Worlds of Natural History

From Aztec accounts of hibernating hummingbirds to contemporary television spectaculars, human encounters with nature have long sparked wonder, curiosity and delight. Written by leading scholars, this richly illustrated volume offers a lively introduction to the history of natural history, from the sixteenth century to the present day. Covering an extraordinary range of topics, from curiosity cabinets and travelling menageries to modern seed banks and radio-tracked wildlife, this volume draws together the work of historians of science, of environment and of art, museum curators and literary scholars. The chapters are framed by an introduction charting recent trends in the field and an epilogue outlining the prospects for the future. Accessible to newcomers and established specialists alike, *Worlds of Natural History* provides a much-needed perspective on current discussions of biodiversity and an enticing overview of an increasingly vital aspect of human history.

H. A. CURRY is the Peter Lipton Senior Lecturer in History of Modern Science and Technology at the University of Cambridge.

N. JARDINE is emeritus Professor of History and Philosophy of the Sciences at the University of Cambridge.

J. A. SECORD is Professor of History and Philosophy of Science at the University of Cambridge.

E. C. SPARY is Reader in the History of Modern European Knowledge at the University of Cambridge.

Advance praise for *Worlds of Natural History*

This massive, comprehensive and extremely rich collection of chapters features a stellar cast of contributors who have created a worthy sequel to *Cultures of Natural History*. From its elegant introduction to its colourful chapters and provocative afterword on the continuing vitality of natural history in the twenty-first century, this book fascinates and instructs. Dazzled by its contents, readers will have a difficult time deciding which compartment in this cabinet of curiosities to open first. This is scholarship in the history of science at its finest.

Bernard Lightman, York University

This volume offers a cornucopia of new approaches to writing the history of natural history from the Renaissance to today. With attention to shifting epistemologies and material cultures, it situates ancient traditions of collecting, classifying and preserving nature in relation to the modern biological and Earth sciences. In our present era of vanishing biological diversity, the authors consider the lessons of the past for the future of both elite and popular scientific institutions, from seed banks to museums and zoos.

Deborah R. Coen, Yale University

Worlds of Natural History comes as close as is humanly possible to living up to its title. The chapters illuminate almost every aspect of the vast enterprise of natural history, from collecting, networking and voyaging to preserving, image-making and classifying. Its sites are as various as the Renaissance apothecary's shop and the contemporary genetics lab; its locales criss-cross the globe. This book crystallises decades of historical scholarship, and is the single best introduction to the topic.

Lorraine Daston, Max Planck Institute for the History of Science, Berlin

"L'Insurrection de l'Institut Amphibie". ————— The Pursuit of Knowledge

Worlds of Natural History

EDITED BY

*H. A. CURRY, N. JARDINE, J. A. SECORD
and E. C. SPARY*

CAMBRIDGE
UNIVERSITY PRESS

CAMBRIDGE
UNIVERSITY PRESS

University Printing House, Cambridge CB2 8BS, United Kingdom

One Liberty Plaza, 20th Floor, New York, NY 10006, USA

477 Williamstown Road, Port Melbourne, VIC 3207, Australia

314–321, 3rd Floor, Plot 3, Splendor Forum, Jasola District Centre,
New Delhi – 110025, India

79 Anson Road, #06–04/06, Singapore 079906

Cambridge University Press is part of the University of Cambridge.

It furthers the University's mission by disseminating knowledge in the pursuit of
education, learning, and research at the highest international levels of excellence.

www.cambridge.org
Information on this title: www.cambridge.org/9781316510315
DOI: 10.1017/9781108225229

© Cambridge University Press 2018

First published 2018

A catalogue record for this publication is available from the British Library.

ISBN 978-1-316-51031-5 Hardback
ISBN 978-1-316-64971-8 Paperback

Contents

EPILOGUE

Colour plates can be found between pages 286 and 287

Plates

Figures

Notes on contributors

SAMUEL J. M. M. ALBERTI is Keeper of Science and Technology at National Museums Scotland, and Honorary Professor in the Centre for Environment, Heritage and Policy at the University of Stirling. He trained in the history of science and medicine. After teaching at the University of Manchester he was Director of Museums and Archives at the Royal College of Surgeons of England (which includes the Hunterian Museum). His books on the history of collections include *Nature and Culture: Objects, Disciplines and the Manchester Museum* (2009); *Morbid Curiosities: Medical Museums in Nineteenth-Century Britain* (2011); and *The Afterlives of Animals: A Museum Menagerie* (2011).

KATHARINE ANDERSON is Associate Professor, Department of Humanities, York University (Toronto), with graduate appointments in history, humanities and science and technology studies. She is the author of *Predicting the Weather* (2005) and other articles on the history of meteorology. She is interested in voyage narratives and ocean sciences in the nineteenth century, and recently co-edited *Soundings and Crossings* (2016) with Helen M. Rozwadowski. Her current project is a book on scientific expeditions and the ocean environment in the 1920s and 1930s.

MITCHELL G. ASH is Professor Emeritus of Modern History and Speaker of the multidisciplinary PhD programme, 'The Sciences in Historical, Philosophical and Cultural Contexts' (supported by the Austrian Science Fund) at the University of Vienna, Austria. He is author or editor of sixteen books and over 150 articles and review essays addressing the social, cultural and political relations of the sciences in the nineteenth and twentieth centuries, the history of the human sciences, and the cultural history of human–animal relations. His edited book, *Mensch, Tier und Zoo* ('Humans, Animals and Zoos') was published in 2008.

ETIENNE BENSON is Assistant Professor in the Department of History and Sociology of Science at the University of Pennsylvania, where he teaches environmental history and the history of technology. His research interests include the use of information technology in Earth and environmental sciences, human–animal relations in urban and industrial spaces and environmentalism as a social movement and as a mode of explanation. He is the author of *Wired Wilderness: Technologies of Tracking and the Making of Modern Wildlife* (2010).

DANIELA BLEICHMAR is Associate Professor of Art History and History at the University of Southern California. She is the author of *Visible Empire: Botanical Expeditions and Visual Culture in the Hispanic Enlightenment* (2012) and *Visual Voyages: Images of Latin American Nature from Columbus to Darwin* (2017), and co-editor of three books on science and empire, cross-cultural collecting and the circulation of objects in the early modern world.

HELEN ANNE CURRY is the Peter Lipton Senior Lecturer in History of Modern Science and Technology at the University of Cambridge. Her book *Evolution Made to Order* (2016) brings together the histories of plant breeding and technological innovation to offer a new account of biotechnological aspirations in the mid twentieth century. Her current research charts the histories of modern-day plant exploration and conservation, focusing in particular on the creation of long-term seed storage facilities, or seed banks, as tools of industrial agricultural production and biodiversity conservation.

REGINA HORTA DUARTE is Full Professor of History at the Universidade Federal de Minas Gerais, Brazil. She is the author of *Activist Biology: The National Museum, Politics, and Nation Building in Brazil* (2016). She also has published articles in international journals such as *Latin American Research Review, Environmental History, Journal of Latin American Studies, Luso-Brazilian Review, Isis* and several Brazilian journals. She is a founding member of the Sociedad Latinoamericana y Caribeña de Historia Ambiental (SOLCHA). Her current research interests focus on the environmental history of zoos in Latin America, in the first half of the twentieth century.

FLORIKE EGMOND is a Dutch historian living in Rome. Since 2004 she has worked for Leiden University as a researcher in projects funded by the Netherlands Organisation for Scientific Research (NOW) on the Flemish naturalist Carolus Clusius, visual and textual cultures of early modern natural history and the historiography of aquatic creatures. She publishes widely on early modern natural history and its overlaps with the history of collecting, medicine, science, communication, and court and aristocratic culture. Her recent monographs include *The World of Carolus Clusius: Natural History in the Making,*

1550–1610 (2010) and *Eye for Detail: Images of Plants and Animals in Art and Science, 1500–1630* (2017).

JIM ENDERSBY is Reader in the History of Science at the University of Sussex. He is a historian of evolution, classification and botany, with a particular interest in the cultural impact of Darwinism and evolutionary ideas. He has written several books including *Imperial Nature: Joseph Hooker and the Practices of Victorian Science* (2008); *A Guinea Pig's History of Biology* (2009); and, most recently, *Orchid: A Cultural History* (2016). He is currently working on a history of the public culture of early twentieth-century biology, tentatively entitled *Mutants, Moths and Midwives: Forgotten Histories of the Genetic Century*.

ROBERT FELFE is Professor for Art History of the seventeenth and eighteenth centuries at Universität Hamburg and member of the Research Group *Naturbilder/Images of Nature*. In 2000 he received his PhD in art history. His thesis discusses natural sciences and book illustration in the period about 1700: *Naturgeschichte als kunstvolle Synthese. Physikotheologie und Bildpraxis bei Johann Jakob Scheuchzer* (2003). His recent book (habilitation) investigates art and nature in the context of early modern collecting: *Naturform und bildnerische Prozesse. Elemente einer Wissensgeschichte der Künste im 16/17. Jahrhundert* (2015).

PAULA FINDLEN is Ubaldo Pierotti Professor of Italian History at Stanford University where she has directed the Suppes Center for the History and Philosophy of Science. Her publications include *Possessing Nature* (1994); *Merchants and Marvels* (with P. Smith, 2002); *Early Modern Things* (2013); and most recently a collaborative volume, *The Paper Museum of Cassiano dal Pozzo: A Catalogue Raisonné. Series B – Natural History, Parts IV and V. Birds, Other Animals and Natural Curiosities* (2017). She has been working on Agostino Scilla's contributions to understanding fossils in the seventeenth century and his surviving collection at the Sedgwick Museum, Cambridge.

NICHOLAS JARDINE is Emeritus Professor of History and Philosophy of the Sciences at the University of Cambridge. His books include *The Scenes of Inquiry* (revised, 2000), *La Guerre des Astronomes* (3 vols., with A. Segonds, 2008), and *Christoph Rothmann's Treatise on the Comet of 1583* (with M. Granada and A. Mosley, 2014). His current book project is 'On Histories of the Sciences'.

ERIC JORINK is Teylers Professor at Leiden University and senior researcher at the Huygens Institute (Royal Netherlands Academy of Arts and Sciences). He was Andrew W. Mellon Visiting Professor at the Courtauld Institute of Art, University of London (2012–13). He has extensively published on early modern scientific culture, specifically on the connections between art, science and religion, including *Reading the Book of Nature in the Dutch Golden Age, 1575–1715* (2010) and *Art and Science in the Early Modern Low Countries* (with B. Ramakers, 2011). He is currently working on a biography of Jan Swammerdam.

LEAH KNIGHT is Associate Professor of Early Modern Non-Dramatic Literature in the Department of English Language and Literature at Brock University in St Catharines, Ontario. She is the author of *Reading Green in Early Modern England* (2014) and *Of Books and Botany in Early Modern England: Sixteenth-Century Plants and Print Culture* (2009), both winners of the British Society for Literature and Science annual book prize.

SACHIKO KUSUKAWA is Honorary Professor of History of Science at the University of Cambridge and Fellow in History and Philosophy of Science at Trinity College, Cambridge. She has published articles on Conrad Gessner, and is author of *Picturing the Book of Nature: Image, Text, and Argument in Sixteenth-Century Human Anatomy and Medical Botany* (2012).

NATALIE LAWRENCE completed her doctorate on the early modern natural histories of exotic monsters at the University of Cambridge. She is fascinated by bizarre and boundary-crossing natural forms and by our relationships with them. Her next project will be a book on the making of images of exotic and extinct beasts and their symbolic roles. She delights in building her own curiosity collection with *naturalia* of a more modest kind than the spectacular contents of many early modern *Wunderkammern*, though, much to her partner's chagrin, one cabinet is certainly never enough.

JUNG LEE is Assistant Professor at the Institute for the Humanities, Ewha Womans University, Seoul. She is the author of several articles including 'Invention without Science' and 'Mutual Transformation of Colonial and Imperial Botanizing'. She is currently preparing a book, tentatively titled *Rooted Plants, Connected Modernity: Botanizing in Japanese Colonial Korea (1910–1945)*.

IRIS MONTERO SOBREVILLA is a Mellon Post-Doctoral Fellow at the Cogut Institute for the Humanities at Brown University. Her research focuses on the interactions between scientific and indigenous knowledge in contexts of colonisation. Her first book will examine the longue durée relationship between hummingbirds and humans in the Americas and its transatlantic resonance. The first iteration of this project as a Cambridge dissertation received an honourable mention for the 2017 Young Scholars Prize of the International Union of the History and Philosophy of Science and Technology.

STAFFAN MÜLLER-WILLE is Associate Professor in the History and Philosophy of the Life Sciences at the University of Exeter (England). His research covers the history of the life sciences from the early modern period to the early twentieth century, with a focus on the history of natural history, anthropology and genetics. Among more recent publications is a book co-authored with Hans-Jörg Rheinberger on *A Cultural History of Heredity* (2012) and two co-edited collections on *Human Heredity in the Twentieth Century* (2013) and *Heredity Explored: Between Public Domain and Experimental Science, 1850–1930* (2016).

KÄRIN NICKELSEN is Professor in History of Science at Ludwig Maximilians University Munich, Germany. Her research focuses on the history and philosophy of biological sciences in modernity, especially the study of plants in both natural historical and experimental contexts. She is the author of numerous papers and books, including *Draughtsmen, Botanists and Nature: The Construction of Eighteenth-Century Botanical Illustrations* (2006) and *Explaining Photosynthesis: Models of Biochemical Mechanisms, 1840–1960* (2015), both of which received international awards.

LYNN K. NYHART is Vilas-Bablitch-Kelch Distinguished Achievement Professor of History at the University of Wisconsin, Madison. Her main research interests lie in the history of modern (post-1789) biology and relations between popular and professional science. She is co-editor (with Scott Lidgard) of *Biological Individuality: Integrating Scientific, Philosophical, and Historical Perspectives* (2017); they are working on a history of concepts of biological part–whole relations in the nineteenth century. Her 2009 book *Modern Nature: The Rise of the Biological Perspective in Germany* analyses the prehistory of German ecology in popular and museum science of the late nineteenth and early twentieth centuries.

MILES OGBORN is Professor of Geography at Queen Mary University of London. He is the author of *Spaces of Modernity: London's Geographies, 1680–1780* (1998); *Indian Ink: Script and Print in the Making of the English East India Company* (2007); and *Global Lives: Britain and the World, 1550–1800* (2008). He is currently writing a book on the historical geographies of speech and slavery in Barbados and Jamaica which, among other things, investigates talk about plants.

BRIAN W. OGILVIE is Professor of History at the University of Massachusetts Amherst. He is the author of *The Science of Describing: Natural History in Renaissance Europe* (2006), as well as articles and chapters on natural history and science in the early modern period. His current research focuses on insects in early modern art, science and religion.

CHRISTOPHER PLUMB received his PhD from the University of Manchester. He is the author of *The Georgian Menagerie: Exotic Animals in Eighteenth-Century London* (2015), and a co-author of *Zebra* (2018).

VALENTINA PUGLIANO is a Junior Research Fellow of Christ's College, Cambridge, and a former Wellcome Trust Research Fellow at the Department of History and Philosophy of Science of Cambridge University. She is currently working on two books: the first based on her doctorate, completed at Oxford University, on the role of apothecaries and pharmacies in the development of early modern natural history; and a second on the diplomatic medicine of the Venetian Republic in the eastern Mediterranean and the Levant, and its intellectual exchanges with Mamluks and Ottomans, *c.*1400–1730.

SADIAH QURESHI is a Senior Lecturer in Modern History at the University of Birmingham. She is interested in modern histories of race, science and empire. Her first book, *Peoples on Parade: Exhibitions, Empire and Anthropology in Nineteenth-Century Britain* (2011), was joint winner of the Sonya Rudikoff Award for best first book published in Victorian Studies in 2011 (awarded 2013). In 2012, the Leverhulme Trust awarded her a Philip Leverhulme Prize for Medieval, Early Modern and Modern History in recognition of her outstanding and internationally recognised research. She is currently working on her next book on extinction in the modern world.

JOANNA RADIN is Associate Professor of the History of Medicine at Yale, where she is also affiliated with the departments of History, Anthropology, American Studies and programmes in History of Science and Medicine and Ethnicity, Race and Migration. She teaches feminist and indigenous science and technology studies, and history of biomedicine, anthropology and global health. She is the author of *Life on Ice: A History of New Uses for Cold Blood* (2017) and co-editor, with Emma Kowal, of *Cryopolitics: Frozen Life in a Melting World* (2017).

SANDRA REBOK is a historian of science and an author. She worked for many years at the Spanish National Research Council in Madrid and has recently finished a two-year Marie Curie Fellowship at the Huntington Library. Her research and publications focus on Humboldt, Atlantic history, exploration voyages and transnational scientific collaborations during the nineteenth century. She is the author of *Jefferson and Humboldt: A Transatlantic Friendship of the Enlightenment* (2014), and she has just finished her new book *Humboldt's Empire of Knowledge: From the Royal Spanish Court to the White House*. She has also curated several exhibitions in the field of history of science.

MORGAN RICHARDS is an Honorary Associate in the Department of Gender and Cultural Studies at the University of Sydney. She is author of numerous journal articles on the history of wildlife documentary in Britain, with a focus on Sir David Attenborough and the BBC Natural History Unit. She is at present working on a book on the history of wildlife documentary in Britain.

ANNE SECORD is an editor of *The Correspondence of Charles Darwin* and an Affiliated Research Scholar in the Department of History and Philosophy of Science, University of Cambridge. Her research has focused on popular, particularly working-class, natural history in nineteenth-century Britain, and on horticulture, medicine and consumption in the eighteenth century. She has produced a new edition of Gilbert White's *Natural History of Selborne* (2016), and is completing a book that explores social class, observation and skill in nineteenth-century natural history.

JIM SECORD is Professor of History and Philosophy of Science at the University of Cambridge, Director of the Darwin Correspondence Project, and a professorial fellow at Christ's College, Cambridge. He has published many books and articles on the history of the sciences during the late eighteenth and nineteenth centuries, including *Visions of Science: Books and Readers at the Dawn of the Victorian Age* (2014) and *Victorian Sensation: The Extraordinary Publication, Reception and Secret Authorship of Vestiges of the Natural History of Creation* (2000). He is writing a short book on current issues facing history of science.

SUJIT SIVASUNDARAM is Reader in World History at the University of Cambridge and Fellow at Gonville and Caius College. His work on natural history includes *Nature and the Godly Empire: Science and Evangelical Mission in the Pacific, 1795–1850* (2005) and the study of animal–human relations in colonial contexts, for instance in a special issue on 'Non-human empires' in *Comparative Studies of South Asia, Africa and the Middle East* (2015). His research also focuses on island history, as in his book *Islanded: Britain, Sri Lanka and the Bounds of an Indian Ocean Colony* (2013) and a current project on the age of revolutions in the Indian and Pacific Oceans.

EMMA SPARY is Reader in the History of Modern European Knowledge at the Faculty of History, University of Cambridge. Her research concerns the history of natural history, medicine, chemistry and food in the 'long eighteenth century' in Europe, particularly France and its colonies. Her monograph *Utopia's Garden* (2000) concerned the formation of the national museum of natural history in Paris, the Muséum national d'Histoire naturelle. She has also published numerous articles on French natural history and two further monographs on the history of diet, *Eating the Enlightenment* (2012) and *Feeding France* (2014). Her current project investigates drug taking in the reign of Louis XIV.

MARY TERRALL is Professor of History at the University of California, Los Angeles (UCLA). She has published widely on the history of the sciences in the eighteenth century. She is the author of *The Man Who Flattened the Earth: Maupertuis and the Sciences in the Enlightenment* (2002) and *Catching Nature in the Act: Réaumur and the Practice of Natural History in the Eighteenth Century* (2014).

ANNA TOLEDANO studies natural history collecting in eighteenth-century Spain and Spanish America. She is a PhD candidate in the Department of History at Stanford University, where her current research investigates the processes by which natural things become objects of material culture. She is also a museum professional and has developed content for exhibitions and collections at the New York Botanical Garden from 2012 to 2015.

Acknowledgements

Thirty years ago, a group of students and staff at Cambridge founded a weekly lunchtime seminar, the Cabinet of Natural History, for informal discussion of the history of natural history and the environmental sciences. Since that time the Cabinet has continued to thrive. Although the conjunction of this anniversary with publication of this book was completely unplanned, it is appropriate, for this volume (like its predecessor *Cultures of Natural History*) is impossible to imagine without the inspiration of the Cabinet's lively sessions.

We are grateful to Edwin Rose for research assistance; to Lynn Nyhart, Sujit Sivasundaram and the anonymous referees for Cambridge University Press for criticism and suggestions; and to our colleagues in the Department of History and Philosophy of Science and the Faculty of History for support, friendship and encouragement. Our editor, Lucy Rhymer, and her team (particularly Melissa Shivers and Cassi Roberts) have made publishing with the Press a genuine pleasure.

The editors particularly wish to thank Andrew Buskell, Marina Frasca-Spada and Anne Secord for support and forbearance.

This volume would not have been possible without the enthusiasm and patience of our contributors, and it has been a pleasure to work with them.

Introduction

Worlds of history

Natural history, conceived broadly to cover all quests for systematic understanding of natural objects – plants, animals and minerals – is vast in scope, both temporal and global. In the West, for example, it reaches back at least to the works of Aristotle and Theophrastus in the fourth and third centuries BCE, and the Chinese tradition is of comparable antiquity.[1] The organised discipline which emerged in Renaissance Europe, while drawing upon these various traditions, took on its own identity through a corpus of texts, characteristic sites of practice and specific textual and iconic traditions. This discipline is the point of origin of *Worlds of Natural History*. Admittedly, it is sometimes seen as having been progressively displaced over the past couple of centuries by fields seen to be more rigorously scientific. But on this score the behaviour of natural history has been reminiscent of that of the old man in *Monty Python and the Holy Grail*, who, about to be cast onto a wagon of corpses, declares: 'I'm not dead'. Contrary to our millennial anxieties, natural history has flourished in recent decades, with ever-increasing funding, media presence and public engagement.[2]

Throughout the period covered by this volume the practices of natural history have been entangled with other enterprises, some extensive – agriculture, commerce, exploration, cross-cultural encounters – some more local – horticulture, hunting, museum display, pursuit of hobbies, gastronomy, and so forth. Accordingly, the history of natural history is closely engaged with many other important and intriguing branches of history. *Worlds of Natural History* celebrates this prospering of natural history itself and its history.

Five hundred years of natural history

Over the past 500 years the practices, theories and institutions of natural history have undergone radical changes; and the past 50 years have seen much innovation in the agendas and methods of its historians. Nevertheless, the history of natural history remains, we believe, a coherent enterprise. Certain shared practices, spaces, traditions and trends run through all the sections of the volume. Among these, material culture looms large, for what characterises much of natural historical endeavour from the Renaissance onwards is the collection, arrangement and representation of natural objects. Most of the chapters in this book concern themselves in one way or another with these practices. Debates over preservation (Felfe, Findlen, Anne Secord, Toledano), over fieldwork or museum practice (Alberti, Benson, Curry), over imagery (Jorink, Kusukawa, Nickelsen, Richards), and over description and grouping (Endersby, Müller-Wille) are parts of a larger debate over collecting and ordering. These practices are more specific to natural history as a discipline than, for example, networking (a practice common to learned communities generally), and more fundamental than, say, the use of collections for the advancement of political or imperial agendas (a practice not necessarily shared by private collections). At the same time, collecting and ordering, preservation and provenance, curiosity and taste, are priorities which natural history has in common with the fine arts.

These preoccupations with the material were not abandoned with the rise of experimentation, nor with the spread of natural history to parts of the world distant from Europe, as the chapters of Terrall and Lee show. It might be possible to identify distinct regimes of natural historical practice: the sixteenth- and seventeenth-century world of princes and polymaths addressed by Egmond, Jorink and Kusukawa appears quite different from the commercial and colonial regime of the later eighteenth century to be found in the chapters of Bleichmar and Plumb; the institutional and public pursuit of nineteenth-century natural history described by Anderson and Nyhart took on a new form with the twentieth-century emphasis on conservation and curation characterising the chapters of Curry and Duarte. These shifts correspond to wholesale changes in the structure and form of the institutions of natural history, as well as in the relationship between learning and power, whether scholarly or governmental. Sweeping transformations such as the emergence of publics, states, the modern university system, the rise of laboratory and field science and the modern environmental movement have all produced new modes of practising natural history.

In saying this we do not seek to imply that the history of natural history can be understood in terms of separate epistemes, paradigms, stages or programmes. Rather, certain themes have borne more fruit for some historical periods than for others. Epistolary networks have emerged as a central focus of the history of the discipline between about 1500 and 1750, while books and paper tools – addressed in this volume by Müller-Wille – have received little attention for the period after 1850. It is only after 1750 that institutions, understood as impersonal spaces within which funded natural historical research could proceed, assume pre-eminence as scenes of enquiry, concurrently with the emergence and diversification of natural historical publics. Yet the basis for the practice of natural history has remained stable: *naturalia* and their representations. Alongside materiality and order, a third enduring criterion, spatiality, might also be seen as characteristic of all natural historical practice, from the national zoos described by Ash to Felfe's account of depicted collections as arguments about order and hierarchy in the natural world.

Several chapters call into question, at least by implication, grand narratives that invoke radical upheavals: the Scientific Revolution; epistemological ruptures (Gaston Bachelard); paradigm shifts (Thomas Kuhn); succession of epistemes (Michel Foucault).[3] The notion of a 'Scientific Revolution' rejecting the authority of the ancients in favour of the testimony of direct observation and experiment does not fit well with Ogilvie's chapter; the Foucaultian leap from a pre-classical episteme of analogies and sympathies to a classical episteme of dispassionate ordering is at odds with the persistence of emblematic and moralising views of *naturalia* noted by Lawrence; many of the later chapters militate against the supposed nineteenth-century displacement of timeless natural historical ordering based on surface characteristics by a modern biological episteme of inwardness and temporal innovation; and Sivasundaram and others challenge the global modernisation of the worlds of natural history. The overall picture that emerges is one of institutional, cultural and national diversity in the development of natural history, a pattern that often shows what Ernst Bloch called 'the non-contemporaneity of the contemporaneous', that is the contemporaneity of past practices and views that seem from our present standpoint as if from different eras.[4] Thus, we have displays by charlatans of wonders and curiosities at the Royal Society at the very time that many of its members were promoting an experimental-philosophical approach to freaks and abnormalities; elaborate codes of moral and sentimental meanings for flowers alongside the nineteenth-century incorporation of many areas of natural history into the new alliance of disciplines called 'science'; and today we see, despite challenges from the botanists, the continued use of Linnaean Latin nomenclature and species descriptions in an age of gene sequencing and cladistic taxonomy.[5]

Trends and turns

When writing our introduction to *Cultures of Natural History* (1996), the precursor to the present work, we inhabited a very different historiographical world.[6] The new cultural history represented historians' embrace of what were being dubbed the 'anthropological turn' and the 'linguistic turn'.[7] In earlier decades, the word 'culture' had generally been used by historians for such elite subjects as canonical literature, classical music or fine art. In its new, anthropological sense, the term was aptly defined by Raymond Williams in *Keywords* as any community sharing a set of significances, whether linguistic, visual or corporeal.[8] Many historians had taken up the implications of this anthropological borrowing to rewrite history in terms of communities, their discourses and their communications. Aspects of this approach, well represented in both *Cultures of Natural History* and the present volume, include 'defamiliarisation' and 'decentring': defamiliarisation being interpretation founded on recognition of the cultural distance of past activities and conceptions from our own, rather than assimilation to present standards and ideas; decentring being the move away from concentration on central and iconic discoverers, authors, texts and settings of the sciences towards critical examination of the full range of their agents, points of view and sites of inquiry. Such decentring is evident throughout this volume, which covers a remarkable range of contributors to natural history: physicians and theologians (Kusukawa); apothecaries and their assistants (Pugliano); informal networks of gardeners (Knight); collectors and dealers (Egmond, Findlen, Lawrence, Pugliano, Toledano); philosophers (Jorink); networks of correspondents (Egmond, Müller-Wille); bureaucrats and entrepreneurs (Müller-Wille); sellers and merchants (Plumb); engravers, draughtsmen, publishers (Nickelsen); morphologists, palaeontologists, ecologists and taxidermists (Alberti); spectators (Alberti, Qureshi); indigenous communities (Curry, Duarte, Montero, Qureshi, Radin, Sivasundaram); anthropologists (Radin); publics (Anderson, Ash, Knight, Nyhart, Plumb, Richards).

The cultural turn of the 1980s and 1990s yielded new accounts of the history of the book which acknowledged the complexity of composition and production, rather than taking texts as direct expressions of authorial intention, and which viewed reception in terms of communities of readers who actively constructed and appropriated, rather than passively receiving, the meaning of texts.[9] At the same time, attention was directed to the modes of communication, persuasion and instruction, both textual and visual, in the sciences.[10] This is of great importance for sound interpretation of past works, given the diversity and (to us) often alien natures of textual and visual conventions prior to more recent standardisation of presentation and illustration in articles, treatises and textbooks. Of special interest in this connection are recent studies, including the contributions of Knight, Ogilvie and James Secord, that have looked into the descriptive and persuasive devices of natural history writing and its relations to such genres as travel and science fiction writing, epic and myth.[11]

The impacts of further 'turns' – the material turn and the spatial turn – are much in evidence in this volume. The material turn had its origins in archaeology and anthropology, and in histories of arts, crafts and everyday life.[12] In the history of science, this turn has led to increased attention to artisanal and day-to-day activities, and to the complex interactions of people, materials, tools and machines in the production and communication of knowledge.[13] Closely associated with this material turn has been the spatial turn, moving from diachronic narratives to synchronic exploration of the pursuit and communication of the sciences in and between diverse sites and settings.[14] Common to both of these turns is the recognition of the hybridity of pursuit of the sciences and the inseparability of cognitive, social, economic (and often commercial and political) activities in the production, consolidation and communication of scientific knowledge.[15]

Especially in the history of natural history, there has been extensive recent study of the intimate links between global communication and exchange of knowledge on the one hand, and exploration, empire and commerce on the other. Here, two domains of research which have sprung into life since 1996 are worth mentioning. Atlantic and Iberian Empire studies, here represented in the chapters of Bleichmar, Ogborn and Rebok, have given rise to a rich body of work uniting the histories of imperialism, colonialism and environment.[16] Works that recognise the dependency of natural history on indigenous sources and informants have taken seriously the standpoints of non-European practitioners and the central role of indigenous knowledge in the formation of European natural history, and this trend is represented by the chapters of Duarte, Montero, Qureshi, Radin and Sivasundaram.[17] A further development, exemplified in the chapters of Egmond, Knight and Müller-Wille, is the greatly increased attention to networks as a model of natural historical practice, in which natural history figures as a collective enterprise, in sharp contrast to earlier emphases on individual naturalists or institutions.

The development of the historiography of the sciences, including natural history, is often presented as a tale of progress through successive 'isms' and progressive 'turns'.[18] Accordingly, positivistic accounts of the accumulation of 'positive' scientific knowledge are said to have been displaced through a turn from internalist to externalist, practice-oriented studies. The practice-oriented sociology of scientific knowledge (SSK) of the 1970s and 1980s, focused on the ways in which social interests have shaped the local construction of scientific knowledge, is seen as being displaced in the late 1980s and 1990s by the exploration of networks of communication and mediation between the agents (both human and non-human) involved in the consolidation of scientific knowledge. This formed a part of what in the mid 1990s many (including ourselves as authors of the introduction to *Cultures*) perceived as a more general cultural turn, characterised, as noted above, by a new historicist sensitivity to the foreignness of past sciences, by recognition of the complexities of the production and reception of their texts, and by 'decentring' the move towards recognition of the full range of participants, sites and tools of the sciences.

Where much of this new cultural history was microhistorical, local both in time and space, the dawn of the new century is widely perceived as marked by a 'global turn' towards macrohistories, extensive not temporally but spatially, concerned with worldwide communication, translation, and cross-cultural interaction in the sciences. As for the present, study of the sciences is enjoying a new incursion from anthropology, the so-called 'ontological turn'. Here what is advocated is not a focus on foundational categories and perspectives on nature, but rather on the things recognised, formed and valued by local communities, their worlds, as evident in their declarations and implicated in their practices.[19] Further, there are many who see all fields of history as strengthened in objectivity and enlarged in temporal and spatial scope through automated linkage and analysis of 'big data'.

There are several objections to conceiving the development of the historiography of the sciences as a series of 'turns'. It can lead to exaggeration, inflating (in the words of Frank Kermode) 'adjustments of normal practice' into 'shocking paradigm shifts'.[20] Further, the antitheses presented in such accounts are potentially misleading. Consider the microhistory vs macrohistory division. Original and constructive works relating to the history of natural history have cut across this division, producing culturally extensive 'middle-sized' histories by working outwards from some single item, exploring the full range of activities and interpretations involved in its production, reception and appropriation: notable examples are Anke te Heesen's *The World in a Box* (2002) and James Secord's *Victorian Sensation* (2000).[21]

The view of historiographical progress through 'turns' also misleads by concealing the rich variety of past approaches. To take a couple of examples from the historiography of natural history, consider two productions from the positivist era: Karl Friedrich Wilhelm Jessen's *Botanik der Gegenwart und Vorzeit in culturhistorischer Entwickelung* ('Botany of the Present and Past in its Historico-Cultural Development', 1864) and Henri Daudin's volumes *De Linné à Lamarck* and *Cuvier et Lamarck* (1926–7). In Jessen's work, we find close attention to the social and institutional settings of natural history and to its links with exploration and commerce; and in Daudin's volumes, past systems of ordering and classification are carefully related to the collecting, horticultural and curatorial activities of the naturalists.

The vision of successive turns is liable to distort not only our perception of the development of the historiography of natural history, but also our practice as its historians. Overcommitment to the latest 'turn' may also lead to accounts that commit anachronism and/or anatropism (that is, misleading application of our categories to cultures other than our own). Thus, while a 'centre of calculation' model has proven useful to describe the operation of certain centralised natural historical networks in the colonial era, it obviously cannot account for all of natural historical practice. For example, work on eighteenth-century Spanish central accumulation of reports, images and specimens gathered by botanical expeditions to colonised territories, and on the co-construction of natural historical knowledge between learned South Asians and Europeans from the seventeenth to nineteenth centuries, fits poorly with a centre and periphery model.[22]

New approaches and methodologies have enriched and diversified the history of natural history; yet, as editors, we hope that they do not consign older themes, such as institutions, systematics, order and the moral or medicinal purposes of collecting, to oblivion. Any foreclosure of alternative viewpoints and approaches, a commitment to external *rather than* internal, global *rather than* local, etc., is genuinely damaging to the historiography of the sciences.[23] Moreover, commitment to new turns, -isms or paradigms has on occasion had a further deleterious effect, leading not to the effective application of new interpretative and explanatory models, but to the spicing up of narratives with buzz words. A formulaic adherence to the language rather than the spirit of such innovations yields results like those described by Grandpa Vanderhof in Frank Capra's romantic comedy *You Can't Take It with You*, when he declares 'When things go a little bad nowadays, you go out, get yourself an -ism and you're in business.'

Controversy over 'turns' can, however, prove fruitful: consider, for example, the reflections on modes of communication in the sciences incited by Bruno Latour's conception of printed works as 'immutable mobiles'. Moreover, fashionable turns may stimulate valuable backlashes, for example, the revival of aspects of the 'history of ideas' in the 'new historical epistemology', of which a splendid example – with much relevance for the history of natural history – is Lorraine Daston and Peter Galison's *Objectivity* (2007).

Moreover, caution is in order in challenging the vision of progressive turns in the historiography of the sciences. For it is all too easy to slip from 'let a hundred flowers bloom' to 'anything goes', an unfortunate slippage, given that there are aspects of formerly and currently favoured approaches that surely do deserve criticism and displacement. For example, while such positivistic accounts of cumulative natural historical progress as Julius von Sachs's *Geschichte der Botanik* ('History of Botany', 1875) and Karl Alfred von Zittel's *Geschichte der Geologie and Paläontologie* ('History of Geology and Palaeontology', 1899) remain mines of useful information, their anachronisms and nationalistic bias in selection and interpretation are to be avoided. And, leaping forward to currently fashionable global studies, these should beware of exclusive focus on worldwide transactions at the cost of inattention to local practices, local worlds and local patterns of communication.

To question the notion of linear progress through successive turns is by no means to deny that recent developments in the cultural history of the sciences have greatly enriched the discipline. With a few exceptions, manifest, for example, in the natural historical works of Jessen and Daudin mentioned above, the history of science long remained an isolated discipline, a peripheral didactic adjunct to the sciences and little involved with other branches of professional history. But the past thirty or so years have seen a major shift, with greatly increased engagement with mainstream history, as is abundantly evident from the proliferation of studies that link the history of natural history to the histories of agriculture, exploration, commerce, politics, art and collecting, and public entertainment. *Worlds of Natural History* bears witness to this new diversity in its four Parts, devoted respectively to 'Early modern ventures' – the Renaissance world of exploration and enterprise; 'Enlightened orders' – the systematising projects of the long eighteenth century; 'Publics and empires' – the imperial world of Victorian natural history; and 'Connecting and conserving' – natural history's very recent past.

Where next?

To wind up this Introduction we offer, in the light of our pleasant experience as readers of the chapters of this volume, some predictions and recommendations.

Let us start with safe predictions. In line with the current ontological turn, we envisage much further study of the skills and ways of life associated with naturalists' interactions with the specimens and equipment of natural history. We further foresee the continuation of profitable engagements between the history of natural history and the histories of exploration, commerce and empire. Further linkages that hold great promise include: art-historical study of geological, botanical and zoological illustrations; study of the ties between the practices of natural history, local history and antiquarianism; links between natural history and the changing interactions of humans with other animals; exploration of the connections between the history of the environment and the development of natural history. Given that the pursuit of natural history has been coeval with the transformation of the Earth through agriculture, deforestation and climate change, the latter seems an especially fruitful field.

Equally safe is the prediction that 'big data' collection, linkage and analysis hold immense potential for the history of natural history.[24] Caution is, however, in order: scientistic declarations on the capacity of big data analysis to render history an objective science are to be resisted; and due reflection is needed on the ways in which the traditional skills of close scrutiny and interpretation of sources, both textual and visual, may be effectively combined with generalisation through big data analysis.[25] But there can be little doubt that the history of natural history will profit greatly from big data collaborative enterprises in the fields of communication and reception, of commerce, and of agriculture and environmental change.

More tentatively, we suggest that there may be a resurgence of temporally extended 'big picture' accounts. Helpful though big data analyses will be for such accounts, they pose serious problems of narrative structuring. Despite their narrative convenience, triumphal stories of natural historical progress are subject to well-known objections, as are dramatic tales of successive epochs, epistemes, research programmes, and paradigms. More promising are thematically based long-duration narratives, covering, to take just a couple of examples: the grounding of natural historical practices in notions of objectivity, as in Daston and Galison's *Objectivity*; and the shaping of natural historical discourse by dominant metaphors, as in Donna Haraway's *Crystals, Fabrics and Fields* (1976).[26] To be hoped for too is more extensive re-engagement with the content of past natural histories. Here we have in mind not so much rehearsals of past discoveries and theories, as in traditional didactic and positivistic histories, but rather accounts of the changing priorities and bones of contention among natural historians. Such studies would explore the ways in which new agendas have transformed the practices of natural history, and how new practices have, in turn, transformed old agendas and created new ones.[27]

Yet more ambitiously, we may hope for big pictures that will relate the changing practices, agendas and forms of communication and education in natural history to changes in power structures and relations. Here we have in mind not the Eurocentric regimes and epistemes of Foucault and his devotees, but rather the powers exerted through exchanges of goods, technologies, forms of governance and ideologies. There is, indeed, literature on which such big pictures may draw: for the early modern period, relating natural history to exploration and trade; for the modern 'age of science', placing it in the context of global cross-cultural and international linkages.[28] In this connection C. A. Bayly's *The Birth of the Modern World 1780–1914: Global Connections and Comparisons* (2004) is of especial interest for its treatment of modernisation – in science, technology, forms of government and political ideologies – not as a 'Triumph of the West', but in terms of international exchange, appropriation and adaptation.[29]

Finally, let us indicate a handful of further topics closely tied to the history of natural history and deserving of further pursuit. A crucial issue, we believe, is that of the relations between nature and artifice in collections and specimens, a key consideration in light of the growing interest in 'heritage' which is facilitating the digitisation of material objects and an expanding view of natural history museums as sites of cultural encounter as well as natural truth.[30] Accordingly, museums have come to be reimagined as anthropological spaces for the symbolic re-presentation of nature, producing the kind of artefact referred to by Alberti in this volume as 'Museum nature'. The space of the natural history collection is a palimpsest overwritten countless times by new accounts of nature, new agendas and new priorities. Historians can tease out many stories: there is not just one itinerary around the gallery.

As already noted, there are solid grounds for challenging the simplistic view that over the past 200 years natural history has been progressively displaced by science. Denise Phillips's *Acolytes of Nature* (2012) is richly revealing of the complex and varied ways in which the social and institutional identities of natural history responded to the consolidation of natural science as a disciplinary alliance in the German Lands. There is room for much further work on the interactions of natural history with other disciplines and pursuits in the 'age of science'. One such topic is that of the relations between 'professionals' and 'amateurs', including artisan naturalists, members of local natural history societies, antiquarians and what have come to be called 'citizen scientists'.[31] Another is the loss and decay of materials in natural history collections resulting from the post-World War II shift from museum- and classification-oriented natural history to laboratory-based teaching and research, followed by the recent recovery of care for specimens and respect for taxonomists occasioned by such factors as cladistic taxonomy and the flourishing of biodiversity studies.

Bound up with multiplicity of participants and changing technologies of communication is the crucial issue of authority and testimony, addressed in the chapters of Anderson, Egmond, Lawrence, Montero Sobrevilla, Ogilvie and Richards. Remarkable changes in the criteria for assessment of testimony are well documented, and there is room for much further exploration of the ways in which such shifts have related to changes in the types of persons involved in natural history and their modes of communication.[32] Allied to this is the trust invested in standardised objects, instruments, descriptions and images, a topic extensively investigated by historians of the exact sciences, but deserving of much further study by historians of natural history.[33]

The pursuit of natural history has long been involved with its own history: in establishing the validity of names and the provenance of old specimens, in using textual and visual evidence for past states of the environment and distributions of species, etc.[34] Moreover, many natural historians have written histories designed for didactic, justificatory and polemical purposes.[35] For example, Sachs's *Geschichte der Botanik*, mentioned above, serves all three functions, being openly didactic in its presentation of discoveries, while denigrating Linnaean taxonomy and promoting Sachs's own field of plant physiology. The long-standing and diverse uses of history by practising naturalists deserve much further study.

As the contributions to this volume show, from its early modern formation to the present, natural history has undergone numerous shifts in agendas, sites, tools, participants and practices. Likewise diverse, as indicated earlier, and manifest in the contributions, are the profitable ways of uncovering its past. Natural history inhabits many worlds; so too does its history.

I Early modern ventures

1 Visions of ancient natural history

Ancient books were essential sources for Renaissance naturalists: above all, Aristotle on animals, Theophrastus and Dioscorides on plants, and Pliny the Elder's encyclopaedic *Natural History*. Equally important, though, were shared visions of how the ancients studied nature. Like the Renaissance understanding of the ancient world in general, naturalists' visions of ancient natural history were grounded in sources, but on those solid foundations they projected their own desires and practices. Not the least of these was their conviction that there was an ancient discipline called 'natural history' with a coherent history, an idea that they projected on what were, in fact, several distinct projects for investigating the animal and plant world. Some of those projects had a continuous tradition, such as the practice of agriculture and herbal medicine, but others – notably the philosophical zoology and botany of Aristotle and Theophrastus – were local, short-lived enterprises. Like humanists more generally, as Renaissance naturalists studied antiquity, they idealised it.

Aristotle and Solomon, founders of natural history?

'Alexander the Great', wrote the Swiss physician Conrad Gessner around 1550, 'burning with desire to know the nature of animals, delegated this work to Aristotle, the master of every subject, and ordered a thousand men to assemble everything that was hunted, fished, and kept domestically in the lands of Greece and Asia, so that he should remain ignorant of nothing born there'. When Aristotle finished his studies and presented fifty books to Alexander, Gessner continued, he was rewarded with a kingly sum: 700 gold talents according to some authors, 800 according to others. In Gessner's day, the scholars Robert Cenalis and Guillaume Budé agreed, 100 Attic talents were 600,000 crowns: thus, Aristotle received 4 or 5 million crowns.[1]

That was thousands of times what Gessner himself received as the city physician of Zurich, his post while he was compiling his own *Historia animalium* ('History of Animals'). But Alexander got his money's worth: the great warrior's name was remembered above all, Gessner claimed, for being Aristotle's patron. Gessner returned to the theme a few years later in his history of oviparous quadrupeds (1554), and again in his history of aquatic animals (1558). In that volume, dedicated to none other than the Holy Roman Emperor Ferdinand himself, Gessner not so subtly hinted that Ferdinand ought to play Alexander to Gessner's Aristotle.

Gessner was far from the only writer to portray Aristotle as the head of a vast research project supported by a royal fortune. In his preface to John Gerard's *Herball* (1597), the physician Stephen Bredwell cited the example negatively as well as positively:

Admirable and for the imitation of Princes, was that act of Alexander, who setting Aristotle to compile commentaries of the brute creatures, allowed him for the better performance thereof, certaine thousands of men, in all Asia and Greece, most skilfull observers of such things, to give him information touching all beasts, fishes, foules, serpents and flies. What came of it? A book written, wherein all learned men in all ages since do exercise themselves principally, for the knowledge of the creatures. Great is the number of those that of their owne private, have laboured in the same matter, from his age downe to our present time, which all do not in comparison satisfie us. Whereas if in those ensuing ages there had risen still new Alexanders, there (certainly) would not have wanted Aristotles to have made the evidence of those things a hundred fold more cleered unto us, than now they be.[2]

In Bredwell's understanding of history, great achievements required munificent patrons; the renewal of natural history in the sixteenth century owed much to patrons such as the Holy Roman Emperor Ferdinand, Cosimo de' Medici and King Henri II of France.

In one way or another, Gessner, Bredwell and others were repeating Pliny the Elder: Gessner's account was lifted, nearly word for word, from Pliny's *Natural History* (8.17.44), with additions from Athenaeus's *The Learned Banqueters*. Two centuries later, the comte de Buffon would refer back to the Aristotle–Alexander story in support of his belief that Aristotle's empirical approach to taxonomy was superior to the rational method of Buffon's rival Linnaeus.

Despite its good classical pedigree, the story was most certainly a fabrication. Aristotle's works reveal that he knew quite a lot about animals, especially those of the eastern Mediterranean.[3] Yet his accounts of the fauna of the Near East are sketchy and rely on second-hand reports. Moreover, Alexander was engaged in a campaign of rapid conquest; he had neither the bureaucracy nor the time to send specimens back to the Lyceum in Athens. Pliny was projecting backwards to Aristotle's time the sort of systematic quest for exotic animals driven by Roman Imperial games in the Flavian Amphitheatre (the Colosseum).[4]

Renaissance naturalists had little reason to doubt Pliny and Athenaeus. The story of Aristotle and Alexander fitted the idealised vision of classical antiquity that characterised the humanist movement of Renaissance Europe. Humanists appealed further to Old Testament sacred history: to Adam, who gave every creature its name, but whose perfect knowledge was lost in the twin catastrophes of the Fall and the Flood, and to Solomon, who 'spake of trees, from the cedar tree that is in Lebanon even unto the hyssop that springeth out of the wall: he spake also of beasts, and of fowl, and of creeping things, and of fishes'.[5] In his scientific utopia *The New Atlantis*, the English natural philosopher Francis Bacon claimed that the people of the fictional island of Bensalem had preserved Solomon's natural history. In reality, Solomon's wisdom was lost to the moderns, but his example, alongside Alexander's, suggested that not only kings but also God himself smiled on natural history.

Humanist naturalists and the ancient past

For Gessner, Bredwell and other participants in the humanist movement of the European Renaissance, classical antiquity was both a beacon and a standard. Already by the early fifteenth century, scholars steeped in the classical *studia humanitatis* – grammar, rhetoric, poetry, history, and ethics – spoke of the 'rebirth of literature' in the works of Petrarch, Boccaccio and other fourteenth-century writers. By the sixteenth century, literary scholars (Polydore Vergil), art historians (Giorgio Vasari), anatomists (Andreas Vesalius) and naturalists (Conrad Gessner) had elaborated this idea of rebirth into a more comprehensive historical scheme. The arts and sciences had flourished in ancient Greece and Rome, from the age of Homer to that of Augustine. In the Middle Ages, the twin forces of barbarism and superstition had ruined them; and in the modern age, they were being restored to their ancient glory. This restoration had been made possible by the scholars who recovered ancient texts, restored them to their pristine condition, replaced erroneous medieval translations and fatuous medieval commentaries with accurate translations and interpretations, and used them as models – not to imitate slavishly, but to emulate.

Needless to say, this vision of antiquity was tendentious.[6] Classical texts had not vanished during the Latin Middle Ages: the very existence of medieval translations and commentaries attests to their availability and use. Nonetheless, the scheme, which we find repeated over and over in sixteenth- and seventeenth-century histories of the arts and sciences, reveals how important ancient texts and models were to Renaissance thinkers. Natural history is no exception. Modern natural history was born out of the attempt to understand ancient books on plants and animals, and to compare·their claims with what naturalists observed themselves. Even as the actual claims of ancient texts diminished in importance, certain ancient authors and their texts – above all, Aristotle – continued to be taken seriously. From the fifteenth through the eighteenth centuries, natural history was profoundly shaped by these visions of classical antiquity.

Renaissance thinkers also shared a broad vision of sacred antiquity. In this vision, Adam, who gave all creatures their proper name, had been imbued with knowledge of all the arts and sciences. Much of this knowledge, and the perfect language in which he had expressed it, was lost with the Fall, or perhaps with the confusion of languages at Babel, but divine revelation kept its spark alive. Moses possessed a portion of it, as did Solomon. And in Egypt, some of this knowledge was passed on to the Egyptian Hermes Trismegistus ('Thrice-great Hermes'), and eventually, during his studies in Egypt, to the Greek philosopher Plato.[7] John Gerard cited Adam as 'the Herbarist' of the Garden of Eden, and noted that Solomon, wisest and most royal of kings, in 'his lofty wisdome thought no scorne to stoupe unto the lowly plants'.[8] The engraved title page of Carolus Clusius's *Rariorum plantarum historia* (1601) gave pictorial form to this idea: underneath the Tetragrammaton (the Hebrew YHWH), four figures represented ancient knowledge of plants: Adam and Solomon above, Theophrastus and Dioscorides below (Figure 1.1).

The English herbalist John Parkinson explicitly contrasted the pagan and Christian stories in his address 'to the courteous reader' in his 1629 *Paradisus Terrestris* ('Earthly Garden'):

Although the ancient Heathens did appropriate the first invention of the knowledge of Herbes . . . some unto Chiron the Centaure, and others unto Apollo or Ae[s]culapius his sonne; yet wee that are Christians have out of a better Schoole learned, that God, the Creator of Heaven and Earth, at the beginning when he created Adam, inspired him with the knowledge of all natural things (which successively descended to Noah afterwards, and to his Posterity).[9]

Ancient sources

Though Renaissance humanists identified both Christian and pagan origins for natural history, the actual texts they had at hand derived almost exclusively from the latter – and not only those that modern historians identify as works of natural history. All kinds of texts, from travel narratives to lyric poetry, provided a trove of facts, some of them more dubious than others, about animals, plants and minerals. But certain texts also introduced methods for studying nature and

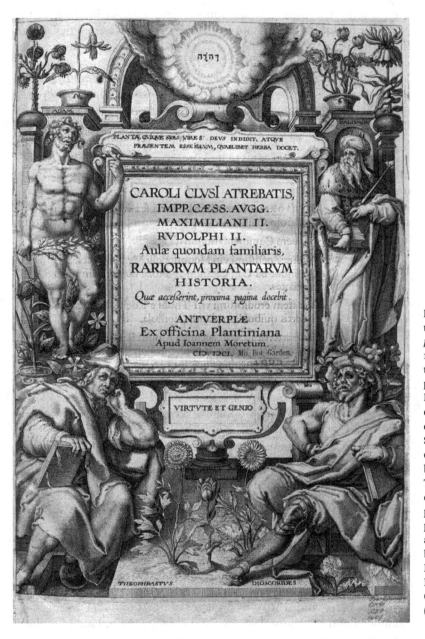

Figure 1.1 Engraved title page to C. Clusius, *Rariorum plantarum historia* (Antwerp, 1601). Under the Divine Name, the engraving depicts the principal ancient sources of Renaissance botany, imagined or real: Adam, who gave all creatures their names; Solomon, who wrote about all creatures great and small (in books that did not survive); Theophrastus, author of books on the history and causes of plants; and Dioscorides, who produced the most important ancient work on medical botany. Image from the Biodiversity Heritage Library. Digitised by Missouri Botanical Garden, Peter H. Raven Library (www.biodiversitylibrary.org).

provided models for naturalists as they wrote their own books of natural history.

Despite what Renaissance naturalists claimed, the classical world had no discipline or genre of 'natural history'.[10] Gessner and his fellow humanists did not consciously set out to deceive their readers. But they interpreted their ancient sources and models in light of the practices for investigating nature that they themselves had developed from the 1490s through the 1550s: identifying plants and animals; collecting them or their parts; describing in image and word their forms, places, habits and medicinal virtues; noting their cultural associations and uses; and trying to make some sense out of the underlying natural order that many of them vaguely perceived. The story of the birth, decline and rebirth of natural history was elaborated in the sixteenth century. The late fifteenth-century humanist Giorgio Valla, for instance, had no separate category for natural history in his encyclopaedic overview of secular learning, *De rebus expetendis et fugiendis* ('Things to Seek and to Avoid', published posthumously in 1501). Following his ancient sources, Valla discussed plants and animals in three separate parts of his work: the books on natural philosophy, agriculture and husbandry, and medicine.

The texts that Renaissance naturalists retrospectively fused into one tradition came, in fact, from several ancient literary genres.[11] There were works called *Inquiries into Animals* (by Aristotle),[12] *Inquiries into Plants* (by his disciple Theophrastus), *On Plants* (attributed to Aristotle but written by someone else), *On the Nature of Animals* (by Aelian), and of course *Natural History* (by Pliny the Elder). But these were not part of a single literary tradition. Aristotle's and Theophrastus's works were part of a philosophical enterprise. Aristotle's *Inquiries into Animals* provided material for better understanding their generation, anatomy and motion; Theophrastus's *Inquiries* were accompanied by another work on *The Causes of Plants*. Aelian's book was a collection of animal lore. And Pliny's work was an enormous compilation of material, mostly drawn from other writers, on everything from geography to the medicinal and magical properties of gemstones, including books on animals and plants.

Beyond these works of natural philosophy and encyclopaedic compendia, there were three further ancient genres that contained material on natural history: medical texts, agricultural treatises and works on hunting and fishing. Of the first, the most important was *On Medicinal Substances*, written by the Greek physician Dioscorides of Anazarbos. Its five books contained hundreds of descriptions of plants, animals and minerals that could be used as 'simple medicines' or compounded into more elaborate drugs. The second included short works by Xenophon, Cato the Elder and Varro, as well as more substantial treatises by Columella and Palladius. And the third consisted of a handful of texts, some of them existing only in fragments, by Xenophon, Nemesianus and two different writers named Oppian.

Renaissance naturalists plumbed the depths of all these texts, but they also scoured other classical works for references to nature. In his 1552 book *De differentiis animalium* ('On the Distinguishing Characteristics of Animals'), the English scholar Edward Wotton cited 214 authorities, though many were indirect citations by way of other sources.[13] Of those, 191 were ancients, mostly Greek and Roman sources, though the Bible and Zoroaster also appeared in Wotton's list. Seven or eight were medieval, and the rest were fifteenth-century humanists or Wotton's contemporaries. And even that enumeration is misleading. Among the ten authorities who accounted for over a third of the index's 432 lines, there were eight ancients. Pliny was cited on 160 of the volume's 220 folios, Aristotle on 131, Galen on 90, Dioscorides on 50, Oppian on 48, Athenaeus on 39, Varro on 37 and Ovid on 34. Only two moderns were in the top ten: Theodore Gaza, cited on 69 folios, and Ermolao Barbaro, on 34. Gaza had translated Aristotle's works on animals into Latin, while Barbaro had written an important commentary on Pliny: even Wotton's modern authorities were cited because they were experts on antiquity.

The uses of antiquity

How did early modern naturalists use these ancient sources? We can identify three broad ways – interdependent, but distinct – in which ancient texts shaped the daily practice of Renaissance naturalists. First, they edited ancient texts, translated them from Greek into Latin, and commented on obscure or controversial passages. Second, they extracted every last fact that could be mined from their pages and organised them in new compilations. And third, they used certain classical works, but not all, as models for their own new compositions.

The earliest humanist engagement with ancient sources for natural history was in translations, textual editions and commentaries. In the 1440s and 1450s, George of Trebizond and Theodore Gaza, rival refugees from Byzantium, produced Latin versions of Aristotle, Theophrastus and other ancient philosophical and medical writers, making their works available to scholars in the Latin West.[14] Gaza's translations became the basis for the first printed editions of many of these authors; the Venetian presses of Aldus Manutius soon followed, around the turn of the sixteenth century, with the first printed Greek editions.

Texts that referred to strange or obscure phenomena were particularly subject to 'corruption' through imperfect transcription, so textual critics like the Venetian Ermolao Barbaro and the Florentine Marcello Virgilio produced 'corrections' to received text in an attempt to ensure that the vitally important medical information contained in the works of Pliny, Dioscorides and others was as accurate as possible.[15] And while the earliest commentaries were philological, focusing on textual problems and comparing manuscripts, later commentaries included factual criticism and, often, substantial new material alongside the ancient text. Indeed, the immensely popular 'commentaries' on

Dioscorides by the Italian Pier Andrea Mattioli ended up dwarfing the original work of the Greek military physician.

Meanwhile, compilers like Leonhart Fuchs, Conrad Gessner and Ulisse Aldrovandi, armed with new editions and translations of ancient texts, scoured those works for all the facts they could provide on the names, anatomy, physiology, behaviour, uses and cultural significance of plants and animals. In works such as Fuchs's *Notable Commentaries on the History of Plants* (1542), Gessner's *Historia animalium* and Aldrovandi's many folio volumes (published from 1599 until well after his death) on the history of trees, metals, birds, quadrupeds and monsters, ancient works were carved up and rearranged in ways their authors would never have imagined. In these weighty tomes, readers could find everything known to ancient and contemporary writers on creatures ranging from the 'antalope' to the 'zibet or civet cat', the first and last alphabetical entries in Edward Topsell's *Historie of Foure-Footed Beastes* (1607), a work that was largely a translation of Gessner's Latin.

Aldrovandi's 1602 tome *De animalibus insectis libri septem* ('Seven Books on Insects') provides an example of the compilers' methods. In his general introduction to insects, Aldrovandi touched on the nature of their coitus and generation, which was sometimes from a fertilised egg, and sometimes spontaneous, the result of the Sun's action on decaying organic matter. In support of these diverse views, he proffered a wealth of ancient sources and some modern commentators: along with extensive quotations from Aristotle's work *Generation of Animals*, he also cited Theophrastus, Pliny, Galen, St Augustine of Hippo, the fifth-century bishop Eucherius of Lyon and the sixteenth-century writer Julius Caesar Scaliger.[16] Earlier, in the dedicatory letter to the Duke of Urbino, Aldrovandi had quoted the *Hieroglyphics* of the late antique writer Horapollo on how the scarab reproduced. But Aldrovandi supplemented ancient texts with his own observations – sometimes extensively; as he noted with surprise, ancient natural histories never mentioned dragonflies and damselflies, despite their ubiquity.

Even as they extracted facts from ancient texts, Renaissance naturalists also used ancient works as models for their own prose compositions. The most influential of these texts was Dioscorides's work *On Medicinal Substances*. Each of Dioscorides's chapters contained the names of a plant, animal or mineral substance; its description; the place and time where it could be found (unless it was available only from resellers in the marketplace); and its medicinal properties. This model was adopted by Renaissance botanists as the basis for their own descriptions of transalpine plants that were completely unknown to the ancients. The more expansive works of natural philosophers such as Aristotle and Theophrastus were also emulated in works like Andrea Cesalpino's *De plantis* ('On Plants', 1583).

Turning away from antiquity

While sixteenth- and early seventeenth-century writers emulated ancient works and incorporated their contents, by the middle of the seventeenth century, those works were largely superseded. Ancient texts continued to provide justifications for studying nature. In the introduction to his *Anatomical Exercises on the Generation of Animals* (1651), the English physician William Harvey praised 'the ancient philosophers, whose industry even we admire', for their 'unwearied labour and variety of experiments'. Their knowledge was limited and their claims sometimes erroneous, but like the moderns who emulated them, they, 'following the traces of nature with their own eyes, pursued her through devious but most assured ways till they reached her in the citadel of truth'. Aristotle provided Harvey with a robust theory of knowledge: drawing on the *Physics*, the *Posterior Analytics* and the *Metaphysics*, Harvey argued that all knowledge came from sensory experience, and thus, knowledge of the generation of animals must be drawn from 'experience, i.e. from repeated memory, frequent perception by sense, and diligent observation'.[17]

But as naturalists and anatomists like Harvey turned toward the kinds of problems that had exercised Aristotle and Theophrastus – attempting not only to enumerate, describe and classify animals and plants but also to explain how and why they were generated and lived – the actual claims of ancient authors no longer attracted much attention. When Jan Swammerdam discussed the method to be adopted in natural history in his *Historia insectorum generalis* ('General History of Insects', 1669), his references were Harvey, Descartes and Robert Boyle, not the ancients. When John Ray justified the publication of his new *Historia plantarum* ('History of Plants') in 1686, he did so because many years had elapsed since the publication, in 1640 and 1650 respectively, of Parkinson's *Theatre of Botany* and Johann Bauhin's *Historia plantarum universalis* ('Universal History of Plants').

Enlightenment naturalists did not reject the Renaissance idea that the ancients had been the founders of their tradition. Nor did they, in general, declare a radical break between the ancients and their own time, as Galileo, Descartes, and other defenders of the New Sciences had done in the seventeenth century. Authors continued to appeal to the symbolic authority of antiquity. In the first volume of his immensely popular work, *Le Spectacle de la nature* ('The Spectacle of Nature', 1732–50), the French naturalist and clergyman Noël-Antoine Pluche offered an engraved frontispiece depicting the plants and animals of the world being brought to King Solomon that he might describe them (Figure 1.2). They read the ancients – whose works, after all, were relatively slim, even Pliny's *Natural History* – but they did not linger on them.

When they did discuss the ancients, it was often critically. In his *Mémoires pour servir à l'histoire des insectes* ('Memoirs to Serve a History of Insects', 1734–42), the French academician René-Antoine Ferchault de Réaumur promoted the memoir, a detailed account of particulars, as best suited to his subject: 'Had Aristotle written his History of Animals following this plan, we would have learned much more from it. It contains many facts, and had he told us which ones he had witnessed himself, they would merit our belief; but he did not provide us a way to distinguish them from the others.' In his estimation, Pliny and Aelian, who based their works on Aristotle, were no better, and the organisation of Aristotle's book, too, was poor.[18]

Réaumur regretted that Aldrovandi, Gessner, Moffett and other Renaissance naturalists had spent so much time studying the ancients instead of nature itself. 'Nature opens, finally, even the eyes of those who are only looking to verify what they read in Aristotle and Pliny', but only after they 'gradually lost – perhaps even too much – the respect owed to the ancients'. Observers like Johannes Goedaert and Maria Sibylla Merian who could not read Latin were, in that respect, at an advantage. Indeed, the first step in studying the history of insects was to dispel the fables with which the ancients had surrounded the subject.[19]

Réaumur's rival, the comte de Buffon, appears to be an exception. In his *Natural History: General and Particular*, published in 44 volumes from 1749 to 1804, Buffon cited ancient writers hundreds of times, especially Aristotle.[20] In the lengthy discourse 'On how to study natural history' in his work's first volume, he proclaimed, 'it seems to me that Aristotle, Theophrastus, and Pliny, who were the first naturalists, were also the greatest in certain respects. Aristotle's history of animals is perhaps still the best work we have on the subject.' And, he added, Alexander's support made the whole thing possible.[21] Buffon's praise of the ancients, though, followed immediately after his attack on his arch-rival Carl Linnaeus and the Swedish naturalist's disciples. He magnified the former to diminish the latter. The ancients, Buffon claimed, had a broad grasp of nature (like Buffon himself), whereas the Linnaeans (like Réaumur) focused on minutiae. When Buffon turned to specific claims made by the ancients, he could be highly critical, such as their appeals to spontaneous generation: 'Most of the species that the ancients believed to be generated from decaying matter in fact come from an egg or a worm, as modern observers have verified.'[22] The nature and context of his citations suggest that ancient

Frontisp. del. Tom. 1.

Et disputavit super lignis, a cedro, quae est in Libano, usque ad hissopum, quae egreditur de pariete: et disseruit de jumentis, et volucribus, et reptilibus, et piscibus. 3.l.de'Regi 4:33.

Figure 1.2 The world's creatures being brought to Solomon for his natural history. Engraved frontispiece to N.-A. Pluche, *Lo spettacolo della natura*, vol. I (Venice, 1786, copy after the original French version from 1732). The artist has imagined Solomon in an eighteenth-century natural history institution, perhaps the Jardin et Cabinet du Roi (Royal Garden and Cabinet) in Paris. Image from the Biodiversity Heritage Library. Digitised by Smithsonian Libraries (www .biodiversitylibrary.org).

authority functioned as rhetorical confirmation of Buffon's own approach to the subject. If the ancients agreed with them, he approved; if not, he corrected them. Nature, not antiquity, was his guide.

Conclusion

Though the content of Enlightenment natural history owed little to the ancients, even toward the end of the eighteenth century we find echoes of the Renaissance vision of ancient natural history. In the third (1775) edition of his *Dictionnaire raisonné universel d'histoire naturelle* ('Universal Methodical Dictionary of Natural History'), Jacques-Christophe Valmont de Bomare offered an engraved frontispiece depicting Adam naming the animals (Figure 1.3). And the reader who turned to the first page of the text would find Adam's counterpart, Aristotle – or at least a figure who could be taken to be Aristotle – standing in a suspiciously Edenic setting and noting down his observations with an assistant at his side (Figure 1.4). The latter engraving echoes a fifteenth-century miniature depicting Aristotle as the 'scribe of nature'.[23]

The vision of a continuous tradition of natural history, with Aristotle and Solomon as its ancient founders, has proven surprisingly durable. In a recent, insightful study of Aristotle, *The Lagoon*, the biologist Armand-Marie Leroi argues that the ancient Greek did nothing less than invent science itself. Modern biologists are less likely to think of Solomon as their ancestor. But it is striking that practising scientists, like Leroi or Ernst Mayr (in his monumental *The Growth of Biological Thought*, 1982), still look back to ancient Greece as if Aristotle and Theophrastus were engaged in the same enterprise as a modern biologist. Why should this be so?

Part of the answer lies in a tendency to define disciplines by their objects, not their methods and traditions: to think that anyone who studies a particular object is engaged in the same enterprise, regardless of how they do it. The eighteenth-century Swiss naturalist Albrecht von Haller took this point of view in his *Bibliotheca botanica* ('Botanical Library', 1771–2), a work promising to review every writing on botany 'from the beginning'. In fact, Haller went beyond his promise, beginning with the ancient Druids, Chinese and Egyptian sages whose writings, like Solomon's, survived only in hearsay. In this view, Aristotle counts as a founder of natural history because he collected animals, studied the differences between them, classified them into broad groups, dissected them and proffered an explanation of how and why they functioned. From this perspective, the fact that his word '*historia*' meant simply 'inquiry' and that he considered himself a *physikos* (inquirer into nature), not a naturalist, are unimportant.

Adduxit ea ad Adam, ut videret quid vocaret ea.

Genese Chap. II.

Figure 1.3 Adam naming the animals. Engraved frontispiece to J.-C. Valmont de Bomare, *Dictionnaire raisonné universel d'histoire naturelle*, vol. I (Paris, 1775). Even as natural history had less and less to do with actual ancient sources and methods, the myth of Adam's complete mastery of the natural order remained potent. Image from the Biodiversity Heritage Library. Digitised by Smithsonian Libraries (www.biodiversitylibrary.org).

Car. Eisen del.   N.ᵉ de Launay sculp.

Figure 1.4 In this engraving, two ancient Greeks – perhaps Aristotle and Theophrastus – describe the animals. Engraving in J.-C. Valmont de Bomare, *Dictionnaire raisonné universel d'histoire naturelle*, vol. I (Paris, 1775). Enlightenment iconography continued to reach for the classical as well as scriptural past. Unlike Adam, whose knowledge is divinely inspired, the Greek investigators are writing down their observations in a book. Image from the Biodiversity Heritage Library. Digitised by John Adams Library at the Boston Public Library (www.biodiversitylibrary.org).

A second reason lies in how subjects were taught in ancient, medieval and Renaissance schools. When texts had to be copied by hand, and were thus relatively rare and expensive, a typical way to teach was by commenting on an authoritative text.[24] To Gessner, Bredwell and other humanists, it was natural to think that the authors of texts used to teach natural history, such as Aristotle, Dioscorides and Pliny, were part of a continuous tradition of practice. We know that their manuscripts were copied and commented, and that they were sometimes used as reference works (especially in medicine), but we also know that, with a few exceptions, their expansive inquiries into nature were not emulated. Renaissance naturalists, though, failed to appreciate this – aided by the fact that, as William McCuaig pointed out, Renaissance scholars often had only a vague appreciation of how profoundly ancient societies had changed; most tended to think of 'antiquity' as a whole.[25]

But there is a third reason. The Renaissance vision of antiquity was powerful because its creators inserted ancient writers, and their texts, into their own milieu. Renaissance naturalists created a new way to study nature, one based on experiencing nature, collecting its specimens, describing them carefully, and cataloguing them, and then they read the ancients in light of those new practices. Humanists had learned that in many ways, antiquity was very different from their own age. But hard-learnt as that lesson was, they often forgot it in practice. Their natural history was collaborative, based on extensive travel and correspondence, often involving objects brought at great expense from faraway lands. It was natural for them to think that Aristotle and Pliny had done the same, just as it seemed natural for Mayr and Leroi to imagine Aristotle as a twentieth-century scientist in ancient Greek garb.

But it is too simple to conclude that Renaissance naturalists – and their Enlightenment and modern successors – were just being bad historians of natural history. They were first of all practitioners, not historians, and they engaged ancient texts because they were, in fact, useful. Aristotle, Theophrastus, Dioscorides, Pliny, and even Aelian and the poets, provoked their readers. They offered unusual claims and, in some cases, powerful theories. Their works, read through the lens of Renaissance and Enlightenment practice, offered models for investigating nature and provocative claims to investigate, confirm or debunk. Ancient writers were not part of a continuous tradition of natural history. But in the Renaissance vision they were, and that vision inspired generations of naturalists.

Further reading

Bolgar, R. R., *The Classical Heritage and its Beneficiaries* (Cambridge, 1963).

Enenkel, K. A. E. and Smith, P. J. (eds.), *Early Modern Zoology: The Construction of Animals in Science, Literature and the Visual Arts* (Leiden, 2007).

French, R., *Ancient Natural History: Histories of Nature* (London, 1994).

Grafton, A., Most, G. W. and Settis, S. (eds.), *The Classical Tradition* (Cambridge, MA, 2010). See especially the articles on 'Botany', 'Natural History', and 'Zoology'.

Hornblower, S., Spawforth, A. and Eidinow, E. (eds.), *The Oxford Classical Dictionary*, 4th edn (Oxford, 2012).

Huxley, R. (ed.), *The Great Naturalists* (London, 2007).

Leroi, A.-M., *The Lagoon: How Aristotle Invented Science* (New York, 2014).

Monfasani, J., 'Aristotle as scribe of nature: the title-page of MS Vat. Lat. 2094', *Journal of the Warburg and Courtauld Institutes*, 69 (2006), pp. 193–205.

Nauert, C. G., Jr, 'Humanists, scientists, and Pliny: changing approaches to a classical author', *American Historical Review*, 84 (1979), pp. 72–85.

Nauert, C. G., Jr, *Humanism and the Culture of Renaissance Europe*, 2nd edn (Cambridge, 2006).

Ogilvie, B. W., *The Science of Describing: Natural History in Renaissance Europe* (Chicago, 2006).

Perfetti, S., *Aristotle's Zoology and its Renaissance Commentators* (Leuven, 2000).

Reynolds, L. D. and Wilson, N. G., *Scribes and Scholars: A Guide to the Transmission of Greek and Latin Literature*, 4th edn (Oxford, 2014).

Sarton, G., *Appreciation of Ancient and Medieval Science during the Renaissance, 1450–1600* (New York, 1955).

2   Gessner's history of nature

Conrad Gessner's *Historia animalium* ('History of Animals', 1551–8) is one of the best-known publications in Renaissance natural history. It is a good example of how a study of nature for a relatively new audience drew on a wide range of cultural resources of the period – humanism, printing, collecting and commerce. As a form of knowledge practised by the ancients, natural history required for its revival proficiency in the classical languages (Hebrew, Greek and Latin), philology, historical sensitivity and source criticism.[1] Gessner possessed all these skills and was furthermore able to draw on the vibrant culture of printed books of his time. He was a voracious reader of books, as well as editor and author of more than seventy titles. In 1545, he published *Bibliotheca universalis* ('Comprehensive Library'), a list of all the books that had ever been written by authors since ancient times. Together with information, drawings and objects gathered from his correspondents and by himself, knowledge garnered from books constituted the foundation of his study of natural history.[2] His *Historia animalium* was published in four volumes (addressing viviparous quadrupeds, oviparous quadrupeds, birds, and fish and aquatic animals) and ran to a total of more than 3,300 folio pages with woodcut illustrations. It included animals mentioned by classical authors (Cicero's 'alces', for example), those well known to Europeans (cats, dogs, cows, horses, etc.), recently discovered exotic animals (armadillo, guinea pig, turkey, etc.), as well as unicorns and sea monsters. It would be misleading to charge Gessner with credulity or gauge him by modern standards of zoology, since his project was expansive, covering everything that was written about an animal, including its uses in fables, poetry, proverbs and emblematic literature.[3] In order to understand what kind of knowledge Gessner's *Historia animalium* embodied, this chapter will discuss how this work differed from the study of animals at universities, what kind of knowledge it encapsulated, and what was involved in bringing together such a knowledge.

Study of animals at universities

The development of Renaissance natural history depended on highly educated scholars, and yet it was a field that developed largely outside academia, since universities did not recognise it as a distinct branch of knowledge.[4] By the sixteenth century, it was well established that the Aristotelian scheme of knowledge differentiated between the causal investigation of 'philosophia' (tackling the 'why') and a descriptive and comprehensive field of 'historia' (tackling the 'what') that provided the basis of philosophy. Natural history as such was not taught at universities, either as preparatory to natural philosophy or as a distinct subject. Aristotle's books on animals were known throughout the Middle Ages, but if they were read at all in the arts curriculum, it was towards the end of the arts course, and they attracted academic commentaries only sporadically.[5]

This is not to say that animals were not part of the study of natural philosophy. They were included, for example, in natural philosophical textbooks that emerged in the first half of the sixteenth century, which summarised Aristotelian concepts and topics instead of commenting on Aristotle's argument line by line. One such textbook, Gregor Reisch's popular *Margarita philosophica* ('Philosophical Pearl', 1503) described minerals, plants and animals as 'mixtures' made out of the four elements (air, water, earth and fire). Animals were divided (as they were in the Bible) into flying, swimming or crawling creatures, with discussion of their manner of generation, but without any extended discussion of different species and their distinguishing characteristics.[6] Morphological variety of plants, for example, was noted as evidence of the Creator's munificence, but not something to spell out in detail. This was indicative of the limited scope for discussing variation or morphological differences within Aristotelian philosophy, because external features such as the colour of plumes or the shape of hooves that could be shared with other species could not constitute an essential definition of a species.[7]

The variety of animals, plants and minerals were studied more closely in medical faculties as part of therapeutics, as they constituted naturally occurring medicines, *materia medica*.[8] In medieval universities, the medicinal effects of substances were commonly studied through compilations from Dioscorides, Galen and Aristotle, such as Avicenna's *Canon*, which included sections on medicinal materials and how to combine them, Serapion's *De simplicibus* ('On Medicinal Simples'), which grouped plants by the strength of their primary qualities, or Sextus Placitus's *De medicamentis ex animalibus* ('On Medicines Made from Animals'), which listed ailments each animal could cure. Such studies were rarely a focal point of the training of learned physicians, partly because of their emphasis on theoretical matters, and partly because of the increasing reliance of physicians on apothecaries, as discussed by Pugliano in Chapter 3, this volume.

The Renaissance enthusiasm for studying directly the original ideas and practices of antiquity led to publications of the Greek texts of Galen, Dioscorides and other medical authors, edited by humanist physicians who also offered Latin translations made afresh from the Greek texts. A greater awareness of a variety of classical authors and newly (re)discovered works by well-known authors generated philological debates about, and comparisons of, medicinal material discussed in different works, while an increasing appreciation of classical rhetorical and dialectical methods led the physician Leonhart Fuchs, for example, to use morphological features to identify the medicinal plants discussed by Dioscorides and Galen.[9]

Because of their familiarity with *materia medica*, university-educated physicians were well equipped to publish books on animals, minerals or plants by the middle of the sixteenth century.[10] Not all such publications were aimed at a medical audience, however: William Turner's *Avium præcipuarum, quarum apud Plinium et Aristotelem mentio est, brevis & succincta historia* ('A Short and Succinct History of the Principal Birds Noted by Pliny and Aristotle', 1544) collated the names of birds mentioned by Aristotle and by Pliny the Elder and provided English or German equivalents, without discussing any medicinal uses. Gessner himself said that he first came to the study of plants, animals and minerals as part of his medical training in therapeutics.[11] What he ended up producing, however, was neither natural philosophy nor medical knowledge as taught at universities.

The audience of *Historia animalium*

Gessner's *Historia animalium* did not represent an institutionalised academic discipline and was certainly not 'Aristotelian' in the strictest sense, because it showed no interest in preparing material for causal investigation. It did draw, however, on the Aristotelian sense of *historia* as a comprehensive form of knowledge, both in the animals to be discussed and the type of information about each animal to be included. Aristotle's study encompassed all living beings in terms of where they lived, their actions, habits, modes of reproduction, and appearance. Gessner too aimed for comprehensiveness (though his work on insects was published only posthumously) and grouped animals by where they lived and the manner of their reproduction. Yet it would not be helpful to categorise Gessner's work as a new, specialist knowledge of 'zoology', as he addressed explicitly a wider Latinate audience of 'philosophers, physicians, grammarians, philologists, poets and all those studying languages'.[12] For Gessner, these constituted the 'well and freely educated' man who would take pleasure in the contemplation of animals.[13] Gessner was here probably drawing on a distinction made in the opening lines of *Parts of Animals*, where Aristotle discussed two types of proficiency: scientific knowledge (*episteme*) that understood matters in a demonstrative way and educated competence (*paideia*) that judged matters in a probable manner.[14] Like Aristotle, Gessner believed that a man who was educated well could judge what he saw and heard not in a demonstrative manner, but by relating each of those things to principles that he had grasped for himself from the repeated use of particulars.

If some of his audience thus had the capacity to judge matters only in a probable way, Gessner nevertheless claimed that knowledge (*scientia*) of animals, plants and metals was certain, or at least more certain than those concerning meteorology or more 'subtle' matters, because the former were closer and 'better known to us'.[15] This was another Aristotelian point: in contrast to divine and unperishable things, Aristotle argued that 'we have better means of information, however, concerning things that perish, that is to say, plants and animals, because we live among them; and anyone who will but take enough trouble can learn much concerning every one of their kinds'.[16] Though by no means a recognisable university discipline of natural philosophy or medicine, Gessner's project on animals was thus configured from several Aristotelian elements, rather than marking a clean break from them.

In the preface to the first volume of *Historia animalium*, Gessner presented his work as useful to many livelihoods.[17] It was of course helpful to medicine, since both medical knowledge and skill could be extended by the study of animals: cures found for animal diseases might be applied to human diseases, trials of new or dubious drugs should be carried out on beasts rather than humans, certain animals could be used to master basic dissection techniques, and parts of animals could be used for medicines. But Gessner pointed out that the book was also beneficial for others. Cooks could learn which parts of animals could be prepared for food and how, fishermen could identify fish to eat for themselves and to sell on, while cowherds and producers of cheese, butter and other things could learn how to cure their flock. An animal's hide, wool or hair could be used by tanners, furriers, curriers and shoemakers to make clothes, purses, saddles and other goods. Cows, oxen and horses could till the land, carry burden, help in construction, and take part in battle. Dogs could guard homes, protect cattle and love humans. Animals also taught moral lessons. Even if no immediate benefit or profit followed, Gessner emphasised that the study of animals would lead to an appreciation of their Creator: each history of an animal was a hymn to God.

As *Historia animalium* was to offer such varied information for each animal, it was ordered alphabetically so that it could be used like a dictionary, though this of course meant that discussion of taxonomy was limited. Different types of information on animals were grouped under alphabetical headings: its name in various languages (A), where it lived and how it differed according to where it lived (B), its manner of living and habits (C), its character, sympathies and antipathies (D), its use other than for food or medicine (E), its use as food (F), as medicine (G), and its literary uses, for example in poems, pictures, proverbs or metaphors (H). As noted by Ashworth, the last section on literary and philological uses was often the longest, reflecting the historical and contemporary interest in the symbolic meanings of animals.[18] These subdivisions guided readers with different interests to locate the piece of information they were looking for promptly.

Although Gessner claimed that his *Historia animalium* could benefit anybody who used animals for their living or was interested in them, the work is unlikely to have been bought by cowherds or cobblers – the four-volume set would have cost just under seven florins, and a coloured set was priced at nineteen florins. Gessner's own annual income from his teaching and medical practice was 142 florins in 1554, so an uncoloured set was about a twentieth of his annual income, and a coloured set was equivalent to a month and a half's salary.[19] This was, rather, a lexical work for a general, Latinate, well-educated audience of some affluence.

The making of *Historia animalium*

What did it entail to create such an all-compassing study of animals in the middle of the sixteenth century? Gessner was in a good position to gather everything that was ever written about animals because of the work he had done for his *Bibliotheca universalis*. In the first volume of his *Historia animalium*, Gessner listed 251 extant and 18 no longer extant titles on animals.[20] Of the 251 extant books, 3 were written in Hebrew, 68 in Greek, 164 in Latin (by classical and contemporary authors, as well as translations of Arabic authors), 8 titles in German, 5 in French and 3 in Italian. About 70 per cent of these works were asterisked, which signalled that Gessner had used all of their descriptions of animals. These were authors who had specifically dealt with animals or intended to convey specific information on animals in part of their work. The works of Aristotle and Albertus Magnus on animals were asterisked, but those by Avicenna or Averroes were not, because the knowledge of the latter two was derived from Aristotle, and whatever additions they offered had been picked up by Albertus. Gessner also explained that he had not excerpted every mention of an animal by historians, poets or others, unless it related to some knowledge about the animal itself. He cited extensively and verbatim from works he deemed reliable, as was the case with the names, descriptions as well as woodcuts from Guillaume Rondelet's *De piscibus marinis* ('On Marine Fishes'). Such borrowings were meticulously acknowledged each time. The extent of copying was such that the title page of the fourth volume of *Historia animalium* on fishes indicated that it contained the works of Rondelet and Belon.[21] Gessner thus worked with all known studies on animals. These were prioritised for content, parsed for relevant information, and cited under the appropriate headings. In effect, this added a historical dimension to the study of animals.

Gessner's ambition for comprehensiveness meant not just gathering historical descriptions of animals from books written since antiquity, but also finding information about new or exotic species by actively cultivating and managing correspondents. He effectively used material prefaced to his publications to elicit contributions, and suggested that letters be sent to him through the mercantile networks at the market cities of Antwerp, Venice, Lyon and Frankfurt.[22] Interest in natural objects was not the exclusive domain of university-educated physicians – princes, merchants, apothecaries and others were keen to collect rare or exotic natural objects of various kinds. The wider cultural network of trading such objects is reflected in *Historia animalium* when Gessner mentions prices: a good-quality hide of a leopard was worth six to seven French gold coins; the bird of paradise was valued at 800 thalers; and a 'horn' of a unicorn was priceless.[23]

Contributors to Gessner's *Historia animalium* were listed at the beginning of each volume and also credited in the text. Many of them were physicians; some were theologians, jurists or humanist scholars; a few were apothecaries, politicians, printers, merchants or surgeons. In the text, Gessner noted assiduously their profession or other qualifications: for example, Ulisse Aldrovandi was described on one occasion as 'a very excellent man in medical matters as well as in the history of plants'.[24] This was not just an elaborate public acknowledgement to induce others to write to Gessner, but also an indication that the source of information was reliable.[25] Albrecht Dürer, 'that very excellent painter whose books on picturing are extant', whose print of the rhinoceros Gessner reproduced, and Lucas Schan, 'the most attentive painter and fowler in Strasbourg', whose pictures of birds Gessner took on trust, were the only two artists named and deemed reliable. In contrast, the picture of an elk that was sent in by an unnamed painter was specified as having been certified as true by several eyewitnesses.[26] Not everybody was named, but Gessner still offered some indication of their reliability by describing them, for example, as 'my erudite friend' or 'my nobleman friend'. He also noted that he learnt from some 'trustworthy' country-folk about the hibernating habit of dormice. This suggests that

Gessner had indeed consulted 'the learned, the unlearned, citizens, foreigners, hunters, fishermen, fowlers, shepherds, and all kinds of people', but that he was also careful to indicate the reliability of their description or images

Gessner was one of a number of university-educated physicians who had a strong sense of history.[27] His study of animals had a 'historical depth' in its engagement with ancient authorities. He collated and compared usages of names of an animal among ancient authors, as he did with 'pardalis' (Greek for panther), 'pardum' (Latin for male panther), 'panthera' (Latin for female panther), 'namer' (Hebrew) and the more recent 'leopardus' (which he determined as a different animal). He sought to establish the identity of animals described by classical authors such as the beasts in the Hercinian forest described in Caesar's *On the Gallic War*; and he also combed through earlier descriptions to determine whether an animal was new or not – a monstrous monkfish discovered recently in Norway was included under the heading of Pliny's mermen.[28] Gessner cited passages verbatim from different authors, even when they were quite similar, because such repetition made statements more reliable; this was a strategy well known in civil history.[29] There was thus a historical dimension to Gessner's textual practice, in that anything that was ever written about living beings had to be gathered, parsed and collated, and new knowledge was to be gauged against a historical genealogy of descriptions of animals. This is a process that has been identified as a form of 'learned empiricism'.[30]

Many entries were accompanied by a woodcut image, conceived as a sort of 'paper menagerie', to which readers had permanent access without the effort or frightening prospect of visiting an actual menagerie.[31] In this sense, pictures were meant to function as substitutes for the objects they depicted, as discussed by Felfe in Chapter 11, this volume. For example, exotic birds Gessner had not seen were described from drawings (see Plate 1). The images in *Historia animalium* were not always drawn from first-hand observation, however. Several of them were copied from earlier publications, manuscripts or prints, just as texts were copied out of earlier works. Gessner acknowledged that his pictures were of varying quality, but did not always replace defective ones with more reliable ones. He often juxtaposed images of the same animal or compared his woodcut to others printed elsewhere, and delivered judgements on pictures, such as 'not so pleasing', 'false', 'not good enough', 'good enough', 'less accurate', 'more accurate' or 'very elegant' (Figure 2.1). He approached images of varying quality just as he worked with texts of variable reliability and informational content – they had to be collated, compared and evaluated.[32]

Images played an additional role in enabling Gessner to 'compile' a new animal. For instance, an image of a toucan was compiled out of

Figure 2.1 'Sepia', C. Gessner, *Historia animalium*, vol. IV, p. 1024. Gessner remarked that in this image made in Venice, the cuttlefish had more legs than the eight that Rondelet had assigned to it, 'no doubt due to the negligence of the painter'. This is one of several examples where Gessner decided to include an image that was made for him which was less accurate than those found in other publications. Zentralbibliothek Zurich, NNN 48 | F.

a beak sent to him and a textual description in André Thevet's *Les singularitez de la France antarctique*.[33] Belon's image of the armadillo was corrected to show shorter feet in light of the animal's carapace, tail and claws that Gessner had received.[34] Objects from his own collection were also represented in order to supplement his discussion, though few of these objects have survived. Images thus helped to compile an animal where neither text, object nor drawing alone provided full information, and became an important part in the process of 'learned empiricism'.

Nature's shapes and colours

A history of animals for a well-educated audience thus required humanist proficiency and the ability to access books, drawings and objects through a network of correspondents and others. The work also required the willingness of a publisher to print not just the text but also invest in the making of woodcuts. The woodcuts generated for *Historia animalium* were recycled in what was essentially a pictorial album of animals, *Icones* ('Icons', 1553, 1555, 1560), which was a common way for printers to optimise their investment. In *Icones*, the woodcuts were rearranged into a grouping that was not alphabetical: the viviparous quadrupeds were divided first into two groups, tamed and wild; the tamed animals were divided further into those with horns and cloven hooves and those without horns and with uncloven hooves; the wild animals were grouped into horned animals; large animals without horns; medium-sized animals without horns; and, finally, small animals without horns, such as hares, rabbits and mice. *Icones* was thus a different kind of publication in that it grouped animals by their external features. Images could function as part of humanist compositional practice, but they could also acquire a life of their own, to be adapted and adopted in different contexts.

The printer Froschauer sold both the volumes of *Historia* and of *Icones* in coloured versions at a higher price. Colour production in printed books remained a technical challenge for centuries, and colouring was mostly done by hand. Gessner assured his readers that the colouring would be done after an exemplar in the printer's shop, though he privately expressed his misgivings over the sloppiness of the work done.[35] Colour was nonetheless an important element in Gessner's world of nature, and his text indicates that he wrote with a coloured copy in mind. He named one parrot 'Eryt[h]roxanthum' because it had yellow ('xanthos' in Greek) plumage within a red ('erythros' in Greek) coat (see Plate 2), and another 'Erythrocyanum' because it had some blue ('cyanos' in Greek) features on its wing in an otherwise red plumage.

The hand-coloured woodcut was close to what Gessner conceived as constitutive of morphology, with lines and colours.[36]

Animals were not the only things in nature that Gessner studied. He was also interested in plants and minerals, and hoped to publish on both topics, though he only managed to do so for the latter. For his unfinished history of plants, he left a large number of drawings, of plants that grew in his garden, of parts of plants that were sent to him from correspondents, and of dried plants that were kept in a separate herbarium. These were drawn in crisp outlines with watercolour. His annotations on these drawings indicate that he consulted many printed books on the topic, recorded the date and names of donors of the plants, and sometimes waited several years to get a picture of all parts of the plant.[37] He hoped to publish his history of plants without the philological section that made his *Historia animalium* voluminous, but the way Gessner approached plants was similar to the way he worked on animals.

This was also the case in his last book on minerals, gems and stones, *De rerum fossilium, lapidum et gemmarum maxime, figuris et similitudinibus liber* ('A Book on the Shapes and Similarities of Things Dug up, of Stones and especially Gems', 1565). Gessner again drew extensively on other publications on the topic as well as on objects that he received from correspondents. He grouped stones, gems and minerals by means of lines and shapes, and emphasised the need for the woodcuts to be coloured, because of the difficulty of distinguishing stones and minerals without colour. In the preface, he gave the strongest justification of why attending to the outward forms of natural objects mattered: the images and shapes of natural objects were like hieroglyphic marks, but truer than the pictograms made by the Egyptian priests because they had been impressed on them by nature. Indeed, he described some stones with images (we would call them fossils) as having been 'depicted by the wonderful skill of nature'.[38] Attention to external features was what united Gessner's study of animals, plants, minerals and fossils, as they were all part of God's creation.

Humanist skills, mastery of the classical tradition, conversance with contemporary scholarship and access to scholarly and mercantile networks were important elements in histories of nature by Pierre Belon, Hippolyto Salviani, Edward Wotton and Ulisse Aldrovandi, which addressed a wide audience that included, but was not confined to, university-educated physicians.[39] Gessner's work on animals typified the skills, resources and effort required for study in the period, and also set a model for others to follow and develop. Aldrovandi, for example, pursued further the idea of nature as painter or even an engraver in his work on fossils.[40] Assembling parts of an animal from a variety of sources became a common way to visualise an unknown animal, including monsters. Gessner's images were copied and recopied in subsequent works of natural history, as attention to morphology became an important feature of knowing nature.

Further reading

Blair, A., *Too Much to Know: Managing Scholarly Information before the Modern Age* (New Haven, 2010).

Egmond, F., *Eye for Detail: Images of Plants and Animals in Art and Science, 1500–1630* (Chicago, 2016).

Kusukawa, S., 'The role of images in the development of Renaissance natural history', *Archives of Natural History*, 38 (2011), pp. 189–213.

Leu, U. B., *Conrad Gessner (1516-1565): Universalgelehrter und Naturforscher der Renaissance* (Zurich, 2016).

Pinon, L., 'Conrad Gessner and the historical depth of Renaissance natural history', in G. Pomata and N. Siraisi (eds.), *Historia: Empiricism and Erudition in Early Modern Europe*, (Cambridge, MA, 2005), pp. 241–67.

3   Natural history in the apothecary's shop

Apothecaries – those makers and retailers of remedies active in most European towns – constituted the second largest group of enthusiasts of early modern natural history after physicians and students of medicine. Historically, however, this category of artisans and the importance of their medical work for natural history have remained behind the scenes. Contemporary botanists like Leonhart Fuchs, Otto Brunfels and Rembert Dodoens were the first to be dismissive, depicting apothecaries as mere herb gatherers, ignorant of the true science of plants, prone to errors and frauds and only wedded to profit.[1] The long-standing reluctance of historians of science and ideas to address the world of craft and commerce did little to alter this image. When scholars influenced by Marxism and sociology eventually began to revaluate the significance of craftwork between the 1940s and 1960s, their studies privileged those arts – mechanical, military and of instrument-making – deemed most relevant to the scientific revolution.[2] Similarly, the sophisticated investigations into the culture of curiosity and collecting which, from the 1980s onwards, uncovered the creativity of courts, universities and scientific academies, treated markets and shops principally as feeding a new taste for the accumulation of goods and news, and providing another hunting ground for the natural philosopher.[3] The apothecaries' lack of university education and the menial nature of their trade was seen to limit them to the role of specimen suppliers and assistants to better-educated and more genteel naturalists.[4]

Nowadays, it is clear that the collective enterprise that was natural history could not have developed without the active contribution of tradesmen, craftsmen and barely literate go-betweens. On the Italian peninsula, apothecaries were undoubtedly those artisans who most influenced the renascent study of nature. Masters in the manipulation of flora and fauna, both local and remote, they were perfectly situated intermediaries who could open up the world of natural materials to the scrutiny of a new generation of *curiosi*. The craft of pharmacy, in turn, contributed material skills essential to the practice of natural history, from techniques of specimen preservation and transportation to the training of the senses for fieldwork. Crucially, pharmacy also facilitated the study of nature locally in neighbourhoods and towns, providing a unique site where scientific interests could be explored collectively and without the hierarchical rigidities of other places of learning: the shop.

Pharmacy, in other words, provided many of the practical tools to implement that new discourse of empiricism and learning by getting one's hands dirty, which became associated with natural history and its study of particulars. Apothecaries actively contributed to this discourse. Rather than go-betweens, many became full-fledged participants in natural history. If some remain anonymous to this day, others, like Francesco Calzolari of Verona or Ferrante Imperato of Naples, achieved fame in their lifetime for their field-work and ownership of the largest museums of *naturalia* in Europe. They were the period's most distinguished naturalists, alongside Conrad Gessner, Pietro Andrea Mattioli, Carolus Clusius, Joachim Camerarius, the Bauhin brothers and Ulisse Aldrovandi. These men (for hardly any women were involved) left behind a wealth of records – including hundreds of letters to naturalists, essays on rare specimens and local floras in print – which reveal apothecaries as the missing link in the history of natural history.[5]

Intermediaries

The histories of pharmacy and natural history are closely intertwined. Before embracing all forms of nature, natural history developed out of an interest in medicinal substances, the subject of prestigious treatises by Graeco-Roman authors 'rediscovered' by north Italian humanists in the late fifteenth century. In order to make sense of the descriptions of *materia medica* contained in these texts, scholars took to visiting pharmacy shops, which regularly used costly drugs from the eastern Mediterranean and Levant – where the ancients' plant-lore had originated – alongside local herbs. What began as textual scholarship soon led to debates on the usefulness of this classical knowledge for contemporary therapeutics. Theophrastus's *On Plants*, Dioscorides's *On Materia Medica* and Pliny's *Natural History* became guides for investigating the stock of pharmacies, and for emancipating apothecaries' practice from the errors into which it had allegedly been led by reliance on medieval Arabic epitomes and folk herbals. Medical students were encouraged to relearn the act of compounding that had been left to 'lesser hands', and thus to follow Galen's injunction that a doctor master all instruments of his profession.[6]

This origin strongly influenced the development of natural history in Italy. By the mid sixteenth century, training at the apothecary's shop had become common for physicians, creating the basis for a partnership between university-educated and artisan practitioners that would be influential throughout the century. In turn, the focus on the medicinal in nature, and the classical corpus describing it, continued to dominate the interests of Italian naturalists.

Up to the late 1540s we learn of the involvement of apothecaries in *res herbaria* primarily indirectly, through the treatises of humanists and physicians. In these early herbals, apothecaries appear as anonymous suppliers of materials, yet remain marginal to the discussion forming around them. Emblematic is Antonio Musa Brasavola's *Survey of All the Simples in Use in Pharmacy Shops* (1537), a fictional dialogue between a vigorous young scholar (Brasavola) and an elderly artisan (Senex Pharmacopola) who meet in a field outside the university town of Ferrara to identify the flora growing at their feet, helped by a copy of Dioscorides.[7] Through their exchange, Brasavola's readers were introduced to a new way of studying nature which integrated reading with fieldwork; they also found in Pharmacopola's age and subservience a transparent metaphor for the decrepitude of knowledge in the shops and the necessity of upgrading it with scholarly help.

Only from the 1550s, when apothecaries began to document these activities themselves, did the extent of their brokerage become clear. At this time, a change in the nature of their involvement can be detected. A telling example is provided by Giovan Battista Fulcheri who, between the 1560s and 1570s, became a close aide of the celebrated Bolognese naturalist Ulisse Aldrovandi, first while training at the Coral pharmacy in Venice, then in his native Lucca. Over the years, Fulcheri sent Aldrovandi the head of a monstrous fish from the Adriatic and coloured stones received from Naples, reported on the difficulties of finding the astringent herb poterion in the Venetian lagoon during winter, compiled a catalogue of the Pisan garden of Father Michele Merini, which the priest was recalcitrant to do, and attempted to rescue a set of miniatures pawned by an anonymous Venetian gentleman for Aldrovandi's collection of specimen drawings.[8]

By the mid sixteenth century, in other words as natural history consolidated into a field of study remote from philosophical speculation and dependent on first-hand examination of specimens, apothecaries had become indispensable interlocutors. Wielding unparalleled access to *naturalia* and the commercial and medical communities that made use of them, they were masters of a practical dimension of natural knowledge and its market for which the university scholar with his book learning was poorly equipped. Their multiple material skills proved especially vital for the culture of collecting that flourished in conjunction with natural history.

This material competence had its origin in the craft of pharmacy or *speciaria*. This was a 'very mercantile' profession 'with much science' according to one famous survey of occupations in Renaissance Italy,

Tommaso Garzoni's *La piazza universale* (1587).[9] Based in humoral theory, sixteenth-century pharmacy restored health by uniting the disparate virtues of spices, herbs, flowers, seeds, leaves, roots, bark, resins, clays, minerals, gemstones, metals, fish and meat, animal fats and oils. Some of these ingredients could be gathered from nearby fields and cultivated in gardens. Yet Italian pharmacy was especially renowned for its complexity and reliance on expensive imports from China, India and the Levant. Its practitioners thus regularly liaised with Venetian, Florentine and Genoese merchants active between the Mediterranean ports and the Asian caravan routes, and with German and Baltic traders responsible for transporting precious stones, metallic ores and amber south of the Alps. These networks could be leveraged to supply the scholar with specimens not found domestically, and also to update him on the state of the market for natural commodities. Marco Fenari, of the Two Moors pharmacy in Venice, was Aldrovandi's informant for cargoes from Crete, for example.[10] Apothecaries also served as reputable links to suppliers of lower status, such as charlatans, fishermen and plant gatherers (*herbolari*), often illiterate men and women who scoured local fields and rivers and sold their catch or harvest to town retailers. To satisfy Aldrovandi's desire for Adriatic fish, Marco Orselini turned to 'Alvise the *herbario* or herb collector' of Venice, one of many individuals who were excluded from the naturalists' community, despite their meticulous knowledge of flora and fauna.[11]

As intermediaries for naturalists, apothecaries were relied upon to provide not only specimens, but also technical know-how about their identification and nomenclature. Thanks to their boundary position among constituencies of knowledge, they held the key to reconciling folk names with pharmacy Latin and the classical humanist terminology. 'Artemisia', explained Giorgio Melichio, 'is called by the common people the herb of the Virgin.'[12] They were also authoritative cataloguers of nature, able to supply precise cues about the taste, smell and texture of an object, and to create a full sensory label for scholars accustomed to verbal description and the observation of morphological traits. This expertise was the outcome of years of formal apprenticeship required by the guild, during which apothecaries learnt to classify the qualities of ingredients using their senses, and to compound them using mortars and ovens (Figure 3.1). They were instructed by example, but also expected to acquire basic Latin in order to memorise the contents of herbals and of the new editions of the ancients consulted by medical students.

For collectors aspiring to immortalise nature in a room, the apothecaries' training supplied an additional set of lessons: how to select specimens in the wild, and, once gathered, how to preserve them and circulate them to colleagues. Testifying to the incompetence of the average scholar in this matter, in 1562 the physician Gregorio

Figure 3.1 Five elder apothecaries in a pharmacy, with an apprentice in the background working with pestle and mortar. *Hortus sanitatis* or *Herbarius zu Teutsch* (Augsburg, 1496). Wellcome Library, London.

Cantarini mourned his inability to send the fish he had attempted to dry in ash to Aldrovandi, because they had all rotted, 'among which the demise of the French shrimp truly upset me'.[13] This lack of manual skill in dealing with perishable *naturalia* stunted the growth of a collection. While every naturalist hoped to receive specimens that were whole and fresh (if not living), these more often arrived in pieces, mouldy or dried up and discoloured with age. Animals were even less tractable, and when it came to rare botanicals or exotic creatures brought back from the Americas, aspiring virtuosi regularly faced unknown material armed at best with inaccurate descriptions. These challenges contributed to the age's fondness for drawings from life and the development of archiving tools like *herbaria*.[14]

Apothecaries could instruct as to both where and when to pick simples. Trees were more vigorous on sunny and windswept mountains; herbs were best picked in spring, just after the morning dew, as this ensured they remained fresher and more easily observed with their proper colour and form and, if necessary, copied on paper. Apothecaries could also teach how to season items for shipping: adding some lichens to wet soil for living plants, for example, or sending samples in jars filled with vinegar. They had learnt to equip their shops with dark and cool rooms appropriate for storage and had perfected preservation techniques. Roots and flowers should be hung in the shade in a cool room or dried in the gentle heat near an oven; animal skins and parts instead required substances borrowed from the kitchen like salt, alum and wormwood. Different ingredients demanded different storage containers: glass jars for flowers, leather sachets for seeds, wooden boxes for roots.[15] The ultimate testament to the import of this craft knowledge for the collecting community can be found in the early cabinets of curiosities, which borrowed heavily from the furniture, design and partitioning principles of the pharmacy shop (Figure 3.2).

Practitioners

It goes without saying that not all apothecaries became naturalists, and for those who did the process was not linear. While the craft gave them a basic competence, they had to reconfigure themselves and the commercial *materia medica* they worked on as epistemic actors and objects. They also had to reach beyond their ordinary clientele and patronage networks and forge for themselves an authoritative role in a social setting that embraced a different value system and espoused the overall aim of documenting nature rather than making use of it.

The first step consisted of transforming one's knowledge of medicinal ingredients into a wider botanical literacy. Such 'improvement' was pursued outside the guild's structures, and benefitted from the new pedagogical facilities of Renaissance Italian towns, including a dynamic print market in the vernacular, vibrant universities, and institutional investment in capturing the secrets of nature. Francesco Calzolari's career took off when, rather exceptionally for an artisan, he decided to audit Luca Ghini's (1496–1556) lectures on

Figure 3.2 Pharmacy interior. J. de Renou, *Antidotarium dogmaticorum vetus renovatum* (Paris, 1608), frontispiece. Wellcome Library, London.

materia medica at Bologna in the 1550s. Instituted by Italian, French and German universities to instruct a new generation of medical practitioners, such lectureships quickly became networking opportunities for aspiring naturalists across Europe. It was here that Calzolari met Ulisse Aldrovandi. Of genteel birth and destined for an academic career, Aldrovandi was instrumental in furthering Calzolari's learning, recommending new publications on *res herbaria*, and in strengthening his scholarly standing, introducing him to key figures like Pietro Andrea Mattioli. Calzolari continued to build his profile by making regular trips from his Veronese pharmacy to the new botanical gardens at the universities of Padua, Bologna and Pisa, and befriending their keepers, Luigi Anguillara and Melchiorre Guilandino.[16] It is no coincidence that most of the artisan naturalists that we know of inhabited urban centres that provided such stimuli for further education: major port cities like Venice and Naples, through which rare imports passed; court towns like Mantua, Florence and Rimini, where fashions for expensive medicaments and luxury items flourished; and university centres like Padua and Verona, with active medical faculties and colleges of physicians.[17]

The returns from such an education were many, for a tradesman. Familiarity with current discussions on *materia medica* made it possible to prepare many important classical remedies that had been abandoned for lack of knowledge or availability of key ingredients. This change is most apparent in the Renaissance revival of theriac and mithridate, two prodigious antidotes against poisons and serious illnesses devised in the ancient world and popularised by Galen, which required eighty-odd ingredients of global provenance and often uncertain appearance, such as Indian amomum and Himalayan costus (Figure 3.3). It was frequently during their efforts to source the ingredients of theriac that sixteenth-century Italian apothecaries discovered natural history. In turn, the successful manufacture of these antidotes came to signal the apothecary's botanical expertise and his newly forged connections to the 'Republic of Naturalists'. (Throughout the period it became common even for apothecaries and physicians indifferent to the study of nature to resort to the expanding natural historical network for their professional work, requesting from naturalists recipes, rare ingredients, and newly printed pharmacopoeias.) A reputation for botanical expertise increased traffic in the shop by widening the master's clientele among the patrons who counted – namely, famous physicians, and the nobility and high clergy who were themselves beginning to dabble in natural history. It was Calzolari's fame as an investigator of *naturalia* that led Mattioli to commission theriac and his patented oil of scorpion from him, to be dispensed at Archduke Ferdinand II's court in Prague.[18]

At stake, however, were not just enhanced competence and greater profit, but also identity. Calzolari and his colleagues were cognisant of doing something new, *other* than their trade. Giovanni Pona introduced himself as someone 'who, always, before any other pleasure put the delight of investigating not only the names and forms of natural things, and of plants especially, but also their faculties'.[19] Aligning their rhetoric with that of other naturalists of the peninsula, apothecaries

Figure 3.3 List of approved ingredients for the recipes of theriac and mithridate employed in the seventeenth century at the Pharmacy of the Two Moors, once run by Marco Fenari. Courtesy of the State Archive of Venice, Giustizia Vecchia, b.211, loose sheet.

drew attention to their efforts in herborising and collecting, and described themselves as those who, 'delighting in simples', inspected nature for leisure. In doing so, they claimed participation in the community of *semplicisti*, as Italian naturalists called themselves, and distanced themselves from potential accusations of mindless labour.

This commitment was measured by their sustained pursuit of membership of the epistolary network, the pre-eminent form of exchange and community creation in sixteenth-century natural history, from which ordinary artisans and suppliers were excluded. Apothecary *semplicisti* shared this community's ethic of honour, friendship and servitude, and its economy of obligation. So, while remaining an important broker of specimens thanks to his stock and mercantile contacts, the apothecary *semplicista* was careful to transform the economic value usually attributed to his services of supplier into intellectual value. He no longer provided commodities paid by coin, but gifts that he expected to see reciprocated in pursuit of the common project of cataloguing nature – 'You shouldn't have paid' a mortified Fenari wrote to Aldrovandi, after being compensated for the plants sent.[20] From a supplicant, moreover, the master turned into a patron, vouching for local acquaintances before important naturalists, and taking responsibility for wayward associates like the young Fortunato Serafini, whom Fulcheri introduced to Aldrovandi and later had to excuse when Serafini began skipping Aldrovandi's lessons on *materia medica* at Bologna on account of a clash with lectures on surgery.[21]

Apothecary *semplicisti* further signalled their departure from their professional role by studying items beyond the common medicinal stock and engaging in collecting. Pona took notes on fossils, Egyptian papyri and Roman coins. Ornamental plants like the Spanish sunflower grew in Fulcheri's garden. Mythological birds like the legless *avis paradisiaca*, and monstrous beasts like crocodiles or two-headed lizards, graced the ceilings of both Calzolari's and Imperato's theatrical cabinets (as Figure 11.1 shows for the cabinet of Imperato). Responding to the period's fascination with the wondrous, these repositories aimed to document the totality of nature and were kept accordingly in the apothecary's home, separately from stock remedies and ingredients. Such collections confronted the artisan with new notions of object hierarchy, of aesthetics of display, and of pedagogical communication. They emphasised how the study of nature demanded a different sensibility towards materials from those employed in the daily operations in the workshop. Pharmacy was in the business of fragmenting and mixing; natural history prized whole items, carefully conserved.[22]

If apothecaries engaged in the staple activities that were coming to define natural history, their craft background ensured that such involvement remained distinctive. Three aspects deserve to be singled out. First, for most sixteenth-century Italian apothecaries, the study of nature continued to mean primarily the study of medicinal plants, especially those described by the ancients. Indeed, the restoration of the classical canon of *naturalia* located in the wider Mediterranean region continued to be privileged over the new American exotica and over zoological interests. Second, apothecaries placed a high premium on experience and on what they had learned through practice. They stressed their work of cultivation and herborisation first hand, taking pride in their ability to disambiguate the mysteries of nature through their senses, and mocking those who could not identify items without a label on their box. It was their time in the field that gave them the right to speak. The persona was that of the indefatigable explorer, in Calzolari's words: 'I devoted my life to searching the true simples across mountains, and plains, valleys and shores.'[23] Third, these two concerns were married in apothecaries' publications, which distinguished the *semplicisti* from their colleagues in the marketplace, and consisted of short treatises on theriacal ingredients, local floras and catalogues of their own cabinets of *naturalia*. Both Calzolari and Pona, for example, described the vegetation of Mount Baldo, renowned for its plant diversity; they exhorted those wishing to herborise on the mountain to read Dioscorides beforehand, and then let themselves be guided to the locations where the specimens could be plucked. These field manuals were meant to showcase these artisans as at once in dialogue with the textual canon of *res herbaria*, and also masters of local knowledge that could only be acquired by saddling a horse and going up the mountain oneself.[24]

The pharmacy shop

Where apothecaries differed the most from the average naturalist was in the way they lived natural history at the local level. Just as university and court influenced the interests and activities of scholars and gentlemen, and the rhetoric in which these were cast, so too the apothecaries' involvement in natural history cannot be dissociated from their work as guildsmen and traders in the Renaissance town. This work not only provided them with a competence that could be spent on pursuing *res herbaria*, as outlined earlier, but also embedded them in specific social relations that gave meaning and pattern to their scientific practice. The epistolary network, which scholars have long identified as the primary social and intellectual setting for sixteenth-century naturalists, was only one dimension through which these artisans lived natural history. An equally important setting was the local town in which each apothecary was rooted, as shopkeeper, healer, neighbour and kin. This setting not only created responsibilities – the Florentine Stefano Rosselli was often kept from his pastime by the need to manufacture theriac, or cure an important patron; but also opportunities – Pona gained access to the forms and hidden meanings of the emblems adorning Verona's *Accademia Filarmonica* thanks to Monsignor Antonio Nichesola, a compatriot and renowned antiquarian.[25] The disembodied web of correspondents was complemented by a local community built through interpersonal interactions. At the centre of this local network stood the pharmacy shop. This was probably the greatest asset that botanically minded apothecaries had to offer to natural history, and they could also draw upon it in pursuit of their own natural historical studies. Though generally overlooked by scholars of today, the shop was both a crucial hub for an urban science based on collaboration and conviviality, and a resource for pursuing natural history. Here, would-be virtuosi congregated weekly with a variety of individuals, especially physicians and apothecaries, but also customers and passers-by. Some shared their interest in nature, others were indifferent to it, but had knowledge and contacts to offer, or simply some gossip to pass the time.[26]

Early modern pharmacies were spacious, elegant and crowded. Announced outwardly by colourful signs ranging from umbrellas and eagles to saints and stylised moral virtues, inside they comprised functional backrooms for cooking and storage and a large reception room. This was typically furnished with expensive wooden panels and cabinets, a large counter, benches to sit on, and shelves supporting richly decorated majolica jars manufactured in the famous potteries of Castel Durante and Montelupo. A painting of the Virgin often blessed the proceedings, reminding the master of the charity owed to the poor. All was designed to create a welcoming environment for medical practice. Here physicians held consultations and patients waited for their remedy to be prepared, attended by several trainee apothecaries and the master, whom the manuals advised to 'be orderly and clean in his dress, and very civil, with neat, beautiful hands, perfumed, and well dressed'.[27] Although pharmacies were known for their greater gentility compared to other medical establishments, notably barbershops, their custom of selling on credit ensured that carpenters, butchers and other labourers with the right credentials could mingle there with the well-to-do and their servants. The shop's hold on urban consumption was tightened by its sale of a variety of non-medicinal products like paper and ink, soap and cosmetics, and specialist products for fellow craftsmen: pigments and dyes for artists; ambergris, musk and myrrh for perfumers; wax for funerals or celebrations. Bigger pharmacies served as post offices and offered facilities for letter writing.

The influx of people resulting from this trade turned pharmacies into one of the most prominent spaces of male sociability outside the home in early modern Italy. Gentlemen and nobles regularly visited to chat about current affairs, read newsletters from nearby cities and gamble on political outcomes. Spies and diplomats included them in their information network, while craftsmen gathered there after work to play cards and chess – illegally – and debate the reformed religion. Questioned by the Venetian State in 1606, Lepido Barbaran described his shop in Vicenza as frequented by 'noble and non-noble persons . . . We talk of the happenings of the world, of the wars and uprisings currently taking place . . . the discussions are varied as are the opinions'.[28] The pharmacy was thus a centre of political and religious culture, including dissent. Apothecaries were often deemed by the Catholic Inquisition to have sown unrest, since their shops hosted clandestine meetings for aspiring Lutherans and Evangelicals, and because they used their literacy to evangelise other parishioners.

The scientific sociability that centred on pharmacies capitalised on the shop's facilities, but also on its pre-existing entrenchment in the local community. For the apothecary, the shop acted first and foremost as a networking tool. It is among these visitors that the master interested in *res herbaria* found his first interlocutors. For Fulcheri, these included the Vanni family of apothecaries from whom he rented his shop, and Lucca's men of letters, like the philosophy professor Flaminio Nobili. It was from neighbouring artisans that he first drew expertise for his new activities – one of the first artists Calzolari approached to commission drawings of *naturalia* was Giovanni Caroto, none other than the eldest son of the apothecary who kept shop at the other end of the square from Calzolari.[29] Scholars have argued that the birth of a scientific community around natural history was predicated on the expression of specific social boundaries that privileged education and social status.[30] While this consideration holds for the epistolary community, it overlooks how far most naturalists were embedded as producers and consumers in the commercial context and built many of their relationships there. Indeed, the most numerous and productive botanical friendships for apothecaries seem to have involved the same town physicians who formerly excoriated them in print. This was a function, as we saw, of the importance assumed by *materia medica* for the medical practitioner, but also of the collaborative nature of early modern medicine. After leaving university, most physicians attached themselves to specific pharmacies to fulfil their prescriptions, and spent hours there weekly recruiting patients and chatting with colleagues. It was Luigi Leoni, physician and keeper of Pisa's botanical garden, who introduced Rosselli to Aldrovandi, praying the latter to make the apothecary known among his friends and give him choice specimens because 'he and I share everything'.[31]

For aspiring local naturalists, the shop's appeal lay in offering at once an archive of *naturalia*, a testing room with useful equipment, and a meeting place where like-minded individuals could discuss flora and fauna. Inspecting the stock was the core activity, encouraged by the masters, who increasingly began to create specimen collections on the side. Naturalists started to include pharmacies in their travels, often on the recommendation of other virtuosi. The Veronese physician Gian Battista Olivi observed of Calzolari's museum, housed above his shop: 'in this theatre we can inspect, examine, and smell many things worthy of notice'.[32] This unveiling of specimens was preceded, accompanied and followed by comparisons between their appearance and written descriptions, and discussions of their healing properties and the etymology of their names. While naturalists with extensive contacts conducted many of these conversations via letter, seeking opinions from far and wide, apothecaries and local virtuosi held them in the pharmacy. There they continued the shop habit of leisurely discussion, debating natural history alongside other topics, more serendipitously and fragmentarily but no less effectively than on paper. The physician Giovan Battista Cavallara declared himself 'transported ... from medical cases to philosophical questions' by the collection of the Mantuan apothecary Filippo Costa.[33]

The shop provided raw materials, and room to examine them, for enthusiasts who lacked the resources to build a microcosm of nature like Aldrovandi's. In Naples, for example, Fabio Colonna used Imperato's collection of plants and fossils for his studies.[34] The availability of equipment and the active operation of the pharmacy – where remedies were prepared variously by baking, boiling, grinding, distilling and oxidating, and with the aid of furnaces, chopping tables, pestles and mortars – made it a space that could be adapted more conveniently than an oak-panelled study to host demonstrations of compounding drugs, or extravagant and potentially messy experiments. Tests on the inflammability of asbestos, by setting fire to its 'ropes', drew in regular audiences in Calzolari and Imperato's shops during the 1570s (Figure 3.4).[35]

Reading was also a shared activity in the shop. Pharmacies' role as the neighbourhood post office was repurposed for natural history, creating a busy traffic as local naturalists visited to exchange correspondence with other virtuosi, often pausing to read it with the master and others present. In 1554, the Veronese noble Gentile della Torre entered Calzolari's shop to deliver a gift to the apothecary, a herbarium sent by Luca Ghini, and spent the afternoon there 'to see the plants together with messer Francesco'.[36] There is also evidence that pharmacies hosted readings of whole works, including those penned by fellow apothecaries. Cechino Martinelli's booklet on amomum was read in the Mantuan pharmacy of Antonio Bertioli with local physicians, for example. Inventories confirm that, alongside recipe collections, pharmacies held at least two or three books on *materia medica* to be consulted at need, a staple being Mattioli's *Commentaries on Dioscorides*, the book that for much of the sixteenth century defined which natural objects warranted study, and who belonged to the emerging community of naturalists. Reading Mattioli in the pharmacy allowed local virtuosi to participate in this discussion.

Natural history in the shop, thus, was a science that was spoken and heard, not just written down. Its exchanges complemented the virtual and disjointed conversations that we find in naturalists'

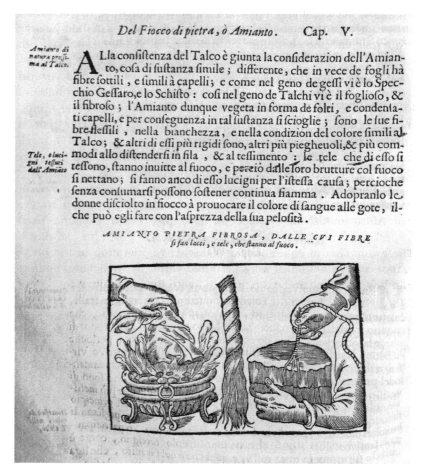

Figure 3.4 Experiment with asbestos. F. Imperato, *Dell'Historia Naturale* (Naples, 1672), p. 592. Author's collection.

correspondence and anticipated the sociable science of seventeenth-century English coffee houses and eighteenth-century French salons. Shop natural history was also a communal endeavour: the shop enabled naturalists to meet, converse and engage in scientific activities together.

To an extent, there is a case to be made for seeing the pharmacy as a pedagogical site, performing a role like that of the university in introducing potential virtuosi to the field, yet far more socially inclusive than the latter. Trainee apothecaries had access to the master's books and drawings of *materia medica*, and could profit from his contacts and from the conversations held within the shop. In Venice there is evidence of multigenerational botanical pharmacies, like the Doctor, the Bell and the Ostrich, where an interest in *res herbaria* was passed on from master to apprentice.[37] Poorer virtuosi too could use their social credit with the master to take a look at his library and collection and discuss their content. Indeed, the pharmacy's commercial nature made it inherently more accessible than other sites of learning emerging at this time, such as the curiosity cabinet, the scholar's library and the astronomical observatory – for instance, it did not require any letter of introduction. This certainly seems to have worked in whetting the appetites of medical students before they left for university, and of gentlemen who did not study medicine, or who had already finished their academic education. The pharmacy probably constituted the main stage of daily scientific practice, not only for the artisans who managed it, but also for many local naturalists who had only limited contact with the intellectual community outside their town and their region.

Conclusion

By the time Cardinal Sforza Pallavicino observed in 1664 that 'Philosophy lives in the shops and the countryside, as well as in books and academies', his words were not a programmatic statement. Rather, they reflected a well-established situation.[38] This chapter has shown that apothecaries were not just bystanders, but contributed to making natural history a distinct discipline. As both brokers and participants, apothecaries inflected and supported the material practices of naturalists, and ensured that the study of medicinal ingredients remained a current concern. By following these artisans' stories and examining the places they worked in, we can better understand how natural history developed as a collective enterprise, and uncover a side of it that went on outside learned epistolary networks. This urban science, for which the pharmacy was a central institution, illustrates that markets and shops were not just sites from which the natural philosopher could extract knowledge, but they were hotbeds of practice in their own right. In turn, apothecaries called into question the long-standing dichotomy between scholar and craftsman, learned and non-learned. Because of their position in the city (wealthy artisans catering to the health and sweet tooth of the citizens) and the pharmacy shop's nature as a hub of sociability, they should be seen as crucial conduits for a dialogue across different social and intellectual circles.

Further reading

Egmond, F., *The World of Carolus Clusius: Natural History in the Making, 1550–1610* (London, 2010).

Fahy, C., *Printing a Book at Verona in 1622: The Account Book of Francesco Calzolari* (Paris, 1993).

Findlen, P., *Possessing Nature: Museums, Collecting and Scientific Culture in Early Modern Italy* (Berkeley, 1994).

Harkness, D. E., *The Jewel House: Elizabethan London and the Scientific Revolution* (New Haven, 2007).

Nockels-Fabbri, C., 'Treating medieval plague: the wonderful virtues of theriac', *Early Science and Medicine*, 12:3 (2007), pp. 247–83.

Palmer, R., 'Pharmacy in the Republic of Venice in the sixteenth century', in A. Wear, R. K. French and I. M. Lonie (eds.), *The Medical Renaissance in the Sixteenth Century* (Cambridge, 1985), pp. 100–17.

Pugliano, V., 'Specimen lists: artisanal writing or natural historical paperwork?', *Isis*, 103:4 (2012), pp. 716–26.

Pugliano, V., 'Pharmacy, testing and the language of truth in Renaissance Italy', *Bulletin of the History of Medicine*, 91:2 (2017), pp. 233–73.

Shaw, J. and Welch, E., *Making and Marketing Medicine in Renaissance Florence* (Amsterdam, 2011).

4   Horticultural networking and sociable citation

In his 1667 account of the newly formed Royal Society of London for Improving Natural Knowledge, Thomas Sprat celebrates the fact that 'rarities' were 'every day given in, not only by the hands of the learned, but from shops of mechanics, voyagers of merchants, ploughs of husbandmen, [and] gardens of gentlemen'.[1] A network like that which Sprat describes – that is, an assemblage of sometimes far-flung collectors, united by their common goal of contributing specimens of *naturalia* and *artificialia* to the Royal Society – has recently been identified as a kind of early modern 'public': the sort of small-scale association scholars have begun to construe as a precedent for the more encompassing public sphere of the Enlightenment.[2] Such informal collaborations often thrived before their formal counterparts (like the Royal Society) were founded. These earlier social formations were composed of people who were not necessarily aware of each other's individual existence, but regarded themselves and one another generally as contributors to a common cultural end through the development of shared interests and expertise. In various arenas, including natural history, printed publications acted as media for the formation and reinforcement of such publics. In early modern botanical books, for instance, citations served as markers and makers of a social network in an emergent and rapidly diversifying field. An actual botanical public, that is, was partly engendered through its public (because published) representation in comments by and about natural historians, herbalists and horticulturalists.

So much becomes clear from a close reading of the first publication of John Parkinson (Figure 4.1), a prominent London apothecary and gardener whose punningly titled *Paradisi in Sole* ('Park in Sun') was a landmark in the history of English gardening literature (Figure 4.2). Few other books dedicated to recreational horticulture, and none larger, were printed in England before it appeared in 1629.[3] In it, Parkinson adopts the loose but recognisable bibliographic apparatus of the herbal encyclopedia, while simultaneously carving out a narrower horticultural public inspired to study plants from something other than a medicinal motive. Parkinson thus minimises his treatment of botanical 'vertues' (their health-giving powers): 'I leave

Figure 4.1 Portrait of author. J. Parkinson, *Paradisi in Sole Paradisus Terrestris* (London, 1629). Dittrick Medical History Center, Case Western Reserve University.

Figure 4.2 A. Switzer, frontispiece to J. Parkinson, *Paradisi in Sole Paradisus Terrestris* (London, 1629). Dittrick Medical History Center, Case Western Reserve University.

the discussing of these and others of the like nature, to our learned Physitians; for I deale not so much with vertues as with descriptions' (p. 398). Similarly, Parkinson distinguishes his treatise from a recipe book when he declares (after elucidating one way to cook spinach), 'I leaue the further ordering of these herbes, and all other fruits and rootes of this Garden' – the section of the book entitled 'The Kitchen Garden' – to 'Gentlewomen and their Cookes [who] can better tell then my selfe' about such matters (p. 496). Parkinson's representation of a segmented market for botanical books both suggests and works to shape a nascent public of recreational gardeners distinct from what his phrasing here suggests were already recognised as their gastronomical counterparts.

As well as selectively adopting conventional elements of botanical description, Parkinson takes up the related herbalist's habit of citing the findings and fieldwork of his fellows – including contributions of both specimens and accounts of their appearance, growth habits and properties. Since both plant materials and botanical know-how stocked Parkinson's famous garden at Long Acre, he acknowledges their sources in the book that he portrays as a bibliographic version of that garden. To move beyond reading such citations as an aspect of Parkinson's self-fashioning, or of the interpenetration of emergent standards for science and civility in early seventeenth-century England, I propose to interpret Parkinson's sociable citation practices in *Paradisi* as a record of the community that made his work possible: a community which this early modern social medium, in turn, reinforced through its acts of acknowledgement.[4]

Parkinson's cultivated public

Parkinson was a founding member of the Society of Apothecaries, who, by 1622, had disowned the governance of his profession in order to cultivate his garden; yet this was hardly to turn from a public to a private role, as is confirmed by his publication of the *Paradisi* by the decade's end. The choice to cultivate one's garden in seventeenth-century England did not, necessarily, signify a retreat from public life – as, by the end of Voltaire's *Candide*, it might. In Parkinson's day, even a private garden grew primarily by cultivating the fellowship of gardeners willing to share slips, seeds and tips for growing them. While evidence survives of a formal society of plant collectors in Norwich by 1631, and others like it crop up elsewhere in later decades, no related society was extant at the time *Paradisi* appeared.[5] In the absence of such a formal arrangement, the paper-based social bonds in books like Parkinson's, abetted by (and often embedding) the manuscript correspondence that often contributed to their composition, were critical in uniting members of a botanical public, as is frequently attested on their pages.

The sociable nature of herbal media may be inferred in brief from a prefatory epistle to Parkinson's second book, the *Theatrum Botanicum* of 1640. In signing off, John Bainbridge identifies himself to Parkinson as '*Your affectionate friend*', despite having opened his letter with a seemingly contrary characterisation: 'Sir, I am a stranger to your selfe'. Bainbridge continues, however, by citing the capacity of Parkinson's books to transmute strangers into friends: 'I am a stranger to your selfe, but not to your learned, and elaborate volumes.'[6] Books enabled such sociability in part through their ventriloquy of authorial voice, as Parkinson appears to have been aware, since, in dedicating his earlier work to the queen, he personifies the *Paradisi* as 'this speaking Garden', and later (in addressing his other readers) casts himself within its pages as the reader's horticultural guide: 'I doe ... play the Gardiner'.[7] Parkinson's diction plays on the book's performative power and suggests its participation in the thriving public theatrical culture of his day. But Parkinson's performance in *Paradisi in Sole* is no soliloquy; rather, his role is one among an ensemble of *dramatis personae* drafted into his representation of a peculiarly English horticultural public.

Although Parkinson cites the standard array of continental sources, for instance, he also imagines his primary audience within English borders, as becomes clear in his first-person plural pronominal references to the English – not just in explicit nods to 'our owne Country' (pp. 382, 385) but with respect to ecosystems seen to march with national borders: 'Gilliflowers, the pride of our English Gardens', for example, are contrasted with 'Out-landish flowers' that are 'Strangers vnto vs' (p. 8). At other times, the dichotomy between outlandish and English plants is both rhetorically and botanically softened: 'Valerians are strangers, but endizoned for their beauties sake in our Gardens' (p. 388). Parkinson writes in a similarly conciliatory mood, and showily gentlemanly mode, in his introduction (textual and horticultural), of the Virginian Silke: 'Lest this stranger should finde no hospitality with vs, being so beautifull a plant, or not finde place in this Garden, let him be here receiued' (p. 444). But the same author at times deliberately ignores plants *not* found in England, and for that reason alone: 'there are two sorts of Licorice set downe by diuers Authors, yet because this Land familiarly is acquainted but with one sort, I shall not neede for this Garden' – that is, his book – 'to make any further relation of that is vnknowne, but onely of that sort which is sufficiently frequent with vs' (p. 533). Parkinson's nationalistic botanising partly derives from the alleged supremacy of English soil and its 'culture': tulips, for instance, are 'made Denizens in our Gardens, where they yield vs more delight, and more increase for their proportion, by reason of the culture, then they did vnto their owne naturals' (p. 65). Both the plants and their delighted public are better, by this account, for roots in English soil. Parkinson even claims 'our English red Artichoke' degenerates when grown elsewhere, because 'our soyle and climinate hath the preheminence to nourish vp this plant to his highest excellencie' (p. 520). Such botanical parochialism can, unsurprisingly, prove uglier, such as when Parkinson explains of 'The Spaniards blush Beares eare' that its flowers are 'of a duskie blush colour, resembling the blush of a Spaniard, whose tawney skinne cannot declare so pure a blush as the English can' (p. 236). In a similarly insular spirit, Parkinson indicates of the 'Ethiopian Mullein' that 'if it please you not, take it according to his Country for a Moore, an Infidell, a Slaue, and so vse it' (p. 383). The frequent personification of plant species shows that their marshalling and cataloguing bore close relation to that of peoples into publics of all kinds.[8]

Despite his frequent turn to a unifying national 'we', Parkinson locates the heart of his horticultural public within London. His identification of the epicentre of English botanical culture becomes clear when he itemises names for gillyflowers which are 'altered as euerie ones fancy will haue them, that carryed or sent them into the seuerall Countries from London, where their truest name is to be had, in mine opinion' (p. 12). As well as privileging urban nomenclature, Parkinson appears to target a local audience when he describes a plant's habitat in such detail that the description appears designed to help those nearby collect specimens, as Parkinson does: 'I gathered diuers rootes for my Garden, from the foote of a high banke by the Thames side, at the hither end of Chelsey, before you come at the Kings Barge-house' (p. 132). Here as elsewhere, Parkinson's colloquial pronoun use suggests that he sees himself as diffusing his knowledge directly from this paper garden to a familiar, neighbourly public for immediate practical use, just as he would in conversation with visitors to his actual garden.

While Parkinson's botanical public is explicitly bounded by geography, he rarely lets Protestant partisanship shape it – perhaps because his dedicatee was the Catholic queen, or because his upbringing in that church lent him tolerance.[9] The limits of such tolerance, though, appear in relation to an illustration uniquely embedded within a paragraph and accompanied by a header set in large type: 'The Iesuites Figure of the Maracoc' (p. 394). Such distinctive layout makes the plant stand out, since most illustrations fill a page crowded with specimens identified in minuscule lettering below (Figure 4.3). Parkinson's distinguishing treatment might appear reverential until, over the page, his prose targets Jesuitical botany for special scorn: 'Some superstitious Iesu-ite would faine make men beleeue, that in the flower of this plant are to be seene all the markes of our Sauiours Passion; and therefore call it *Flos Passionis*: and to that end haue

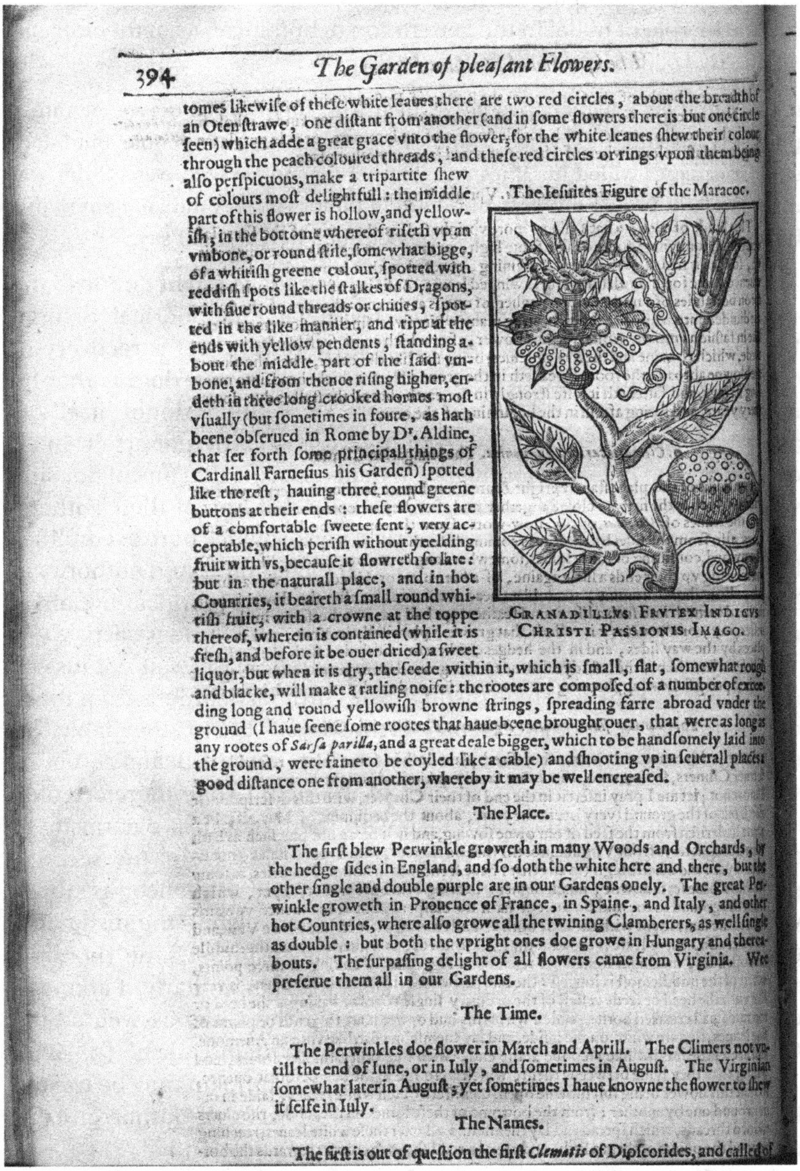

394   *The Garden of pleasant Flowers.*

tomes likewise of thefe white leaues there are two red circles, about the breadth of an Oten ftrawe, one diftant from another (and in fome flowers there is but one circle feen) which adde a great grace vnto the flower, for the white leaues fhew their colour through the peach coloured threads, and thefe red circles or rings vpon them being alfo perfpicuous, make a tripartite fhew of colours moft delightfull: the middle part of this flower is hollow, and yellow- ifh, in the bottome whereof rifeth vp an vmbone, or round ftile, fomewhat bigge, of a whitifh greene colour, fpotted with reddifh fpots like the ftalkes of Dragons, with fiue round threads or chiues, fpot- ted in the like manner, and tipt at the ends with yellow pendents, ftanding a- bout the middle part of the faid vm- bone, and from thence rifing higher, en- deth in three long crooked hornes moft vfually (but fometimes in foure, as hath beene obferued in Rome by Dr. Aldine, that fet forth fome principall things of Cardinall Farnefius his Garden) fpotted like the reft, hauing three round greene buttons at their ends: thefe flowers are of a comfortable fweete fent, very ac- ceptable, which perifh without yeelding fruit with vs, becaufe it flowreth fo late: but in the naturall place, and in hot Countries, it beareth a fmall round whi- tifh fruit, with a crowne at the toppe thereof, wherein is contained (while it is frefh, and before it be ouer dried) a fweet liquor, but when it is dry, the feede within it, which is fmall, flat, fomewhat rough and blacke, will make a ratling noife: the rootes are compofed of a number of exten- ding long and round yellowifh browne ftrings, fpreading farre abroad vnder the ground (I haue feene fome rootes that haue beene brought ouer, that were as long as any rootes of *Sarfa parilla*, and a great deale bigger, which to be handfomely laid into the ground, were faine to be coyled like a cable) and fhooting vp in feuerall places a good diftance one from another, whereby it may be well encreafed.

The Place.

The firft blew Perwinkle groweth in many Woods and Orchards, by the hedge fides in England, and fo doth the white here and there, but the other fingle and double purple are in our Gardens onely. The great Per- winkle groweth in Prouence of France, in Spaine, and Italy, and other hot Countries, where alfo growe all the twining Clamberers, as well fingle as double: but both the vpright ones doe growe in Hungary and there- bouts. The furpaffing delight of all flowers came from Virginia. We preferue them all in our Gardens.

The Time.

The Perwinkles doe flower in March and Aprill. The Climers not vn- till the end of Iune, or in Iuly, and fometimes in Auguft. The Virginia fomewhat later in Auguft, yet fometimes I haue knowne the flower to fhew it felfe in Iuly.

The Names.

The firft is out of queftion the firft *Clematis* of Diofcorides, and called of

The Iefuites Figure of the Maracoc.

GRANADILLVS FRVTEX INDICVS CHRISTI PASSIONIS IMAGO.

Figure 4.3 'The Iesuites Figure of the Maracoc'. J. Parkinson, *Paradisi in Sole Paradisus Terrestris* (London, 1629), p. 394. Mount Allison University Libraries and Archives.

caused figures to be drawne, and printed, with ... thornes, nailes, speare, whippe, pillar, &c. in it' (p. 396). Parkinson includes this depiction for comparison with the correct figure. His condemnation of the Jesuits' disfigurement reveals, however, underlying anxieties about the role of belief in the generation of botanical, as of theological, authority (p. 396):

these bee their aduantagious lies (which with them are tolerable, or rather pious and meritorious) wherewith they vse to instruct their people; but I dare say, God neuer willed his Priests to instruct his people with lyes ... But you may say I am beside my Text, and I am in doubt you will thinke, I am in this besides my selfe, and so nothing to be beleeued herein that I say.

As if realising that the violence of his own language might diminish the credibility of his attack on florid Jesuitical metaphorical diction, Parkinson dithers after his disjunction, only to enact a recovery of confidence in his own credibility, dismissing his own doubts that he might be doubted by others, simply impugning doubt itself as a partisan tactic: 'For, for the most part, it is an inherent errour in all of that side, to beleeue nothing, be it neuer so true, that any of our side shall affirme, that contrarieth the assertions of any of their Fathers' (p. 396). Parkinson's complex rhetorical performance here exemplifies his recurrent concern with establishing the credibility and authority of his peers and himself as authentic members of a sceptical botanical network concerned, foremost, with the evidence of things seen.

Parkinson often addresses a botanical public narrowed not just by nation but by rank when he distinguishes ordinary gardens from those of 'the better sort of the Gentry of the Land' (p. 8). Inevitably, he favours the latter, as when he tells us that 'garden Beans serue ... more for the vse of the poore then of the rich: I shall therefore only shew you the order the poore take with them, and leaue curiosity to them that will bestow time vpon them' (p. 521). Here, the second-person pronoun characterises his presumed audience as those wealthy enough not to know how to eat beans, while the distancing third-person pronoun scorns those willing to invest time (possibly their only asset) in so contemptible a species. Similarly, Parkinson comments of pumpkins that 'the poore of the Citie, as well as the Country people, doe eate thereof, as of a dainty dish' (p. 526). The quiet concluding simile clarifies that the pumpkin may be treated by some as dainty, but is not so in the opinion of author and his postulated audience: not poor pumpkin eaters, evidently.

In such ways, Parkinson patrols the borders of his preferred botanical public. He separates horticultural sheep from goats at one point, for instance, by noting that a particular plant 'beareth no seede that euer I could see, heare of, or learne by any of credit, that haue noursed it a great while; and therefore the tales of false deceitfull gardiners, and others, that diliuer such for truth, to deceiue persons ignorant thereof, must not bee credulously entertained' (p. 392). As with jesuitical Jesuits and pumpkin eaters poor in pocket and taste, Parkinson impugns the credibility of those who proffer hearsay as little better than the 'imposters' who flog ineffective botanicals 'as an antidote against the Plague' (p. 160) or profit from fake botanical wonders, such as 'cunning counterfeit' mandrake roots, carved to look like little men (p. 377). Commercial nursery operators are similarly condemned as a duplicitous and rapacious counter-public to Parkinson's preferred model of honest and genteel botanical exchange: he notes that 'our Nursery men do so change the names of most fruits they sell, that they deliuer but very few true names to any', and he specifically objects that 'Iohn Tradescantes Cherrie is most vsually sold by our Nursery Gardiners, for the Archdukes cherrie, because they haue more plenty thereof' (pp. 571, 574). In seeking private profit – as opposed to the common good Parkinson espouses by enshrining his botanical expertise in print – such figures violate the decorum of Parkinson's public, idealised in his pages as a gentlemen's game of freely exchanged gifts.

Engendering the garden

Yet women also formed an important part of Parkinson's emergent botanical public – not least since the volume is dedicated to his queen, who took extensive interest in renovating the royal gardens.[10] Many less eminent recreational gardens were still closely connected with physic gardens, which provided gentlewomen with the ingredients they compounded for household and community treatments.[11] Parkinson acknowledges 'small diseases as are often within the compasse of the Gentlewomens skils, who, to help their owne family, and their poore neighbours that are farre remote from Physitians and Chirurgions, take much paines both to doe good vnto them, and to plant those herbes that are conducing to their desires' (p. 470). He even credits biblical women, in a manner that sets them on par with ancient and medieval male herbalists, on the virtues of the mandrake (p. 378):

Dioscorides first, and then Serapio, Aiucen, Paulus Aegineta, and others also do declare, they conduce much to the cooling and cleansing of an hot matrix. And it is probable, that Rachel knowing that they might be profitable for her hot and dry body, was the more earnest with Leah for her Sonne Rubens Apples, as it is set downe Genesis 30.verse.14.

Rachel's female descendants seemed to form a kind of sub-public when it came to botanical nomenclature, since Parkinson often highlights plant names used specifically by women in their own botanical dialect. He mentions a lily called '[o]f diuers women here in England ... Lilly of Nazareth' (p. 33), and notes that the colour of one daffodil

'hath caused our Countrey Gentlewomen ... to entitle it Primrose
Peerlesse' (p. 74); he says 'some English Gentlewomen call the white
Grape-flower Pearles of Spaine' (p. 115), and of another hyacinth that it
'doth somewhat resemble a long Purse tassel' and 'thereupon diuers
Gentlewomen haue so named it' (p. 116). While Parkinson sometimes
justifies the basis of women's naming, he elsewhere disparages feminine nomenclatural judgement: 'Some haue thought it to be a yellow
Anemone, that haue looked on it without further iudgement, and by that
name is most vsually knowne to most of our English Gentlewomen'
(p. 294). At other times, he classes women's names as mere nicknames:
'[S]ome of our English Gentlewomen haue called it, The Princes
Feather, which although it be but a by-name, may well serue for this
plant to distinguish it' (p. 234). As in the examples shown, Parkinson
regularly asserts the genteel class of the female members of the botanical public to whom he apportions credit.

While Parkinson acknowledges and tolerates the cultural phenomenon of women's botanical naming, he rarely cites the help of individual women in locating specimens. In the two cases when he does
name women, he is quick to associate their botanising with that of
the eminent plant collector John Tradescant, as though anxious to
offset his reliance on feminine expertise by juxtaposing it with an
authority beyond reproach. One acknowledged female contributor to
his garden is identified as 'a good old Lady, the widow of Sir Iohn
Leveson', who is commended for 'nours[ing] vp' a very rare specimen
in one instance (pp. 603–4), but is primarily identified only as
a property owner when Parkinson reports a specimen 'found neere
the salt Marshes by Rochester, in the foote way going from the Lady
Levesons house thither, by a worthy diligent and paneful obseruer
and preseruer ... Iohn Tradescante' (p. 489). Parkinson honours
another woman's botanical expertise when he acknowledges a plant
sent by 'a courteous Gentlewoman, a great louer of these delights,
called Mistris Thomasin Tunstall, ... who hath often sent mee vp the
rootes to London' (p. 348). Elsewhere he praises the 'industrie' of this
same 'worthy Gentlewoman' in sending another rare plant; as with
Lady Leveson, however, he immediately mentions a second specimen
'sent me by my especial good friend Iohn Tradescante, who brought
it among other dainty plants from beyond the Seas' (p. 389).
Tradescant and his international plant-collecting expeditions are
thus linked paratactically to the women's local contributions, with
the well-known authority of the former underwriting both their expertise and Parkinson's reliance on it.

Parkinson signals by name his debts to many more male colleagues, such as 'the pods of this Rose bay, brought mee out of

Spaine, by Master Doctor Iohn More' or some seeds 'had of Master Iames Cole a Merchant of London lately deceased, which grew at his house in Highgate, where there is a faire tree which hee defended from the bitternesse of the weather in winter by casting a blanket ouer the toppe thereof euery yeare, thereby the better to preserue it' (p. 401). By describing the long-distance travels and horticultural innovations of the men who increased his own botanical knowledge and stores, Parkinson both honours their labour and establishes his own horticultural holdings (and their rarity); he also provides would-be members of the botanical public with models of the conduct required of those in good standing. But as well as acknowledging by name members past and present of his horticultural public, Parkinson actively recruits from the ranks of those he does not yet know by name but anticipates as a component of his audience. After describing one little-known plant, for instance, he claims (p. 420) that it

might easily be knowne, if any of our Merchants there residing [that is, in Turkey], would but call for such a shrubbe, by the name of a Foxe taile in the Turkish tongue, and take care to send a young roote, in a small tubbe or basket with earth by Sea, vnto vs here at London, which would be performed with a very little paines and cost.

Parkinson also appeals for more lively participation from peers of both sexes at home (p. 348), where

many things doe lye hid, and not obserued, which in time may bee discouered, if our Country Gentlemen and women, . . . in their seuerall places where they dwell, would be more carefull and diligent . . . to finde out such plants as growe in any of the circuits or limits of their habitations, or in their trauels, as their pleasures or affaires leade them.

The need for an enlarged horticultural public is portrayed as a consequence of the broad purview of its subject – so broad that, as Parkinson elsewhere suggests, no individual can grasp its entirety: 'many that haue trauelled in the sowing of the seed of Tulipas many years, may obserue each of them to haue some variety that others haue not: and therefore I thinke no one man can come to the knowledge of all' (p. 55). Botanical knowledge requires, instead, the medium of a social network, as well as the network of a social medium like that Parkinson generated in *Paradisi*.

Horticultural capital

As in earlier herbals, Parkinson's discussion of botanical names often reveals those of their discoverers and disseminators as well:

plant names themselves acted as social media in miniature. For instance, Parkinson notes of one flower that '[s]ome call it *Narcissus Caparonius*, of the name of the first inuentor or finder thereof, called Noel Caperon, an Apothecary dwelling in Orleance, at the time he first found it' (p. 44). Parkinson similarly promulgates a botanical name embedding that of its disseminator when he notes that 'The yelow Italian Daffodill of Caccini' refers to 'that honourable man that sent it' abroad, and adds that this distinction 'is most fit to continue' (p. 90). Elsewhere, Parkinson honours the first to send not a botanical specimen but an illustration of it, by expressing his preference for the artist's choice of botanical name: '[i]t is of some called *Lilium Persicum,* the Persian Lilly: but ... I had rather with Alphonsus Pancius the Duke of Florence his Physitian, (who first sent the figure therof vnto Mr. Iohn de Brancion) call it *Corona Imperialis*' (p. 28). Parkinson himself indulges in the honorific naming of a carnation: 'the most beautifull that euer I did see was with Master Ralph Tuggie, which I must needes therefore call Master Tuggies Princesse' (p. 312). But such personal accolades could contribute to the redundant multiplication of names that was a most vexing feature of early modern botany: nicotine was thus named 'of one Nicot a French man, who seeing it in Portugall, sent it to the French Queene, from whom it receiued the name of *Herba Regina*' (p. 364). The duelling citational principles yield nomenclatural fog but also reveal how a plant's audience could be as significant as its author, depending on the relative cultural capital of each.

Among the dozens of personal names embedded in Parkinson's catalogue of plants, the most frequently cited is John Tradescant's.[12] Parkinson frequently identifies Tradescant with a variant on the phrase 'my very louing and kinde friend', but in one case does so in a context that rarefies their friendship still further, botanically and socially: 'There is likewise another sort of these male Mandrakes, which I first saw at Canterbury, with my very louing and kinde friend Iohn Tradescante, in the garden of the Lord Wotton, whose gardiner he was at that time' (p. 378). Here, the secondary citation of a high-ranking figure endorses the credibility of the first. In another instance, Parkinson's elaboration of Tradescant's pedigree shows how thoroughly an individual's stock of cultural capital could be exploited in such citations. In discussing 'The soon fading Spider-wort of Virginia, or Tradescant his Spider-wort', Parkinson (p. 152) indicates that

for it the Christian world is indebted vnto that painfull industrious searcher, and louer of all natures varieties, Iohn Tradescant (sometimes belonging to the right Honourable Lord Robert Earle of Salisbury, Lord Treasurer of England in his time, and then vnto the right Honourable the Lord Wotton at Canterbury in Kent, and lastly vnto the late Duke of Buckingham) who first receiued it of a friend, that brought it out of Virginia, thinking it to bee the Silke Grasse that growth there, and hath imparted hereof, as of many other things, both to me and others.

Parkinson's initial lofty praise of Tradescant is followed by parenthetical substantiation of Tradescant's authority through his succession of big-wig bosses; this résumé is conveniently inserted before a phrase that goes all but unnoticed after this tide of celebration of the plant named for Tradescant, 'who first receiued it of a friend, that brought it out of Virginia'. It is thus not, in fact, Tradescant to whom 'the Christian world is indebted' – but he receives the credit, with his name embedded in that of the plant as well as in the *Paradisi*.

Tradescant did not only supply Parkinson and others with specimens, however; he also disseminated his knowledge of them, as is shown in Parkinson's attribution of a description to Tradescant when he presents his own account of the Pine Parsnep as being 'as Iohn Tradescante saith', adding that Tradescant 'hath giuen me the relation of this, and many other of these garden plants', such that 'euery one is a debtor' to him (p. 508). Parkinson's acknowledgement exposes how his botanical public, like his book, took shape through shifting relations of debt and credit derived from exchanges of plants and descriptions of them. Parkinson's related purpose in *Paradisi* is, therefore, to disseminate not only plants or botanical knowledge, but also his own trust, a form of credit extended to those he cites favourably and subtracted from those he suspects. His own belief is not, moreover, currency he trades lightly, as his excoriation of the Jesuits' Maracoc confirms. Parkinson always prefers to witness what he reports, and is alert to the need to bolster his own credibility, as becomes clear when he notes of a rare type of rosemary, 'few haue heard thereof, much lesse seen it, and my selfe am not well acquainted with it, but am bold to deliuer it vpon credit' (p. 425). He generally hesitates to substantiate hearsay in this way, and his nervousness in doing so is shown when he caps his description of the plant by repeating his caveat: 'This I haue onely by relation, which I pray you accept, vntill I may by sight better enforme you' (p. 426). Parkinson's distinct preference for eyewitnessing shows the emergent standards of Parkinson's public closely tracking those later upheld by the Royal Society.

Given his sceptical mindset, it is unsurprising that Parkinson's citations of fellow plantsmen are not exclusively honorific but record disagreement, not only with heretical sources or local or low-ranking figures but even with renowned continental authorities. At times, Parkinson notes his discord with the illustrious herbalist Carolus Clusius, for instance, as in this conflated expression of praise and blame in which Parkinson identifies but still lets stand an error by the same: 'Clusius hath set forth this plant among his *Ornithogala* ... and although it doth in my minde come nearer to a *Hyacinthus*, then to *Ornithogalum*, yet pardon it, and let it passe as he doth' (p. 139). Parkinson also goes to considerable pains to avoid overtly criticising Clusius in the following example (p. 148):

Clusius hath set downe, that it was reported, that there should be another Liliasphodill with a white flower, but we can heare of none such as yet; but I rather thinke, that they that gaue that report might be mistaken, in thinking the Sauoye Spider-wort to be a white Liliasphodill, which indeed is so like, that one not well experienced, or not well regarding it, may sone take one for another.

Parkinson here distances Clusius from his published error by directing attention to 'they that gaue the report' to him; he thus invokes an anonymous botanical public in order to excuse a high-ranking member. In another instance, he muffles a dispute with Clusius in a parenthetical phrase buried mid-sentence (p. 45):

[T]he seede of a *Media Tulipa* did neuer bring forth a *Praecox* flower (although I know Clusius, an industrious, learned, and painfull searcher and publisher of these rarities, saith otherwise) so farre as euer I could, by mine owne care or knowledge, in sowing their seede apart, or the assurance of any others, the louers and sowers of Tulipa seede, obserue, learne, or know.

His admission of Clusius's dissent from his perspective (and thus his from Clusius's) is preceded and to some extent superseded by his reverential praise. Parkinson couches his critique of other erring allies in similar fashion, such as when he notes of a plant sporting a proliferation of names (pp. 96–7):

I receiued the seede of this Daffodill among many other seedes of rare plants, from the liberality of Mr. Doctor Flud, one of the Physitians of the Colledge in London, who gathered them in the University Garden at Pisa in Italy, and brought them with him, returning home from his trauailes into those parts, by the name of *Martagon rarissimum*, (and hauing sowne them, expected fourteen years, before I saw them beare a flower, which the first yeare that it did flower, bore foure stalkes of flowers, with euery one of them eight or ten flowers on them) which of all other names, doth least answer the forme or qualities of this plant.

Parkinson's lengthy parenthetical account of his hard-won success in persuading the plant to flower springs up after his acknowledgement of Flud's gift and his impressive credentials; through this syntactical diversion, Parkinson separates his praise of Flud from the nomenclatural critique that is the real point of the sentence. He thus twists his prose into rhetorical knots in order to prioritise civility and gratitude while still respecting the botanical facts. Something similar occurs in another slowly unravelling account: 'assuredly I haue been informed from some of my especiall friends beyond Sea, that they haue a double white *Clematis*, and haue promised to send it; but whether it will be of the climing or vpright sort, I cannot tell vntill I see it: but surely I doe much doubt whether the double wille giue any good seede' (p. 392). The 'assured' confidence of his initial phrase is vitiated by sentence's end with his equivocally 'sure' doubts about his 'especial' friends' specific claims; perhaps his doubts explain why Parkinson does not name the friends he impugns more than honours in this citation.

One contemporary basis of scholarly citation lies in evolving ideas of intellectual property, and a proprietary relationship to knowledge-making contributes to Parkinson's citation practice too. In describing what he calls 'The greatest double yellow bastard Daffodill, or Iohn Tradescant his great Rose Daffodill', he notes that it 'belongeth primarily to Iohn Tradescant, as the first founder thereof, that we know, and may well bee entituled the Glory of Daffodils' (p. 102). Parkinson's syntax blurs his final referent: is the daffodil the glory of its kind, or is Tradescant? Both possibilities pay off for Parkinson, since immediately following Tradescant's Daffodill is 'The great double yellow Spanish bastard Daffodill, or Parkinsons Daffodill' (p. 103); by ordering the presentation of these plants just so, Parkinson associates his authority with that of the resplendent Tradescant. The next entry, moreover, is for 'The double English bastard Daffodill, or Gerrards double Daffodill' (p. 103): Parkinson's horticultural catalogue threatens to turn into a catalogue of horticulturalists, with this entry referring to the John Gerard (1545–1612) who claimed to cultivate over a thousand species in his Holborn garden. Parkinson justifies naming this plant after him by saying that 'Mr. Gerrard first discouered it to the world, finding it in a pore womans Garden in the West parts of England, where it grew before the woman came to dwell there' (p. 104). The nameless poor woman's claim to be the 'first discouere[r]' of the plant is neatly purged by Parkinson's implication that its flourishing pre-dated her cultivation; moreover, she did not disseminate knowledge of it 'to the world' – that is, to the botanical public – as did Gerard, who therefore gets the credit.

Conclusion

These examples suggest the variety of ways in which an early modern book about plants also embedded a matrix of information about a community in formation. That community was not situated in any single geographic site, nor did it meet at regular intervals: instead, it found its being at the nexus of plants and texts about them, in an age characterised by Parkinson as 'more delighted in the search, curiosity, and rarities of these pleasant delights, then any age I thinke before' (p. 45). The map in his *Paradisi* of relations among herbalists, horticulturalists and their bigwig employers would have been more clearly legible to Renaissance readers than to us; and through the rhetorical pressure of his citational practices, Parkinson both reflected and shaped the relations that he planted in the minds of readers. Members of the horticultural public whom he documented were thus not only among the book's audience, but also contributing authors, sharing botanical specimens, descriptions, illustrations, names and knowledge along with credibility through their webs of association. As Paula Findlen has written in the context of Mattioli's commentary on the ancient herbal of Dioscorides, such books were 'collective projects whose success relied on the cooperation of many', so that a single herbal could act as a 'visible location in which to define who really belonged to the emerging community of naturalists'.[13] *Paradisi in Sole* is, as this chapter has attempted to summarise, a similarly viable site for excavating such a community, since people like Parkinson, as well as people Parkinson liked (in the lingo of today's defining social medium), shaped an English botanical public in part by transcribing the bonds they made in the course of making their gardens, as well as their books about them.

Further reading

Findlen, P., 'The formation of a scientific community: natural history in sixteenth-century Italy', in A. Grafton and N. Siraisi (eds.), *Natural Particulars: Renaissance Natural Philosophy and the Disciplines* (Cambridge, MA, 1999), pp. 369–400.

Harkness, D. E., '"Strange" ideas and "English" knowledge: natural science exchange in Elizabethan London', in P. H. Smith and P. Findlen (eds.), *Merchants and Marvels: Commerce, Science, and Art in Early Modern Europe* (New York, 2002), pp. 137–62.

Laroche, R., *Medical Authority and Englishwomen's Herbal Texts, 1550–1650* (Farnham, 2009).

Parkinson, A., *Nature's Alchemist: John Parkinson, Herbalist to Charles I* (London, 2007).

Robinson, B. S., 'Green seraglios: tulips, turbans and the global market', *Journal for Early Modern Cultural Studies*, 9:1 (2009), pp. 92–122.

Swann, M., '*The Compleat Angler* and the early modern culture of collecting', *English Literary Renaissance*, 37:1 (2007), pp. 100–17.

Willes, M., *The Making of the English Gardener: Plants, Books and Inspiration 1560–1660* (New Haven, 2016).

Yale, E., *Sociable Knowledge: Natural History and the Nation in Early Modern Britain* (Philadelphia, 2016).

5 European exchanges and communities

The long sixteenth century saw a particularly explosive growth of expertise on living nature, coinciding with and stimulated by increasing numbers of exotic plants and animals arriving in Europe and the beginning of a systematic exploration of the European flora and fauna. A satisfying explanation has so far not been found for why 'green fashion' (this strong fascination with living nature in all its manifestations) emerged when it did, or why it spread so quickly, both geographically and socially, throughout Europe. But exchanges of objects, images and textual information played a crucial role in its rapid expansion, forming the basis for the formation of crisscrossing and partly overlapping networks, in which expertise on living nature was shared and shaped in the course of this century. These networks of collectors of *naturalia* (many of whom also collected other items), and naturalists (most of whom developed specialist knowledge in their field, while by no means all of them collected) grew out of practices of exchange, collecting and related activities, rather than vice versa. It is in, and by means of, these exchanges and shared practices that notions about what aspects of living nature should be investigated and how this should be done developed. In other words, exchange helped to generate a European circle of more or less professional naturalists, who always remained embedded in the wider networks of *curiosi*, collectors and practical experts, as well as the locations, instruments, research methods (for example, field work and experimentation), research collections (for example, botanical gardens or image collections), and standards of textual and visual reporting of natural history.[1]

This chapter will first consider exchanges of *naturalia* in the sixteenth century, starting with the objects, living and dead, continuing with images – visual representations of *naturalia*, especially drawings – and ending with textual exchanges occurring before publication. Lastly, it will focus on the networks of expert and increasingly professional naturalists of the period *c.*1560–1610: did they form a *community* of naturalists?

Collecting and exchanging *naturalia*

As green fashion spread in Europe in the course of the sixteenth century, and *naturalia* collections grew, both elite collectors and budding naturalists strove for increasingly complete surveys of the natural world. The collector's aim of completeness can scarcely be distinguished from the naturalist's aim of encyclopaedic coverage of a particular domain of nature. Surveys could take the form of living collections in gardens and menageries, dried or otherwise preserved *naturalia* in cabinets, drawings or printed images, or descriptions in handwriting or print. Ideally, a collection of *naturalia* would contain all these elements; often it formed part of, or was combined with, a larger collection comprising works of art, antiquities, weapons and so on. Very few European *Kunst- und Wunderkammern* of this period lacked *naturalia*, but some collectors specialised in *naturalia*.[2]

Most collectors and naturalists acquired plants and animals via commercial transactions. Nearly all, however, also relied at least in part, and sometimes almost exclusively, on non-commercial networks of exchange for access to new items or information about how and where these could be obtained. The circulation of many rarities did not depend upon monetary exchange: many European wild plants that circulated were of purely botanical interest, apparently without any commercial value. Whether commercial or not, exchanges of natural history objects used the same routes and means of transportation as other types of goods – road and water, both within Europe and between continents.

The intriguing presence in southern Germany of what are thought to be the earliest European representations of the American potato, sunflower and tomato should probably be understood in terms of this geography and transportation infrastructure. The drawings form part of a huge collection of some 1,500 original plant drawings, created by the naturalist Leonhart Fuchs in Tübingen between around 1535 and 1560; they are roughly contemporaneous with drawings of American plants in the less famous but equally large image collection of the Venetian naturalist-patrician Pietro Antonio Michiel (see Plate 3). The Fuchs drawings of the potato, tomato and sunflower are lifelike and brightly coloured. They are presumably made after living plants that grew in Fuchs's own garden. But how did Fuchs in Tübingen – remote from any European port connected with the New World – obtain Central or South American plants or their seeds? And why did these early representations of Americana emerge in the collection of a south German naturalist rather than in Antwerp, Seville, Lisbon or in some major courtly collection? The most likely answer so far emerges from a particular facet of early European expansion politics: the role of the Welser family of bankers and merchants, from Augsburg and

Nuremberg, in Central America and what is now Venezuela. The Welsers did not merely operate as merchants, but were formal power holders on behalf of the Spanish king in Venezuela from 1528 until the 1550s, and ran a crucially important factory in Santo Domingo which regulated and supervised the flow of objects and information between the Welser domains in the New World and their home base in Southern Germany. It seems likely that collectors' items and *naturalia* from various locations in the New World were shipped to Southern Germany via Santo Domingo and Antwerp alongside ordinary merchandise, reaching the Welsers – and presumably Fuchs – overland or along the Rhine.

From the very early sixteenth century onwards, gift exchanges between members of ruling families (in particular the Habsburgs and the Medici), both within Europe and between Europe and Asia, involved precious animals such as elephants, rhinoceroses, monkeys, parrots and civet cats. Especially from the 1550s onwards, the most important princely collectors in Europe employed expert assistants, agents and naturalists to acquire rare and novel *naturalia*. Such experts fulfilled a large range of functions, from middlemen in transactions to scouts for rarities, plant hunters in the wild, garden consultants and contact persons who arranged exchanges with other aristocratic collectors. In the 1540s, for instance, Duke Cosimo I de' Medici in Florence had exotic plants imported, studied books on exotic and unusual animals, founded the Pisa botanical garden, and surrounded himself with living and painted wild animals – his menagerie held live lions, tigers, wolves and bears – in a conscious emulation of both the papal animal collection and the Roman Emperor Augustus's display of intriguing animals in his villas. For about two decades, from *c.*1578 onwards, his successor Francesco I de' Medici employed the naturalist Giuseppe Casabona to obtain new plant material for the private and university gardens of the Medici, either by means of exchange with international collectors and naturalists, or via expeditions to regions known for their varied flora: Casabona's field work ranged from the Alps and Apennines to Crete. In the Iberian peninsula, Hans Khevenhüller acted from 1574 to 1606 as representative of the central European Habsburgs, procuring exotic animals and *Kunst- und Wunderkammer* objects for their menageries and collections, finding specialised staff, and supervising transport, to name only a few of his many duties. These expert agents relied in their turn on exchanges via their personal networks for information about interesting *naturalia*.[3]

Princely and aristocratic natural history collections of the sixteenth century were not always intended merely for display, self-fashioning or contemplation: like university collections and gardens, they also functioned as sites of knowledge and nodes in European exchanges of information about nature. Many courts had resident or visiting naturalists, like Pietro Andrea Mattioli, at the court of Cardinal Bernardo Cles in Trento in the 1520s to 1530s, then at the Habsburg court in Prague from around 1550 to 1566; Alfonso Panzio at the Este court in Ferrara in the 1570s; Carolus Clusius at the Habsburg court in Vienna in the 1570s; or the two Jean Robins, father and son, at the royal court in late sixteenth- and early seventeenth-century Paris. The Prague court of the Habsburg Emperor Rudolf II (1583–1612), with its enormous collections of living, dead, painted and sculpted rare animals and plants, has been described as a 'highly interesting cultural centre where not only artists of many specialities but also a string of scientists from different countries gathered to work under optimal conditions'.[4]

The importance of Habsburg court circles for nature study and exchanges is also evident in the introduction of new plant species from the Balkans and the Middle East to Western Europe, especially during the 1550s to 1570s. European diplomats operating between Brussels, Vienna and Constantinople during a long period of warfare with the Turks returned with plant species previously unknown in Europe, which quickly became fashionable. Foremost among these were tulips and other bulbs, but other plants introduced into continental Europe in this way were the horse chestnut and lilac. Ladies at the Viennese, Madrid and Brussels courts were particularly active in organising plant and seed exchanges, for instance between the royal gardens of Aranjuez, the private gardens of the court aristocracy in Brussels and Malines, and those of the Viennese elite. During the 1570s, aristocratic ladies in Vienna even sent their private servants to Constantinople or used diplomatic couriers in order to obtain more botanical rarities. In a similar manner, Canon Levenier in Bordeaux sent out a nephew to Constantinople and other places in the Levant in 1602–3 with specific instructions to obtain rare *naturalia* that could be grown in his botanical garden. Levenier also had personal contacts with the New World, via a friend in Mexico who sent him information about trees and fruits. Levenier was equally interested in the European flora, employing his personal staff and local villagers to 'comb' entire mountain flanks in the Pyrenees for rare flowers that could be transplanted to his gardens, or exchanged for other rarities from fellow collectors and naturalists.

Gift exchanges were also at the heart of the Europe-wide networks of many naturalists, of whom Conrad Gessner and Carolus Clusius are two of the better-known names. In Clusius's case, these exchanges continued for at least half a century, from the late 1550s to 1609, and expanded geographically, so that his network eventually ranged from Naples to Norway and from Seville to England and Hungary.

Indirectly, his lines of exchange also extended outside Europe. For instance, Clusius asked a friend in London, the apothecary James Garet, who specialised in exotic plants and spices, to contact Francis Drake, or members of his entourage, with a view to obtaining new exotica from Drake's voyages. A few correspondents, like the physicians Onorio Belli from Crete and Tobias Roels from Middelburg, sent both seeds and complete botanical surveys to Clusius, who planted the former in the Leiden botanical garden and published the latter as appendices in his own works under the original author's name. Belli sent seeds to many others besides Clusius; by the late sixteenth century the living results of Belli's fieldwork on Crete could be admired in the botanical gardens of both Leiden and Padua, and in many gardens of private collectors in the Veneto. Generally, Clusius gave seeds, bulbs, plants and gardening advice in return for newly discovered plants and plant products, seeds, cuttings, bulbs or written and visual information about plants and animals, especially exotic ones. Other exchanges of gifts were immaterial, but no less valuable: Clusius mentioned many donors, patrons and intermediaries by name in his printed works; Gessner operated in a very similar way. The personal presence of members of exchange networks in sixteenth-century natural history publications not only served to emphasise and convey the authoritative character of the information, but also strengthened the communities of exchange and made them more visible.

There were no barriers between the circuits of princely collectors and those of naturalists, as is only to be expected, since many naturalists moved continually in and out of the sphere of courtly patronage. In the 1560s to 1570s, the court physician Alfonso Panzio in Ferrara, for instance, was a key contact in the transmission of Egyptian and other Middle Eastern plants and plant products to his patrons in Ferrara, Italian naturalists, plant collectors in the Southern Netherlands and Clusius. During the third quarter of the sixteenth century, selections of seeds from the Spanish royal gardens arrived with some regularity in the gardens of male and female plant collectors linked with the Brussels court, via diplomatic couriers. The sunflower made its way into print precisely through such a line of exchange: from Spain via court-connected gardens in Malines, to the Flemish botanist and court physician Rembert Dodoens, who published the first European illustration of one in 1569. Particularly for rare plants, plant collectors and naturalists appear to have kept track of such filiations, not only in recognition of the chain of obligation linking donors and recipients in their many-stranded and often international exchanges, but also because of an interest in the plant's descent and heredity.

Image circulation

Illustrations in printed works formed only a part of the huge body of visual material, comprising maps, tapestries, engravings, oil paintings, drawings, frescoes, wood carving, embroidery, sculpture and other decorative arts, that conveyed information about living nature in early modern Europe.[5] As is well known, many representations of *naturalia* in the decorative arts – from exotic animals on Flemish *Verdure* tapestries to plants and animals on frescoes – were copied after or inspired by illustrations in printed works that had a huge readership in the whole of Europe. Conrad Gessner's *Historia animalium* (4 vols., 1551–8) and its vernacular editions were a particularly widely used source. Examples of animal decorations based on Gessner's illustrations can be found, for instance, in Jacopo Zucchi's frescoes with birds painted in the Roman Villa Medici during the 1570s, but also in needlework by Mary Stuart and on the early seventeenth-century painted wooden ceiling of the Long Gallery at Earlshall Castle, a sixteenth-century tower house near St Andrews in Scotland.[6]

For many other representations of plants and animals in the decorative arts, no printed sources have been identified, however. Some were based on direct observation. Others were copied from other non-printed images. The complex filiation of image sources directly reflects the intricate ways in which images travelled and were exchanged. Some 80 Brazilian birds are depicted on the ceiling panels of the central hall in the hunting lodge of Hoflössnitz near Dresden, built between 1648 and 1650 on the orders of Johann Georg I, Elector of Saxony (Figure 5.1). These paintings are based on original bird drawings made in Brazil by the Dutch painter Albert Eeckhout, now part of the Libri Picturati collection in Kraków.[7]

Sixteenth-century European collections of images of *naturalia* in the form of albums on paper or parchment contain hundreds, even thousands of original coloured drawings. Such drawings and albums varied greatly in style, quality, status and function. Some were high-quality showpieces, commissioned by princes and aristocrats, like Jacopo Ligozzi's high-definition paintings of fish, birds and plants, made for the Medici, or the plant and animal drawings in the Libri Picturati, made for the southern Netherlandish nobleman Charles de Saint-Omer during the 1560s. Others were much more modest, such as the Dutch fish merchant Adriaen Coenen's home-made encyclopaedic fish albums of the 1570s to 1580s, or the herbals with plant drawings that were made all over Europe during the fifteenth and sixteenth centuries in the service of health, herbal medicine, and the study of plants. Very large image collections were also created by naturalists such as Conrad Gessner, Felix Platter,

Figure 5.1 Ceiling panels with paintings of Brazilian birds in the hunting lodge of Hoflössnitz near Dresden, built between 1648 and 1650 on the orders of Johann Georg I Elector of Saxony. Photo by the author.

Ulisse Aldrovandi, Leonhart Fuchs and Pietro Antonio Michiel. Most of these would never have come into being without intensive international exchanges of images.

Conrad Gessner, for instance, was involved in a complex web of exchange of images: he not only reused many fish illustrations printed earlier by his French colleague Guillaume Rondelet, but also received large numbers of animal drawings from friends all over Europe, quite a few of which eventually served as models for the printed illustrations in Gessner's *Historia animalium*. One of the chains of exchange in which Gessner was involved extended from Rome via Zurich to Meissen and Torgau not far from Dresden during the 1550s. In Rome, the Dutch physician-naturalist Gysbertus Horstius commissioned original drawings of fish and marine animals from local painters and shared them with visiting physicians from beyond the Alps, one of whom arranged carriage to Gessner in Zurich (see Plate 4). Subsequently, further copies were made, or multiple copies were shared. A regular two-way exchange of images occurred between Gessner and the naturalist Johannes Kentmann in Meissen: their respective albums contain many nearly identical drawings of fish, while further copies have been traced in other German and Dutch collections.[8]

Like printed illustrations, paintings or drawings of *naturalia* crossed vast geographical distances as they were copied and circulated many times. They also moved back and forth between media and genres, fulfilling many different functions from decoration to documentation. Often few if any traces of these exchanges of visual materials survive in correspondence or other textual evidence; they have to be reconstructed from comparisons between images.

Correspondence and the development of a discipline

Besides images and the relatively small numbers of extant *naturalia* from this period, collections of handwritten letters constitute our most important source for investigating exchange networks among naturalists and collectors in early modern Europe. By no means all of this material has been explored, let alone compared or linked in terms of personal connections or content. Even a brief inspection of letters to and from Aldrovandi, Clusius, Gessner and members of their networks shows enormous diversity in language, style and content, often within one body of correspondence, occasionally even within a series of letters by the same person. Naturalists of the sixteenth century were quite capable of producing both formal open letters about *naturalia* in Latin, inspired by the model of the humanist letter and functioning almost as publications, and chatty letters in the vernacular, or a combination of Latin and the vernacular, that mixed detailed plant or animal observations with gossip, personal health issues, politics, colour experimentation, garden planning and herbal remedies.

It is very hard to generalise, therefore, but three characteristics appear to have been more common in correspondence between naturalists, collectors and other *curiosi* of the sixteenth century than in the Republic of Letters of the seventeenth and eighteenth centuries. The first is the regular use of vernacular languages alongside Latin. The second, which helps to explain the first, is the social diversity of the exchange networks, reflecting the wide spread of green fashion in Europe. Such networks, as Pugliano's chapter shows, included apothecaries as well as princes, merchants as well as physicians, and men as well as (mainly elite) women. The latter generally corresponded in French, almost never published, but were often acknowledged as experts in medical botany, garden layout and the cultivation of rare flowers. Finally, information exchanges by letter were often pragmatic in character and only occasionally took the shape of formal 'research reports' in Latin.

Sometime between the 1520s and the 1580s, correspondence concerning *naturalia* in the vast and heterogeneous world of *curiosi*, collectors and naturalists came to delineate and shape a more limited and socially much more homogeneous community of men (and not women) for whom the study of nature became their main interest. They found common ground in their effort to identify, name and inventory nature. In a seminal article, Richard Palmer has phrased it as follows: 'Associated with all these developments was the emergence of a large, but closely-knit community of botanists and natural historians, international in scope and spanning the religious divisions which followed in the wake of the Reformation. These were men united by common aims, and linked by frequent correspondence.'[9]

While many of these professional naturalists possessed medical knowledge or qualifications, and the whole project of inventorising nature should also be seen in the context of a revived interest in classical antiquity and humanist philology, the influx of practical knowledge into natural history remained very important indeed in following centuries. From at least the 1530s onwards, personal exchanges with local informants during fieldwork in mountainous zones (the Alps, Dolomites, Apennines, Jura or Pyrenees), several Mediterranean islands, and extensive areas of the Middle East were often essential to the discovery and identification of new plants and animals. Relative openness and the sharing of information were also relevant *within* the circles of professional naturalists. All those who participated in the shared research project benefitted from an easy exchange of information, and naturalists were well aware of the fact. Natural history differed in this respect from, for instance, alchemy, where secrecy generally offered more advantages. Bitter rivalries did, of course, occur among naturalists, but seem to have manifested themselves more often in accusations of incorrect identification and naming than of withholding information.

Underpinned by exchanges of objects, images and texts, and stimulated by the challenges posed by the correct identification of increasing numbers of new plants, a whole series of practices came to form part of a shared scientific attitude among naturalists: precise documentation, detailed description and depiction, the value of first-hand evidence, double-checking and other forms of verification, the repetition of tests during experimentation, and the development of a specialist idiom.[10] Many examples can be found in the letters and writings of Clusius. In the early 1570s, he received images of the East Indian betel, areca and coconut palm from Alfonso Panzio in Ferrara.

Clusius was, however, not convinced of their correctness, and did not include the images in his editions of Garcia da Orta's work on tropical plants, until the mathematician-cartographer Fabrizio Mordente from Salerno assured him of their correctness during a long conversation in the Viennese court in the late 1570s. Mordente had lived in Goa and was known as an observant and reliable witness. In later years, Clusius continued to verify his information about the betel via correspondence with the French royal physician Nicolas Rasse, and personal interviews in Amsterdam with both Asians and Europeans who had been in the East Indies. In another case, Clusius's attempts to corroborate the existence of a red daffodil (which did not exist, but was illustrated) involved an intricate exchange of letters between 1597 and 1605 with the Duke of Arenberg in the Southern Netherlands, the plant collector Matteo Caccini in Florence, the physician-naturalist Simón de Tovar in Seville, the Brussels plant collector Jean de Boisot and Canon Levenier in Bordeaux (Figure 5.2).

Many of the practices listed here were by no means the exclusive prerogative of a few top naturalists with a humanist education, however. Numerous letters by apothecaries and aristocratic flower collectors who had no Latin refer, for instance, to their experimentation in the garden, in particular with colour variation, over periods of years and sometimes decades. Their references imply some kind of systematic note-taking to keep track of the results. While the medical knowledge shared by many naturalists was undoubtedly one of the important cultural roots of the developing scientific methodology of sixteenth-century natural history, it was certainly not the only one. Much of the methodology of natural history and its links with practice-based knowledge still needs further exploration.

The international exchange community

To what extent did the criss-crossing and partly overlapping personal networks of exchange of the sixteenth century constitute true communities – or even one virtual European community – of naturalists?[11] It seems far less useful to ascertain whether they fit various modern definitions of a community or network, than to establish how such networks were perceived and used at the time. Naturalists generally did not refer to themselves as a community, and appellations equivalent to 'Republic of Letters' were seldom used. Yet a strong indication of a shared research interest and an awareness of the importance of collaboration emerges from their published works, correspondence and further exchanges. Communication at a distance between naturalists who never met in person is particularly significant in this respect. Introductions of one naturalist by letter as a 'friend of a friend' were common, and indicate awareness of each other's reputation. Intensive

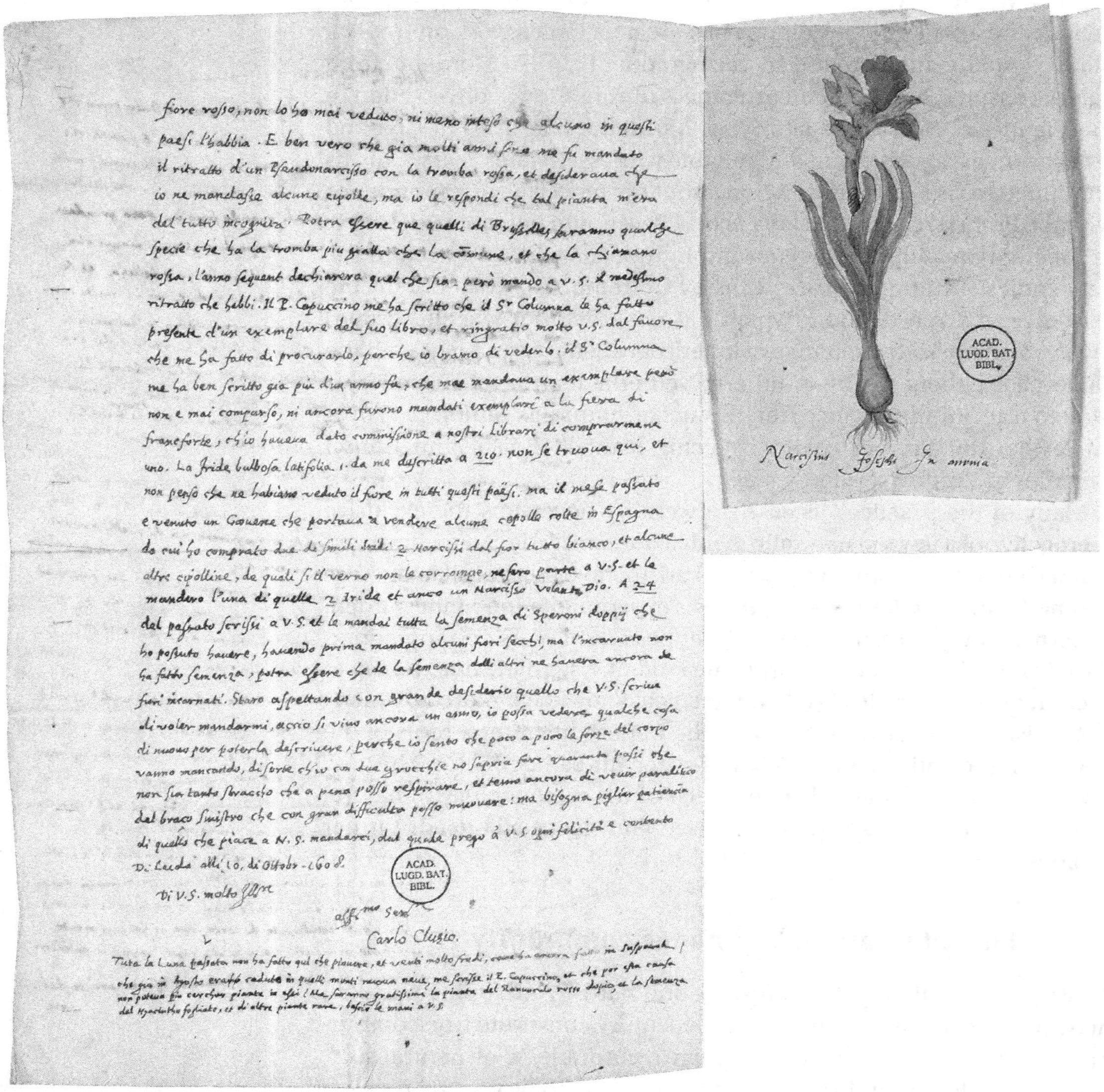

Figure 5.2 Drawing of a red daffodil attached to a letter from Carolus Clusius in Leiden to the Italian naturalist Matteo Caccini in Florence (10 October 1608). Leiden University Library, shelf mark BPL 2724a.

international exchanges lasting for years and even decades sprang up between naturalists and collectors who never met in person. A well-connected naturalist in one European country could certainly find out (especially via informal exchanges) who was important in the field in other parts of Europe, and what was being discovered. By the second half of the sixteenth century we may certainly imagine a 'cloud' of exchanges concerning *naturalia* that stretched from one end of Europe to the other, and reached out to other continents.

Within this cloud, more or less dense zones may be discerned, depending on how many personal networks criss-crossed and overlapped. Among the denser zones and hubs of the second half of the sixteenth century were courts like Prague, Vienna, Brussels, Madrid, Munich and Florence, where major collectors and experts congregated in a single location; some ports, like Venice, Antwerp and Lisbon, where exotic *naturalia* arrived; and some university centres like Padua, Pisa, Montpellier and Basel. There were also privately organised salons and informal societies. Much remains to be clarified, however, concerning the changing geography of natural historical exchange in Europe. Early in the sixteenth century, parts of southern and central Europe – Portugal, Spain, Italy, Austria-Hungary, parts of Germany, the south and west of France – were far more important in terms of natural history exchanges than peripheral countries like England or Holland. The growth of Atlantic trade, and the increasing relevance of Antwerp and, later in the sixteenth century, of the Dutch ports and London, does not necessarily imply, however, that natural historical exchange in or with the Mediterranean area diminished. Notions of a European centre and periphery may obstruct understanding rather than help it, and it is probably much more relevant to compare regions rather than states.

In organisational terms, it was typical of the exchange networks of the sixteenth century that nearly everyone fulfilled multiple roles. Clusius might nowadays be best known as an author and naturalist, but he was also a middleman, the patron of younger naturalists, a courtier and a client of his own princely patrons, a friend and a donor as well as receiver of information, *naturalia*, texts and images. The multiple roles and many-stranded connections of sixteenth-century exchange networks make it almost impossible to speak of diffusion and reception of knowledge, since it is often unclear who distributed, who received, and who acted as middleman. Important new plants and animals could as easily be discovered by local *curiosi* as by famous naturalists. Nonetheless, some particularly well-connected persons acted as nodes between the criss-crossing personal networks, and their roles deserve further investigation. During the 1570s to 1590s, for instance, the aristocrat, humanist, naturalist and bibliophile–collector Giovanni Vincenzo Pinelli in Padua created one of the best private libraries in Italy, and acted as patron, mentor and intermediary for many Italian scholars and scientists, including the young Galileo. He put his collections at the disposal of what was to all intents and purposes an informal private academy – a centre of erudition based at his home – which he coupled with a European network of exchange.[12]

In the late sixteenth century, informal meetings with the specific purpose of discussing living nature were also held at a much more modest social level, but very little is known as yet of this phenomenon. The physician Tobias Roels, for instance, was a member of a circle of local *curiosi* in Middelburg, Zealand, during the 1590s and early 1600s. They met regularly and discussed exotic and other rare *naturalia*, comparing the information which they gathered from their various contacts within and outside Europe. This circle included apothecaries and physicians, a minister, flower painters, and merchants. Although it consisted of local *curiosi*, its reach was definitely international. Roels himself had travelled in Germany and Italy, and maintained contacts with naturalists in those parts. Ships arrived in Middelburg with *naturalia* from the West African coast, and members of this circle interviewed travellers about living nature in the tropics. The professional shipping contacts of the merchants in this circle furthermore underpinned regular exchanges of *naturalia* with a similar group of *curiosi* in Bordeaux.

Exchange obstructed

It would be misleading to speak about connections and exchanges without at least briefly mentioning obstructions and gaps in the transmission and exchange of information about living nature. Both incidental, temporary barriers and far more structural ones blocked exchanges. Latin has been called the *lingua franca* of the world of learning, but by no means all those with expert knowledge of nature in Europe had Latin. The frequent use of vernacular languages in correspondence about *naturalia* indicates a clear awareness of the existence of valuable expertise outside the Latinised world. Major publications on plants and animals in French by Pierre Belon, Flemish by Rembert Dodoens, Italian by Pietro Andrea Mattioli and Luigi Anguillara, German by Hieronymus Bock, English by John Gerard, as well as the many vernacular editions of Gessner's works, show that both authors and publishers consciously addressed a large non-Latin readership.

To what extent wars impeded communication about and exchanges of *naturalia* is unclear. War with the Ottoman Empire seems to have stimulated exchanges between Western Europe and Constantinople rather than stopping them, judging by the case of plant imports. Nor do the fall of Antwerp and the subsequent closure of its port by the emerging Dutch Republic seem to have adversely affected the town's position as a major centre for collectors. On the other hand, Philip II of Spain's policy of secrecy, influenced by international rivalry and imperial competition, definitely impeded the circulation of knowledge from and to the Iberian peninsula. Its results are visible, for instance, in decade-long gaps in the correspondence between Clusius and his Spanish fellow naturalists, and in the non-circulation in Europe of the major results of the 1570s Hernández expedition to New Spain (Mexico).[13] Only major gaps in the transmission of information can explain why three of the largest collections of botanical drawings of that century, created by Conrad Gessner, Leonhart Fuchs and Pietro Antonio Michiel, disappeared from sight (and were apparently largely forgotten) relatively shortly after the deaths of their creators, only reappearing centuries later.

Conclusion

Focusing on what may be regarded as typical of the exchanges in natural history during the long sixteenth century, the following characteristics should be mentioned in particular. The rapid spread of green fashion throughout Europe manifested itself in widespread collecting of *naturalia* and the study of living nature, and in a very wide geographical range of information exchanges about *naturalia* in the form of objects, images or textual information. These characteristics were reflected in the social constellations in which these exchanges took place and which they helped to shape. The social composition of the circles involved in sixteenth-century nature study was highly diverse, ranging from princes to merchants and apothecaries. This diversity was reflected, for instance, in the regular use of vernacular languages alongside Latin, the relative openness towards information based on practical experience, and the lack of barriers, or rather the overlap, between circles of elite collectors and those of expert naturalists.

Whether the criss-crossing networks of information exchange in Europe together formed *one community* of naturalists remains a moot point. They were not, or only rarely, organised in any formal sense, and a single denomination for such a European community of naturalists is lacking. Yet students of nature in different parts of Europe recognised each other's expertise, were aware of each other's discoveries and publications (often via informal channels of information), clearly shared a sense of purpose, and recognised that sharing information was, on the whole, generally to everyone's advantage. Within the patterns formed by the criss-crossing of personal networks, denser zones of information exchange can be discerned, while certain nodal figures linked networks. The relative lack of formal organisation of these expert networks is matched by the fact that virtually no sixteenth-century nature expert can be pinpointed as occupying *only* one role, whether as donor, agent, collector or naturalist. Nearly all received, as well as creating or distributing, information; many acted as middlemen.

There was seemingly no abrupt transition between sixteenth-century networks of exchange and the scientific societies and academies of the seventeenth century, such as the Roman Accademia dei Lincei, the later Florentine Accademia del Cimento and the English Royal Society. Exchanges appear to have been gradually channelled and formalised through official meetings and journals. This transition, which deserves further investigation, can be seen as part of a process of specialisation, in the course of which the domain of natural knowledge was to some extent fenced off by boundaries in the form of the jargon and social exclusivity of the official societies, and which at the same time entailed an elevation of the status of natural history and natural philosophy. But, as the history of the following centuries shows, natural history as a discipline was never closed off completely from 'amateurs', nor did the journals and meetings of scientific societies that took over parts of professional information exchange replace personal correspondence and exchange networks.

Further Reading

Bethencourt, F. and Egmond, F. (eds.), *Correspondence and Exchange in Europe, 1400-1700* (Cambridge, 2007).

Davids, K., 'Dutch and Spanish global networks of knowledge in the early modern period: structures, connections, changes', in L. Roberts (ed.), *Centres and Cycles of Accumulation in and around the Netherlands during the Early Modern Period* (Zurich and Berlin, 2011), pp. 29–52.

Dupré, S., De Munck, B., Thomas, W. and Vanpaemel, G. (eds.), *Embattled Territory: The Circulation of Knowledge in the Spanish Netherlands* (Ghent, 2016).

Egmond, F., *The World of Carolus Clusius: Natural History in the Making* (London, 2010).

Egmond, F., *Eye for Detail: Images of Plants and Animals in Art and Science, 1500–1630* (London, 2017).

Findlen, P., *Possessing Nature: Museums, Collecting and Scientific Culture in Early Modern Italy* (Berkeley, 1994).

Findlen, P., 'The formation of a scientific community: natural history in sixteenth-century Italy', in A. Grafton and N. Siraisi (eds.), *Natural Particulars: Nature and the Disciplines in Renaissance Europe* (Cambridge, MA, 1999), pp. 369–400.

Impey, O. and MacGregor, A., *The Origins of Museums: The Cabinet of Curiosities in Sixteenth and Seventeenth-Century Europe* (Oxford, 1985).

Kusukawa, S., 'The sources of Gessner's pictures for the *Historia animalium*', *Annals of Science*, 67:3 (2010), pp. 303–28.

Kusukawa, S., *Picturing the Book of Nature: Image, Text and Argument in Sixteenth-Century Human Anatomy and Medical Botany* (Chicago and London, 2012).

MacGregor, A., *Curiosity and Enlightenment: Collectors and Collections from the Sixteenth to the Nineteenth Century* (New Haven, 2007).

Ogilvie, B., *The Science of Describing: Natural History in Renaissance Europe* (Chicago, 2006).

Olmi, G., '"Molti amici in varii luoghi": studio della natura e rapporti epistolari nel secolo XVI', *Nuncius. Annali di Storia della Scienza*, 6:1 (1991), pp. 3–31.

Palmer, R., 'Medical botany in northern Italy in the Renaissance', *Journal of the Royal Society of Medicine*, 78 (1985), pp. 149–57.

Pérez de Tudela, A. and Jordan Gschwend, A., 'Renaissance menageries. Exotic animals and pets at the Habsburg courts in Iberia and Central Europe', in K. Enenkel and P. Smith (eds.), *Early Modern Zoology: The Construction of Animals in Science, Literature and the Visual Arts* (Leiden and Boston, 2007), pp. 419–47.

Smith, P. H. and Findlen, P. (eds.), *Merchants and Marvels: Commerce, Science and Art in Early Modern Europe* (New York, 2002).

6   Making monsters

The word 'monster' derives from the Latin, *monstrare*, to demonstrate, or *monere*, to warn. The *Oxford English Dictionary* defines a 'monster' as 'something extraordinary, a prodigy, a marvel', and 'monstrous' as 'deviating from the natural order'. In the early modern period, monsters were seen as deviations from 'normal' nature that could reveal God's plan and nature's inner workings, or singular instances that acted as omens. Monsters could take many forms. They were commonly hybrids with characteristics that crossed accepted natural boundaries. This involved exaggeration, proliferation or absences of normal body parts, such as deformed births with excess limbs or multiple heads, races of one-footed *sciapodes* and dramatically undersized pygmies.[1]

Many kinds of monsters were represented in different types of early modern publications, including cheap pamphlet literature, illustrated compendia of medical monstrosities, such as *Des Monstres et des Prodiges* ('On Monsters and Prodigies', 1573) by French surgeon Amboise Paré, and natural histories, such as the *Monstrorum historia* ('History of Monsters', 1642) of the Bolognese collector Ulisse Aldrovandi. The material, often taken from humanist works, combined several classical traditions of monstrosity. Aristotle's *On the Generation of Animals* cast monsters as deviations from natural development, often through the excessive influence of the female imagination on the foetus. Cicero's *On Divination* described monsters such as deformed foetuses as individual, supranatural occurrences, interpreted as prodigies or omens. Finally, Pliny's *Natural History* described whole populations of deviant human forms such as dog-headed *cynocephali* or headless *blemmys*.[2]

These Plinian races and other monsters were often placed at the shifting margins of the known world, or *ecumene,* on maps, from medieval *mappaemundi* to early modern charts. As geographical knowledge developed with exploration and colonial activity in the early modern period, the medieval *ecumene,* previously bounded by impassable ocean and an uninhabitable 'torrid zone', was rapidly expanded. The locations of these monstrous races shifted also, retreating out of reach to the edges of maps or to the unknown centres of continents. Monsters almost always originated *elsewhere*, either from distant places or local but inaccessible locations.[3] Such intangibility made travellers' tales of monsters impossible to falsify, but it also explained, for some scholars, the generation of these strange creatures in nature: the extreme conditions of unknown and distant reaches of the globe would produce similarly extreme humans, animals and plants.[4]

Many exotic beasts that first reached Europe from the late fifteenth century were deemed to be 'monsters'. They were often seen as adulterated or degenerate forms, inferior to God's original creations that had populated the Garden of Eden. New World species were sometimes interpreted as Old World species affected by dramatic climates: the whiteness of the polar bear resulted from the loss of the brown colour of other bears in freezing Arctic conditions. Similarly, the startling whiteness of the ivories of walruses and elephants was produced by the extremes of the 'coldest weather' and the 'heat of heaven' in their respective regions of origin.[5] Sometimes new monsters were the result of miscegenation, the illicit coupling of 'Godly' beasts. For example, according to some authors, the union of the tortoise and the hedgehog in Noah's Ark had produced the armadillo.

This chapter examines how and why exotic creatures were made into monsters in early modern European natural histories. It focuses on the making of one particular monster, the bird of paradise, that arrived in Europe in the early sixteenth century as trade skins. Unlike other birds, these creatures were thought to live floating perpetually in the skies without legs or wings. The first two sections outline the reasons why new creatures had to be assembled by naturalists, and the early constructions of these spectacular birds in natural histories and cabinets. The third and fourth sections examine the dramatic reworking of these creatures in the early seventeenth century by the Flemish Professor of Botany at Leiden University, Carolus Clusius, and the patchwork nature of the images that he constructed. The fifth section highlights the long chains of exchange through which Clusius's materials passed, though he was attempting to present 'direct experience' of the birds. The final two sections explore the ways in which such composite monsters could be used for emblematic functions, and how their chimerical forms made them very malleable so they could be used for different symbolic purposes.

Assembling parts

New animals and plants did not arrive in Europe as 'monsters'; they rarely even came as complete specimens or living creatures. Rather, most were brought in as preserved or decaying objects, collected by travellers and sailors and transported for many months on merchant ships from European colonies such as Batavia or New Spain. The trade routes taken by ships were determined by practical and economic concerns, so these things rarely moved across the globe by the most direct paths. As a result, the conditions in which exotic natural objects were collected and transported were often less than ideal. Housing specimens in transit and preserving them from decay was very difficult, so the commercially valuable or most easily preserved parts of creatures were most likely to be transported, such as walrus tusks or armadillo carapaces.[6]

After arriving in Europe, these things entered flourishing markets for items of *naturalia* or circulated within the elite networks of gift exchange between scholars, collectors and patrons. Objects were displayed in cabinets of curiosity, broadcast in pamphlet literature and described in detail in more formal publications. Physical material was often accompanied by travellers' tales of exotic natures, which contained both eyewitness information and material collected from natives in colonial locations. This information could remain closely associated with the objects, but was sometimes dissociated.[7]

The mutilated or incomplete forms of these specimens and the accounts of exotic beasts encouraged reconstruction by naturalists, who often had to actively assemble unknown creatures from disparate body parts and pieces of transported information. They gathered specimens from their own collections, correspondents and secondhand accounts, as well as using familiar referents from classical authorities, previous accounts and known creatures. The results were chimerical European constructions, often very different to the living animals and plants: they were actively made into things that played roles in European cosmologies. They often became 'monstrous', boundary-crossing or uncategorisable things, deviations that were in fact used in defining boundaries and structuring the world.[8]

New creatures arrived in Europe as physical objects, but it was the printed images and textual imagery relating to them that circulated most widely. New beasts existed primarily in the pages of early modern publications. The construction of one 'monster' offers a rich case study of such construction: the birds of paradise or *manucodiata* in the *Exoticorum libri decem* ('Ten Books of Exotic Things', 1605) by the naturalist Carolus Clusius. This volume was the first European work to deal exclusively with the natural history of 'exotics', and the extensive material Clusius presented made his book authoritative for over a century. Clusius worked on the *Exoticorum* intensely despite his failing health, keeping unusually close control over its production throughout the publishing process. He continued to gather new material and accounts, to be added in appendices if the relevant chapters had already been printed.[9]

The birds of paradise were not creatures new to Europe at the time that Clusius described them; they first appeared in printed natural histories in the mid sixteenth century. Clusius did not entirely construct these 'monsters' from new material, but reconstructed them into beasts very different from previous representations. Looking closely at this process of reconstruction and the details of Clusius's account reveals the hybridity of the materials he used and the hybridity of his mode of description, which shifted seamlessly between the empirical and the mythical, literal and symbolic, objective and subjective. Using heterogeneous sources and multiple descriptive styles produced beasts that were concomitantly monstrous and chimerical.

The early history of the birds of paradise

The first examples of birds of paradise in Europe were probably five skins brought to the court of Emperor Charles V of Spain in 1522 as part of a cargo of spices and other marvels from the East Indies. These fabulous skins were shrunken and wingless as a result of preservation methods used by New Guinean tribal hunters who prepared them, causing their beaks and gorgeous plumes to be disproportionately exaggerated. Bird of paradise skins had been part of extensive Asian trade networks for at least 5,000 years before Europeans reached the region in the late fifteenth century. Europeans explored the East in search of precious commodities such as cinnamon, cloves and nutmeg, and in the hope of rediscovering the fabled Eastern 'terrestrial paradise'. Birds of paradise were significant in Asian and Islamic mythologies and were associated with the spices with which they were traded, though the living birds were virtually unknown to anyone outside New Guinea.[10]

The ship's chronicler of the voyage on which these plumed treasures had been collected, Antonio Pigafetta, described the skins: 'These birds are as large as thrushes ... legs slender like a writing pen, and a span in length; they have no wings, but instead of them long feathers of different colours, like plumes.'[11]

The Spanish court secretary, Maximilianus Transylvanus, wrote another, more widely circulated account of the voyage based on interviews with sailors, whose reports included Islamic Asian material:

The kings of Marmin ... saw that a certain most beautiful small bird never rested upon the ground nor upon anything that grew upon it; but they sometimes saw it fall dead upon the ground from the sky. And as the Mahometans, who travelled to those parts for commercial purposes, told them that this bird was born in Paradise . . . they call the bird *Mamuco diuata* [Bird of God] . . .[12]

The first bird of paradise skins to arrive in Europe were rapidly assimilated into aristocratic *Wunderkammern*. Most other examples brought to Europe through the sixteenth century lacked legs as well as wings, and they remained scarce and fetched high prices on the open market.

Some ten or eleven significant scholarly descriptions of the birds of paradise were published between 1550 and 1600. An elaborate body of imagery was developed around the birds in natural histories, books of emblems, natural philosophies and other print forms: they became angelic, perpetually floating creatures that could not land, represented by bizarre dried skins to which authors rarely had personal access. These ideas were based in part on the Islamic imagery published by Transylvanus that told of the paradisial birds that never landed, as well as on their value as exotic objects traded from the East, apparently originating from a rediscovered terrestrial paradise. They were named the *manucodiata*, a Latinisation of the Malay name *Mamuco diuata* ('birds of God').

The legless nature of most of the skins circulating around sixteenth-century Europe was used to support these angelic images. However, it also made these birds monstrous, because they deviated from the 'normal' avian type as described by classical authors. Aristotle, for example, maintained that 'no creature is able only to move by flying', birds being two legged and winged creatures.[13] The leglessness of the *manucodiata* was a focal point for most of the natural historical descriptions of them: these emphasised the monstrous deviation both from the expected presence of limbs and from contact with both earth and sky made by typical birds.

The most influential sixteenth-century account of the *manucodiata* was in the *Historiae animalium* (1555) of the Swiss naturalist Conrad Gessner, in which he brought together as many sources as he could acquire to form a comprehensive history of the birds. He included previous natural historical accounts as well as material from correspondents who owned skins. A contact in Padua described the small plumed birds found in the Indies that he thought might well be the *rhyntace* of Plutarch or *phoenix* of Herodotus. Another correspondent in Augsburg sent a report and image of his specimen, which Gessner used to produce a woodcut (Figure 6.1). Gessner discussed the Islamic imagery and ideas about the life histories of the birds, such as the possibility that they were kept effortlessly suspended by their haloes of plumes.[14]

20   Auium

Paradifi auis, uel Paradifea, ex Nouo orbe, noftri tantum
fæculi fcriptoribus commemorata.

ITALICE Manucodiata: quod uocabulum Indi-
cum uel Noui orbis eft, ubi Mamuco diata, id eft Aui-
cula Dei nominatur.

GERMAN. Paradyßuogel/ Luffuogel.

Figure 6.1 '*Manucodiata*'. Illustration from C. Gessner, *Icones animalium*
(Heidelberg, 1606), p. 20. This woodcut of a legless bird of paradise skin was
produced after a specimen owned by an acquaintance of Gessner's in Augsburg.
Reproduced by kind permission of the Syndics of Cambridge University
Library, M.13.31.

The most extensive treatment of all of the available information on the legless *manucodiata* was Ulisse Aldrovandi's *Ornithologiae* (1599). Aldrovandi had several skins in his collection, and differentiated four species of *manucodiata,* each depicted in woodcuts drinking dew and floating amongst clouds. However, Aldrovandi argued that the birds could not possibly live on dew alone and conjectured that their 'sturdy beaks' were 'very fit to strike insects'.[15] This natural historical imagery was widely used in other print genres, most notably in books of emblems, such as Joachim Camerarius's *Symbolarum et emblematum* (Nuremberg, 1596) (Figure 6.2). The birds, with their perpetual motion, were emblems of spiritual ascension, lofty thinking and restless, mercurial thought.

Remade monsters in the *Exoticorum*

Carolus Clusius's *Exoticorum libri decem* was produced at the same time as the balance of European power shifted in the East Indies. Whereas in the sixteenth century there had been a crown-led Portuguese and Spanish hold on spice trading, in the seventeenth century the commercially financed operations of the Dutch East India Company (Vereenigde Oost-Indische Compagnie (VOC), chartered in 1602) took prominence.[16] The VOC commissioned voyages to take stock of the natural capital in these areas as part of its colonial operations. As a result, northern European towns such as Amsterdam and Enkhuizen became centres of accumulation for material from Southeast Asia. These Dutch expeditions to the Indies brought more bird of paradise skins to Europe, some of which still had their legs attached. They were accompanied by reports of accounts from Malay traders of how the New Guineans prepared the skins.

In 1601, Clusius heard of bird of paradise skins with legs that had arrived in Amsterdam, but was frustrated in his attempts to see them. They had been rapidly sold to one of the richest and most prolific collectors of the day, Rudolph II, for his fabulous collection in Prague.[17] Clusius had to obtain assurance from the broker of the specimens, Johan de Weely, that the skins had possessed legs. This information caused Clusius to hurriedly produce a section on the *manucodiata* as a last-minute appendix. He admitted that he had, presumably like the majority of his well-educated readers, held the 'erroneous opinion' that the birds lacked legs and that Pigafetta's account was false, as informed by reputable sources such as Aldrovandi and personal experience of seeing legless skins. He was now convinced, however, on the strength of new material, that the birds of paradise did possess legs.[18]

Figure 6.2 Emblem showing a bird of paradise accompanied by the adage '*Terre commercia nescit*'. J. Camerarius, *Symbolorum & Emblematum centuria tertia* (Mainz, 1668), XLII, p. 86. Reproduced by kind permission of the Syndics of Cambridge University Library, Hhh.1123.

This first-person shift in opinion framed Clusius's account of the birds of paradise. As he displayed the particulars that convinced him of a changed perspective, he simultaneously constructed a new natural history of the birds, weaving different types of sources together within the narrative of his own quest for knowledge. Presenting accounts as if from direct, first-person experience was a method commonly used in travel narratives in the early modern period. The style of 'autoptic' narratives lent veracity to otherwise potentially unreliable sources such as travel tales.[19] Similarly, the autoptic presentation of Clusius's account facilitated his refashioning of the *manucodiata* into his own, legged monsters to present to the reader, using diverse modes of expression and sources that were not always quite as he presented them. He was creating something new by skilfully crafting heterogeneous materials into things that were at once empirical and also potent emblems.

Assembling the *manucodiata*

Clusius began by referring to the philological history of the birds, in the tradition of Renaissance natural histories, but in a truncated fashion. He reiterated that the *manucodiata* were 'unknown to the ancients' and described by many recent authors. He referred specifically only to Aldrovandi's account, praising it as the clearest work on the birds. This was not a commentary or reworking of old material; Clusius quickly shifted to set himself in opposition to these previous works, writing that Aldrovandi and 'all the rest who have talked about this bird ... judge it to lack feet'.[20] Though he was providing evidence against the leglessness of the birds of paradise, this monstrous feature still acted as the focus of Clusius's account. It was a point of contention: the presence of legs, as the inverse of the abnormal state, in itself became monstrous. These were exotic creatures that had transgressed the traditional attributes of being a bird, and this monstrousness remained, despite the addition of legs.

Like Gessner, Clusius drew on his wide correspondence network to gather material on the birds. This included figures from a wide range of geographical locations and social strata, from colleagues at Leiden University, collectors and scholars further afield, to dealers in *naturalia*, merchants and sailors. The scholarly correspondence networks through which objects and information moved and were exchanged were a central feature of natural history in this period, and Clusius had an especially extensive set of connections from diverse social backgrounds. In particular, he had close contact with several naturalists, scholars and collectors who actively helped him to gather material. Using a specimen owned by his colleague, Pieter Pauw at Leiden University as well as a specimen of a new kind of bird of paradise, a 'king bird', owned by Emmanuel Swerts, Clusius could write first-hand descriptions of specimens and have plates produced for the *Exoticorum* (Figure 6.3).

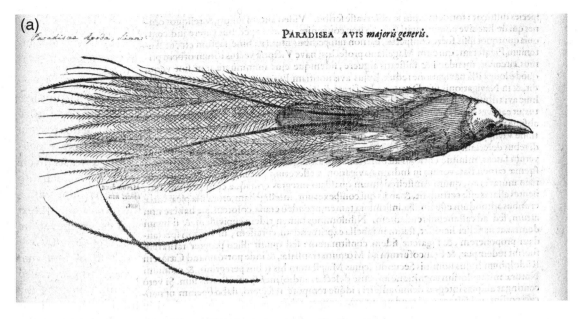

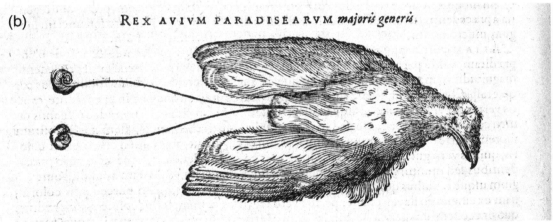

Figure 6.3 Plates from C. Clusius's *Exoticorum libri decem* (Leiden, 1605). (a) '*Paradisea avis*', after Pieter Pauw's legless specimen, p. 360. (b) '*Rex avium*' after Emmanuel Swert's 'king bird' specimen, p. 362. Reproduced by kind permission of the Syndics of Cambridge University Library, L.2.6.

Just as exotic objects stood in for experience of distant places that could not be experienced, certain kinds of description could stand in for experience of unseen objects. About a third of Clusius's account was taken up with the painstakingly detailed depictions of these two specimens, using rich and vivid language:

From the throat right up to the breast, the feathers … [are] saturated with green colour so elegant and splendid, as in the throat of a wild male duck, that none more elegant may be seen; the feathers covering the breast are very fine, but long and very soft, of a dark rufous colour, looking like nothing so much as silken threads …[21]

His words made readers into 'virtual witnesses' of these physical objects. Giving 'vivacity' to an account through copious descriptive details was another important way to imbue a text with the impression of reality, like the autoptic perspective. A variety of descriptive techniques, such as *enargeia* (vivid verbal recreation) and *ekphrasis* (lively description of visual objects), were drawn from classical scholarship and were used both to imply credibility and to give a description with a lifelike quality. Such description was used in many other types of text and was an especially distinctive feature of travelogues, whenever the aim was to convey the credible experience of something absent.[22]

Such 'painting' with words was a powerful rhetorical tool that transmitted not only the visual impression of something but also the subjective experience of it to the reader, by communicating the effect of viewing the object on the writer. Aside from describing their beautiful plumes, Clusius repeatedly stated how the 'form of the bird', even as a mutilated skin, could engender 'infatuation' in the viewer. In the early modern period, visual experiences of things could create 'wonder', either because of their mode of display, as in the *Wunderkammer*, or the mystery of the objects themselves.[23] Clusius's description of these specimens, while empirical, was also a way of making these birds into 'wonders' that could evoke awe and amazement in viewers.

Such language gave the impression that the skins were supranatural objects, beyond the ordinary creations of nature. They were not only vividly coloured, with feathers of 'blood red', 'dark black' and 'golden', but they also seemed almost artisanal creations, from materials like 'silken thread' and 'shoemaker's cord'. Blurring the boundaries between the natural and artificial was a popular motif of early modern curiosity collections, which often contained objects such as nautilus shells set in elaborate metalwork, or gilded Seychelles coconuts. The interplay between divine creativity and the artisanal expertise of man was a great source of fascination. Many other kinds of 'monsters' in collections, such as 'dragons' formed from artfully dried rays or 'mermaids' made from joining various animal body parts and other materials were also popular: part deceptions, part playful jokes.[24] Clusius's *manucodiata* description played on this aesthetic elision between art and nature, and evoked their exoticism. They were still monstrous wonders from the East, even if not the perpetually floating creatures previous naturalists had described.

Chains of exchange

It was a cruel irony that Clusius never saw a bird of paradise skin with its legs on. Indeed, in order to show the erroneous nature of previous naturalists' interpretations of mutilated legless skins, and to make a truth-claim about the legged state of the birds, he had to assemble a chimerical legged bird himself. Critically, though he used the legless specimens as physical tokens to represent the legged creatures he was constructing, the verbal images of these specimens stood for the experience of the examples he had not seen himself. He only had second-hand accounts of legged specimens, from de Weely and sailors from the voyage bringing them to Europe. In his letter to Clusius, de Weely described how:

The bird of paradise was in every respect like the vulgar sort, somewhat flat . . . it had two large feet like those of a sparrow hawk or harrier, that looked unseemly and ugly . . . The leg was dried and looked ugly too, so that the Indians very sensibly cut off the feet together with the leg, for it is the ugliest part of the bird, and in my opinion they all have similar feet.[25]

This second-hand description was Clusius's only assurance that what he had heard about the legged state of the new skins was true. He relied on the good faith and expertise of de Weely as assurance that the knowledge he imparted was reliable.

Clusius rejected the Islamic imagery assimilated by other authors from the first voyage bringing bird of paradise skins to Europe but did use accounts from sailors. These contained information several times removed from direct sources, and would not usually have been deemed 'trustworthy' reports by early modern scholars: 'I am assured that the sailors who brought back these words, though they have not visited the islands in which those birds are born and live, understood from those from whom they purchased them that all are endowed with feet and walk and fly like other birds.'[26]

The sailors' accounts led Clusius to include a new set of tales about the birds of paradise. This was the idea of the 'king bird', which was seen to lead the other birds, 'thirty or forty in flocks'. Such was their loyalty to 'their King or Captain' that if it happened 'to be killed or fall down, the rest that are in the flock fall with him, and yield themselves to be taken, refusing to live'. The greater and lesser birds of paradise each had their peculiar 'King'. The 'Lesser King' was black and 'less handsome', the 'Greater King . . . was a very rare one', a 'little bird' and brilliantly coloured. [27]

Clusius also included material that he probably acquired from the unpublished diary of Jacob van Heemskerck, who captained the first Dutch voyage to the Indies from 1599 to 1601. Heemskerck relayed information, from an Ambonese captain, that the birds of paradise were hard to find, located only on certain islands, and only flew in certain winds. Natives of the places where the birds dwelt caught them by trickery: hunters waited until after one bird had 'tasted' the water from a pool and shown it to be safe for the other birds, then poisoned the water to catch the flock that descended to drink.[28] Clusius related this information with the proviso that it was a 'fable', yet he still included it, making it part of the new images of the birds. While rejecting earlier travellers' images of the *manucodiata*, Clusius incorporated a new set of travellers' tales into his own version of the creatures.

Colonial monsters

Monsters often had long lives as they circulated between different publications. That Clusius was simply making new monsters even as he appeared to 'normalise' them becomes clear when considering the fate of the legged birds of paradise. Clusius's new imagery featured in subsequent seventeenth-century natural histories, developed further and repurposed for specific symbolic roles. In particular, the birds he depicted acted as an image of the East, not as a bountiful paradise as they had in the sixteenth century, but of the increasingly problematic colonial interactions there. Monsters were often used to embody shifting European relationships with distant places and peoples. Colonial locations such as Southeast Asia were regions both full of promise and full of potential danger, commercially lucrative and politically fraught. The natural historical constructions of exotic plants and animals from such locations were loaded with meaning.

In the *Historiae naturalis et medicae* of the VOC physician Jacobus Bontius, the birds of paradise were described as sturdy-legged birds of prey, quite unlike the angel-like entities in sixteenth-century natural histories:

Birds of paradise, formerly considered by many to have no legs, are found here abundantly . . . It is untrue, however, that those paradise-birds have no legs, or feed on air, as they hunt small birds, like finches, with their curved, sharp nails and devour the same immediately, like other birds of prey, and it is untrue that they are only found when dead, for they roost on trees and are killed with arrows by the natives.[29]

Similarly, in the *Ornithology* (1678) of Francis Willughby and John Ray, the birds of paradise were rapacious carnivores with fierce talons, classed with the 'cowardly and sluggish, lesser and exotic' rapacious 'land birds'. The contentious legs became the defining feature of the new birds of paradise, as monstrous hunter's apparatus.[30] These depictions placed even more emphasis on the legs of the *manucodiata* than Clusius had done: they turned the legged birds into emblems of colonial conflict. The new interpretations of the birds' physical form, nature and place in creation were linked to and symbolic of changing European relationships with the birds' region of origin.[31]

Even from the sixteenth century, the birds of paradise had represented the riches to be gained in the Indies: 'The golden birds that ever sail the skies' described in the *Os Luciadas* (1572) of the Portuguese poet Luis Vaz de Camões symbolised the riches of the exotic southern regions in the Pacific. The ruthlessly mercenary interests of the Portuguese and Spanish ships reaping spice harvests were echoed by de Camões's description of the birds of paradise:

> From bower to bower, on busy wings they rove;
> To seize the tribute of the spicy grove.[32]

The shift from footless and angelic images to legged and carnivorous creatures can also be tied to the changing European image of the Indies from a fabled paradise to that of an infernal and hostile region. Contributing to this were violent encounters with native peoples resulting from explorations around the 'Southland'.[33] The cosmological fall from grace implied by becoming legged, terrestrial birds, no longer the angelic creatures perpetually floating in the heavens, was mirrored by a shift in the way that the birds' region of origin was perceived.

Depictions of the legged *manucodiata* reflected the increasingly complex and ambivalent colonial relationships in the Indies through the seventeenth century. Attempting to monopolise Moluccan spice production in the late seventeenth century, the Dutch systematically destroyed whole islands of clove trees: commercial success rested on battling an intractable exotic nature as much as political control. These measures caused considerable bloodshed and destruction of the spice production in some Moluccan islands. The VOC's spice trading venture came to involve increasing violence and antagonism with local populations as tensions rose, though it remained profitable until the end of the eighteenth century. Bontius's sturdy-legged *manucodiata*, that could 'tear and devour' other birds with their 'crooked and very sharp claws' came to signify a region that was no longer impossibly bountiful, but dangerous and threatening.[34]

Myriad monsters

Monsters often existed in plural forms in print and images, they were malleable and they served multiple purposes. An important demonstration of the hybrid and emblematic nature of European images of the *manucodiata* is the fact that the images of legless birds did not disappear. The ideas of earlier naturalists, described and dismissed by many authors, were not rapidly or indeed universally rejected. Even well into the seventeenth century the nature and form of the *manucodiata* were debated in natural histories and collection catalogues, such as Jan Jonston's *Historiae naturalis* (1657).[35] The printed images in the works of Gessner, Aldrovandi and Clusius were all used to represent the birds in later natural histories, including Jonston's, as well as in many other genres of publication, such as books of heraldry, theological pamphlets and books of *mirabilia* (Figure 6.4).

The persistence of these printed images was both the result of the symbolic value of the things they depicted, and the desire of printers to save money on producing new plates for publications.

Similarly, while depictions of birds of paradise in artists' images shifted in parallel with natural histories, especially in northern Europe, this was by no means ubiquitous. For example, the birds were shown, legged and winged, with other terrestrial birds in paintings such as Peter Paul Rubens and Jan Brueghel the Elder's *The Garden of Eden with the Fall of Man* (c.1615).[36] Yet many painters depicted the birds as floating, angelic creatures, as in Roeland Savery's *Landscape with Birds* (1627).[37] Even when made terrestrial, the birds of paradise were set in Edenic scenes, amongst other 'birds of Eden' such as the hoopoe, as in Frans Snyder's *Concert of the Birds* (1629).[38]

There were multiple emblematic roles for the birds of paradise, and they existed in multiple forms, both in natural histories and across other media. Just as the physical objects existed with and without legs in different collections across Europe, a variety of virtual *manucodiata* existed in Europe. How the birds might be represented depended not only on the empirical experience of an author or the specimens to which they might have access, but on the other pieces of imagery and information they used to construct their account, and, crucially, on the underlying aims of the author.

Conclusion

Exotic monsters did not arrive in Europe; rather, they were made, often by naturalists such as Clusius. Highlighting the potential monstrosity of new beasts was both a way of dealing with

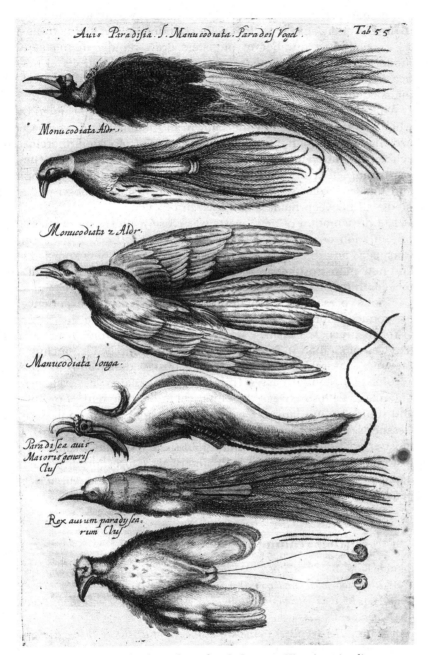

Figure 6.4 Various birds of paradise, after C. Gessner, *Historia animalium*, U. Aldrovandi, *Ornithologiae hoc est de avibus historia* and C. Clusius, *Exoticorum libri decem*. J. Jonston, *Historiae naturalis, de avibus libri VI* (Amsterdam, 1657), tab. 55. Image from the Biodiversity Heritage Library. Digitised by Smithsonian Libraries (www.biodiversitylibrary.org).

problematic or shocking characteristics and a way of accentuating their worth, because monsters were commercially valuable things. As a result, the number of monsters grew rapidly in the early modern period. Once constructed, virtual representations of new beasts, especially monsters, circulated far more widely than specimens could do.

What Europeans defined as monstrous or not was determined by what was included within the known and familiar and what was excluded. The geographical marginalisation of monsters therefore echoed the inherently marginal nature of monstrosity itself.[39] They were things that bridged geographical distances between Europe and unseen exotic regions, just as natural historical descriptions mediated between readers and unseen objects.

The presentation of new beasts was closely intertwined with the presentation of the author themselves. The *Exoticorum* presented Clusius as an arbiter of the exotic and purveyor of new knowledge and new monstrosities. In remaking the *manucodiata*, Clusius distanced himself from earlier encyclopedic accounts such as Gessner's, presenting his material as if from personal experience, to produce a specific, new image composed of discrete tropes.

Yet in practice, Clusius's text was less distinct from older texts like Gessner's than it appeared. His array of sources also included previous natural histories, personal observations and indirect accounts that had passed through long chains of exchange. The relationships between the empirical and emblematic, objective and subjective in Clausius's work were certainly similar to those in Gessner's. The textual conventions of natural histories, even the most erudite ones, were far more fluid than is usually assumed by historians.[40] In particular, the chimerical natures of the beasts depicted in these natural history texts mirrored the chimerical patchwork of the texts themselves.

Further reading

Campbell, M. B., *The Witness and the Other World: Exotic European Travel Writing, 400–1600* (Ithaca, 1988).

Cook, H. J., *Matters of Exchange: Commerce, Medicine and Science in the Dutch Golden Age* (New Haven, 2007).

Egmond, F., *The World of Carolus Clusius: Natural History in the Making, 1550–1610* (London, 2010).

Johns, A., *The Nature of the Book: Print and Knowledge in the Making* (Chicago, 1998).

Margócsy, D., *Commercial Visions: Science, Trade and Visual Culture in the Dutch Golden Age* (Chicago, 2014).

Mason, P., *Before Disenchantment: Images of Exotic Animals and Plants in the Early Modern World* (London, 2009).

Mittman, A. S. and Dendle, P. J. (eds.), *Ashgate Research Companion to Monsters and the Monstrous* (Farnham and Burlington, 2013).

Platt, P. G. (ed.), *Wonders, Marvels, and Monsters in Early Modern Culture* (London, 1999).

Ritvo, H., *The Platypus and the Mermaid* (Cambridge, MA and London, 1997).

Schaffer, S., Roberts, L., Raj, K. and Delbourgo, J. (eds.), *The Brokered World: Go-Betweens and Global Intelligence, 1770–1820* (Sagamore Beach, MA, 2009).

7   Indigenous naturalists

In the autumn of 1576, an unlikely group of collaborators undertook
a prolonged writing retreat in the Franciscan Real Colegio de la Santa
Cruz in Tlatelolco, in what is now Mexico City. They were producing
a natural history of New Spain as part of a longer general history that
described the beliefs, customs, institutions, social organisation and
past events of the Mexica, one of the indigenous peoples of Central
Mexico, commonly called Aztecs.[1] The group working on the natural
history was composed of a variety of informants, among them healers,
animal keepers in Moctezuma's 'zoo', merchants reporting on natural
species from across the fallen Aztec Empire, and hunters, fishermen
and feather-workers describing their practices.[2] All this information
was gathered in visual form by a team of twenty-two indigenous artists
proficient in the native pictorial tradition and acculturated in
European iconography.[3] It was also described textually in Nahuatl by
a group of four 'grammarians', indigenous scholars distinguished in
the use of their native language. These scholars had trained in Spanish
and Latin in the Colegio, which had been founded in 1536 to educate
indigenous noblemen as humanists. Lastly, the Nahuatl texts were
translated into Spanish by the head of the whole operation,
Bernardino de Sahagún (1500–90), a Franciscan friar who arrived in
Mexico in 1529 and was set to compile the first comprehensive compendium of the things of New Spain, frequently known as the 'Aztec
encyclopaedia'.[4]

Modelled on Pliny's *Natural History* (79 CE), the *Historia general de
las cosas de la Nueva España* ('General History of the Things of New
Spain'), also known as the *Florentine Codex* (*c*.1575–7), is a twelve-book
illustrated manuscript, now bound in three volumes and housed at the
Biblioteca Medicea Laurenziana in Florence. Like Pliny's work,
Sahagún's compilation is a descriptive narration with an encyclopaedic scope, concerned with all things that are found in the world, that is,
all of nature, including humans. Both works are historical, in that they
cover the expansion of particular societies, the Roman and Mexica
respectively. But unlike the *Natural History*, which, according to Roger
French   presents   Pliny's   world   as   'anthropocentric   and

anthropomorphic', the *Florentine Codex* offers a window onto a world where hierarchies among beings operated in a far more fluid way.[5]

Relationships between the Mexica and the physical world were reciprocal in nature, often taking cyclical temporalities and attributing agency to animal, vegetal and mineral entities. We know this from varied types of evidence. The Mexica did not record their ideas about nature in a descriptive way before the Conquest, but they did produce elaborate narratives intertwining the human and non-human domains, which were transmitted through oral traditions and depicted in painted manuscripts, murals and objects.[6] In these narratives, we find evidence of observations accumulated over generations, and of the creatures and phenomena that particularly attracted the Mexica's attention. Among the many topics treated in these narratives, those of hummingbirds (*Trochilidae*), tiny birds endemic to the Americas, stand out as paradigmatic. This chapter offers a microhistory of these birds to illustrate both the ways in which animal behaviour was studied and used to articulate the ritual life and historical record of pre-Columbian Mexico, and the elasticity of the early modern genre of natural history to incorporate these indigenous notions.[7]

The animals of Book 11: encyclopaedia entries or incarnations of the divine?

Book 11 of the *Florentine Codex*, devoted to the natural world, is the longest and most richly illustrated of the twelve, with 253 folios and 965 paintings, amounting to one-quarter of the total number in the *Codex*. Typically, its entries consist of a Nahuatl description in the right column, and a Spanish translation and one or more images in the left column.

Recording indigenous notions of nature in this manner was no simple exercise, as is suggested by the activities of those in the Tlatelolco workshop in 1576. It is likely that the organisation of the book followed a scheme inspired by a questionnaire, now lost, designed by Sahagún himself.[8] Some scholars have seen Sahagún as a pioneer of anthropological methods, attributing to him motives at once 'ethnological, historical, philological and linguistic', which constituted a mixture that was rare in missionary writers.[9] Others have argued that since the scheme designed by Sahagún was inspired by known encyclopaedic works, it imposed a European Christian imprint on the native information that was being gathered.[10] In actuality, even with its European framework, the content of Book 11 presents native principles in areas germane to natural history, such as naming and classification. But, as Diana Magaloni has discovered, it also constitutes a window onto the

intrinsic relationship between the natural world and the materialisation of divinity as understood by the pre-Columbian Mexica.

To illustrate this aspect of the *Codex*, let us turn to the case of the
hummingbird. Regarding naming and classification, hummingbirds
functioned as a lens through which the Mexica observed and named
the natural world more generally. Hummingbird characteristics, such
as iridescence, shape or dietary preference, were criteria for naming
species in the vegetable, animal and mineral world. An iridescent
stone like the opal, for instance, is called *uitzitziltetl*, or 'hummingbird
stone' because 'its appearance is like the feathers of the hummingbird'; a fish with a pointy beak, probably a swordfish, is called *uitzitzilmichi*, literally 'hummingbird fish', 'because its beak is very pointed,
long, just like a hummingbird's bill'. A dark flower, thorny as the
hummingbird's beak and frequented by hummingbirds is called *uitz-
tecolxochitl* or 'char-thorn flower'.[11] This 'hummingbird lens' is also
depicted iconographically, as shown in Figure 7.1, where the iridescence of the *uitzitziltetl* is represented by a tiny hummingbird emerging from the stone, the shape and function of the *uitzteculxochitl* is
captured through a hummingbird sucking nectar from it, and the bill
of the *uitzitzilmichi* resembles a hummingbird beak. These choices
provide illuminating evidence of the distinct network of associations
operating in the Mexica conception of nature, and the overarching
presence of the hummingbird within it.

As these examples suggest, Sahagún's indigenous informants
recorded valuable insights about their worldview in the form of encyclopaedia entries in the *Florentine Codex,* and this manuscript is
rightly considered the main repository of pre-Columbian customs
produced in colonial Mexico. But the encyclopaedic allure of the
Codex can be misleading. Indeed, to interpret these examples as
mere entries in an early modern compendium overlooks a central
ambition held dear by all the participants in the endeavour: the
management of the supernatural or divine.

Sahagún never lost sight of his primary objective in compiling the
General History of the Things of New Spain: he sought to know the
language and culture of pre-Conquest Mexico in order to design more
effective tools for evangelisation. Nature played a big part in this
project as, true to Franciscan didactics, the 'book of nature' provided
endless examples for guiding spiritual edification. In this context,
adapting the mysteries of the Christian faith to the new environment
was much facilitated when elements from that new environment could
be used to explain the dogma. As Sahagún explained in the prologue to
Book 11:

Knowledge of natural things is no lesser jewel of the evangelical coffer when it
comes to giving examples and making comparisons, and we have seen the

y llaman le Ayztli es la manera del marmor de españa.

¶ Ay vnas piedras preciosas que se llaman Vitzitziltetl, que quie re dezir, piedra que parece al cincon, y es piedra pequeñuela, y blanca: pero la luz, haze la parecer de diuer sos colores, como tan bien haze pare cer de diuersos colores, a la pluma del cincon parece de diuersos colores: esta piedra, segun la diuersidad, de la luz, que le da: esta esto esplicado bien en la letra, tiene hechura como de huir mjea, hallase esta piedra, a las orillas de la mar, entre la arena: y tambien se halla en vn rio que corre por la tierra de totonacapan veen la de noche, porque resplandece a la ma nera de luciernaga, o como vna can delita pequeña que esta ardiendo, y de lexos, no parece sino luciernaga: conocen ser la piedra dicha en que esta que da aquella luz, y no se mue ue es rara y preciosa no lo vsan sino los señores estas parentes o a lo menos, de la color de vna perla muy fina.

chipavac, tehcaltic, cuechtic, achi xoxoxvic.

¶ Vitzitziltetl: injtoca itech qujca in vitzitzili, joan tetl: ipampa in itlachieliz, iuh qujnma vitzitzilin mvito, mmjtoa totozcatleton. In itlachieliz iuh qujnma centzon tli, icpitl, itech moiava xoxotla, iuhqujn tlatla, itech conqujca, ca itech conqujqjztica in tlapalli, inchi chiltic, in xoxoctic, in xiuhtotatl, in tlauhquechol, in amopalli, in tlavitl, in qujltic, etc. çantel atle itechca tliltic: injn mjmjltontli, iuhqujn teil acatl tacepiktontli, conqualton, contepiton; ompa icham, ompa neci intevatonca, il huica axaltitlan, immotta, ioan ompa intotonacapan atoiac, tex calapan in neci: auh caniо valtica in neci, iuhqujn icpiton xotlatica; anoce iuh qujn cande laton tlatlatica, in tlacan valne neci caicpitl: auh in tlacan ieca, in tlacan ietlatlatica, caichoatl in vitzitziltetl, chipavac, atic, nal tic, vellacpitl, mavitztic, mavico tlamavicoltic, xoxotla, pepetla

Figure 7.1 (b) The 'char-thorn-flower', described as dark and thorny like the hummingbird's beak, is depicted attracting the bird. *Florentine Codex*, Book II, MS Mediceo Palatino 220, Biblioteca Mediceo Laurenziana, Florence, fo. 190r. Reproduced by permission of MiBACT. Further reproduction by any means is prohibited.

Redeemer himself using it. And the more these examples and comparisons are familiar to the listeners, and spoken in the words and language commonly used among them, the more effective and advantageous they will be. To this end this book is a thesaurus, made with much effort and labour, which records in the Mexican language the properties and exterior and internal shapes attained by the most commonly known and used animals, fish, trees, herbs, flowers and fruits that exist in this land, where there are copious terms and much language, all very proper and much used, and a delightful subject.[12]

The main purpose of writing a 'natural history of New Spain' for Sahagún, then, was to use the natural world as ammunition in the war against idolatry.

Understanding the shapes idolatry could take in early colonial New Spain proved no small challenge, however, particularly with regard to animals. The testimony of the animal keepers who were recording their experience for Book 11 illustrates this well. The *totocalli* of Moctezuma, literally a 'house of birds', was a large space adjacent to the palace of the *tlatoani*, or main ruler, in Mexico-Tenochtitlan. Various accounts by Cortés and his soldiers described this sumptuous

¶ Ay vnpez enla mar que sella
ma vitzitzilmychi: llamase anxi,
porque tiene el piquillo muy del
gado, como el auecilla, que se lla
ma zinzon, que anda chupando
las flores./.

Figure 7.1 (c) This portrayal of the 'hummingbird fish', perhaps a swordfish, stresses its morphologigal similarity with the bird. *Florentine Codex*, Book II, MS Mediceo Palatino 220, Biblioteca Mediceo Laurenziana, Florence, fo. 62v. Reproduced by permission of MiBACT. Further reproduction by any means is prohibited.

space where all kinds of birds and other animals were kept, fed and bred.[13] But the brief description in the *Florentine Codex* shows that this *totocalli* was very different from the kind of 'zoo' or menagerie typical of contemporary European courts, which served to show the might of the ruler or to amuse him. Here, according to Sahagún's informants, 'the major-domos kept the various birds – eagles, ibis, thrushes, yellow parrots, small parrots, big parrots, pheasants. And all the various artists with their crafts are there too: the goldsmiths, the silver and copper workers, the feather workers, the stonecutters, the green-stone mosaic workers, the carpenters.'[14] Precious animals coexisted with craftsmen, experts in the highest arts described in the *Florentine Codex*.

But why would animals inhabit the same house as goldsmiths, feather workers or stonecutters? Because animals were, among other things, raw materials for some of these arts, and these arts, in turn, were the ways of materialising divinity, which could take animal form.[15] Feathers, skins and pigments, as well as metals and precious stones such as turquoise, obsidian or jade, were used to craft objects of ritual and ceremonial import. These materials were in fact essential in producing 'incarnations' (*teixiptla*) of divinity (*teotl*). The deities in the Mexica pantheon did not 'exist ontologically', as fixed entities with specific names and physical attributes. Rather, divinity was 'called forth' by the creation of an incarnation.[16] The incarnation of the god Huitzilopochtli, or 'Hummingbird of the Left (of the Sun, meaning the South)', in Figure 7.2, for example, is produced of jadeite and takes the shape of a human figure with the hummingbird device after which it is named at its back. This co-production of divinity between animals and craftsmen, announced by their coexistence in the *totocalli*, is also evoked in many images in Book 11 – including this depiction of a hummingbird, in almost identical style to the jadeite *teixiptla* of the god.

Given the importance of animals for the materialisation of divinity, altering the relationship between the Mexica and animals was promptly recognised as a way to destabilise their belief system. What happened to the *totocalli* during the Conquest campaign illustrates this with great clarity. Cortés himself recounted that:

seeing as how the inhabitants of the city were rebelling and showing such determination to die and defend themselves . . . and so that they would feel it more, I set fire to the big houses of the main square . . . and others next to them, where Moctezuma had all the types of birds available in this land; and although it was hard for me to do that, I decided to burn them, because it would hurt them more.[17]

Perhaps it was quite intentional that after the fall of the city in 1521, the place occupied by the *totocalli* became the site where the Convent of Saint Francis was built, making it the spot of the first Christian enclave in the continental Americas. It was here that the first Franciscan friars in charge of the evangelisation started their campaign, most famous among them Toribio Benavente Motolinía, a pioneer in the appropriation of local nature for didactic examples and comparisons. Considered in this context, we can see the indigenous informants' contribution to the *Codex* as a way of ensuring that elements of their worldview were granted at least some role in one of the foundational documents of the new regime.

The fact that the first Christian site in the Americas rested atop the former house of birds attests to a persistent challenge for Franciscan evangelisers. Many Aztec gods had animal manifestations, including eagles, jaguars, serpents, rabbits, deer, bats, dogs, quetzals and, of course, hummingbirds. This fact greatly elevated the relevance of Book 11 for the Franciscans. Recording indigenous accounts of the natural world would make evident cases where the natives were still attributing divine qualities to inferior creatures. The fact that animals were called 'gods' by the natives was deeply troubling to Sahagún. In his mind, they were oblivious to the great chain of being, a hierarchy in nature in which creatures possessed a particular place in an

Figure 7.2 Stone and paper incarnations of divinity. Jadeite figurine of Huitzilopochtli, 6.7 cm (copyright Musée du Quai Branly/Art Resource, NY, photo Patrick Gries and Benoît Jeanneton), and '*vitzitzili*' or 'hummingbird', *Florentine Codex*, Book 11, MS Mediceo Palatino 220, Biblioteca Mediceo Laurenziana, Florence, fo. 24r (detail). Reproduced by permission of MiBACT; further reproduction by any means is prohibited.

organised ladder ascending towards divinity.[18] As Sahagún explained, 'This work will also be very timely to make [the natives] understand the value of all creatures, so that they do not attribute divinity to them, because they were calling any creature that was eminent either in a positive or negative way *teotl*, which means "God".'[19]

If animals could be 'gods', and this was a notion particularly opposed by Sahagún, what strategies did indigenous intellectuals and artists use to convey the qualities of animals in the natural history they related to the missionary? And what did they do when the animal in question was the hummingbird, the *teixiptla* of Huitzilopochtli, the main *teotl* in the Aztec pantheon at the time of the arrival of the Spanish? The main strategy was disguise. And disguise could take a variety of forms. To explain, the following section illustrates two disguises that often worked in tandem: the use of Nahuatl temporal frames and visual cross-referencing.

Why was the hummingbird god a hummingbird?

Because of his pre-eminence at the time of the conquest, Huitzilopochtli – from '*huitzilin*' (hummingbird) and '*opochtli*' (left or left-handed) in Nahuatl – has remained somewhat of a puzzle for specialists. He was a newcomer to the pantheon, coinciding with the rise of the Mexica, and very visible in Mexico-Tenochtitlan. His statue dominated the main temple in the city at the arrival of the Spaniards, and Cortés had it removed immediately after seizing the capital. This was the first of many acts of targeted iconoclasm against Huitzilopochtli. With few surviving material representations, his identity over the centuries has been fluid and evasive, with no clear account as to his human or divine origins, much less his hummingbird attributes.

The most widely known and accepted description of Huitzilopochtli, recorded by Sahagún early in his research in the town of Tepepulco, around 1559, narrates the deity's extraordinary birth when his mother, Coatlicue, was impregnated through a feather bundle that fell from the sky. Huitzilopochtli was born as a fully attired warrior ready to fight his older siblings, Coyolxauhqui and the Huitznahua, whom he killed or exiled. The pioneer linguist and anthropologist Eduard Seler interpreted this story as the triumph of Huitzilopochtli (the Sun), born of the Earth (Coatlicue), over the Moon (his sister Coyolxauhqui) and the stars (the Huitznahua).[20] In this interpretation, Huitzilopochtli is primarily an astral god, representing the cosmic battle between day and night. His hummingbird attributes are never clearly mentioned in this story. According to Seler, 'there is . . . not much to say about the hummingbird nature of the god. The sources do not divulge what the lovely, tender bird has to do with the terrible god of war.'[21] Yet there must have been good reasons for the Mexica to deify this tiny bird. It is precisely Book 11 of the *Florentine Codex* that shows how the empirical observation of hummingbirds helped embody ideas relating to the seasonal cycles and to the longer history of the Mexica, in particular their migration from the northern land of Aztlan to Central Mexico, which in turn became the centre of Huitzilopochtli's cult.

The rendition of hummingbirds in Book 11 is one of the richest animal entries in the entire *Codex*. To begin with, as we see in Plate 5, there is a vivid illustration of hummingbirds in what seem to be various stages of their life cycle. From left to right may be identified, first, a young hummingbird atop its nest, waiting to fly; then a still hummingbird, colourless and shrunken, hanging motionless from a tree branch in what is surely a representation of the phenomenon of torpor, the state of nightly or post-flight hibernation that humming-birds can briefly enter; followed by another hummingbird in the fore-ground, sucking nectar from a red flower; and last, a group of four hummingbirds on the wing.[22] In the different stations in which the hummingbirds are presented, there is a sense of time passing, perhaps even seasons, from the animal's birth to its slumber or 'deadening', its reconstitution through food and, lastly, to its active state. Time emerges from both the image and the accompanying text as an import-ant variable in the bird's behaviour. The text in the Spanish column describes how the hummingbird

renews itself every year at the time of winter; they hang by the beak from the trees; hanging there they dry up and moult; when the tree grows green again, it revives, and grows feathers once again; and when it starts to thunder for rain, then it awakes and moves and resuscitates.[23]

According to the description, the depicted hummingbird 'renews itself' (*renuévase*) every year 'at the time of winter' (*en el tiempo del invierno*), presenting a significantly altered temporality to the nightly or post-flight torpor.

An entirely different temporal frame for this 'renovation' is given in the Nahuatl column. Here, the beginning of this phase is sig-nalled by the term *in tonalco*, which according to a Nahuatl–Spanish dictionary contemporary to the *Codex* meant 'summer, the time when it does not rain' (*estio parte del año ... el tiempo que no llueve*). In the Nahuatl version, the 'renovation' happens during the dry season, not in the winter. What happens to the hummingbird is also different in the Nahuatl column. Whereas the Spanish version talks about 'renovation' ending in 'resurrection', clearly a phrasing with Christian connotations, the Nahuatl account states that when it does not rain the hummingbird 'dries itself' (*mooatza*). *Mooatza* (or *mo-vatza*) in Nahuatl means 'to get drained, dried out or thin'.[24]

Placing the chronology of torpor within the dry season rather than the Spanish 'winter', and the drying of the hummingbird as the main phenomenon, invites a different reading of the image altogether. In Central Mexico the dry season starts in November and intensifies through April, when drought can become extreme. In the Mexica calendar, this moment of extreme dryness called for a particular feast named, precisely, *Toxcatl* or 'Dryness'. As the *Tovar Codex*, another ethnographic compendium contemporary to the *Florentine Codex* explained,

The fourth month was called *Toxcatl* which means the dryness or sterility of the land, and was celebrated around April, because by then [the Mexica]

had great desire for water; because in this land the rains start at this time, they exposed and paraded the weapons and trophies of the god Huitzilopochtli.[25]

So, according to the ritual calendar, the feast of 'Dryness' was offered to Huitzilopochtli, the hummingbird god, to invoke rain. At this time, his weapons and trophies were exhibited, perhaps before putting them to rest as the rainy season arrived with its demands of working the land.

Describing torpor as starting '*in tonalco*', the Nahuatl texts signals us to recognise the Mexica division of the year into two main seasons, rainy and dry, both marked by rituals. In Book 2 of the *Florentine Codex*, devoted to the ritual feasts, Sahagún explained that three feasts were offered to Huitzilopochtli during the ritual year. Besides the April feast of Toxcatl, there were Tlaxochimaco, meaning 'Giving of Flowers', and Panquetzaliztli, meaning 'Raising of Banners'. During Tlaxochimaco, celebrated between mid July and mid August, Huitzilopochtli would be offered the best flowers of the year. Panquetzaliztli, the last and most important feast, celebrating him as a swift warrior deity, occurred between mid November and mid December, just when the dry season starts in the valley of Mexico. That is, Huitzilopochtli's presence in the ritual calendar began as the dry season was ending and the rains arriving, and ended at the start of the next dry season. From this we can see that the other hummingbird activities depicted in Plate 5, namely feeding and flying, inspired the remaining feasts devoted to Huitzilopochtli. To understand the hummingbird images as an evocation of the feasts is also to see them as part of Mexica history, since these rituals memorialised specific episodes in Mexica annals.

Continuing the analysis of this image of hummingbirds in Book 11, we can now turn to an overlooked feature at the beginning of the visual narrative. To the left of the newborn hummingbird in its nest there seems to be a gratuitous symbol of a broken tree. This motif refers to a different timeline: the rise of Huitzilopochtli as the main god in the Mexica pantheon. Numerous pictorial and textual sources depict the migration of the Mexica people from a northern land called Aztlan to Central Mexico, where they founded Mexico-Tenochtitlan. This migration occurred roughly between the year 1-Flint (estimated in some sources as 1138 CE) and 2-House (1325 CE). The journey from Aztlan was initiated, according to these annals, by a 'talking hummingbird' that invited the Aztecs to start their southbound voyage, revealing itself as a tutelary deity (Figure 7.3).[26] In the town of Quauhuitzintla, upon recognising the sign of a broken tree, Huitzilopochtli tells the Mexica that they should part with the other

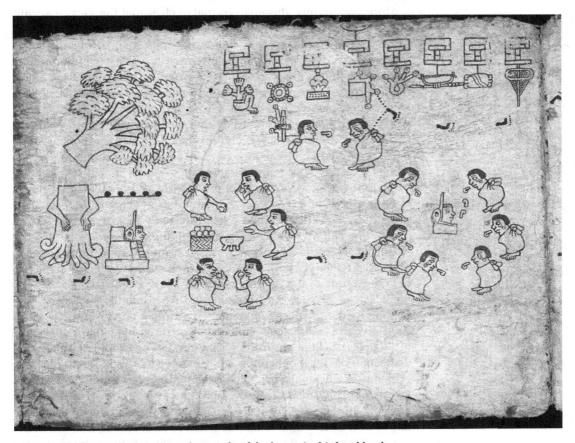

Figure 7.3 The Mexica are born as the people of the hummingbird god by the broken tree omen. Huitzilopochtli gives instructions to proceed with the migration on their own, as signalled by speech scrolls. The Mexica adopt their name, a warrior identity and Huitizlopochtli as their tutelary deity. *Tira de la peregrinación* or *Boturini Codex*, Museo Nacional de Antropología, fo. 3. Reproduction authorised by the Instituto Nacional de Antropología e Historia, Mexico.

seven tribes with whom they were travelling and continue the voyage alone. He also tells them they will change their name from Aztecs – or the people from Aztlan – to Mexica. Lastly, this is the moment when they become warriors, as this is the place where they receive their weapons, the bow and arrow. According to these annals, the broken tree represents a stop along the migration where the Mexica become a separate people and a warrior people under the influence of the hummingbird as tutelary deity. Reading this image as a representation of the Mexica migration from Aztlan to Mexico-Tenochtitlan, it is easy to see the indigenous agency behind this visual resource.

More importantly, this analysis makes clear that the Mexica keenly observed the migration patterns of hummingbirds, breeding in the north and wintering in the south, year after year, developing two intertwined narratives: one of the presence and absence of hummingbirds according to the seasons, and one of their own migration. Along the migration route, the hummingbird god takes various shapes. It is alternatively passive and active, saving energy and engaging in combat. This extraordinary ability to alternate between extreme energy saving and extreme activity is closely related to the migration cycles of hummingbirds, which the Mexica represented as attributes of the god Huitzilopochtli. This representation of the god is also inspired by Mexica observation of the phenomenon of torpor. Watching hummingbirds dart and feed in sustained bursts of activity, then sleep so profoundly that they seemed dead, translated in Aztec visual narratives into an alternatively resting and fighting deity.

Other sources, both pre- and post-Conquest, support this connection. The *Codex Azcatitlan* richly illustrates this frequently depicted double attitude of the hummingbird. In one example (see Figure 7.4), the Mexica, carrying the hummingbird as a bundle (top right), are passing by the town of Tzonpanco. Their quiet voyage is interrupted by a battle where Huitzilopochtli is shown armed and fighting alongside his people, labelled 'Mexican'.[27] That torpor was a trait relevant in the deification of Huitzilopochtli is confirmed by looking again at the only extant tridimensional *teixiptla* of the god from pre-Columbian times in Figure 7.2. This jadeite figurine, only 6.7 cm high, presents the god's hummingbird device on his back, strikingly resembling the bird in torpor in the *Florentine Codex*, as we have pointed out, and the hummingbird bundle in the *Codex Azcatitlan*. When Sahagún's informants described the hummingbird's seasonal drying and reconstitution in the *Florentine Codex*, they linked the seasonal cycle of migration to the phenomenon of torpor because they saw them as connected.

Completing the cycle suggested in the various depictions of the southward Aztec migration guided by the hummingbird god, there is also a rich account of Aztec emissaries travelling northward, back to the land of Aztlan, to visit their homeland and to enquire about Huitzilopochtli's mother, Coatlicue. This trip is recorded as occurring during the reign of Moctezuma I (1440–69), a period of consolidation of the Aztec Empire and Mexico-Tenochtitlan as the dominant city in Central Mexico. However, what they learned during the trip foretold the likely fate of a civilisation at its zenith: a fall. In an eloquent passage recorded by the Dominican Diego Durán, a contemporary of Sahagún, Moctezuma's emissaries transform into birds and other animals upon reaching Coatepec near Tula, in order to reach the northern land of Aztlan. When they get to Aztlan, they ask if Coatlicue, Huitzilopochtli's mother, is still alive. She is, and says that since he left she has been in

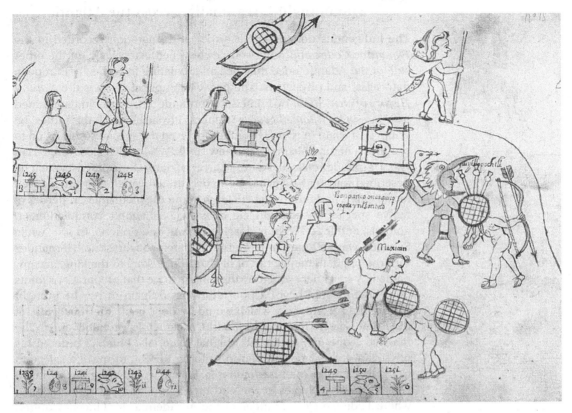

Figure 7.4 Torpid and active hummingbird god. Huitzilopochtli alternatively resting carried as a bundle (top right) and attired as a warrior fighting (centre right) along the migration from Aztlan to Mexico-Tenochtitlan. *Codex Azcatitlan*, MS Mexicain 59–64, Bibliothèque nationale de France, fo. 8r. Reproduced by courtesy of the BnF.

mourning, waiting for his return. She then proceeds to recount what the hummingbird god told her when he started the southward migration, announcing the cyclical nature of his journey:

Oh my mother, I shall not tarry, I shall soon return after I have led these seven barrios to find a dwelling place, where they can settle and populate the land that has been promised them. Once I have led them there, once they are settled and I have given them happiness, I shall return. But this will not be until the years of my pilgrimage have been completed. During this time I shall wage war against provinces and cities, towns and villages. All of these will become my subjects. But in the same way that I conquered them they will be torn from me. Strangers will take them from me, and I shall be expelled from that land. Then I shall return, then I shall return here because those whom I subjected with my sword and shield will rise against me.[28]

Hummingbird seasonality across the Atlantic

The indigenous notions about torpor and migration recorded in the *Florentine Codex* and other sources travelled extensively on the other side of the Atlantic once they began circulating in works by European naturalists and physicians. The prime example of these is the *Natural History of New Spain* by Francisco Hernández.[29] After being appointed his Majesty's *Protomédico*, the 'General Physician of all the Indies' by Philip II of Spain in 1570, Hernández received the special commission to produce a natural history of the new lands. While in Mexico from 1571 to 1577, Hernández collected a broad array of natural specimens, which he described and had depicted by native draughtsmen. He interviewed local healers and interpreters to learn about the medicinal properties of native species. In addition, he consulted Sahagún's compilation on animals, and followed him closely in many descriptions. In 1576, while Sahagún was still hard at work in the Tlateloco workshop, Hernández sent the original manuscript of his *Natural History* to the king, accompanied by specimens and drawings, in the hope that preparations for its publication would begin. Upon returning to Spain in 1577 he brought with him a second copy which would be deposited, after his death, at the Jesuit Colegio Imperial in Madrid. Meanwhile, probably due to the massive scope of Hernández' original materials, Philip II ordered his physician, the Neapolitan Nardo Antonio Recchi, to make a selection from Hernández's manuscript in 1580. Recchi would bring this abridged version back to Naples at the end of his appointment in El Escorial, where Federico Cesi, founder of the Accademia dei Lincei, acquired a copy. The end result was *Rerum medicarum Novae Hispaniae thesaurus* (1651), the best-known edition of Hernández's work.

In the *Thesaurus*, Hernández borrowed heavily from Sahagún's account of hummingbird torpor, but added two interesting elements. The first was an emphasis on witnesses. Paving the way for the treatment of torpor as a natural historical fact, Hernández affirmed that the phenomenon should not be doubted, for it 'is ascertained on the word of the most worthy men and confirmed on the testimony of writers in more than one place'.[30] Hernández was referring to the missionaries who had described the phenomenon before him, specifically Motolinía, Durán and of course Sahagún. The testimony of three different 'eyewitnesses' to the phenomenon of torpor, based upon accounts by indigenous peoples, provided Hernández with enough evidence to report the fact authoritatively. The fact that his sources were Spanish, men of faith and proficient in Nahuatl meant, for him, that they were indeed qualified witnesses: what he called 'most trustworthy men'.

The second theme Hernández introduced to his account of hummingbird torpor was the notion of experiment. He concludes thus:

This is not an idle tale, or a thing open to doubt, for this bird has been more than once kept attached to the stock of a tree [*stipiti*] inside a chamber [*cubiculum*], and when it had hung, as it were, dead for six months, at which time Nature had appointed, it revived, and being let go flew away into the neighbouring fields.[31]

Here Hernández invokes the testimonies of Motolinía and Durán, who both claimed to have seen torpid hummingbirds 'with their own eyes'. But the way in which Hernández presented this information, urging his readers to trust the description because the phenomenon had been studied under controlled conditions, was a powerful use of rhetoric indeed. His rendering suggests that observing torpor in the artificial setting of 'a chamber' confirmed what his witnesses had reported they had seen in the open.[32]

Hernández's account in the *Thesaurus* circulated widely in other natural historical treatises and encyclopaedias throughout the seventeenth century. With different degrees of confidence, physicians in New Spain and Europe reported and commented on what the *protomédico* Hernández reported as happening to hummingbirds for half the year. One example is Francisco Ximénez, a medical practitioner originally from Aragon, who worked in the Hospital of the Santa Cruz in Huaxtepec, New Spain. Through unknown channels, Ximénez received a copy of Recchi's abstract of Hernández's work, using this for his *Quatro libros de la naturaleza y virtudes de las plantas y animales de la Nueva España* ('Four Books on the Nature and Virtues of the Plants and Animals of New Spain', 1615). His careful eye, together with his unique position as another medical practitioner in New Spain, made Ximénez the perfect commentator, someone qualified either to support the *protomédico*'s claims, or to contradict them in the light of his own observations. The chapters on animals in the *Quatro libros* mostly follow Hernández, but there is an important addition to the description of hummingbirds:

The hummingbird does not die [in] the temperate lands of this New Spain, where there are flowers all year round, only in harsher climes, where it finds no food due to the cold.[33]

Merely by noting his own alternative experience, Ximénez subtly corrected Hernández's account without openly accusing him of error.

Back in Europe, a Fellow of the Royal Society in London also had a say on Hernández and torpor. This was John Ray in his influential *Ornithology of Francis Willughby* (1678; in Latin, 1676). Considered the first scientific natural history of birds, the collaborative work offered a classificatory system based on morphology and departed from the emblematic model that privileged antiquarian and philological approaches. Acknowledging that seasonal hummingbird torpor was, as he put it, 'taken for an undoubted truth', Ray resisted the Hernandian rendition of the phenomenon. He did not doubt

Hernández simply because of some prevalent anti-Spanish prejudice – in fact, he also doubted other Englishmen like John Josselyn, who, in his *New England's Rarities Discovered in Birds, Beasts, Fishes, Serpents, and Plants of that Country* (1672) had similarly supported humming-bird long-term torpor. By the time Ray was editing Willughby's book, there was evidence which contradicted the claims about seasonal deadening and revival. For instance, following Dutch naturalists Piso and Marcgraf, Ray reported that in Brazil hummingbirds were seen 'all year long in great numbers'. Here, as in the case of Ximénez, the fact that hummingbirds were perennial in other latitudes raised questions about their sustenance.

The issue of hummingbird feeding habits provoked such interest after Ray's remarks that it was included in the Royal Society's research agenda in 1693, when the physician and plant anatomist Nehemiah Grew, its former secretary, published a query in *Philosophical Transactions*. He wanted to know whether hummingbirds fed exclusively upon flower nectar or also upon insects. To answer this, Grew suggested 'the bird should be open'd' and then it would be clear if it had entrails 'fitted only for liquids' or 'the same stomachs and guts as other birds'.[34] Grew's question, published in 'the world's first science journal', attests to the endurance of the indigenous perspective on hummingbird cycles across geographic and epistemic distance. For these critiques begin to expose hummingbird seasonal torpor not as a universal truth, but as the product of a particular group of people: the team in Sahagún's workshop in Tlalelolco.

Indigenous natural histories

Why did the team of intellectuals, painters, grammarians and scribes decide to disguise the ritual and historical attributes of Huitzilopochtli within their description of hummingbirds in the *Florentine Codex*? The most likely answer is to make a bid for their transmission. The team knew that Sahagún's purpose in compiling a thesaurus of nature was Christian conversion, and so they did not reveal all the complexity of their deities' cult, particularly that of their tutelary god, either in Book 1 on the gods or in Book 2 on ritual life. Instead, they conveyed their way of understanding the hummingbird god in Book 11, an unexpected place in the Aztec encyclopaedia. Furthermore, they employed their own Nahuatl terminology and their pictorial conventions, which, as we have seen, were not commensurable with the role played by images in the European textual tradition. Mexica images were not illustrations to texts but, rather, whole worlds of meaning that had the potential of summoning rituals and histories, alternative temporalities and even divine entities. The painting of hummingbirds in Book 11 is in fact a vivid answer to Sahagún's questionnaire about the animals of New Spain: one in which the Mexica restored their own historical agency. But it is also a portal to a disappearing world, captured in the most comprehensive compendium of the new colonial order.

By the time the Tlatelolco team was finishing the natural history, the prospect of their world disappearing was more tangible than at any time since the Conquest. A new epidemic had begun three months before, striking the indigenous population with unprecedented fury. As Sahagún described in the middle of Book 11:

Now, in this year of 1576, in the month of August a general and great plague began, which already continues for three months. Many people have died, die, and every day more are dying and I do not know how long it will last and how much damage it will do … from the time it began until today, the 8th of November, the number of dead has always gone increasing: from ten [to] twenty, from thirty to forty, from fifty to sixty and to eighty. And from here on I do not know what will be.[35]

In the midst of so much death and uncertainty about the future, a sense of shared urgency must have permeated the workshop at Tlatelolco. For Sahagún, it must have seemed crucial to finish the natural history before there were no witnesses left to what the Mexica knew about nature. For the team, it must have been imperative to express their indigenous way of conceptualising nature in a genre that would endure. And so the Mexica group working on the *Florentine Codex* in the autumn of 1576 adapted the early modern genre of the encyclopaedia, directly modelled on Pliny, to register their notions about hummingbirds and the place of this bird in their history.

This is, if we think about it, the ultimate natural history. Not only is it a history of nature in narrative form, as in the Plinian tradition; it is a history where nature, in the form of an animal, plays the main role in the history of a people, through the mouthpiece of a god created to represent them both.

Further reading

Boone, E. H., 'Incarnations of the Aztec supernatural: the image of Huitzilopochtli in Mexico and Europe', *Transactions of the American Philosophical Society*, 79:2 (1989), pp. i–iv and 1–107.

Cañizares-Esguerra, J., 'The colonial Iberian roots of the Scientific Revolution', in *Nature, Empire and Nation: Explorations of the History of Science in the Iberian World* (Stanford, 2006), pp. 14–45.

Few, M. and Tortorici, Z. (eds.), *Centering Animals in Latin American History* (Durham, 2013).

León Portilla, M., *Aztec Thought and Culture: A Study of the Ancient Nahuatl Mind* (Norman, OK, 1990).

Magaloni Kerpel, D., *The Colors of the New World: Artists, Materials and the Creation of the Florentine Codex* (Los Angeles, 2014).

Montero Sobrevilla, I., 'The slow science of swift nature: hummingbirds and humans in New Spain', in P. Manning and D. Rood (eds.), *Global Scientific Practice in an Age of Revolutions, 1750–1850* (Pittsburgh, 2016), pp. 127–46.

Norton, M., 'The chicken or the *Iegue*: human–animal relationships and the Columbian Exchange', *American Historical Review*, 120:1 (2015), pp. 28–60.

Russo, A., 'Plumes of sacrifice. Transformations in sixteenth-century Mexican feather art', *RES: Anthropology and Aesthetics*, 42 (2002), pp. 226–50.

Wolf, G., Connors, J. and Waldman, L. A. (eds.), *Colors between Two Worlds: The Florentine Codex of Bernardino de Sahagún* (Florence, 2011).

8   Insects, philosophy and the microscope

One day in around 1590, the artist Joris Hoefnagel presented his magnificent album 'The four elements', containing 277 water-colours of living creatures, to the Holy Roman Emperor Rudolph II.[1] Between 1572 and 1585, Hoefnagel had painted delicate images on vellum of around 1,000 animals, ranging from the elephant to the squirrel, from the whale to the civet cat. The section 'Terra' presented four-footed beasts and reptiles; 'Ignis' insects; 'Aier' birds and amphibians; and finally 'Aqua', unsurprisingly, aquatic animals. Framed by golden ovals, each page showed related creatures, often embellished by Biblical verses like Psalm 104, 'O LORD my God, thou art very great.' Although most of the representations were based on earlier images, some published by Conrad Gessner, they evoke liveliness, giving the impression that they were done from life. The overall atmosphere is one of serenity, harmony and beauty – making the observer forget that the album was executed during the years of religious upheaval in the Low Countries, a time when thousands were executed, and many people moved abroad. Hoefnagel had left Antwerp permanently after devastating plundering by Spanish soldiers in 1567, settling first in London, along with friends like the botanist Matthias de Lobel and the cartographer Abraham Orthelius. Later, he would be employed by princely courts in Germany.

The most striking characteristic of Hoefnagel's album is the presence of many species of insects. This was a remarkable novelty. In earlier works of art and devotion, such as psalters, a butterfly had occasionally been depicted in the margins. Generally, however, insects were neglected, even despised. Hoefnagel, however, painted butterflies, beetles, flies and an occasional ladybird, as well as creatures that, seen from a modern perspective, belong to different orders, such as spiders. In the dominant Aristotelian concept of nature, insects represented the lowest stage of being.[2] In this hierarchically ordered universe, angels and man were the highest of God's creatures, followed by the mammals, birds and the rest. The rather unspecified category of *insecta* was the lowest stage, alongside fungi, snails, snakes and

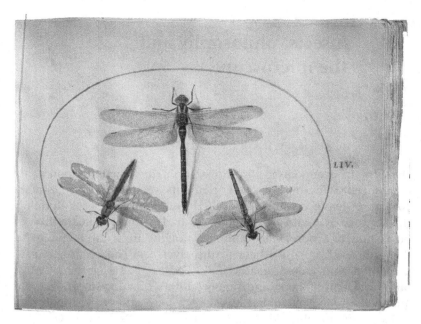

Figure 8.1 Three dragonflies. Watercolour and gouache on vellum; the wings are real dragonfly wings, glued to the vellum. J. Hoefnagel, *The Four Elements*; section II, 'Ignis', plate LIV. National Gallery, Washington, DC. Gift of Mrs Lessing J. Rosenwald. Inv. nr. 1987 20.5.55.

the like. These were thought to be the result of spontaneous generation – that is, to originate from decaying organic matter. On hot summer days in the kitchen, one can still see the logic behind this idea.

In Hoefnagel's album, insects were moved from the margins to the centre of the pages. His delicate paintings were based on his own meticulous observations. Hoefnagel's images are still praised by entomologists, and most of the creatures are easy to make out. His artistic style and his elaborate use of shadows create the illusion of life – the creatures seem to crawl on the vellum or be ready for take-off. There is play with artistic conventions: on plate LIV, Hoefnagel painted the delicate bodies of three dragonflies – to which he glued *real* wings, thereby blurring the distinctions between reality and representation, *ars* and *scientia* (Figure 8.1).

Hoefnagel's study of insects marks a shift in attention of artists and scholars alike towards the world of the smallest of creatures. Between *c.*1560 and 1740, men like Hoefnagel, Thomas Penny, Thomas Moffet, Robert Hooke, Jan Swammerdam and Antonie van Leeuwenhoek observed, described and depicted all kinds of small and low life. Earlier, prior to his premature death in 1565, Conrad Gessner had also studied insects, leaving notes which would be elaborated upon by Penny and Moffet. After 1620, aided by the newly invented microscope, scholars entered upon a previously unexplored domain, demonstrating that a frog, a louse or even mould could be as fascinating as the mightiest lion. This was, as many contemporaries observed, a 'new world'.

This fascination for the small has some interesting characteristics. First, this interest antedated the introduction of the microscope by at least half a century. It was this previously existing field that prompted the use of the microscope, rather than vice versa. Moreover, this was a predominantly (although not exclusively) Protestant affair, mainly (but, again, not exclusively) manifesting itself in the Low Countries and in England. Religious and philosophical factors created a fertile intellectual climate for the study of insects. Protestants embraced the concept of the 'Book of Nature', the idea that God made himself known to mankind by his creation.[3] Neostoicism and Cartesianism, which both explicitly rejected a hierarchy of beings, seem to have been of great importance here as well. These philosophical systems referred to 'laws of nature', and argued for underlying order and uniformity. By stressing the ontological equality of all creatures, great *and* small, both provided a philosophical framework for elevating the 'low' and 'imperfect' creatures, and for making them worthy of attention as well as contemplation. Thus, the use of the microscope in the seventeenth century was only one manifestation of a much broader tendency, of which the life-casts of toads, lizards and insects by Wenzel Jamnizer and Bernard Palissy (both Protestants), or the emergence of the new artistic genre of the still-life painting around 1590, were related expressions. In this chapter, we will focus on some of the key figures in the study of the 'small world', as well as discuss some of the underlying causes. As we shall see, this study required not only a sharp eye and optical tools but, most of all, a conceptual framework.

Religion and the eye of the beholder

Hoefnagel was trained as a scholar. His interest in nature must be seen against the backdrop of both the Reformation and the rise of Neostoicism. Although often associated with morals and ethics, Neostoicism was an all-encompassing philosophy, in which nature played a crucial role. Seen as the result of God's creative power, *natura* had to be studied and contemplated. Nature consisted not of the four Aristotelian elements but of 'pneuma': the generative principle that organised and connected the individual and the cosmos. Nature was uniform; there were no 'higher' or 'lower' beings. Within the context of Neostoicism, the works of Seneca, Cicero and Pliny gained new relevance, as they stressed that *everything* in *natura* was the manifestation of a higher being. Overtly rejecting

Aristotelian notions of hierarchy, Pliny had famously stated that, while people are often filled with wonder by large elephants and lions, 'really Nature is to be found in her entirety nowhere more than in her smallest creations' ('*rerum natura nusquam magis quam in minimis tota sit*').[4] In Neostoic circles, '*natura*' was interpreted as 'God'. Hoefnagel operated in circles where such ideas were voiced, most notably by Justus Lipsius. Nature was seen as a religiously neutral common ground between Catholics and Protestants. Amidst the iconoclasm and atrocities of the Dutch Revolt, gardening and the study of plants and animals could unite Christians in their love for God, and give peace of mind to the individual.

The rise of Protestantism had already put a new stress on the study of nature. Philip Melanchthon and Jean Calvin emphasised that nature showed the greatness of God, and that it was man's duty to honour the Creator. In 1561, the Protestants of the southern Low Countries stated in their confession:

We know him [God] by two means. First, by the creation, preservation, and government of the universe, since that universe is before our eyes like a beautiful book in which all creatures, great *and* small, are as letters to make us ponder the invisible things of God ... Second, he makes himself known to us more openly by his holy and divine word.[5]

Here, once again, a powerful incentive to contemplate even the smallest of God's creatures was given. Hoefnagel and many of his friends, like Christofel Plantijn and Carolus Clusius, subscribed to these ideas. Hoefnagel himself made abundantly clear that he considered nature in general, and small creatures in particular, as manifestations of God's creative power.

The many biblical quotations in the *Four Elements* manuscript could only be read by a few; however, the same basic message was transmitted to a larger audience via a collection of engravings Hoefnagel published in 1592, the *Archetypa* series.[6] Here, once again, beautiful and stunningly detailed ensembles of tiny creatures were framed in a religious context with explicit biblical references, like Psalm 9: 'I will tell of all thy wonderful deeds, O Lord.' Hoefnagel stressed that the power of God is seen everywhere, but nowhere more clearly than in the tiniest of his creations (Figure 8.2). This was a theme echoed by later students of insects. As John Ray put it, a century after Hoefnagel, in *The Wisdom of God* (1691):

If man ought to reflect upon his Creator the glory of all his works, then ought he to take notice of them all, and not think any thing unworthy of his cognizance. And truly the wisdom, art and power of almighty God, shines forth as

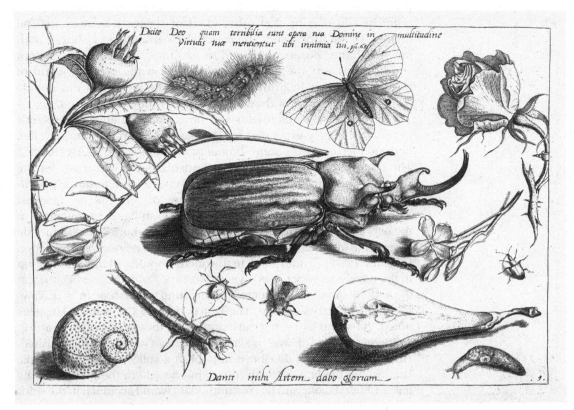

Figure 8.2 Elephant beetle and other small creatures. Engraving from *Archetypa studiaque patris Georgii Hoefnagelii* (1592), Pars I, 1. The creatures are framed in an emblematic context; the upper quotation is from Psalm 66: 'Say to God, How terrible are thy deeds! So great is thy power that thy enemies cringe before thee.' The lower line refers to Hoefnagel himself, witnessing God's Creation: 'To him who gave me my skill I shall give glory.' Rijksmuseum, Amsterdam.

visibly in the structure of the body of minutest insect, as that in that of a horse or elephant: Therefor God is said to be, *maximus in minimis*.[7]

Hoefnagel operated from what has been called 'the emblematic worldview': creatures were not only interesting per se, but above all as nodes in a web of myriad references and symbolic meanings. A butterfly, wondrous in itself, was also seen as a reference to the Resurrection, since the winged creature was believed to arise from the dead caterpillar. On similar grounds, the stag beetle was presented as a reference to Christ. Gessner's studies of insects were presumably framed in an emblematic context as well; and the book published in 1602 by that other giant of sixteenth-century natural history, Ulisse Aldrovandi, *De insectibus*, certainly was.[8] Based both on first-hand observation and a stunning reading of ancient sources, this was the first scholarly work on the broader world of insects to appear in print. The rather crude woodcuts did not do justice to the beautiful drawings on which they were based.

The attitude towards insects was now rapidly changing. Gessner's notes were hugely expanded by Thomas Penny and Thomas Moffet, both connected to the same humanist London group as Hoefnagel.[9] The manuscript, today held in the British Library, was ready for the printer by 1598, but would only appear in 1634, as *Insectorum sive minimorum animalium theatrum* ('Theatre of Insects or Smallest Animals').[10] Around 1600, insects also started to figure prominently in the new genre of still-life paintings. The Antwerp-born artist Jacques de Gheyn II, a follower of Hoefnagel, was one of the pioneers in this genre. De Gheyn also provides a link to the microscope. As a kind of follow-up to the telescope, which was invented in 1608, Cornelis Drebbel, then at the court of James I, constructed a compound microscope in 1620. News soon spread across Europe. The Dutch virtuoso Constantijn Huygens noted: 'It really is as if you stand before a new theatre of nature, or are on a different planet.'[11] Huygens's friend de Gheyn was equally astonished. Huygens tried to get de Gheyn to publish a collection of microscopic studies, but the latter's death in 1629 put an end to the project. In the 1660s, Robert Hooke and Jan Swammerdam would act much in the spirit of Huygens's idea.

The order of nature

In 1628, René Descartes settled in the Dutch Republic.[12] His mission was to explode the Aristotelian system and replace it with his own. Indeed, he became the most important – although not the only – advocate of a new, mechanistic philosophy of nature. Nature, Descartes claimed, did not consist of four elements, but of tiny little particles, colliding with each other. Nature was essentially just matter in motion: a huge clockwork, obeying fixed laws. There was no ontological distinction between comets, clouds and animals; nor between big or small, rare or ordinary. Animals were just automata; man only differed from them because he had a rational capacity.

Cartesianism did put crucial issues, earlier touched upon by the Stoics, on the agenda. These included Descartes's idea that there was no fundamental distinction between the 'lower' and 'higher' beings, for example. Moreover, Descartes's conception of the body as a piece of engineering stimulated research into anatomy and physiology. Processes such as respiration and the circulation of the blood became increasingly studied, often by vivisecting dogs, frogs and other animals, which by some were now taken to be machines. The Cartesian conception of matter – the *res extensa*, the tiny particles filling the Universe – made scholars curious whether they could be observed with the microscope. Cartesian, and more generally mechanistic, notions provided a conceptual framework as well as an agenda for research. In Italy, Marcello Malpighi, who was working in the mathematical and empirical tradition of Galileo and knew the work of Descartes, embarked upon systematic research into the pulmonary system. Pioneering both the use of microscopes and new techniques of preparation, and aided by a mechanistic conceptual framework, Malpighi published a seminal study on the working of the lungs in 1660, mostly based on frogs, in the course of which he discovered capillaries and thereby confirmed Harvey's theory of the circulation of the blood. This was the first anatomical publication in which the microscope was used. Being not only an able anatomist, but also a talented draughtsman, Malpighi did the illustrations for the book himself, thus mapping unknown territory and developing new representational techniques.

Other issues gained attention as well. Of great importance in this context is the issue of generation and reproduction.[13] Descartes had hinted at conceptions of the human body as a machine, but had been conspicuously silent on an obvious question: if the body is a machine, where do baby-machines come from? This was to become a major issue. In England, William Harvey published his highly influential *De generatione animalium* ('On the Generation of Animals') in 1651, based on empirical research on deer and claiming that 'all animals are generated after the same manner from an egg-like rudiment' – a challenging concept that Harvey, however, did not develop in any detail. The issue of the generation of insects was not addressed in the book, although it attracted increasing attention. Already in 1635 the Dutch still-life painter Johannes Goedaert had started to follow the development of caterpillars into butterflies, and the life cycles of other insects as well. Offering a highly symbolic view of nature, his work was published in three richly illustrated volumes in Dutch and Latin from 1660–9, followed by translations into French and English.[14] Although a patient observer, Goedaert did not use the microscope, nor does he seem to have been much aware of the philosophical discussions of his time. Order, and the laws of nature, were now increasingly the buzzwords. Although hinting at some kind of metamorphosis of the caterpillar into the butterfly, Goedaert basically endorsed the theory of spontaneous generation. He noted, for example, that, of two identical caterpillars, one would transform into a beautiful butterfly, while from the other a swarm of 'flies' emerged. He took this for a divine miracle.

Others also were intrigued by similar problems. One of them was Johannes Hudde. Highly influenced by René Descartes, Hudde was one of the most talented mathematicians of his time, and would later become burgomaster of his native Amsterdam. In 1657, he announced plans to start studying the problem of generation and procreation with the aid of the microscope. He developed a single-lens microscope which was easier to handle than the compound microscope. (It was basically the same design that van Leeuwenhoek would later make famous.) Hudde generously shared his ideas and instruments with, amongst others, Jan Swammerdam.

The same interest in the underlying laws of nature is also to be seen on the other side of the Channel. Francis Willughby, a very early associate and Fellow of the Royal Society, interested both in mathematics and natural history, had started studying insects in the late 1650s, together with his friend John Ray.[15] Willughby was referred to at length in Ray's *Catalogus plantarum circa Cantabrigiam nascentium* ('A Catalogue of Plants Growing around Cambridge', 1660), significantly enough in a passage describing the anomalous generation of a certain caterpillar. Willughby continued his researches on insects, but, due to his untimely death in 1672, published little; most of his observations were, however, included in John Ray's posthumous *Historia insectorum* ('History of Insects', 1710).

Robert Hooke and Jan Swammerdam

Other scholars in the circle of the Royal Society took an interest in insects and the problem of generation. Christopher Wren made detailed drawings of a louse and the wing of a fly with the aid of a microscope. Other fellows discussed 'an experiment of the production of bees out of dead bullocks'.[16] The Society also published a book that became iconic: Robert Hooke's *Micrographia*, that appeared in January 1665.[17] Based on meticulous observations, and a series of lectures for the Society, Hooke demonstrated the power of the microscope in a visually stunning way. Many sources testify to the immediate impact the work made. 'Good figures', Christiaan Huygens wrote to Johannes Hudde, 'Flea and louse the size of a cat.'[18] Hudde, in return, wrote that he regretted not being able to understand English – a broader continental problem at that time. Being a trained draftsman, Hooke depicted mould, cork, feathers and the like, in astonishing detail (Figure 8.3). The things and creatures were mostly depicted out of context, leaving the observer wondering about size and relative proportions, and astonished about the tranquillity and sheer beauty of the micro-world. Using various representational techniques, including an advanced use of shadows, the engravings created the illusion that doing microscopic

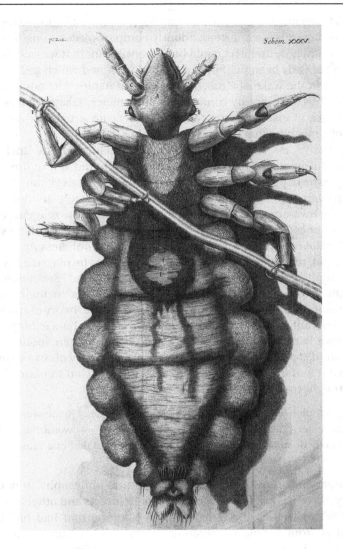

Figure 8.3 Louse. Engraving from R. Hooke, *Micrographia* (1665), schema 35, facing p. 212. Hooke depicted the creature in alienating detail; the elaborate use of shadows made it look even more impressive.
The reader had to fold out this engraving, which measures approximately 52 × 20 cm, and presented the creature almost 200 times as big as in real life. Courtesy Museum Boerhaave, Leiden.

observations was as easy as peeping through a hole. Insects only entered the book in the thirty-fourth of the fifty-nine observations. Hooke worked from a mechanistic standpoint, comparing insects to tiny little engines, and often praising God's ingenuity:

Nature does not onely work Mechanically, but by such excellent and most compendious, as well as stupendious contrivances, that it were impossible for all the reason in the world to find out any contrivance to do the same thing that should have more convenient properties. And can any be so sottish, as to think all those things the productions of chance?[19]

Here, Hooke made a point which from now on would be made repeatedly: as nature is such a tremendously complex piece of engineering, only the almighty architect could be responsible for it. It was a message that had already been put forward by the Stoics, and which gained new currency in the wake of a mechanical view of nature. The cosmos was governed by laws, not by contingency and chance. Like produces like. In the case of insects, it was rather their delicate structure than their symbolic meaning that pointed to God.

Impressive as Hooke's book was, and although he included a beautiful image of the egg of a silkworm, he did not address the thorny issue of procreation. This was a problem that the young Leiden student Jan Swammerdam was starting to research at that time. Although Swammerdam has had bad press, due to the religious crisis he experienced in 1674–5, when he abandoned his research for a short time for the religious community of the mystic prophetess Antoinette Bourignon, recent research has demonstrated that he played a key role in the study of insects and in debates on generation.[20] Swammerdam brought the ideas of Descartes, Hooke and others to their logical conclusion: if nature is stable and uniform, then the theory of spontaneous generation is untenable, as it leaves room for chance. Moreover, as nature is ruled by laws given by God, to believe in spontaneous generation is outright atheism. This theory was 'a very obtuse error', 'a falsification of the natural wonders and truths of God in nature', and 'the straight road to atheism'.

For if generation were at random, man could be generated in the same way: as some have been rash enough to write: although God is as wondrous in both; for the body of an animal is as wondrous in its way as that of a human in its way.[21]

Swammerdam transferred Descartes's natural philosophy to natural history, through novel investigations of both insects and other forms of 'low life', as well as the female body – subjects that had had little attention from scholars operating in the Aristotelean framework. During his Leiden years (1661–7), and in close collaboration with his friend Nicolaus Stensen (Steno), he did pioneering research. Starting with the axiom that spontaneous generation is impossible, he demonstrated that all insects come from eggs.

Most insects were traditionally believed to have no internal anatomy. By closely observing and occasionally dissecting some, Swammerdam showed that this was wrong. He recognised four types of insect development, in accordance with the number of stages involved. The louse, for example, comes from a nit and goes through a process of steady growth, basically not changing its outward appearance (Figure 8.4). The butterfly, however, comes about through a radical process of transformation. Swammerdam took great pride in his discovery, achieved through a complicated process of preparation, that parts of the future butterfly were present in the caterpillar. Not only did he reject the idea that the caterpillar dies, and that from its remains a butterfly emerges, he also mocked the symbolic use of metamorphosis to refer to the Resurrection. In dissecting what was taken to be the king of the beehive, Swammerdam discovered, much to his astonishment, that the creature unmistakably possessed a female reproductive system and egg masses – it was a queen! Here again, Swammerdam both expressed his astonishment at the bee's wondrous anatomy, and ridiculed the ancient metaphorical use of the beehive as a model of patriarchal society under the divine rule of a king. Swammerdam marked the transformation from the emblematic worldview to what in the eighteenth century would be called 'the argument from design'.

Like Hooke, Swammerdam was a skilled draughtsman, and he was very conscious of his own talent and that of his predecessors, amongst whom he explicitly mentioned Hoefnagel. Some of his original drawings are signed with a proud *'delineavit auctor'* – the author has drawn this himself. Swammerdam, obviously able to read English, was well aware of the work of the Royal Society, to which he referred. Moreover, he entered into a visual dialogue with Robert Hooke. For example, whereas Hooke had depicted the water gnat devoid of any context, Swammerdam presented the creature as it appeared under the surface of the water, both life sized and when magnified about fifteen times. Most of Swammerdam's observations in these years were done with the naked eye, or with only a slightly magnifying microscope, for which he acknowledged Hudde. Occasionally he dissected an insect, but most of his observations were devoted to the change in the outward appearance of the creatures. In other words, he carried out his observations starting from a particular philosophical and religious standpoint, and not vice versa.

Dissecting insects

While Swammerdam was finishing the manuscript of his first book on insects in 1668–9, two major publications on the same subject appeared: Francesco Redi's *Esperienze Intorno alla Generazione degl'Insetti* ('Experiments on the Generation of Insects') and Marcello Malpighi's *De bombyce* ('On the Silk-worm'). Redi, employed by the Medici court, had earlier published a book on vipers, reflecting the court's interest in medicine and poison. Coincidentally or not, shortly after Swammerdam's friend Steno joined him at the court, Redi took a great interest in the theory of spontaneous generation.[22] Redi developed an experiment that was as expedient as it was

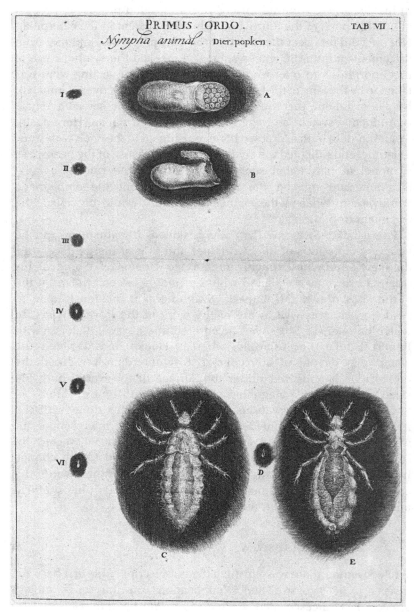

Figure 8.4 Louse. Engraving and etching from J. Swammerdam, *Historia generalis insectorum* (1669). Like Hooke, Swammerdam chose a fold-out, but he represented his insects both at life size and slightly enlarged. Swammerdam also showed the subsequent stages of development (i–vi). He followed the same template with the three other orders he identified. Courtesy Leiden University Library.

revolutionary. He put carefully selected sorts of meat and fish in vessels of glass, covering each sort with fine gauze so that only air could enter. The same sorts of meat and fish were put in control jars that were left open. After the process of decay started, in the open jars maggots showed up, which subsequently turned into flies. No insects appeared in the closed jars. In elegant prose, Redi described his discoveries, stressing that all life comes from life. To the treatise were added descriptions and images of small insects such as lice, fleas and ticks, done with a slightly magnifying microscope. Like Hooke and Swammerdam at this time, Redi did not do many dissections.

Marcello Malpighi did, however. In the wake of Hooke's *Micrographia*, the Royal Society, especially its secretary Henry Oldenburg, kept a keen eye on the subject. Contact between Oldenburg and Malpighi began in 1667. Oldenburg's importance can be hardly underestimated: he directed Malpighi's interests and connected them in very subtle ways to those of the Society, its Fellows and its correspondents. The result was Malpighi's research on the silkworm, involving spectacular dissections. Partly driven by the economic importance, rather than the pious connotations of the silkworm, Malpighi looked into the internal anatomy of the creature, developing new techniques for preparation and dissection. No one had done this before, and there were no clear clues of what to expect, what to see or how to depict it. Malpighi's drawings were pathbreaking with regard both to content and to style.[23] An original drawing, now at the archive of the Royal Society, shows Malpighi's use of colour in order to depict organs and structures – a characteristic lost in the translation from drawing to engraving. However, much to the frustration of his readers, he did not give details of the way he operated, nor of the microscopical techniques he used. The manuscript earned Malpighi a Fellowship, and when the book appeared in July 1669, its amazed readers could not only see, in the forty-eight highly detailed engravings, the way a moth formed out of a caterpillar, but also the extremely complex structure of the creature, no less advanced than that of the so called 'higher' beings.

Swammerdam included long references to both Redi and, at the last minute, Malpighi in the book he would publish in November 1669, the *Historia generalis insectorum*. Despite its main title, it was published in Dutch, and translated into French and Latin in 1682. Although in the short run this work had less impact, it conveyed basically the same message as those of Redi and Malpighi: spontaneous generation does not occur and insects have a very complicated internal anatomy. Moreover, it made more general claims about order and the process of procreation. Decades later, while reworking Willughby's and his own researches on insects into a book, John Ray, in a letter to a friend, announced that he wanted to follow the method of the *Historia* of Swammerdam, 'which seems to me the best of all. It would be long to describe it, and therefore I refer you to the book.'[24]

Excited by Malpighi's book, from 1670 on Swammerdam increasingly focused on anatomical dissections and microscopic observations (Figure 8.5). With some interruptions, most notably his stay in Bourignon's religious sect, he developed the framework published in the *Historia* over the next years. Contrary to what is often stated, after his return from the sect, he did not drop his studies, but did his most pioneering research, including the famous dissection of the louse, in which, amongst other things, he described its male and female genital systems. Shortly before he died in January 1680, he finished the manuscript of what was perhaps the most coherent and impressive study of insects of the age. However, the *Biblia naturae* ('The Book of Nature') would only be published in 1737–8, thanks to Herman Boerhaave. In the meantime, another Dutchman had acquired more lasting fame: Antonie van Leeuwenhoek.

Van Leeuwenhoek began his researches somewhere around 1670, using a single-lens microscope that he improved on the Hudde design.[25] His talent was spotted by the physician Reinier de Graaf, once a friend of Swammerdam, but by this time a bitter enemy. In an evident attempt to introduce a competitor, de Graaf recommended van Leeuwenhoek to the Royal Society. Swammerdam's opinion of van Leeuwenhoek was that he was 'opiniated and very barbaric in his reasoning, not having studied'.[26] Indeed, van Leeuwenhoek had a minimum of education, did not know any foreign languages and had but a skimpy knowledge of the scientific literature. However, with his matchless microscopes (those that have survived magnify up to 250 times), he studied practically everything that came or crept his way: rainwater, a piece of wood, blood, hair, all kinds of living beings and even his own excrement. Like Malpighi before him, he was to a certain extent guided by the Royal Society. Van Leeuwenhoek never wrote a coherent book, but instead noted his observations in letters, mostly addressed to the Royal Society and published in the *Philosophical Transactions*. Among his most spectacular discoveries were the human and animal spermatozoids and the microorganisms that had been totally unknown and inconceivable until then. As is well known, it was only after a long process of repeated observations and written testimonies that the Society recognised these as facts.

It almost goes without saying that van Leeuwenhoek also scrutinised insects. His first two contributions to the *Transactions* already contained descriptions of bees and lice.[27] In succeeding decades he was to study mites, mosquitoes, fleas, wasps, butterflies and scorpions. Like Swammerdam, he was an outspoken opponent of the theory of spontaneous generation. Van Leeuwenhoek became an international

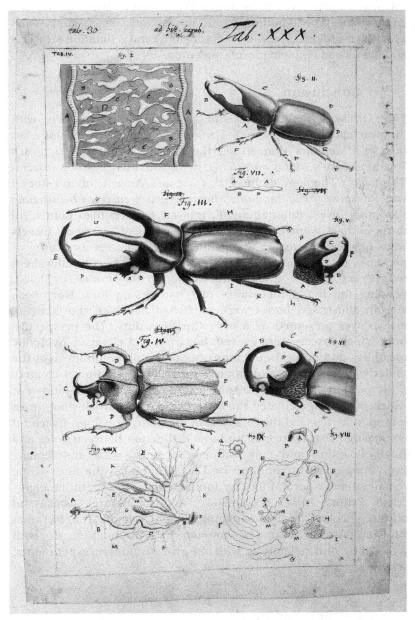

Figure 8.5 Various beetles. Original drawing by J. Swammerdam, from the manuscript of the *Biblia naturae*, executed in 1677 but only published in 1737. Like Hoefnagel – whose work he knew and highly praised – Swammerdam was much aware of his talent as a draftsman. Contrary to Hoefnagel, the beetle is not framed in an emblematic context. Now it is the creature's delicate anatomy, including the female and male genital system, depicted below, that points to God. Courtesy Leiden University Library.

celebrity who received visits from a procession of princes, scientists and *curieux*. The reporting of his microscopic observations dealt a blow to traditional ideas about the anatomy and generation of insects and the related emblematic interpretation.

Conclusion

The 1740s were the apotheosis of the fascination for insects, with the publication of Réaumur's *Mémoires pour servir à l'histoire des insectes* ('Memoirs Concerning the History of Insects', 1734–41), Swammerdam's *Biblia naturae* (1737–8), Friedrich Lesser's *Insecto-Theologia* ('Theology of Insects', 1742) and August Johann Roesel von Rosenhof's *Der monatlich-herausgegeben Insecten-belustigung* ('Monthly Insect Entertainment', 1746–61) as their high points. All these authors were Protestants. In the wake of Linnaeus, insects were now considered as a separate class, and seen as worthy or perhaps, on account of their size, complexity, beauty and sheer number of species, even more worthy to be studied in detail. Echoing the initial motivation that had driven Joris Hoefnagel, lavishly illustrated books were published in which the study of insects was presented as a truly Christian duty. The perspective from which insects were viewed, however, had changed. Symbolic meanings and biblical quotations became less important than the underlying regularity and minute structure of all things created. An important factor here was the emerging mechanical worldview, which not only postulated the existence of tiny particles but, most importantly, rejected the Aristotelian hierarchy of beings. Instead it emphasised that there was no basic distinction between great and small, beautiful and ugly, ordinary or rare, male and female. Everything in nature obeyed fixed laws. Although the microscope remained an important tool for fundamental research in the eighteenth century, it was most of all a tool for natural history and physico-theology in the emerging Enlightenment. The fact that studying insects became so popular had as much to do with philosophy and religion as with the rise of modern experimental science.

Further reading

Bennett, J., Cooper, M., Hunter, M. and Jardine, L., *London's Leonardo: The Life and Work of Robert Hooke* (Oxford, 2003).

Birkhead, T. (ed.), *Virtuoso by Nature: The Scientific Worlds of Francis Willughby FRS (1635–1672)* (Leiden, 2016).

Egmond, F., *Eye for Detail: Images of Plants and Animals in Art and Science, 1500–1630* (London, 2017).

Jorink, E., *Reading the Book of Nature in the Dutch Golden Age, 1575–1715* (Leiden, 2010).

Kaufmann, T. D., *The Mastery of Nature: Aspects of Art, Science and Humanism in the Renaissance* (Princeton, 1993).

Meganck, T., *Erudite Eyes: Friendship, Art and Erudition in the Network of Abraham Ortelius (1527–1598)* (Leiden, 2017).

Neri, J., *The Insect and the Image: Visualising Nature in Early Modern Europe, 1500–1700* (Minneapolis, 2011).

Ruestow, E., *The Microscope in the Dutch Republic: The Shaping of Discovery* (Cambridge, 1996).

Wilson, C., *The Invisible World: Early Modern Philosophy and the Invention of the Microscope* (Princeton, 1995).

II Enlightened orders

9   The materials of natural history

In 1734, in the midst of transforming all the specimens in his cabinet into a lavishly illustrated book which redefined what it meant to catalogue a collection, the Amsterdam apothecary Albert Seba (1665–1736) proudly declared that his generation had materially advanced the science of natural history: 'Our century outstrips all the preceding ones in that people see with their own eyes and handle with their own hands the things that people describe.'[1] Seba's famous portrait at the beginning of the first volume of his *Locupletissimi rerum naturalium thesauri accurata descriptio* ('Accurate Description of the Very Rich Treasury of Natural Objects', 1734–65) reveals a naturalist not simply immersed in the materials of natural history but presenting himself in full command of a wide variety of wet and dry specimens (Figure 9.1). Holding a sealed glass container of a snake preserved in spirits, perhaps the tropical *Ninia sebae* (redback coffee snake) which eventually bore his name, Seba gestures to a table strewn with shells and minerals, containing an open book of natural history illustrations on top of loose sheets, presumably containing all the preparatory materials for his virtual cabinet. Behind him are large cases, their shelves filled to capacity with specimen jars, with an elegantly articulated piece of coral and shells perched above. We are no longer in the Renaissance cabinet of curiosities, but have been invited to enter a simulacrum of an early eighteenth-century natural history collection.

Seba's cabinet was the self-conscious product of increased access to exotic nature. The strategic location of his shop in Amsterdam's harbour gave him the opportunity to supply people departing on ships sponsored by the Dutch East India Company (Vereenigde Oost-Indische Compagnie (VOC)) with everything they needed to collect on his behalf in exchange for his medicines, a decent price for choice items and the opportunity to join his expansive network of collectors. The scale of Seba's enterprise was impressive; but he was also the beneficiary of practices that evolved over several generations, beginning with Dutch physicians such as Bernard Paladanus of Enkhuizen and Carolus Clusius, recruited to Leiden

ALBERTVS   SEBA, ETZELA OOSTFRISIVS
Pharmacopoeus Amſtelaedamenſis
ACAD. CAESAR. LEOPOLDINO CAROLINAE NAT. CVRIOS. COLLEGA XENOCRATES DICTVS,
SOCIET. REG. ANGLICANAE, et ACAD. SCIENTIAR. BONONIENSIS INSTITVTVS SODALIS.
AETATIS LXVI. ANNO CIƆIƆCCXXXI.

Figure 9.1 D. Finnin, portrait of Albertus Seba, frontispiece for his *Locupletissimi rerum naturalium thesauri accurata descriptio*. Rare Book Division, Department of Rare Books and Special Collections, Princeton University Library.

from Rudolph II's Prague to help establish the university botanical garden founded in 1593. Together they put Dutch natural history cabinets on the map, signalling a new phase in the entanglements between scientific collecting and the growth of European overseas empires. Clusius boldly hung the remains of a manatee calf, acquired from Dutch sailors in Amsterdam in 1600, above the entrance to the Leiden garden. Naturalists from long-standing collecting cultures, such as Italy, took note of the new-found richness and diversity of Dutch collections, as they had previously done with the Portuguese and the Spanish; others began to consider how to harness the personnel aboard ships, especially surgeons and apothecaries, to observe and collect on their behalf.[2] Within a generation or two, a surprising influx of natural curiosities grew into a steady stream of artefacts, transforming the humanist fascination with correlating words with things into a full-fledged commercial and empirical enterprise. Specimens were no longer precious rarities but ubiquitous commodities in constant motion in a market with fluctuating prices.

Seba managed to acquire enough materials to create not one but two major collections, selling the first at a handsome profit to Peter the Great in 1716. He was well known for his care in acquiring the best specimens which is why every major naturalist, including Linnaeus (who named the redback coffee snake after him in 1758), visited his cabinet. His collection benefitted from ongoing relations with other collectors – physicians, surgeons, apothecaries, natural philosophers and numerous people from many different walks of life who shared his passion for acquiring and displaying natural specimens, or simply profited from them. Their collective investment in Seba's project allowed him to create, sell and remake his cabinet. He belonged to a generation of ambitious naturalists enthralled with the possibilities of collecting. The end result was a rich and complex material culture of natural history, accompanied by a growing interest in how to collect, display and preserve nature.

Very few of Seba's specimens survive today. When we study early natural history collections, we confront a paradox of studying a subject richly described and catalogued on paper, illustrated in images, bound in books, but now materially impoverished for the simple reason that nature is often fragile and difficult to preserve. Some portion of these collections nonetheless survives. Integrating this material into our understanding of the historical evolution of natural history is one of the current challenges and opportunities presented by this subject. By paying greater attention to the difficulties which early modern naturalists overcame in order to bring nature into their homes, we gain a better understanding of how they built and maintained their cabinets. This chapter foregrounds the question of preserving nature and the status of surviving specimens to reconstruct the stages by which naturalists learned to work with objects between the sixteenth and eighteenth centuries.

Creating the Renaissance archive of nature

Only a tiny portion of the natural artefacts collected before 1650 can be identified today in contemporary natural history museums or other historical repositories. They are overwhelmingly dried plants, fossils, earths and minerals, enlivened by the presence of some fish, animal skins, skeletons, entomological remains and occasionally an entire animal. These fragile sixteenth- and early seventeenth-century specimens include Felix Platter's herbarium in Basel (Figure 9.2), Ulisse Aldrovandi's 'theatre of nature' in Bologna, and the remaining artefacts of John Tradescant the Elder and the Younger's collection, originally in South Lambeth but part of the Ashmolean Museum at Oxford since 1683. All of these collections have been restored in the past few decades and made more accessible to a public eager to see the cabinet of curiosities reborn. They also provide materials for future research, as curators and historians of science discover the advantages of working together to understand what kinds of histories we can write from surviving artefacts.

In the early decades of the sixteenth century, physicians interested in medical botany began to discuss the best techniques to preserve plants. During the 1530s, the *hortus siccus* was born. Part of a new-found empiricism that inspired an interest in preserving nature for scientific investigation, the goal of drying plants and pasting them on paper containing useful information was not simply to study them in winter, but also to create a botanical archive for purposes of identification and comparison. If Luca Ghini – who taught medicinal simples in Bologna and Pisa, where he inaugurated the earliest university botanical garden in 1543 – did not literally invent the idea of pressing dried plants in books, as the most important teacher of his generation, he helped to popularise this practice. Four of the earliest surviving herbaria belonged to Ghini's disciples; the largest one, containing almost 5,000 specimens in fifteen volumes, was the creation of the greatest Renaissance collector Aldrovandi. Improving upon the original idea, Aldrovandi claimed that in 1553 he was

the first in Europe to discover the way to dry green plants between paper strips, reducing them to the form and shape so that they seemed depicted, dried and glued forever to the papers, which is very useful to those who cultivate these sciences who otherwise forget them because of the difficulty of this arrangement.[3]

(a)

(b)

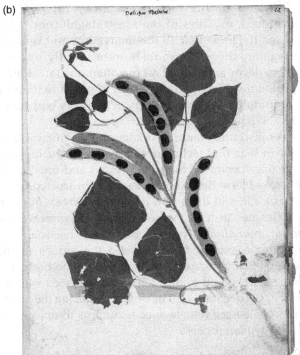

Figure 9.2 (a) A botanical illustration by Leonhart Fuchs (1542) bound alongside (b) a corresponding specimen of *Dolichus phaseolus*, the common bean (now known as *Phaseolus vulgaris*) from Felix Platter's herbarium. Burgerbibliothek Bern, ES 70.7 (86) and (87).

The goal was to preserve as much of the living plant as possible. At mid century, the Swiss naturalist Conrad Gessner called these books of plants an *herbarium*.

Preserving nature in books became a canonical feature of Renaissance natural history. By 1571, when Matthias de Lobel and Pierre Pena recalled how Ghini affixed dried plants to paper to complement his notetaking and illustrations, such practices were commonplace.[4] The herbarium was not always inert, however. In 1605 Clusius described his technique of soaking dried plants in order to bring forth their colours in preparation for having them illustrated.[5] Such practices foregrounded the importance of specimens as instruments for understanding of nature. They prefigured passionate debates in the mid seventeenth and eighteenth centuries regarding different methods of transforming natural objects into useful museum artefacts.

The famous 1599 engraving of the Neapolitan apothecary Ferrante Imperato's museum elegantly depicts shelves filled with books of plants, stuffed birds perched on top of cabinets, drawers containing samples of earths, minerals and fossils, animals and fish hanging from the ceiling. Such images capture the passion for possessing nature, but cannot reveal the *actual* state of a collection since the effect is timeless. Instead, we can consider the limited shelf life of many Renaissance artefacts by listening to discussions of the problems of preservation. When the Flemish physician and librarian Samuel Quiccheberg described an ideal museum for Duke Albrecht V of Bavaria in his *Inscriptiones; vel, tituli theatri amplissimi* ('Inscriptions, or Titles of the Most Magnificent Display', 1565), he expressed the hope that all the 'marvelous and very rare animals' in his imaginary collection would be miraculously 'by conservation kept free from decay and dried'. All too often, the fate of many specimens resembled the description of a bird of paradise in the 1607–11 inventory of Rudolph II's *Kunstkammer*: 'in a very bad state and eaten by moths'.[6] What was to be done?

Sixteenth-century naturalists knew that nature was fragile. They were already in search of techniques to maximise the longevity of far more challenging specimens than the pressed and dried plant. The French apothecary Pierre Belon enthusiastically instructed his readers how to open, gut, salt and dry a bird in his *L'histoire de la nature des oiseaux* (1555); Gessner instead focused on how to preserve the skins in his *Historia animalium* (1551–8), but then he was a furrier's son. The ubiquitous crocodile illustrated the virtues of mummification, which entailed a complete evisceration of the body, typically leaving only the skin behind to be stuffed and seasoned with ingredients that did not easily decompose. The dream of making the animal body last an eternity challenged Renaissance naturalists to unlock the secret of this ancient Egyptian process.[7]

Figure 9.3 Balloonfish and pufferfish specimens that survive in Ulisse Aldrovandi's natural history collections in Bologna. Courtesy of Università di Bologna, Museo di Palazzo Poggi, Bologna, Italy.

After visiting many collections before formulating his advice on how to create a museum, Quiccheberg understood how hard it was to preserve specimens. Acknowledging the difficulties of keeping animals intact, he envisioned a room devoted to 'parts of larger animals and even of small ones', preceded by a room containing life castings and followed by a room filled with skeletons, bones and artificial body parts. This threefold approach to knowing animals offered different perspectives on the same creature. Quiccheberg planned another room containing 'seeds, fruit, vegetables, grain, and roots', adjoining a room filled with plants, their flowers and their roots, tree branches, boughs, bark and wood; he insisted that all samples be 'dried, genuine, and selected'.[8] This was a well-curated *archive of nature*, a phrase used increasingly by Aldrovandi, whom Quiccheberg visited in Bologna.

Aldrovandi's methods of preserving animals – 'either in some instances their individual parts or else dried and stuffed in their entirety or at least depicted in still lives (as he tended to do in the case of fish)' – provided a concrete example for Quiccheberg's readers of how to keep nature in the museum (Figure 9.3).[9] Taking Aldrovandi's method of studying the elk as an example – he complemented the hoof in his possession with a portrait of the complete animal – we see how he frequently combined a material fragment with a paper specimen. He also experimented with methods for preserving fish, though it was the Roman physician Ippolito Salviani who was probably the first to treat fish like plants by gluing them to paper. As early as 1553, Renaissance naturalists speculated whether it might be possible 'to find a way to preserve dried fish in their own shape, like one does with these herbs'.[10] The emergence of a scientific collecting culture encouraged naturalists to develop new methods of preparing specimens, including the invention of fabulous creatures such as the basilisk and the hydra which existed on paper but not in nature.

Nature bottled, boxed and preserved

As natural history cabinets grew in size and scope, experiments in preparing natural objects began to proliferate. In the early seventeenth century collectors such as the Roman virtuoso Cassiano dal Pozzo perfected the idea of a 'paper museum' to counterbalance the limits of preserving nature.[11] Late seventeenth- and eighteenth-century specimens, both wet and dry, offer opportunities to understand how techniques evolved, even as challenges remained. The rare surviving examples of early modern cabinets such as John Woodward's seventeenth-century fossil collection, or the eighteenth-century cabinets of Jean Hermann in Strasbourg, Clément Lafaille in La Rochelle, Joseph Bonnier de la Mosson in Paris, Lazzaro Spallanzani in Reggio Emilia, or August Hermann Francke in Halle, to name a few well-known examples, often contain only a tiny fraction of the original collection. Yet they preserve far more than any earlier cabinets because early modern naturalists had begun to understand how to give nature a second, artificial life. Seba's prize specimens represent a sea change in techniques of collecting and preservation that locked each object into its proper, hermetically sealed place, so that naturalists might envision an eternity of looking and learning. They were among the first specimens to be displayed in bottles.

In his 1681 catalogue of the Royal Society repository, Nehemiah Grew carefully described a small finch in their collection:

A young LINET which being first embowel'd hath been preserved sound and entire, in rectified Spirit of Wine for the space of 17 years. Given by the

Honourable Mr. *Boyl.* Who, so far as I know, was the first that made trial of Preserving Animals this way. An Experiment of much use.[12]

Presented by Robert Boyle to the Royal Society in 1664, the linnet was an early example of a wet specimen enclosed in a bottle filled with spirits. Boyle wrote approvingly of the 'Balsamic faculty' that made this possible.[13] In the late sixteenth century the French royal surgeon Ambroise Paré (1510–90) recommended the use of brandy as a preservative, but no serious exploration of how to create anatomical specimens that might endure occurred before the 1650s. New techniques of preservation pioneered in the Netherlands were a subject of much discussion among early Royal Society members, who attempted to replicate them. Grew proudly reported that Boyle's linnet was in excellent condition after seventeen years in a bottle.

That linnet no longer exists, unlike the African grey parrot buried with the Duchess of Portland in 1702, considered to be the oldest surviving example of early modern taxidermy, which gave birth to a new kind of dry specimen coated with substances such as tobacco dust, pepper, snuff, camphor and eventually arsenic, in an attempt to ward off insects.[14] Other bottled specimens survive in repositories such as the Hunterian Museum in Glasgow, or the St Petersburg *Kunstkamera* founded by Peter the Great, which famously contains Frederik Ruysch's anatomical *tableaux vivants*, hauntingly photographed by Rosamond Purcell. This last collection may still have some of Seba's reptiles, despite the 1747 fire that destroyed the majority of the specimens. Other Seba specimens became part of the Napoleonic booty shipped to Paris around 1794. A number of his bottled snakes and fish, still sealed with red wax, ended up in Utrecht from where, after passing through several hands, they entered the British Museum in 1867.[15] Curators used the images from Seba's *Thesaurus* to identify them.

The growing interest in comparative anatomy and physiology encouraged a productive intersection between natural history and *Paracelsian chymistry*, as experiments in preservation required knowledge of how different ingredients, combined properly in the laboratory, could maintain animal bodies after death. In 1664, the year that Boyle presented his bird to the Royal Society, the Flemish noble Lodewijk de Bils's proprietary secret of embalming animals was posthumously revealed. It encouraged two ambitious young Leiden medical students, Jan Swammerdam and Ruysch, to experiment further with oil of turpentine, spirits, resin and other preservatives. By 1667, Swammerdam had developed a new method of preparing anatomical specimens with wax injection. Yet ultimately it was Ruysch who became famous for tricking time with art. In 1671, he opened his museum in Amsterdam. Ruysch's boast that his specimens would last for several hundred years has been borne out by the almost 1,000 surviving examples of his anatomical art, many no longer in their original fluid or bottles. Peter the Great paid an extra 5,000 guilders to learn Ruysch's secret of preparing specimens.[16]

The difficulties of preserving nature become amply apparent when we consider the origins of Sir Hans Sloane's famous collection, dating from the fifteen months he spent in Jamaica in 1687–9. Many years later, Sloane bitterly recalled his failure to successfully transport any specimens besides dried plants home to London. The living animals died en route, 'and so it happens to most People who lose their strange live Animals for want of proper Air, Food or Shelter'.[17] Having embalmed the body of his deceased patron, the Duke of Albemarle, for the voyage home, Sloane failed to understand how such techniques might assist him in his quest to fill the ship with natural curiosities. Instead, he invested in a paper museum of Jamaican fauna.

In the next few decades, however, Sloane became increasingly familiar with a wide variety of new ways to collect and preserve nature. The aggressive collecting strategies of his generation inspired printed instructions for agents working on behalf of armchair naturalists who rarely travelled but expected high quality results. John Woodward's *Brief Instructions for Making Observations in All Parts of the World* (1696) may well be the first printed guide 'for the Collecting, preserving, and Sending over Natural things, from Foreign Countries', formalising the kind of remarks made in correspondence since the late sixteenth century. Priding himself on the quality of his specimens, Woodward insisted that his agents select even the most ordinary things carefully, choosing examples that were '*perfect* or *whole*'. Even the most durable specimens should be nestled between papers, labelled to recall their origin and packed carefully in boxes to avoid breakage. Woodward worried about sea urchins' spines, the dirt and moss surrounding roots, how to pin an insect securely, knowing which ones to bottle instead, and the best method to skin and dry large animals, while confining the rarest of the small animals to 'small *Jarrs*, filled with *Rum, Brandy*, or *Spirit* of *Wine*, which will keep them extremely well'.[18] Boyle's tentative experiment with a finch spawned an entire industry of specimen-making.

The more commercially minded naturalists were not as picky. The London apothecary James Petiver did his best to encourage quantity over quality. His *Brief Directions for the Easie Making and Preserving Collections of all Natural Curiosities* (*c.*1700) reduced an entire pamphlet of instructions to a single broadsheet, recommending substitutes for the more expensive spirits, including 'a strong Pickle or Brine of sea Water' seasoned with salt and powdered alum, if all else failed. Any container would do and it could be sealed with just about anything. If a bird could not be preserved intact, Petiver willingly accepted 'Heads, Leggs, or Wings'. He sold premade labels and loaned out equipment. When Petiver went to the Netherlands in 1711 to purchase Paul Hermann's collection for Sloane, his cavalier attitude towards specimens horrified Seba. After returning to London, Petiver indicated that he would be thrilled to accept Seba's seconds – 'whatever you can spare & tho not intirely perfect' – with the plan of dedicating an illustration to the Amsterdam apothecary.[19]

Upon Petiver's death in 1718, Sloane purchased his collection. He indicated his dissatisfaction with Petiver's carelessness with nature and lack of interest in cataloguing things, describing the disorderly collection, piled in heaps, only partly labelled, and 'many of them injured by Dust, Insects, Rain, &c'.[20] Despite these criticisms, Sloane had good reasons for wanting Petiver's collection. In addition to collecting and identifying many otherwise unknown species, Petiver had done for butterflies what Renaissance naturalists had done with plants and fish, preserving them between sheets of mica in his 'Collecting Book' (Figure 9.4) to retain their colour and prevent mould and insects from destroying them.[21] In the long run, the mica sheets also ruined many a specimen, but at the time it was an important innovation. Sloane placed them in wooden boxes, often with glass tops and bottoms, with the edges sealed with paper. The preservation of colour in many different kinds of specimens posed real difficulties for naturalists, but the quest to make nature durable and lasting was by far the greatest challenge that they felt they had to overcome.

Comprehensive collecting toward the modern museum

In 1780, the naturalists Jacques de Favanne and Jacques-Guillaume de Favanne de Moncervelle enumerated 695 collections, mostly French and especially Parisian, in their list of 'the most famous cabinets of natural history in Europe'.[22] As preservation techniques improved into the mid eighteenth century, naturalists sought methods by which to organise their growing collections. The system of binomial nomenclature, developed by Swedish botanist Carl Linnaeus to describe the interrelationships among species, gave birth to a period that defined scientific collecting as the methodical labelling of once living specimens and the logical insertion of them into the biological family tree. Linnaeus considered the museum to be a useful and necessary complement to studying nature in the field. His *Instructio musei rerum naturalium* ('Instruction for a Museum of

Figure 9.4 Hans Sloane labelled and organised the specimens that he received from other collectors, including Petiver's butterflies, which he placed in mica sheets. '*Danaidae* sp., milkweed butterflies in mounts', Image 022851, ©The Trustees of the Natural History Museum, London.

Natural Objects', 1753) defined the ideal, well-lit space in which to scrutinise nature with all the senses, and carefully summarised all the techniques of collecting and preservation then in use, including his own innovations. Both Linnaeus and his disciples who visited Sloane's collection found it disorganised.[23] It failed to meet their criteria for a well-classified museum.

Linnaeus developed his famous system from his own formidable collection. Thanks to James Edward Smith's 1783 acquisition of the collection from Linnaeus's widow, the Linnean Society of London holds his largest intact collection today. Arguably the most significant component is Linnaeus's original herbarium, with more than 14,000 specimens, which he began in 1727. Breaking with tradition, he did not make books of plants. Each page contains a single specimen and was kept unbound in Linnaeus's cabinet, organised according to his sexual system, a technique he taught other naturalists. His *Philosophia botanica* (1751) contained a diagram of his cabinet's organisation (Figure 9.5), and his guide to creating a Linnaean natural history museum referred readers to this demonstration of the best way to organise a botanical collection.[24]

Today, the herbarium is unaltered: the specimens are labelled and mounted as they were in Linnaeus's time. The objects remain united with many of Linnaeus's other personal effects, such as his books and letters. A number of fish – 168 to be exact – insects and zoological specimens survive. Linnaeus furthered the project first begun by Renaissance naturalists when he decided to glue dried fish to paper rather than preserve them in bottles or illustrations. Yet two of the three purpose-built cabinets that housed this fabled herbarium returned empty to Uppsala as a symbolic compensation for the loss of this valuable Swedish patrimony. His forty-three stuffed birds displayed in glass cases have vanished. Smith sold the 2,424 minerals at auction on 1 March 1796.[25]

The obsession with categorisation led to the reform of the familiar cabinet of curiosities. It was as if the 'very *syntax* of the cabinet' had been rewritten in the form of Latinate taxonomy.[26] The newly conceived institution was the natural history museum: the visual exemplar of Linnaean classification and its quest for logic. This eighteenth-century museum was a public institution, meant to spread knowledge to all by rendering the 'book of nature' transparent.[27] During this period Europeans greatly expanded their first-hand knowledge of nature on a global scale, using the museum as a tactile, theatrical and didactic tool for a much broader audience. Guidebooks on how to collect, preserve and present nature began to proliferate. Curators self-consciously embraced the idea that how they arranged specimens

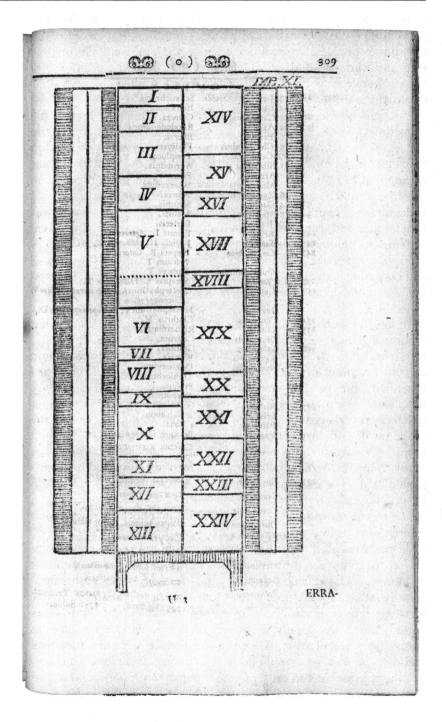

Figure 9.5 Linnaeus's design for his herbarium cabinet in his *Philosophia botanica* (Stockholm, 1751), p. 309. Courtesy of Department of Special Collections, Stanford University Libraries.

greatly influenced how the populace viewed science. The physical ordering of *naturalia* in the museum was now a deliberate process to present a symbolic system dictated by the conventions of taxonomy. The complex rearrangement of nature became the curator's art.[28] A glimmer of the Renaissance cabinet of curios remained inside the pedagogic museum, but the marvel now lay in the configuration as a whole more than the wonder of individual rarities.

Having so many diverse artefacts of natural history under one roof allowed naturalists to do comprehensive research as the foundation for a new encyclopaedia of nature. It also changed the way in which they interacted with objects. The French naturalist Georges Cuvier championed the technique of comparative anatomy, which used the very objects of the museum to draw new empirical conclusions. He interpreted the lineage of species by comparing anatomical structures that served analogous vital functions. In 1796, detailed drawings of a strange skeleton held at the Spanish Royal Cabinet of Natural Sciences in Madrid fell into the young Cuvier's hands. This huge, mysterious creature bearing immense claws was the first fossil skeleton to be put together and mounted in a museum.[29] It found its place within the Linnaean hierarchy thanks to the other objects within Cuvier's museum. The French naturalist published his *Memoir on a Fossil Skeleton*, in which he finally solved the mystery behind the *Megatherium*, a giant sloth (Figure 9.6), and gave it its Latinate name, meaning 'giant beast'. Classifying a fossil animal using binomial nomenclature was unheard of before Cuvier placed the *Megatherium* within this schema. He boldly asserted that these remains did not belong to any species of living animal.[30] Without skeletons of two- and three-toed sloths with which to compare, he never would have come to his revolutionary conclusion.

That same year, the Pennsylvania native Charles Willson Peale bought a private collection of animal and plant specimens which inspired the idea for his own museum. The aptly titled Philadelphia Museum, the first American museum dedicated to displaying objects of natural history, 'helped put American natural history on the international map'.[31] Through a strange chain of events, Peale came to possess a fossil skeleton of a mammoth – actually a mastodon – which he displayed in his museum adjacent to that of a mouse, in order to accentuate its size rather than assert its place in the Linnaean hierarchy. An artist by trade, Peale created unique, naturalistic painted backgrounds for many of his mounted animals. At the end of his gallery lay his greatest achievement: the first habitat model diorama.[32] The medium contextualised the stuffed birds and mammals within a framework that juxtaposed related species, but not in the way that Cuvier had envisioned. In 1771, the comte de Buffon critiqued

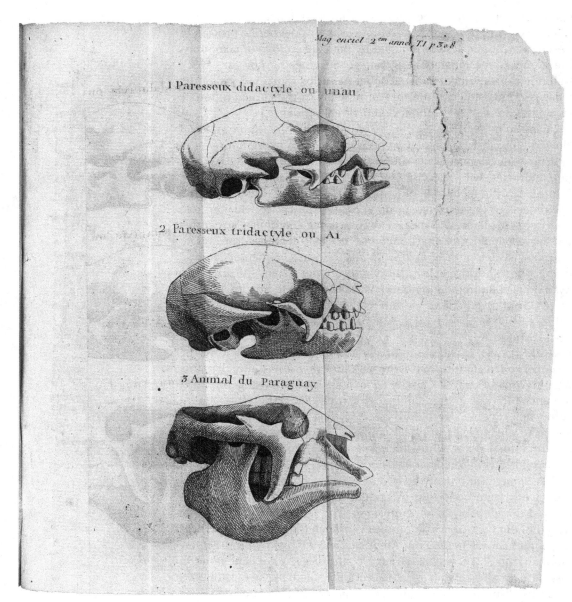

Figure 9.6 Cuvier's comparative analysis of the skulls of the two-toed sloth, the three-toed sloth and the *Megatherium*, respectively from 'Notice sur le squelette d'une trés-grande espèce de quadrupède inconnue jusqu'à présent, trouvé au Paraguay, et déposé au cabinet d'histoire naturelle de Madrid, rédigée par G. Cuvier', *Magasin encyclopédique*, 2:1 (1796), p. 308. Note that the skulls are not to scale. Rare Book Division, Department of Rare Books and Special Collections, Princeton University Library.

the increasingly elaborate methods of arranging specimens in cabinets as being 'nothing more than a still life, inanimate and superficial'.[33] Peale instead saw this as a science that included his own improvements upon eighteenth-century techniques of preservation, including the development of arsenical soap. A sign warned visitors: 'Do not touch the birds as they are covered with arsenic Poison'.[34] The American museological methodology – new like the nation itself – marked a turning point toward the ecological arrangement of specimens in natural history museums today and the abandonment of the early modern cabinet to history.

What's next? Digital collections and beyond

During the nineteenth and early twentieth centuries the taxonomic natural history museum eclipsed the early modern cabinet. The contents of the former cabinet of curiosities were of little interest to curators, who often shuttered them away. Luckily, many such collections embedded within extant museums have experienced a resurgence. In recent decades, curators have become interested in incorporating the histories of these early modern objects into the modern museum. In some instances, they have physically reconstructed the original displays, whether on exhibit or behind the scenes. The Enlightenment Gallery at the British Museum uses many of Sloane's objects to evoke the nature cabinet with which this collection began.

As institutions hoping to flourish in the twenty-first-century information economy, museums with early modern roots must adapt their approaches to objects. Museum professionals as well as historians have embraced new technologies to aid them in reconstituting the cabinet. Consider Hans Sloane's collection: during the past two and a half centuries, the British Museum scattered Sloane's possessions among countless different museum departments and other institutions, as it expanded and restructured. In one instance, a museum professional inadvertently slipped botanical illustrations from Joseph Banks's collection into a set of drawings from Sloane.[35] Now, researchers – including historians, curators and librarians – working on the 'Sloane's Treasures' project are investigating the histories that link these disparate materials to reunite them once more to tell a more complete story and deepen our understanding of provenance research. Similarly, the Sedgwick Museum has digitised the Sicilian painter Agostino Scilla's seventeenth-century fossil collection, allowing texts, drawings and engravings to be viewed next to three-dimensional scans of surviving specimens. Digital technologies are not only altering the ways in which researchers work with artefacts, but also the manner in which the public experiences them. Visitors can take virtual tours of galleries or examine an object from every angle online. Although digitisation is a huge expense for museums – both in equipment and staff time – these massive undertakings are becoming a critical component of the stewardship of natural history objects in the twenty-first century.

What does the future hold for the digital cabinet of curiosities? One might aspire to a comprehensive database of surviving arte-facts, possibly even on a global scale. Virtual reality could bring faraway scholars into contact with fragile objects that resist hand-ling. Augmented reality technology may insert objects scattered among other collections into museum installations, or restore damaged objects to their former glory on a screen. The limits of what one can do with these objects is a boundary for curators and researchers to push; but the possibilities have already reanimated interest in early natural history collections.

Further reading

Bleichmar, D., *Visible Empire: Botanical Expeditions and Visual Culture in the Hispanic Enlightenment* (Chicago, 2012).

Farber, P. L. 'The development of taxidermy and the history of ornithology', *Isis*, 68 (1977), pp. 550–66.

Findlen, P., *Possessing Nature: Museums, Collecting and Scientific Culture in Early Modern Italy* (Berkeley, 1994).

Hunter, M., Walker, A. and MacGregor, A. (eds.), *From Books to Bezoars: Sir Hans Sloane and his Collections* (London, 2012).

Margócsy, D., *Commercial Visions: Science, Trade and Visual Culture in the Dutch Golden Age* (Chicago, 2014).

Mason, P., *Before Disenchantment: Images of Exotic Plants and Animals in the Early Modern World* (London, 2009).

Price, D., 'John Woodward and a surviving British geological collection from the early eighteenth century', *Journal of the History of Collections*, 1 (1989), pp. 79–95.

Prince, S. A. (ed.), *Stuffing Birds, Pressing Plants, Shaping Knowledge: Natural History in North America 1730–1860* (Philadelphia, 2003).

Schiebinger, L., *Plants and Empire: Colonial Bioprospecting in the Atlantic World* (Cambridge, 2014).

Schulze-Hagen, K., Steinheimer, F., Kinzelbach, R. and Gasser, C., 'Avian taxidermy in Europe from the Middle Ages to the Renaissance', *Journal für Ornithologie*, 144 (2003), pp. 459–78.

Spary, E. C., *Utopia's Garden: French Natural History from Old Regime to Revolution* (Chicago, 2000).

Stearns, R. P., 'James Petiver: promoter of natural science, c.1663–1718', *Proceedings of the American Antiquarian Society*, 62 (1953), pp. 243–365.

10   Experimental natural history

Philosophers from Aristotle to Francis Bacon defined natural history in contrast to natural philosophy: the former as a science of accumulating particulars (or facts) and the latter as the search for the general causes of phenomena. In 1751, Jean d'Alembert elaborated on this canonical distinction when he mapped out the branches of human knowledge for one of the most widely read texts of the Enlightenment, the 'Preliminary Discourse' to the massive encyclopaedia he was editing with Denis Diderot. Here, d'Alembert grounded the radical distinction between natural history and natural philosophy on the different aspects of the human mind, separating the sciences of memory (such as natural history) from the sciences of reason.

In this chapter, I look at the breakdown of these canonical distinctions in the practice of Francophone natural history in the eighteenth century. In the circles around the French naturalist René-Antoine Ferchault de Réaumur, natural history became part of '*la physique*'. Réaumur, who spent years on a multivolume natural history of insects, described the stages of the caterpillar's metamorphosis as 'one of the most curious subjects in physical science (*la Physique*)'.[1] The cognate 'physics' does not quite capture the full meaning of this term as it was used in the period; 'natural philosophy' is somewhat better, but still implies something distinct from natural history. French writers often used '*la physique*' to designate science in a general sense, or a properly scientific outlook. This usage endorsed an attitude – leave no stone unturned in the search for understanding; be precise as well as thorough; anything, no matter how apparently trivial, is worthy of sustained investigation – as well as a set of tools and techniques. Experimentation, the hallmark of physical science by the eighteenth century, played a vital role in natural history, blurring the divide between natural history and natural philosophy.

Experimental physics and natural history

In 1753, Jean-Antoine Nollet took up a newly created post as Professor of Experimental Physics at the College de Navarre in Paris. His contemporaries knew Nollet as a maker of elegant scientific instruments and demonstration apparatus, as an expert in electrical phenomena, and as a spectacular scientific lecturer. In his inaugural address, he drew a familiar distinction between experimental physics and natural history:

The object of experimental physics is to understand the phenomena of nature, and to show their causes by proofs of fact. It differs from natural history in that the principal aim of the latter, without accounting for effects, is to give us detailed knowledge of the bodies that make up the universe, to distinguish for us the genera, species, individual varieties, the relations between these things and their different properties. The first of these sciences undertakes to unveil the mechanism of nature, the second gives us, so to speak, an inventory of nature's riches.[2]

Revealing mechanisms or compiling an inventory, determining causes or classifying according to similarities and differences – these were the time-honoured oppositions distinguishing natural philosophy from natural history.

Nollet agreed with d'Alembert that the contrast between causal explanations and the accumulation of particular facts could serve as a schematic point of departure, but as he turned from the abstractions of epistemological categories to the practices associated with making these different kinds of knowledge, Nollet immediately blurred the familiar distinction. Experimental physics and natural history, he went on, 'are so closely tied together that it is almost impossible to separate them. A *physicien* who is not at all a naturalist is someone who reasons at random and about objects he does not know anything about; the naturalist who is not also a *physicien* is using nothing but his memory'.[3] Experiment provides the key to moving in both directions from physics to natural history and back, and to breaking down the Aristotelian categories that no longer applied unproblematically. The practitioner of experimental physics finds his subject matter anywhere and everywhere: 'countryside and city, the elements, the seasons, anything that breathes, vegetates, is born and dies – all these give him what he needs to meditate, to learn, and to profit from'.[4] The distinction between physics and natural history disappeared in this vigorous pitch for experimentation. Nollet claimed the realm of living things for physics; but he also implied that the naturalist should exploit the methods and instruments of experimental physics. To exemplify the ideal practitioner of this kind of science, Nollet invoked none other than Réaumur, citing the naturalist's work on insects for the kind of attentive observation necessary for doing physics.[5] Réaumur was a pillar of the Paris Academy of Sciences, a prolific observer and writer, famous for his experimental work in many areas, including the fabrication of steel and porcelain, thermometry, preservation of natural history specimens, the natural history of insects and birds, and the artificial incubation of chicken eggs. His books on the natural history of insects exposed readers not only to the phenomena of insect lives, but also to an experimental and observational approach to investigating the natural world.

Nollet sometimes called this kind of experimental science an 'art', to evoke the material entanglement of observer and observed. His reflections on the meaning of physics as an approach to natural knowledge point to the key role of experiment in natural history in this period. Paying attention to experimental trials and interventions will alert us to the ways that naturalists engaged with the objects of their attention, as they continually invented techniques and devices to make visible hidden and often mysterious processes. As cabinets filled to overflowing with animal, plant and mineral specimens from far and wide, in nearby laboratories and workrooms and gardens naturalists were pursuing experiments and devising strategies of observation designed to fill in the 'histories' of living things. In turn, these natural histories supported generalisations (sometimes explicitly designated as 'general laws') about such matters as physiology and generation.

Experiments on living plants and animals did not replace collecting, displaying and organising of preserved specimens. But this kind of direct interrogation of nature, using instruments and all five senses as well as the rational mind, opened up new kinds of questions, and the answers became part of the natural histories of particular species. Do all animals reproduce through the copulation of two sexes? What are the properties of materials produced by insects for various purposes? How do temperature variations affect the development of insects or the formation of chick embryos inside eggs? How does a queen bee know how much food to deposit with each egg in its cell of the honeycomb? How can a sea star replace a severed arm or a lizard a severed tail? What role does air play in the physiology of caterpillars, or their chrysalises? These questions, and so many others, could not be answered by simple inspection; naturalists manipulated living things and intervened experimentally in any number of ways. They used their hard-won empirical evidence to uncover the causes of phenomena and the general laws of life.

Though better known for his physics lectures and his experiments with static electricity, Nollet had first-hand experience of natural history. Twenty years before delivering his manifesto on experimental physics, he had worked as an assistant to Réaumur, in a laboratory where they studied the life cycles and physiology of insects alongside the heat capacity of liquids and the properties of materials. In the service of an elaborate research programme on temperature and heat capacity, Nollet built and calibrated thermometers and other instruments, and experimented on the expansibility of alcohols and the melting and freezing temperatures of different substances. He took charge of jars and boxes of living insects, as well as thermometers, furnaces and air pumps. A surviving list of tasks for Nollet, written in Réaumur's hand, includes, alongside measurements of temperatures of various salt solutions and the quantity of air released in mixtures of water and spirits of wine, instructions for 'suffocating insects coated in butter in the air pump, to see where the air emerges from their bodies'.[6]

Réaumur and Nollet studied the physiology of respiration in cater-pillars as a physical problem of the movement of air across mem-branes, just as they studied the solubility of air in spirits of wine. Investigating the freezing temperatures of water mixed with different salts, they used caterpillar 'blood' as a test substance. With none of the properties of other liquids with very low freezing points (alcohol or acids), the caterpillar fluids were 'not at all flammable, with no sensible acidity, and they can nevertheless resist very cold tempera-tures'. They subjected caterpillars to freezing temperatures, rendering their bodies rigid and apparently lifeless, and then revived them with the application of heat. Some species survived freezing and some did not; younger specimens resisted intense cold more effectively. In some kinds, even dead caterpillars, when brought to temperatures colder than ice water, retained their suppleness. 'The kinds of blood, the liquids that circulate in the vessels of different kinds of caterpillars, are thus related to each other in the same way that spirit of wine or a very strong brandy is related to a weak brandy'.[7] Measurements on caterpillar fluids treated as physical materials in the laboratory were integrated into the natural history of insects, with experiments on the effect of air temperature on the timing of metamorphosis, or the function of transpiration in this process. The study of chrysalises led directly to further experiments using temperature and humidity to control the preservation of chicken eggs.[8] In the same laboratory, experiments on insects provided data for physical science research (measurements of specific heats and the physical chemistry of liquids), while the methods of experimental physics fed into the natural his-tories of insects.

The conformation of animals and plants – size, shape, colour, inter-nal structures and so on – provided the baseline for the descriptions familiar to readers of natural history books. A closer reading shows that much more than looking, dissecting and measuring went into these 'histories'. Naturalists attempted to decipher the life cycles and behav-iours of their subjects, as well as their interactions with other species and the distinctive properties of the materials they produced (silk, wax, excrement, coloured liquids and so on). Microscopes and scalpels served as obvious tools of the trade, but so did thermometers, air pumps, ovens, furnaces and distillation apparatus. Take the example

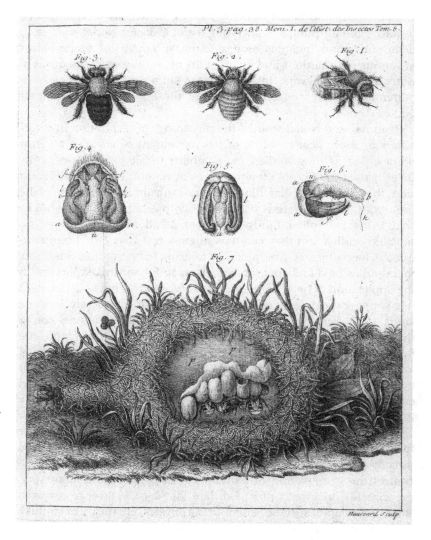

Figure 10.1 Three different species of bumblebee (*bourdon*), dissected male organs of common bumblebee and moss nest with interior exposed to show wax and honeycomb. R.-A. Ferchault de Réaumur, *Mémoires pour servir à l'histoire des insectes* (Paris, 1734–42), vol. VII, plate 3. Huntington Library, San Marino, California.

of the bumblebee (*bourdon velu*), a species that constructs its nest on the ground from moss fronds to shelter the hive (Figure 10.1).

Réaumur started his investigation by enlisting the help of labourers mowing nearby fields to gather hundreds of these nests, with their living inhabitants. A systematic study of this bumblebee – one of the many species of similar nest-building insects that occupied Réaumur's attention for years – required moving the insects wholesale from their natural habitat into a space in or near the house, where the nests could be opened, the various substances analysed, and the insects themselves dissected, manipulated and depicted. Describing the structure of the nest was only one aspect of the bumblebee's natural history; to understand the sequence of its construction, Réaumur removed the top of the nest and scattered the moss, so he could watch the bees rebuild it, strand by strand. He opened other nests to analyse the materials produced by the insects, including the water-proof lining they secreted as a coating for the inside of the moss domes. This substance appeared waxy, but did not behave exactly like normal beeswax when manipulated in the laboratory:

Having made a little ball of this material by rolling it between my fingers, I put it in a coffee spoon which I balanced on hot coals. Even though the ball heated up, it did not liquefy, as a ball of wax would have done in the same circum-stances. When it was heated to a certain point, it ignited, and burned for some time. After the flame went out, a little mass of black charcoal was left. This coal was quite different from ordinary coals – after leaving it for two hours, I found it reduced to a moist powder.[9]

This detail – the attributes of the waterproof material, determined in the laboratory – was only one of the many bits and pieces comprising the complete history of this particular kind of bee. The paste of pollen and wax where the queen deposited her eggs (designated *pp* in Figure 10.1), the waxy cells containing nectar, the ovoid cocoons spun by the larvae before their transformation into adults, the mossy passa-geway leading into the nest, all these were examined in due course, and then represented in images and text. Revealing the operation of the male sex organ, depicted just above the nest, involved manipulat-ing the dissected abdomen to engage the mechanism and inflate the relevant anatomical part, so it could be viewed under the microscope and then sketched. In the top row of images, the queen bumblebee (upper right) is depicted next to two similar bees native to Egypt, drawn from preserved specimens in Réaumur's collection. This single plate (one of forty-seven in this volume) displays several kinds of knowledge at once: comparison with more exotic preserved speci-mens; anatomised sex organs; bees and their materials in situ, exposed to view, with the surrounding vegetation as it would appear in the field. We can read this collection of images as a microcosm of the practice of natural history, with references to collection, dissection, microscopy, drawing and engraving, manipulation of materials and observation of life cycle and behaviour.

In the histories of other species the details vary, of course, but the elements of the experimental and observational approach remain consistent. The materials produced by insects, investigated with the tools of experimental physics, became part of these histories, and led to generalisations across species as well. Silk organs were put into solvents to isolate the silk material and study how it hardened; silk fibres were tested for strength and flexibility and stretch. The shiny surface of certain chrysalises or the hard interior surface of some egg cases displayed a spectrum of the physical properties of insect products which might be formed into long threads or spread into smooth surfaces and left to harden. After testing the properties of many such 'silks', Réaumur decided that 'the liquid silk is nothing but a kind of varnish; if the caterpillar, ... while the silk is still sticky, covers a smooth surface with it instead of pulling it into long threads, this surface will become varnished'.[10]

Practical applications of natural history

Naturalists turned their attention and their tools to the study of familiar denizens of field and forest such as the bees and caterpillars discussed above. For insects, especially, understanding the habits and life cycle of common species was a prerequisite for the potential application of the knowledge gained by observation and experiment. Réaumur's examination of the tinea (or clothes moth) is a typical example of his experimental approach, as he studied these small but destructive household pests and looked for ways to discourage or exterminate them. The moths deposit their eggs on furs and woollen fabrics – upholstery, curtains, bedding, carpets, clothes – where the worm-like larvae feed on the fibres, which they also use to make the tubular structures where they live. In the process, the pests leave the cloth peppered with holes. The naturalist started by enclosing moths in glass jars with squares of woollen cloth, where the insects obligingly deposited their eggs. A few weeks later, these had hatched into little caterpillars that could be observed 'working to clothe themselves', constructing cylindrical sheaths of wool fibres, held together with the caterpillar's own sticky silk.[11] The 'worm' stuck out its head to bite off fibres that it worked into first one end of its sheath and then the other. Curious about how the creature manoeuvred within its woollen tube, Réaumur intervened, cutting off the ends of the cylinder to make it shorter than the caterpillar. 'By gently pressing one end of the sheath, I forced the tinea to move a bit towards the other end, and then with scissors I cut off the part I had forced it to abandon.' Thus exposed, the insect worked to rebuild its home in full view. The observer then moved the insect onto cloth of a contrasting colour, so he could see how the new fibres entered into the construction (see Plate 6):

We force it to show us all its procedures, by constraining it to dress itself from scratch ... I poked a small stick ... into one end of its sheath, then pushing the

stick little by little, I chased it out of its sheath ... In various similar experiments, the tinea always preferred to make itself a new garment rather than to
go back into the one it had been chased from, even though it had cost it so
many months of work.[12]

The strategy of manipulating the insect and its surroundings is
emblematic of the experimental approach evident in all of
Réaumur's work.

Having established the sequence of the life cycle of this clothes
moth, Réaumur set out to test various methods for protecting household goods from its ravages. Common practices like shaking out furs
or beating rugs could be effective, but only after all eggs had been laid
and before the larvae had attached themselves to the fabric in preparation for metamorphosis; choosing the right season for this task
depended on knowing the insect's life cycle. More effective treatments would require deterrents or poisons, which had to be tested in
the laboratory. So Réaumur raised thousands of moths in captivity
and distributed them in groups of twenty to glass jars containing
pieces of fabric treated with various substances. Noticing that
untreated animal fibres were never attacked by the tinea, he experimented with the effects of lanolin, rubbing woollens with an
untreated lamb fleece or dipping them in hot water where virgin
wool had been washed. (This treatment rendered the woollens unap-
petising to the tinea, and protected the fabric quite effectively.) Other
jars contained swatches of woollen serge dipped in an infusion of
tobacco, solutions of salt, soot and other substances, or aromatic
plants. Inspired by a country remedy using pine cones, he tried
various methods of exposing the caterpillars to turpentine fumes.
This method killed the insects quite dramatically, as did exposure
to tobacco smoke.[13] The same jars where he had watched the tinea
making their sheaths, and where moths were enclosed to produce the
next generation of experimental subjects, served as the testing
grounds for potential poisons.

Having found two substances lethal to the tinea, he experimented
with how to treat or prevent infestations on a larger scale. In the
vignette illustrating this natural history (Figure 10.2) we see not the
laboratory where the experiments were done, but a storeroom with
domestic servants putting the results into practice: one is beating a fur
muff; another is sweeping the insects off a rug. At the table two men are
laying sheets of paper treated with oil of turpentine on a tapestry.
At the back of the room, a fifth servant is placing a smoking brazier
filled with tobacco into an armoire to fumigate its contents.[14]
The image reinforces the connection between the experiments and
their application.

MEMOIRES

POUR SERVIR

A L'HISTOIRE

DES INSECTES.

Figure 10.2 Techniques for destroying clothes moths. Vignette, R.-A. Ferchault de Réaumur, *Mémoires pour servir à l'histoire des insectes*, 6 vols. (Paris, 1734–42), vol. III. Wellcome Library, London.

Experimenting with generation

Nollet found a wide public for his *Leçons de physique expérimentale* ('Lectures on Experimental Physics', Paris, 1745–8), a print version of his lecture courses and a showcase for his scientific instruments. In a lecture on the shapes of microscopic bodies, he focused his microscope on salt crystals, then turned to samples of liquids made from plant and animal materials: vinegar, oyster liquor and infusions of grains, leaves and flowers (Figure 10.3). He described the shapes and motions of different kinds of microscopic beings visible in different preparations, then launched into a tirade against spontaneous generation, or the production of life from unorganised matter. For naturalists in the circles around Réaumur and Nollet, spontaneous generation was anathema, a throwback to the days of irrational superstition. This was a current issue in the 1730s, as

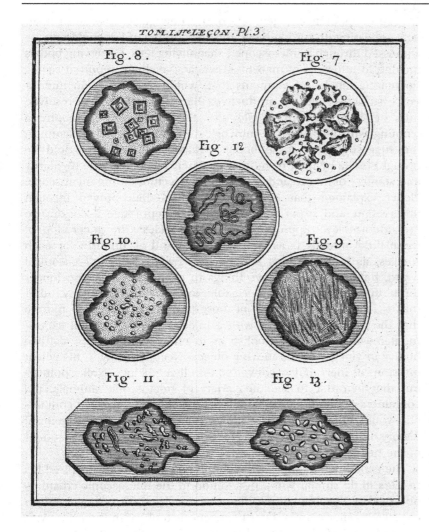

Figure 10.3 Microscopic views of different solutions and infusions. J.-A. Nollet, *Leçons de physique expérimentale*, 6 vols., (Paris, 1745–8), vol. I, plate 3. Wellcome Library, London.

evidenced by a Jesuit attack on Réaumur after he ridiculed claims about the spontaneous production of maggots in rotting meat.[15] 'Modern' naturalists were committed to the principle that generation – the reproduction of living forms – could not be equivocal or random. Animals always and everywhere produce offspring of their own kind, so that nature is consistent and regular in its operations, and therefore intelligible to naturalists. Réaumur went so far as to label this one of the fundamental 'principles of the natural history of insects'.[16]

In the 1740s, observations of infusions of meat and grain, especially by the English microscopist John Turberville Needham, revived the spectre of spontaneous generation in a new guise. Instead of maggots appearing on rotting meat, this new drama of life could only be witnessed through a microscope's magnifying lens. Observing bodies moving in infusions of macerated wheat, Needham reported seeing fluctuating tangles of filaments that 'would swell from an interior Force so active, and so productive, that even before they resolved into, or shed any moving Globules, they were perfect Zoophytes teeming with life, and self-moving'.[17] To naturalists like Réaumur, the suggestion that these bodies, whatever they were, could be brought to life by their own 'interior Force' challenged the regularity and stability of nature and its laws. In his lecture on the microscope, Nollet explained that 'modern naturalists' had proved through observation and experiment that insects reproduce according to the uniform laws appropriate to their species: 'the generation of these little animals is as well regulated, and as uniform for each species, as it is for lions and horses'. Even in the microscopical world, nature does not leave things to chance. 'We can no longer think that Nature, so conformable to herself, adopts any other means for the multiplication of those that are so extremely minute that they can be seen only with a microscope.'[18] A careful experimenter would not be so foolish as to understand the tiny eel-like bodies in vinegar or the moving objects Needham saw in his wheat infusions as inert matter converted into living beings. Nollet pointed out that in order to see the variety of microscopic animals, the containers of liquid had to be left uncovered; tightly sealed containers would not produce any life at all. The experimental natural history of insects, according to Nollet, carried lessons for the student of the microscopic world.

The experimental approach championed by Réaumur inspired his readers to do similar work. In addition to the microscopic creatures observed by Needham, experiments on novel and surprising phenomena like regeneration and parthenogenesis took off in the 1740s, with an explosion of discoveries by naturalists in Switzerland, France, England and the Netherlands. Charles Bonnet in Geneva tried to clarify the apparently asexual process of reproduction in aphids, after reading Réaumur's report of his own inconclusive results; Pierre Lyonet and Abraham Trembley in The Hague replicated these experiments, circulating their results to each other and to Réaumur in Paris. Female aphids produce living offspring, but they seemed to do this without copulating. Réaumur suspected that they might be hermaphroditic, without the need for fertilisation:

Luckily it was not difficult to imagine experiments that could confirm this suspicion, or destroy it. All one would have to do would be to seize an aphid at the moment it emerged from its mother's body, to raise it and make it live out its days in a solitude where it would not be permitted to communicate with another insect of its kind, or even with any insect of any size whatever.[19]

Réaumur attempted these experiments, but failed to keep the solitary aphids alive long enough to see if they would produce young of their own. Bonnet, and then several others, successfully isolated newborn aphids and watched the next generation emerge, confident that mating had never taken place.[20]

Then Trembley found strange 'organised bodies' in his local pond water. These polyps, as they became known, had green tubular bodies, less than half an inch long, with fine threads like tentacles at one end. They attached themselves to the leaves of water plants, and to the bottom and sides of glass jars that soon filled Trembley's workroom. Their bodies contracted and expanded, sometimes walking inchworm-style, migrating toward sunlight. When cut in two, both parts remained alive and sensitive to light, and each of them gradually grew back its missing half over the course of a few weeks.[21] Trembley spent several years exploring every aspect of the anatomy and physiology of these 'enigmas', eventually finding and describing many other species with similar properties. He reported a litany of unexpected and inexplicable results. One body sectioned into three pieces grew into three fully functioning individuals; the interior and exterior membrane of the body's tube enclosed a green gel-like substance of unknown purpose; the body contracted and expanded in length, causing a change in colour as well as shape; the flared end near the tentacles seemed to be a receptacle, sometimes empty and sometimes full; fully grown polyps produced protuberances that gradually developed into complete bodies that broke off and lived independently. After months of experiments, he was able to confirm that the tentacle end served as the animal's mouth, only visible when it was open to receive food. He watched a mother with a budding offspring share a little eel, which passed from one body to the other. Once he had observed digestion, he used the digestion of eels as a test of whether regenerated bodies were truly complete polyps.[22] In one of his most striking experiments, Trembley turned the body of a polyp inside out, to see whether it would still be able to eat and reproduce. When the tubular animals had been inverted, he strung them up with hog bristles, to keep them from turning back again.

Trembley's startling results with his freshwater polyps inspired Bonnet to look for them in the countryside outside Geneva, to no avail. He did find an abundance of aquatic worms and decided to work with them instead:

If the trial that I was considering succeeded on this worm, easily recognised as an animal, I would have demonstrated that there truly are animals that can be propagated, so to speak, by budding, which would confirm Trembley's beautiful discovery, still in its early stages. Indeed, the experiment succeeded: my worm, divided in two, soon gave me as many complete animals.[23]

He went on to cut them into more and more pieces, up to twenty-six. Most of these grew to some extent and several became complete worms (Figure 10.4).

He tested the effect of different temperatures on the rate of the worms' growth, inspired by the analogy to experiments by the English physician Stephen Hales on the growth of plants.[24] Bonnet wondered about what could be driving their reproduction, and whether the worm's 'vital principle' could be exhausted over time. The generation and growth of organic bodies, Bonnet commented, 'is a very important point of Physics, and still very little illuminated'.[25] At around the same time, independently of Bonnet, Lyonet also noticed that aquatic worms could regenerate when cut. 'What seemed quite surprising to me', he noted,

was that not only does a piece one-sixteenth of the whole worm have the ability to repair its loss, but I have seen one-thirtieth and even one-fortieth of a worm become complete worms. I do not know if these worms copulate as earthworms do, but I have seen them multiply quite often without this expedient, by cutting themselves in two and regrowing.[26]

As this sort of experiment proliferated, and as the news spread around Europe and beyond, what had been inexplicable novelties developed into sustained research programmes of systematic experiments and observations, resulting in the careful natural histories of polyps and worms produced by Trembley and Bonnet. All of these naturalists were committed to investigating the full histories of their subjects experimentally, especially when the initial observations seemed startling or even impossible, just as Réaumur had done with his clothes moths and bumblebees.

Conclusion

Eighteenth-century naturalists filled many volumes of comprehensive natural histories with the results of their experiments and observations. Text and illustrations explained not only the details of structure, function and behaviour, but also the methods employed for investigating them. Natural history – the accumulation of knowledge about all aspects of the animal world – relied on systematic investigation using many of the same tools and techniques being applied in the same period to problems in the physical sciences. The microscope was an essential tool, as were

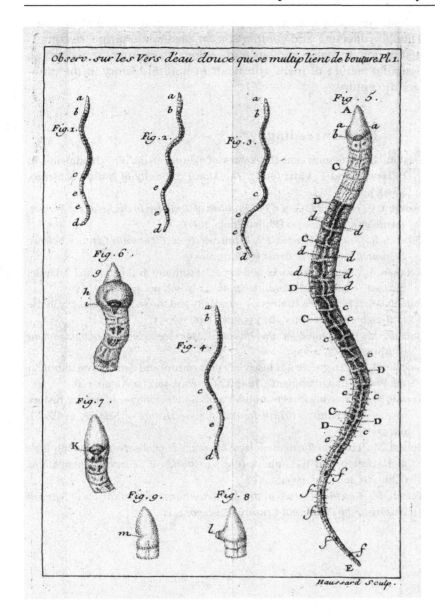

Figure 10.4 Regenerating worms. C. Bonnet, *Traité d'Insectologie; ou observations sur quelques espèces de vers d'eau douce* (Paris, 1745). Wellcome Library, London.

thermometers, air pumps, glassware, furnaces and ice baths. In practice, the development of experimental natural history blurred the traditional distinction between natural history and natural philosophy. From the later eighteenth century onwards, increasingly sophisticated experiments on living things took naturalists further into embryology, the life cycles of microorganisms, animal electricity and other aspects of plant and animal physiology. Though collecting and comparison for taxonomic purposes continued in botany, palaeontology and zoology, experiment remained an essential feature of many sub-fields of natural history in the nineteenth century.

Further reading

Daston, L., 'Attention and the values of nature in the Enlightenment', in L. Daston and F. Vidal (eds.), *The Moral Authority of Nature* (Chicago, 2004), pp. 100–26.

Dawson, V., *Nature's Enigma: The Problem of the Polyp in the Letters of Bonnet, Trembley and Réaumur* (Philadelphia, 1987).

Gibson, S., *Animal, Vegetable or Mineral? How Eighteenth-Century Science Disrupted the Natural Order* (Oxford, 2015).

Kellman, J., 'Nature, networks and expert testimony in the colonial Atlantic: the case of cochineal', *Atlantic Studies*, 7 (2010), pp. 373–95.

Ratcliff, M., 'Trembley's strategy of generosity and the scope of celebrity in the mid-eighteenth century', *Isis*, 95 (2004), pp. 555–75.

Ratcliff, M., *The Quest for the Invisible: Microscopy in the Enlightenment* (Burlington, VT, 2009).

Stockland, E., '"La guerre aux insectes": pest control and agricultural reform in the French Enlightenment', *Annals of Science*, 70:4 (2013), pp. 1–26.

Terrall, M., 'Following insects around: tools and techniques of natural history in the 18th century', *British Journal for the History of Science*, 43 (2010), pp. 573–88.

Terrall, M., 'Frogs on the mantelpiece: the practice of observation in daily life', in L. Daston and E. Lunbeck (eds.), *Histories of Scientific Observations* (Chicago, 2011), pp. 185–205.

Terrall, M., *Catching Nature in the Act: Réaumur and the Practice of Natural History in the Eighteenth Century* (Chicago, 2014).

11   Spatial arrangement and systematic order

The spaces of early modern collections shaped observation even through their material arrangement. A short passage at the start of Jean Bodin's *Colloquium heptaplomeres* ('Colloquium of the Seven', *c.*1590), offers a particularly vivid portrayal of this in a fictional account of a meeting of seven scholars, with diverse religious and philosophical views, at a house in Venice. Their host, one Coronaeus, is described as a collector with a large correspondence network, and a '*pantotheca*' which forms the natural historical core of his cabinet.[1]

The *pantotheca* is a wooden cupboard or cabinet with a square display window measuring six by six feet, about the height of an adult man. This piece of furniture is divided into equal-sized compartments, each devoted to a single natural object, from the fixed stars and planets via the elements to objects from all three natural kingdoms: *mineralia, vegetabilia* and *animalia* (creatures endowed with souls). The objects it contains are displayed in their entirety, represented by a particular part, depicted in images or described in text. The construction and presentation of this display cabinet, Bodin adds, allow optimal visibility of the specimens contained within it. When the curtain that usually covers the *pantotheca* is drawn back, each individual object and the whole collection can easily be encompassed in a single glance. A classificatory order of nature is implied by the compartmentalisation of the cabinet. The 36 by 36 compartments can accommodate 1,296 specimens, which represent the totality of natural things. Each object is presented in its compartment as a discrete unity. Yet the arrangement of the compartments with respect to one another also offers the opportunity to highlight similarities, possible relationships and transformations in nature, by juxtaposing particular specimens.

But the *pantotheca* does not stop short at the claim to universality. The fundamental number of six around which it is designed stands for, among other things, the six perfect bodies with their cosmological symbolism: the six directions; the six harmonies; and the six senses of mankind – sight, hearing, smell, taste, touch and common sense (*sensus communis*). The analogical relationship between the world as macrocosm and man as microcosm, well-known to early modern scholars, is figured as a relationship of correspondence between the ideal space of cosmological order, the material elements of nature and the human body. This correspondence finds an echo in human cognition, in the relationship between the perceptible qualities of the objects in the collection and the six senses of its human makers and observers.

The description of the *pantotheca* as a universal repository both unites and problematises a range of conceptual aspects of early modern collecting. The many *naturalia* it contains serve as a synopsis of the continuity and connections between the cosmos and sublunary nature, from fixed stars to quadrupeds. So it is not surprising to find that the *pantotheca* is also a mnemotechnical instrument. Coronaeus has looked at the various objects in it so often that he has completely memorised their locations.[2] The display cabinet, here, links the practice of collecting explicitly with the *ars memoriae* widely adapted in early modern times from classical authors like Cicero and Quintilian. These mnemonic techniques were based on imaginary spatial systems where every item to be memorised was represented by an image and located in a specified position. To memorise by means of such a method meant visualising the deposited items in the mind's eye according to their proper connections. There were actual projects which manifested this spatial system in a material object, like the memory theatre of Giulio Camillo.[3]

The natural philosophical and methodological properties of this fictitious cabinet are thoroughly conventional, in intellectual historical terms. They suggest a pervasive view of the order of nature as static. Yet at the same time, the text also discloses moments of dynamism: by stressing the mediating role of some objects in particular, the display makes it possible to flag transitions between discrete units of the order. This role is possible because Coronaeus treats his many-sided world-box as an open structure, constantly adding to and thus transforming it. If the *pantotheca* in Coronaeus's house might be taken as an 'ideal type' of the symbolic aspects of natural historical collecting in the early modern period, it is nevertheless important not to overgeneralise. Natural history collecting would develop in divergent directions in the seventeenth century. Collections owned by sixteenth-century scholars and humanists were predominantly sites and resources for their own studies, as well as for exchanges with other naturalists. They could also serve as vehicles for improving social status. In the case of courtly collections, the goals of princely representation were often paramount, and priorities differed. Objects in such collections often displayed a territory's natural resources and potential for princely dominion.[4] As new categories of collection appeared later in the seventeenth century, such as institutional collections, collecting practices diverged further. Professional research, connoisseurship and virtuosity, and individual didactic goals meant that collections came to manifest particular emphases and preferences generating collective conventions of credibility, new methods of studying specimens and new networks of exchange, as discussed in Egmond's chapter in this volume.

If types of collection began to diverge, cultures of collecting continued to show similarities well into the eighteenth century. Sometimes this is apparent in individual collecting practices. John Woodward, for instance, is one prominent example of a type of scholar around 1700 who was a highly regarded collector both of minerals and fossils, and also of works of art and antiquities.[5] His collecting preferences represented just one of many possible combinations within what was understood as a complementary relationship between nature and the arts, in the broad sense. Both were understood as 'works', the product of labour of different kinds. Artificial objects were the product of human invention and labour upon materials, using various technologies. Natural objects were the work of nature, produced by the effective causes of a *natura naturans* – nature doing as nature does – more or less subject to divine agency and a divine plan. Even traditions like physicotheology in the later seventeenth and eighteenth centuries were fundamentally linked to this analogy: their protagonists interpreted every complex structure, beautiful object or function in nature as evidence of design, and thus necessarily as evidence of a divine originator.

Such a conceptual framing was a highly ambivalent requirement in relation to the spatial whole of early modern collections and to the natural historical project of producing taxonomic order. On the one hand, collectors were under pressure to arrange their items in an orderly way. For disciplines like botany or mineralogy, collections were decisive for the establishment of systematic order. Yet interest in rare or wonderful natural phenomena, and the analogy between art and nature, both of which survived into the seventeenth century, formed a counterweight to the great project of classifying.[6] On top of this, there were also works, produced for princely collectors in particular, in which 'raw nature' and artistic shaping intersected, producing hybrid works of art. The contrasts and similarities between human and natural artistry formed a particularly appealing feature of such products, and gave them a metamorphic potential, a downright anti-taxonomic emphasis that challenged any stable order based upon marked separations between objects.[7] In contrast to this dynamic of transformation and change, museum space served in various ways as both medium and instrument of order and knowledge. An important source for reconstructing this structuring role is images, which encompass both a performative aspect, in the sense of gesture and action, and a structuring one, in the sense of classification and arrangement.

Scenographies of the collection

From the late sixteenth century onwards, interior views of collection spaces were often published in connection with printed descriptions of particular collections or theoretical writings about the *Kunstkammer*.[8] One of the earliest examples is a view of Ferrante Imperato's Neapolitan collection, published as the frontispiece to his *Dell'Historia Naturale* in 1599 (Figure 11.1). After 1610, the particular spatial arrangement and design shown in this print would be taken up by Flemish painters. Such gallery paintings show collecting scenes in which the fine arts, especially painting, play a dominant role.[9] These woodcuts, etchings and copperplate engravings only allow limited conclusions as to the actual condition of the collections they depict. Although they convey some aspects of the spatial arrangement, in most cases the interiors have been revised to idealise the arrangement and use of the collection. Images thus show a visual interpretation of the space of the collection. Especially in the case of printed images of cabinets dating from the late sixteenth to mid eighteenth centuries, three distinct categories may be identified.

Typical of the first category is a clearly delineated, box-shaped room, with a throng of objects positioned on shelves, in drawers, on the walls and often on the ceiling. The viewer's gaze can hardly overlook the abundance of individual specimens, but is simultaneously invited to travel over them in succession, discerning an astonishing level of precision and detail even in the more distant objects.

In this category of image, it is also common for actors to be present, inducting the viewer into the use of the collection. That is, the visit is portrayed not as a silent immersion of one viewer, but rather as a sociable enterprise. The viewing of the objects is accompanied by lively discussion and gesticulation; individual specimens may be abstracted from the whole at any moment for detailed inspection. In the frontispiece to Basilius Besler's index of collections, published in Nürnberg in 1616, a young assistant exhibits a large dolphin skull to the collector and a companion. Welcomed into such images, visitors and guides are effectively included among the exhibits. The act of looking at the objects is bound up in the act of sharing the space of the collection. The visual experience is shaped in no small degree by the fact

Figure 11.1 Frontispiece. F. Imperato, *Dell'Historia Naturale* (1599). Herzog August Bibliothek, Wolfenbüttel, 37.2 Phys. 2°.

that the majority of the displayed objects are present to the touch, 'at hand' in a very concrete sense. Books like this, which offered descriptions of particular collections, were used by collectors as works of reference and sources of information, and were recommended as preparation for museum visits.

A second category of representations organises the space of the collection in what seems at first sight a very similar manner to the first. Yet there are two characteristic differences. First, these pictures contain no people. They seem to present the specimens in a three-dimensional manner, evoking an atmospheric combination of still life and interior through the use of subtle gradations from light to dark. This is particularly apparent in the 1622 catalogue of Francesco Calzolari's Veronese collection (Figure 11.2).[10] Spaces without people, like the frontispiece to the *Museum Calceolarianum* ('Calzolari's Museum'), are also organised in fundamentally different ways.

Figure 11.2 Frontispiece. B. Ceruti and A. Chiocco, *Musaeum Francisci Calzeolari Iunioris Veronensis* (Verona, 1622). Herzog August Bibliothek, Wolfenbüttel, 38 Phys. 2° (1).

Instead of showing a more or less flat space, where the figures' interactions run parallel to the surface of the picture, in these images depth is exaggerated. This perspectival enhancement reorganises the relationship between viewer and objects. In the absence of a mediating guide, the reader or observer as individual is connected in a far more immediate, and more specifically visual, manner with museal space. When not embedded in a collective interaction, the objects on show here appear as if directly addressed to the observer. The image no longer commands attention by inviting viewers to participate in a depicted exchange, but rather challenges them to enter its own space, even if only in imagination.

The way the viewer is drawn into the image is exemplified in the frontispiece of Ole Worm's *Museum Wormianum* (1655). The collection space is empty of people; at its centre stands a table whose top, receding into depth, bears the book's title, nothing more. Such tables were common in early modern collections, serving for the study of individual objects, reading and note-taking. Contemporary readers moving in collecting circles would have understood Worm's frontispiece as an invitation to step inside the collection and use the table in just this way. The dominant characteristic of these unpopulated collecting spaces is the erasure of the social conventions associated with visiting a collection, in favour of an invitation to individuals to immerse themselves in the objects' microcosm.

A third type of image is principally distinct from the other two in the interruption of the spatial continuity between the viewer and the space of the collection, often through the insertion of an architectural frame between the space of the viewer and that of the picture. The earliest printed example of this kind is the frontispiece of the *Gottorfische Kunstkammer* ('Gottorf Art Cabinet', 1674). Such frames were a widespread book ornament in the period, only later used for collections, but then becoming very desirable. One of the more ambitious examples is the frontispiece to Georg Everhard Rumph's *Amboinschen Rariteitkamer* ('Amboinese Rarity Cabinet', 1705), engraved after drawings by Maria Sibylla Merian (Figure 11.3). Here the facade emphasises the extension of the collecting space into the depths of the picture, while simultaneously closing off the collection's interior space from the space occupied by the beholder by means of a 'portal'. The fictive portal is often represented as an entrance, allowing entry only to the gaze and the imagination.

Here, human actors reappear on the scene, but they are distinct from the guided group of visitors in my first category. Instead, as for example in Imperato's frontispiece, they figure in an allegorical representation. This image shows a group of men around a table surrounded by cabinets, in animated exchange over some objects. They are to be understood as an ideal company of scholars: dressed in antique or oriental garb, they are clearly distinct from the scantily clad figures situated to the front of the portal arch, who are bringing in baskets and boxes or proffering an array of exotic sea creatures. These servant figures are mostly depicted as inhabitants of the Pacific islands, where Rumph compiled his collection while serving as a colonial official in the service of the Dutch East India Company (Vereenigde Oost-Indische Compagnie or VOC). The scene is flanked by two sculptures representing Terra

Figure 11.3 Frontispiece. G. E. Rumph, *Amboinsche Rariteitkamer* (Amsterdam, 1705). Herzog August Bibliothek, Wolfenbüttel, Nh 2° 7.

and Neptune, providing a framework of mythical polarity between the Earth, as bearer of all human culture, and the untamable seas.[11]

In collection images of this third kind, the space of the museum is often removed from the reader or observer in two senses. First, it appears less immediately accessible, screened off as it is by allegorical figures. Second, it is pictured as the site of an idealised circle which no longer invites the viewer to a communal inspection of the collection. Moreover, these images do not promise direct contact with the objects in the museum. The latter appear in a second order of pictorial space, mostly shut away in tall cupboards. The space of these collections is increasingly charged with meaning: religious or cosmological imagery is painted on the ceiling, or else the heavens themselves are shown curving above the cabinets of the collection, as for example in Levinus Vincent's *Wondertooneel der Nature* ('Spectacle of Nature', 1719).

The majority of printed images of collection interiors produced between the late sixteenth and mid eighteenth centuries hint at an important structural transformation. From the late seventeenth century onwards, space recedes ever further into the depth of the image, often distanced from the gaze and semantically accentuated as a location where the material objects of the collection are sublimated into a higher sphere of knowledge. Simultaneously, a smaller foreground becomes the boundary of the image, a transfer zone between the space of the collection and the onlooker. This accentuation of a narrow spatial layer emphasises the picture as at once medium and tableau of the systematic orders of collected items.[12]

Furniture and spatial order

It was not just a trivial condition of collections that every object included in them had to be allocated its own space. Rather, such placements entailed decisions of the greatest significance. *Where* a particular specimen found a place in the collection as a whole was often of decisive importance in determining *what* it was from then on. In this sense, the arrangement of cupboards and repositories in collections was always the instrument of an order of objects extended in space. An early published description of such an item of furniture is to be found in Conrad Gessner's *De omni rerum fossilium genere* ('About Every Kind of Fossil Objects', 1565). 'Fossil objects' here meant everything that was dug up out of the earth; the book lays out for the reader an encyclopaedic knowledge of stones and earths, which also encompasses issues like their use in pharmacy and the arts. Gessner's work is the first printed book devoted to the *mineralia* to represent its objects not only through written descriptions and commentary, but also through pictures.[13]

The order to be adopted in a collection of stones so as to produce the best effect in a cabinet is addressed at the start of the book. The inventor of this order, Johannes Kentmann, a doctor and collector living in Torgau in Saxony, was one of Gessner's correspondents. The book presents Kentmann's ordering system as an abstract scheme. The various groups of minerals, each allocated a number, are represented in pairs on a plane, in thirteen compartments of a rectangle divided up in tabular form. On the next page, the same numbers appear on the drawers of a cupboard in the same order. The cupboard's construction replicates the ordering scheme, translating it directly into architectural form.

This piece of furniture thus embeds an abstract classificatory order in a material structure which allows the manifold appearances and qualities of stones to be brought together in the form of a set of exemplary pieces. A fairly simple cupboard encloses the individual objects, its compact architecture representing an order which lays claim to completeness. The description of the cupboard as an '*arca*' (ark) underlines this claim to comprehensiveness, as well as the protective role of the cabinet. Both roles allow the Biblical motif of the ark to serve as a leitmotif for natural history collecting, and sometimes even to impart its name to entire collections, such as John Tradescant the Elder's 'Ark of Lambeth', as it was commonly known, open to the public in South London by 1629. When Gessner prefaced his far more heterogeneous juxtaposition of individual essays with the concrete spatial order of objects that was Kentmann's Ark, it suggested to the reader that everything in the pages that followed was supported by a factually grounded order of the things in themselves.

A few years later, Michele Mercati's *Metallotheca Vaticana* ('Vatican Cabinet of Minerals') would display a complex relationship between the cabinets and space of the collection, systematic order and individual specimens. The work was largely written around 1580, when Mercati was a physician at the papal court, responsible for the Vatican's botanical garden and mineralogical collection. At his death, he left behind an unfinished manuscript and a series of copperplate engravings in his own hand, produced for the book. These materials were only published in 1717, with corrections and commentaries. Despite its delayed publication, the book became an important work of reference for the study of stones in the first half of the eighteenth century.[14]

Figure 11.4 Frontispiece. M. Mercati, *Metallotheca Vaticana Opus Posthumum* (Rome, 1717). Herzog August Bibliothek, Wolfenbüttel, Nf 2° 15.

The *Metallotheca* covers a wide spectrum of natural objects, including many from the mineral kingdom. One area of interest were hard-to-explain phenomena like dendrites or petrifactions. But such things also figure in the book as part of a whole collection of objects apparently made by human hand. The most striking examples, here, are antiques like the Laocoon, the Apollo Belvedere and the Belvedere Torso, which appear in the *Metallotheca* among the marble specimens.

The frontispiece extends an open space before the reader's eye, empty of human figures, on a double folio page whose front opens out to reveal a curved portal to the exterior (Figure 11.4). The specimens of the collection are housed in compact compartments in nearly identical cupboards. Each is labelled and identified with a Roman numeral. A table in the foreground lists all the numbers along with the name of the group to which each specimen belongs.

These are divided into two hierarchical categories: the 'metals' in the strict sense, and all other stones – the 'orycta'. The same division is reflected in the spatial arrangement of the cupboards: the metals are placed to the left, all other specimens to the right. In this way, the frontispiece affords a view inside a museum space in which a differentiated order of objects is anchored in, and fixed by, the cupboards it contains.

At the start of every main chapter, the same spatial order is taken up afresh. Alongside a written description of the group of objects being considered in that particular chapter, there is an illustration of the relevant cupboard (Figure 11.5). Although the majestic architectural facades do not really show the cupboards visible in the frontispiece, the way the cupboard facades repeat in the reader's imagination evokes the connection to that space. But the view of an open cupboard positioned at the head of every chapter does more than merely gesture at the space of the collection as a whole. Rather, it serves as an iconic index and map of the arrangement of the objects. Every specimen described is given a 'loculo', that is, a specific location within the compartment. The petrified fish in compartment number fifty-one, for example, are especially attractive pieces (Figure 11.6). Their particular quality is their lifelike appearance, the cause of which was the subject of extended disagreements among naturalists. The *Metallotheca* takes this striking characteristic to generate a category of 'ideomorphic stones', to which this entire cupboard was devoted.

Even the printed book frames the objects of the collection, together with all the information about them, in spatial terms. A work like *Metallotheca* which describes a collection is not just to be read or viewed: the act of reading itself implies, at least in imagination, the act of moving the objects in and out of the collection. This publication, through the labour-intensive artistic endeavour its engravings demanded, adapts and broadens techniques of inventorising used in the *Kunstkammer* of Ambras, Munich or Prague around 1600. And yet the *Metallotheca* simultaneously distances itself from the praxis of inventorising. It thus marks a turning point in the way the space of the collection and its taxonomic order are related to one another. Experimentation with different locations for the specimens and their distribution on the plates at once liberates them from the concrete space of the Vatican's collection, and makes them available for other orderings.

Even in the late period of the *Kunstkammer* and *Naturalienkammer*, the cupboards where specimens were housed continued to play an important instrumental role. They allowed systematic orders of the objects of natural knowledge to be developed and displayed in ever

Figure 11.5 Table (cupboard). M. Mercati, *Metallotheca Vaticana Opus Posthumum* (Rome, 1717). Herzog August Bibliothek, Wolfenbüttel, Nf 2° 15.

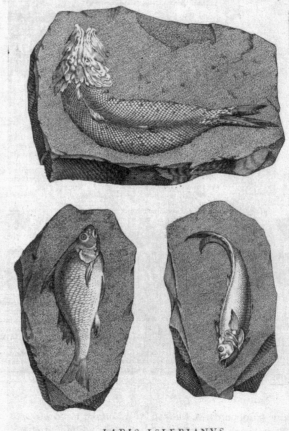

Lapides ΙΔΙΟΜΟΡΦΟΙ.   *319*

L O C U L O LI.
SPINUS MELIBOCI.
C A P. L X I.

Spinus lapis eſt fiſſilis, colore ater, & bituminoſus. Quibuſdam locis ſcinditur ob venas æreas, quas urendo exigunt, quemadmodum dicetur. Hunc verò de hoc figmenta ludentis Naturæ recenſeri debent. Cruſtas enim, quæ ex lapide facilè diſparantur, ipſa ſibi pro tabellis pictorum inſtituit, ima-

LAPIS ISLEBIANVS

gines animalium eis ſuperinducens, non aliis uſa coloribus, quàm lapidis qualitas concedit. Cùm ærofus lapis, ac bituminoſus ſit, aureas pyritæ micas ita diſponit, ut delineamenta animalium omnino reddantur. Neque commodiùs

Figure 11.6 Petrified fish. M. Mercati, *Metallotheca Vaticana Opus Posthumum* (Rome, 1717), p. 319. Herzog August Bibliothek, Wolfenbüttel, Nf 2° 15.

finer detail. In the *Kunstkammer* of Halle's Franckesche Stiftungen, founded in 1741, the display cabinets as a whole manifested the universalising claims of this teaching collection for boarding-school pupils. Here too, the cabinets were alike in construction and design. The facades were adorned with carefully painted architectural elements. The decoration on each cabinet included detail specific to the group of objects it contained. The specimens themselves were visible through the glazed fronts, but could not be touched. One of these cupboards housed the mineral specimens. A number marked out the cabinet containing stones as the first in the sequence. This conformed to contemporary classificatory schemes of the cosmos in which nature was divided into three kingdoms, minerals, plants and animals, progressing from the raw forms of matter up to living beings, capable of movement, feeling and ultimately reason. To devote the first of the cupboards in this collection to 924 specimens of stones was therefore to appeal to established models of the order of the macrocosm.[15] To this order even artefacts in all their variety could ultimately be added, since they were produced by the art of man, who was the summit of the *animalia* and the image of God. Moreover, the three kingdoms of nature were obviously not understood as strict divisions without intermediate forms. Between the collection of stones and the next large group, the *vegetabilia*, was a cabinet containing corals, those sea creatures long held in natural history to partake of both mineral and plant natures.

Pragmatic recommendations like instructions for guides taking visitors around the collection also underscore the fact that the individual divisions of this systematic order corresponded with specific compartments in the cabinet. Guides were under strict orders to put each object removed for individual inspection back in its precise location, and never to risk 'confusion' by taking too many specimens out of the compartments at one time. If the precise arrangement of the specimens in cupboards played an important instrumental and didactic role, such rules of conduct announce that *using* collections – increasingly the property of public institutions – could endanger those goals. In the long term, this was one reason why spaces of scientific work and public viewing in the museum became separated, as Alberti's chapter in this volume argues.

Pictorial practice

In directly mirroring the spatial structure of cabinets, the books of Gessner and Mercator exemplify a particular use of pictorial representation which did not separate images from the space of the collection. Pictures not only played an important role in the use of collection space, they also allowed new modes of working with individual objects liberated from the material context of the museum.

The most ambitious attempt to depict a complete collection in its spatial setting is a series of etchings made in St Petersburg in 1741. These plans and elevations of the new buildings for the city's famous academy show the precise arrangement of the cupboards and the order of the specimens, wall by wall. Numerous items on display can be identified in the miniature representations. The visual indexes of this collection *in situ* are backed up by a corpus originally consisting of over 4,000 drawings in pen and wash, produced to document the collection, to serve as aids for study, and perhaps eventually to be published. These meticulously detailed individual studies depict a substantial portion of the total collection. Together, the two layers of this 'paper museum' offer the possibility of moving back and forth between an overview of the wall-mounted arrangements and the scrutiny of individual specimens, independent of their actual location.[16]

Pictures were essential in creating connections between the overview of the collecting space and the study of individual objects. They might substitute for objects that formed essential parts of a systematic arrangement, but were physically absent. In Coronaeus's *pantotheca*, pictures appeared in the compartments designated for the objects. Clearly, an image not only substituted for an *object* which was absent, but was also a visible placeholder in the space allocated to the *species*. This manoeuvre indeed appears to have occurred in practice. In his 1565 museological treatise, Samuel Quiccheberg revealed that, in his admirably complete display of *naturalia*, the Bolognese collector Ulisse Aldrovandi replaced missing objects with pictures.[17] Visitors to Aldrovandi's collection reported the suggestive illusion produced by these small paintings, which depicted objects lifesize, scarcely distinguishable from the real items surrounding them. Such practices apparently continued over a long period: Johann Daniel Major recommended the use of images in his *Unvorgreifflichen Bedencken von Kunst- und Naturalien-Kammern* ('New Views of Art and Natural History Cabinets', 1674), and in a 1705 introduction to arranging collections, the reader was advised to put paintings of missing objects in the empty spaces, because in that way 'everything would appear full / look pleasing and magnificent / and a stable order will be achieved'.[18]

This particular use of pictures highlights a discrepancy between the theoretical claims and practice of collecting. It was impossible in practice to represent the totality of nature in systematic arrangements using exemplary objects alone, and it was apparently customary to fill the lacunae using pictures. This shows that the space of the collection was far more than merely a container for objects. Rather, it was understood as the material form taken by an uninterrupted, perceptible order that could be represented through both real objects and various media. Even in the absence of a material, deictic relationship with the actual space of the collection, or a close link to a complete set of specimens, a wide range of pictorial representations was produced and used in natural historical collecting. I have already mentioned drawings of individual specimens. Collections of drawings, like the one at St Petersburg discussed earlier, or Cassiano dal Pozzo's *'museo cartaceo'* (paper museum) in Rome, were exceptions, understood even in their own time as collections in their own right.[19] A far more common form of collecting of images of *naturalia*, mostly in small format, were manuscripts and albums containing written notes, drawings, 'nature prints' made using actual specimens and ordinary prints. Images could be arranged and ordered according to the owner's research interest, and over time were often removed and rearranged.

Underlying this approach to images was a general concern with the ordering of individual subjects on a neutral pictorial surface, something which already had a long history by the start of the early modern period. Combinations of text and image, whether drawings, miniatures or printed images, were particularly important media of natural historical knowledge in manuscripts, prints and book illustrations, especially from the fifteenth century onwards. Learned scribes borrowed an image style from antique papyrus to add individual images to manuscript text columns. The codex, on the other hand, offered more possibilities for collating multiple images. Individual pictures could be freely arranged on a single- or double-page spread. The simplest form taken by such arrangements was scattered or clustered images in which exemplary individual images are informally distributed or loosely grouped on the page, often accompanied by titles linking them to the text. An example here is the 'Viennese Dioscorides', a sixth-century Byzantine manuscript.

This principle of arranging images survived well into the eighteenth century; its flexibility was probably its great strength. The pictures could be positioned so as to stimulate particular comparisons or sharpen resemblances or differences. The methodological implications of this type of arrangement for pictorial representations were also very attractive for those working on taxonomic order in the strict sense. It offered an ideal way to isolate whole objects or their parts and clarify the criteria of classification. The possibilities were particularly significant in analytical reductions like that used by Joseph Pitton de Tournefort, a founder of modern botanical taxonomy, for the plates of his *Institutiones rei herbariae* ('Elements of Botany', 1700).

This consistent form of a tableau, however, represented only one strand of making and using pictorial representations, if a very important one. An entirely different way of composing a page can be found in learned writings throughout the Middle Ages, in the form of a grid dividing the page into separate quadrants.[20] This pictorial tradition merged into an increasingly broad practice of collecting, and such images are once again anchored in museum space, in that they depict or represent reified spaces in cabinets into which objects can be placed. They thus clearly refer to material situations. In seventeenth-century paintings of *Kunstkammern*, such as those of Johann Georg Hinz or Domenico Remps, such images became an autonomous genre of still life, which used virtuoso *trompe-l'oeil* effects to draw the viewer into a play of synaesthetic influences between the overall impression of a carefully balanced order, and the apparently real presence of extremely heterogeneous objects. This is no process of distribution and evaluation of fundamentally equivalent elements; rather, these pictures present the viewer with an irreducible diversity of shapes, textures and surface effects, evoking a bewildering array of semantic relations.

But the opposite trend, towards taxonomic models, evidently does not reflect a boundary between the scientific and artistic use of images. A striking example of this is the etchings produced by Martin Lister's daughters Anna and Susanna for his vast *Historia conchyliorum* ('History of Shells', 1685–92).[21] The collector used images primarily as instruments of taxonomic differentiation. Yet the plates in his book captivate with their densely atmospheric still-life quality. The etchings combine a precise linear delineation of form with a subtle modelling of the objects' plastic qualities. They transform the specimens into highly sophisticated graphic representations whilst at the same time evoking a suggestive spatial and material relationship to the beholder. Even prints of collected items thus often bring to mind the spatial arrangement of real collections. This is underlined when their pictorial qualities echo the compelling interior of the frontispiece, as is the case in the etchings in *Museum Calceolarium*, or when topological orders structure the whole book, as in *Metallotheca Vaticana*, or when the reader could identify specific items shown in the frontispiece from accurate descriptions in the book, as in *Museum Wormianum*.[22]

The spatial organisation of natural history collections, the great project of *taxonomia* and the role of images in these two thus highlight opposing trends that acted over the *longue durée*. Representative examples are Gessner's collection of drawings and notes for his *Historia plantarum*, compiled from around 1560 onwards, and Johann Jakob Scheuchzer's *'Klebebände'* or albums, compiled in around 1720 for a dictionary of petrifactions.[23] An increasing quantity of objects and information, organised in topological arrangements, filled collections of art and natural history from the sixteenth century onwards. The analogies between microcosm and macrocosm, between *natura* and *ars*, were natural philosophical concepts that allowed even heterogeneous specimens to be included and experienced within a connected web of meanings. The differentiation of taxonomic orders and systems progressed hand in hand with the redistribution of objects. Models of methodical comparison and systematic arrangement, as suitable forms of natural historical order, alleviated the complexities of the space of the collection. In the process, pictures, as instruments and media, played a pre-eminent role. The scattered or clustered image in particular provided ideal conditions for the mutable arrangement of individual representations, in accordance with defined criteria.

However, opposing tendencies coexisted within this instrumental function of images. Particular aesthetic properties appealed to both artistic amateurs and naturalists. The greater the emphasis of pictures upon the physical existence of objects – sometimes through explicit references to museum displays – the greater the density of details and information, producing pressure to infuse *naturalia* with new forms of significance.

Further reading

Bredekamp, H., *The Lure of Antiquity and the Cult of the Machine: The Kunstkammer and the Evolution of Nature, Art and Technology* (Princeton, 1993).

Daston, L. and Park, K., *Wonders and the Order of Nature, 1150–1750* (New York, 1998).

Findlen, P. and Smith, P. H. (eds.), *Merchants and Marvels: Commerce, Science and Art in Early Modern Europe* (New York, 2002).

Freedberg, D., *The Eye of the Lynx: Galileo, his Friends, and the Beginnings of Modern Natural History* (Chicago, 2002).

Herzog Anton-Ulrich-Museum Braunschweig and Kunstmuseum des Landes Niedersachsen, *Weltenharmonie. Die Kunstkammer und die Ordnung des Wissens* (Braunschweig, 2000).

Kistemaker, R. E., Kopaneva, N. P, Meijers, D. J. and Vilmbakhov, G. V. (eds.), *The Paper Museum of the the Academy of Sciences in St Petersburg c.1725–1760* (St Petersburg, 2005).

Marx, B. and Rehberg, K.-S. (eds.), *Sammeln als Institution. Von der fürstlichen Wunderkammer zum Mäzenatentum des Staates* (Munich and Berlin, 2007).

Ogilvie, B. W., *The Science of Describing: Natural History in Renaissance Europe* (Chicago, 2006).

te Heesen, A. and Spary, E. C. (eds.), *Sammeln als Wissen. Das Sammeln und seine wissenschaftsgeschichtliche Bedeutung* (Göttingen, 2001).

12   Linnaean paper tools

In this chapter, I am going to explore a theme that has recently become a 'hot topic' in cultural studies of early modern science and medicine more generally: the use of ink-and-paper tools, both in script and print, to accumulate, process and communicate information across geographic, socio-political and cultural distances. This is a topic that promises to deepen our understanding of the history of natural history, especially in its 'classical' period, which stretches between the tenth edition of Carl Linnaeus's *Systema naturae* (1758) and Charles Darwin's *On the Origin of Species* (1859).[1] In the first section, I shall provide an outline of the information economy of classical natural history, contending that it was characterised by an increasing heterogeneity, rather than homogeneity, of sources of knowledge. In the second section, I shall argue that the adoption of two information-processing devices that Linnaeus had introduced – namely binomial nomenclature and the so-called hierarchy of taxonomic ranks – gave discursive unity to classical natural history despite this heterogeneity. The third section, finally, will present some examples of how these devices were deployed in the form of paper tools designed for the storage, indexing and exchange of information on plants and animals. Overall, I want to suggest that attention to the material construction and practical deployment of such paper tools can tell us a lot about natural history and its highly dynamic research culture.

The information economy of classical natural history

Late eighteenth and early nineteenth-century natural history experienced social and institutional changes that involved both diversifying and centralising tendencies. On the one hand, rising levels of literacy and the spread of cheap print made it increasingly easier to consume, and contribute to, natural history, resulting in a widening basis of amateur naturalists engaged in specimen collection and epistolary exchange, and the formation of local and regional associations of naturalists.[2] Moreover, demand increased for expert naturalists to fill positions both in the context of state bureaucracies, economic enterprises, and organisations engaged in long-distance trade and colonial administration.[3] These experts both contributed to, and depended upon, the production, exchange and consumption of information on minerals, plants and animals, their properties and their uses. The 'information economy' of natural history began to offer a crucial stepping stone for members of the middling sort and subalterns to enter various occupations and careers of an administrative, brokering or entrepreneurial nature.[4]

Until the mid eighteenth century, the exchange of specimens, letters and publications centred on individuals. Linnaeus in Sweden, Georges-Louis Leclerc, comte de Buffon in France, Hans Sloane and Joseph Banks in England, Herman Boerhaave in the Netherlands and Albrecht von Haller in Switzerland are just a few among the naturalists who personally managed and controlled vast social networks. Along with the letters they exchanged, they received specimens accompanied by written information and drawings, as well as manuscripts and publications.[5] By the early nineteenth century, central institutions that were there to stay had taken over this role, such as the Muséum national d'Histoire naturelle in Paris, Kew Gardens and the British Museum in London and Berlin University with its newly established gardens and collections in Prussia.[6] Individuals associated with these institutions certainly continued to wield enormous power over information flows in natural history.[7] But two important structural features distinguished the 'new' museums from their early modern counterparts.[8] First, they represented collections of collections rather than collections *tout court*. Often starting out with the acquisition of a single large collection, these museums then expanded by acquiring further collections, or by commissioning travelling naturalists to hunt for specimens on a global scale. The most striking example of this is the Muséum national d'Histoire naturelle in Paris, which received a boost from the confiscation of aristocratic collections during the French Revolution, the provenance of which was carefully noted in a catalogue.[9] Second, and concomitantly, museums became increasingly articulated into specialised departments, offering hierarchically organised positions for curators or 'keepers' and various amanuenses who administered its collections. A new generation of 'professional' naturalists emerged, often socialised through participation in long-distance natural historical exploration, during which they collected for their patrons or institutions, and later moving on to curatorial work in metropolitan collections and libraries.[10]

The knowledge networks that underwrote natural history were not just expanding and diversifying in the late eighteenth and early nineteenth centuries. Rather, over the same period, institutions emerged which formed central hubs in these networks and appropriated the knowledge that was generated. This double process of diversification and centralisation turned natural history into an increasingly disparate and tension-ridden field. Numerous conflicts testify to this and, as one might expect from the complexity of the developments just outlined, they did not align with a simple centre–periphery dichotomy. To be sure, tensions existed between imperial ways of practising natural history and their colonial and indigenous counterparts.[11] But such conflicts played out with equal intensity on a regional scale between local naturalists and metropolitan experts, both within European nations and within their nascent colonies.[12] They even occurred on the microscale, within local institutions and associations, where professional and disciplinary standards, as well as social status and authority, were at stake.[13]

To date, 'information economy' largely remains a suggestive metaphor; the conceptual tools for studying the actual economic mechanisms that supported and propelled natural history during its classical period still need to be developed.[14] But the metaphor draws attention to the infrastructures that mediated information exchange in classical natural history.[15] In order to gain a better understanding of these infrastructures, the next section will turn to two innovations that formed the cornerstones of Linnaeus's self-styled 'reform' of natural history – the naming of plant and animal species by 'trivial' names composed of genus name and specific epithet (as in *Homo sapiens*); and their ordering in a hierarchy of taxonomic ranks, by variety, species, genus, order (or family) and class. By the end of the eighteenth century, both of these innovations had been universally adopted by naturalists, and as the few analyses of their reception that exist demonstrate, it was precisely their practical value in communication that made them so attractive to naturalists in their pursuit of ever-enhanced levels of collaboration and information exchange in natural history.[16]

Linnaean paper tools

One of the most astonishing aspects of classical natural history is the success that Linnaeus's taxonomic publications enjoyed in terms of print run, especially if one considers that these were not publications made for leisurely reading or intellectual entertainment, but catalogues filled with names of genera and species, references to earlier literature, short morphological diagnoses and abbreviations designating geographic range and ecological habitat. Linnaeus himself counted twelve editions of his *Systema naturae* between 1735 and 1768 – expanding over that time from a single eleven-page folio volume to four octavo volumes numbering 2,441 pages in total – six editions of *Genera plantarum* from 1737 to 1764, and two editions of *Species plantarum* (1753 and 1762). Moreover, the success of the books far outlived Linnaeus himself. From the late 1760s onwards, but especially after Linnaeus's death in 1778, other naturalists began to publish new editions, translations and adaptations of his works. Some of these continued Linnaeus's own numbering of editions, so there exists a sixteenth edition of the botanical part of *Systema naturae*, which was issued in Göttingen as *Systema vegetabilium* in five volumes from 1825 to 1828, a two-volume ninth edition of *Genera plantarum* (Göttingen, 1830–1), and an aborted sixth edition of *Species plantarum* (Berlin, 1831–3).[17]

The lasting success of Linnaeus's taxonomic work is often explained by claiming that it provided naturalists with the means to communicate unambiguously about plant and animal kinds (Figure 12.1). But what allows for unambiguous reference in modern taxonomy is the type method: the practice of associating taxonomic names with type specimens. And this method was only introduced to natural history towards the end of the nineteenth century.[18] Linnaeus himself, when introducing binary names and the five-tiered hierarchy of taxonomic ranks, advertised quite a different advantage. Traditional or 'legitimate' names, as Linnaeus called them, were composed of the genus name and a diagnostic phrase that spelled out the traits by which the named species differed from other species of the same genus. The 'trivial' or binary name, in contrast, just added a 'single word ... freely adopted from anywhere' to the generic name. The trivial name was thus not only shorter and more easily reproduced; above all, it was more stable, since it did not have to carry any diagnostic meaning, and hence did not have to be changed whenever new species were discovered.[19] In highlighting the advantage of a 'systematic' arrangement by class, order, genus, species and variety, Linnaeus drew on a similar contrast with traditional systems, whose primary function was also diagnostic. Whereas such diagnostic systems, or 'keys', guided naturalists 'along their way' in identifying organisms with known kinds by

Figure 12.1 Frontispiece of the first volume of *Caroli Linnaei ... Systema Naturae*, edited by J. J. Lang (Halle, 1760). The frontispiece shows a statue of Diana, and is a variation of the frontispiece of Linnaeus's *Fauna suecica* (Leiden, 1746), adding a human figure taking notes and pointing to a monkey in the top of the tree to the right. The heading refers to 'numbers and names' (*numeros et nomina*) as essential elements of Linnaean natural history. Courtesy Uppsala University Library.

applying progressively narrower sets of criteria, they could not stake out the 'borders' of these kinds, i.e., circumscribe them as groups or collections of organisms.[20] The identity of the taxa constituting the Linnaean hierarchy was determined not by any particular difference they happened to exhibit with respect to other taxa, but by what they contained, and came to contain – each genus containing a particular set of species, each order a particular set of genera, and so on.

In short, Linnaean names served a mere indexing function, like labels, while the Linnaean hierarchy simply provided a nested set of containers, of 'boxes within boxes'.[21] To gain a better understanding of how they facilitated communication among naturalists, it is useful to look at the role Linnaean names and taxa played in the creation of paper tools – the devices made from paper and ink, whether in manuscript or print, that were employed in practices of extracting, storing and processing written information, including note-taking, listing, cataloguing or tabulating.[22] Up to the early eighteenth century, the predominant forms of scholarly annotation had been marginalia and topically organised commonplace books, that is, media that tended to fix information in relation to a relevant (con)text.[23] The late seventeenth and eighteenth centuries witnessed a transition to more flexible paper tools, like loose files and card catalogues, and to more complex techniques of extracting, rearranging and displaying information, like forms, tables, diagrams and maps, often deployed for highly idiosyncratic purposes.[24]

Linnaeus participated in this transition, experimenting throughout his career with a diversity of annotation and filing systems, various forms of lists and tables, and, towards the end of his life, with paper slips that resemble index cards. In all of these media, Linnaean taxa carved out an allocated paper space – whether on the printed pages of a book, in a handwritten list or table, in the form of a file produced from folded paper sheets, or by cutting paper into small slips of a standard size – that was labelled with the name of a genus or species and then used to collect pieces of information under that name. Since names served as mere indexing devices, the resulting packages of information could be freely extracted from their context, and their contents inserted, or even redistributed, elsewhere, without losing their identity as long as the labels remained attached.[25] As early as 1737, Linnaeus used a remarkable metaphor to explain the role of generic names as mere labels: 'The generic name has the same value on the market of botany, as the coin has in the commonwealth, which is accepted at a certain price – without needing a metallurgical assay – and is received by others on a daily basis, as long as it has become known in the commonwealth.'[26] This metaphor clearly implies that Linnaean names and ranks derived their value by serving as a material vector for exchanging and accumulating information, rather than from any intrinsic meaning. *Species plantarum, Genera plantarum* and *Systema naturae* were designed to serve as templates for communal annotation, whether this took the form of creating a numbered list of named specimens sent to a correspondent, or whether an interleaved copy of one of these works was used to absorb new observations gathered from reading the latest literature, from a letter received, or in the field. Linnaeus himself employed copies of his own publications exactly for this purpose throughout his career (Figure 12.2). He was thus able to churn out one edition after another on the basis of information he received from correspondents and travelling students. There is growing evidence that other naturalists used Linnaeus's

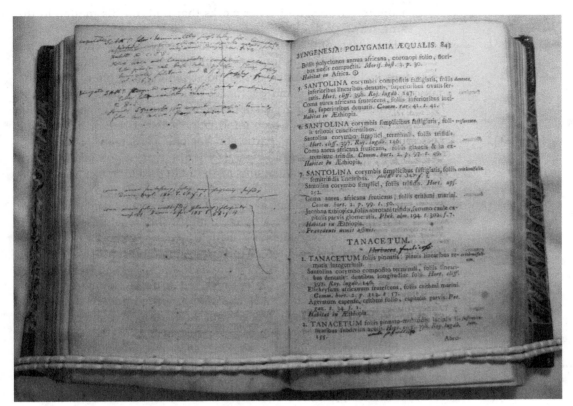

Figure 12.2 Page from Carl Linnaeus's own interleaved and annotated copy of his *Species plantarum* (Stockholm, 1753). Note that each entry for a species in the printed text follows a strict layout and order, and occupies roughly the same space. Each entry starts with a serial number and the genus name (in capitals), followed by a short diagnostic phrase distinguishing the species from other species of the same genus as well as references to works (in italics) that mention the species under that name. The genus name and the diagnostic phrase together form what Linnaeus called the 'legitimate' name of species. The subsequent paragraphs list 'synonyms', that is, alternative legitimate names under which the species was treated in the literature, again with references in italics. Most entries then end with a few short notes, partly employing symbols, containing additional information on geographic distribution, ecological habitat, life cycle, and taxonomic position. On the margin, against the first line of each entry, the specific epithet is noted in italics, which forms the 'trivial' name of each species together with the genus name. In addition, the text is structured by headings naming the genus. The page header spells out the class and order in Linnaeus's sexual system to which the genera treated on the page belong. Linnaeus's annotations were made in the preparation of a new edition of *Species plantarum* and include

(*cont. on next page*)

taxonomic works, and other works that built on his model, in the same way.[27]

The effects of the 'Linnaean reform' are evident in changes in the ways naturalists communicated with one another in letters. Rather than including elaborate descriptions, letters were often reduced to mere lists of names and references to Linnaeus's works, whether in instructing a correspondent or travelling naturalist about the minerals, plants or animals on which information was desired, or in reporting observations from a particular region, garden or collection.[28] The ready availability of Linnaeus's books made it possible to reduce observations to emendations of or additions to those works; the reporting of errors became 'an essential component of the correspondence routine', as Bettina Dietz has argued, and was notably not perceived as an attack on the authority of the addressees of such reports.[29] The overwhelming feedback that Linnaeus's own works generated in this form forced the Swedish naturalist to make recourse to an elaborate system of paper tools towards the end of his life; he would routinely reduce letters received to long, numbered lists of provisional binomial names, and then use the numbers and names to cross-reference the letters with herbarium specimens he had received, with his own manuscript notes on species that, at this stage in his career, he kept on small paper slips of standard size, and lastly with his taxonomic publications, which no longer appeared in the form of grand editions of *Systema Naturae* or *Genera plantarum*, but rather as supplements to these works, entitled *Mantissa*, a Latin word for makeweight, a small or worthless addition serving only to restore balance.[30]

Paper empires

The fact that Linnaeus's taxonomic publications were most useful to naturalists because of the format in which they presented natural

---

Figure 12.2 (*cont.*)   corrections and short additions that are entered directly in the printed text, and entries for new species and synonyms on the facing page. Note that the latter emulate the typographic layout of the printed text, and that the position where they are to be inserted is either directly indicated by their position on the page or with the help of a drawn line. Interestingly, in this case Linnaeus seems to have decided that two synonyms employing the same genus name, 'Coma', are to be distributed into two different genera ('Santolina' and 'Tanacetum'). Linnean Society, London, Library and Archives, Linnaean Collections, call no. BL83. Courtesy Linnean Society of London.

historical knowledge explains one curious aspect of the many posthumous editions. Strictly speaking, these were not new editions or
translations at all, but rather *continuations* of Linnaeus's taxonomic
project, which incorporated new observations.[31] Many of the editors
of these works pointed this out explicitly. Thus, the Dutch physician
and naturalist Martinus Houttuyn stated in the preface to his
Natuurlyke Historie (1761–85) that he had adopted Linnaeus's 'system'
and 'Latin bynames', i.e. Linnaeus's binominal or 'trivial' names. But
he also stated that he had inserted information from publications by
naturalists like Buffon or Jacob Theodor Klein, whose works rivalled
those of Linnaeus in their scope and authority.[32] The Frisian theolo-
gian and philosopher Philipp Ludwig Statius Müller, teaching cam-
eralism and natural science at the University of Erlangen in Bavaria,
made similar remarks in the preface to his German edition of *Systema
naturae*, warning his readers in the very first sentence that they
should 'not expect a translation', and then detailing his sources,
notably Houttuyn's Dutch edition of *Systema naturae*, but above all
the growing number of journal articles in natural history.[33] In the
preface to a supplementary volume, which appeared in 1776, Müller
even asked his readers to report any new discoveries directly to him
by providing, at the least, a short description and indication of the
new species' taxonomic position.[34] Müller unfortunately died before
this supplementary volume appeared in print, but its publisher,
Gabriel Nicolaus Raspe, explained in a short note that he was keen
to continue the project.[35] And indeed, between 1777 and 1779, Johann
Friedrich Gmelin (1748–1804), Professor of Medicine at the University
of Göttingen, edited four more volumes covering the mineral kingdom for Raspe, which were reprinted in 1785. Thirteen volumes covering botany were edited by two other naturalists and appeared
between 1777 and 1788.[36]

One can see from this short sketch that translations and editions
of Linnaeus's *Systema naturae* were products of complex, long-
term paper work, not only because they often built on one another,
rather than directly on Linnaeus's own publications (Figure 12.3),
but also because their authors relied on a wide array of additional
written sources – other general works in natural history, local floras
and faunas, journal articles and letters from correspondents – to
include the latest discoveries. Thus Gmelin acknowledged his
material debt to 129 named naturalists in the preface of his own
'thirteenth' edition of Linnaeus's *Systema naturae* (1788–93).[37]
Müller coined a revealing expression for the unflagging compila-
tory activity that lay behind such works. In advertising his supplementary volume, he emphasised that 'all *Addenda*, *Appendices* and
Mantissae of Herr von Linné have been properly slotted in (*gehörig*

Figure 12.3 Frontispiece and title page of *Des Ritters Carl von Linné Lehr-Buch über das Natur-System* (Nürnberg, 1781). This was a shortened version of Philipp Ludwig Statius Müller's seven-volume *Des Ritters Carl von Linné vollständiges Natursystem* (Nürnberg, 1773–6) prepared by the clergyman Jeremias Höslin (1722–89) 'for everybody, rather than scholars', as it says in the preface. The inscriptions on the large volume and plaque held by a putto illustrate the succession of publications that the book builds on: 'Linnaeus composuit' points to the Swedish naturalist's tenth edition of *Systema naturae*, 'Houttuÿnius explicavit' to Houttuyn's expanded 'translation' of the tenth edition (started in 1761 and still ongoing at the time), and 'Mullerus ad Ed. XII reformavit' to Müller's attempt to provide a synthesis of Houttuyn's edition, Linnaeus's own twelfth edition of *Systema naturae* (1766–8), as well as other works. Courtesy Staatsbibliothek zu Berlin – PK | Abteilung Historische Drucke | Signatur: Le 2019.

eingeschaltet)', and that the same had been done for new species reported by other naturalists.[38]

Einschalten is a verb with overtones of mechanical or bureau-cratic labour, and simply means to insert an object into a pre-existing series of other objects.[39] It thus vividly expresses how easy it had become to compile data on plant and animal species since the Linnaean reform. Linnaean names and taxa empowered even those naturalists who, like Müller in Erlangen, were situated in peripheral contexts to build their own 'paper empires' on the basis of purely derivative literary techniques like extraction, com-pilation or rearrangement of names and accompanying descrip-tions. It is in this context, it seems, that even the more esoteric paper tools Linnaeus had developed, like the index cards he used in later life, began to spread, finding their way, for example, to the first two 'keepers' of botany at the British Museum, Linnaeus's own student Daniel Solander and Robert Brown.[40]

In order to demonstrate in more detail how Linnaean paper tools enhanced exchange, and especially, how they shaped naturalists' perceptions both of their own status and of the order of nature, I want to turn to a final example on a much smaller scale than those discussed so far. In 1768, the German naturalist Johann Reinhold Forster was commissioned to produce a volume on insects for Thomas Pennant's multivolume *British Zoology*. Just two years later, he published a curious first fruit of his labours, entitled *A Catalogue of British Insects* (Figure 12.4). It consisted of a list of slightly over 1,000 trivial names for insect species, neatly numbered consecutively, both overall and within each genus, and structured by headings giving the names of the genera to which the species belonged, which again were consecutively numbered. The overall purpose of the catalogue, as well as the meaning of the abbreviations set against many of its entries, were succinctly explained in the preface:

The author of this catalogue intends to publish a *Fauna of British Insects*; and as he thinks not to set out upon it, till he can offer to the public a work, as little imperfect as possible, and to give no other descriptions than from ocular inspection: he presents his most respectful compliments to all ladies and gentlemen who collect insects, and begs them to favour him, if possible, with specimens of such insects, as they can spare, and which he is not possessed of: for this purpose he has made this catalogue, and put no mark to the insects in his possession; those which he has so plentifully as to be enabled to give some of them to other collectors, are marked with a (*d*); those which he has not, are marked either *Berk.* signifying *Dr. Berkenhout's Outlines of the Natural History of Great*

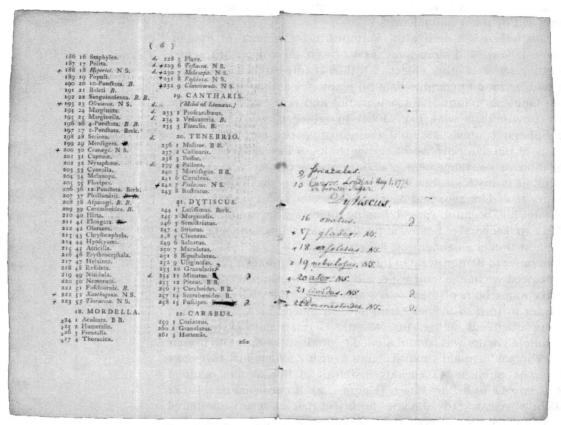

Figure 12.4 Two pages from the author's own annotated copy of Johann Reinhold Forster, *A Catalogue of British Insects* (Warrington, 1770). The printed text lists genera and species of insects, employing Linnaean trivial names. The notes document additional species that Forster came across after publication, many of them marked as new species ('N. S.'), and in one case reporting when and where the species was found: '10. [Tenebrio] Cursor. Londini Aug 1. 1771. in brown sugar'. The latter remark is probably referring to a beetle from Florida that later became known under that name among entomologists and which had been found in a shipment of sugar. See text for further discussion. Courtesy Staatsbibliothek zu Berlin – PK | Abteilung Historische Drucke | Signatur: Lt 12373.

Britain; or *B.* signifying a manuscript catalogue of *British Insects* communicated to the author; or *B. B.* which signifies *Berkenhout*, together with the manuscript catalogue. *N. S.* is put to such insects as have not yet been described by Dr. *Linnaeus*, and are *new species* with new specific names.[41]

At a glance, then, Forster's catalogue informed its readers how many species and genera of British insects were known to him, which of these he possessed in abundance (the 'd' probably standing for 'dupli-cate'), and which he was still looking for.[42] The catalogue was an open invitation to enter into an exchange, and this tactical move was appar-ently successful. The Staatsbibliothek Berlin holds an interleaved copy of the *Catalogue* in which Forster carefully noted additions to his collection, by deleting the abbreviations '*Berk.*', '*B.*' or '*B.B.*', by adding '*d.*', or by noting additional species names on the interleaves, often followed by '*N. S.*' and/or '*d.*'. A note on the flyleaf of this copy states 'Aug. ye 28. 1771. 42 more insects', and a calculation at the very end of the catalogue registers '43 additional Insects' below the 1,004 already listed, and proudly draws up a new sum total of 1,047.[43]

Forster's *Catalogue*, with its extreme reduction of content to species names arranged according to the Linnaean hierarchy, illus-trates the degree to which classical natural history was dominated by concerns with naturalists' position within the 'marketplace' of natural history. It is a document of book-keeping in an almost literal sense, recording credits and debits on the level of Forster's personal collection, and at the same time representing the British insect fauna itself in terms of abundance and scarcity of species. With regard to the latter, Forster's *Catalogue* shows striking structural similarities with what is certainly one of the most intriguing visual representations in late eighteenth-century natural history, the 'genealogical-geographical table of plant affinities' (*Tabula genea-logico-geographica affinitatum plantarum*) that Paul Dietrich Giesecke produced on the basis of notes taken during private lec-tures he received from Linnaeus (Figure 12.5). The table represents the plant kingdom in the form of fifty-eight circles of varying size, distributed over the sheet in an unruly manner, a little bit like an archipelago. The explanations that accompanied the table indeed speak of a 'map', and of the circles as 'provinces' or 'islands', each of them standing for a particular 'natural order' of plants, their sizes corresponding to the number of genera they contained, and their mutual positions expressing relations of 'affinity'.[44] The aims of Linnaeus's speculations about a 'natural' plant system may have been loftier than those of Forster's *Catalogue*, but his manuscript explorations of plant affinities took exactly the same form of num-bered lists structured by headings, and were certainly of equal strategic importance in his dealings with other plant collectors.[45]

Conclusion

It should be emphasised that the evidence that I have used in this chapter about the use of paper tools in natural history is still very

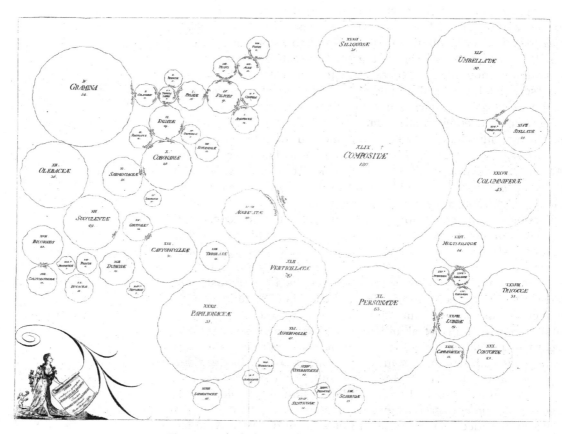

Figure 12.5 'Tabula genealogico-geographica Affinitatum Plantarum' from
Caroli Linnaei Praelectiones in ordines naturales plantarum, edited by
P. D. Giseke (Hamburg, 1790). The circles represent 'natural orders' or plant
families, their size the number of genera they contain, which is also noted in
the centre of each of them, alongside the family's name and Roman numeral.
The relative position of each circle indicates its taxonomic relationship with
other families, sometimes highlighted by inscribing the names of closely
related genera on the inside of circles where they approach each other.
Courtesy Uppsala University Library.

patchy, and that the basis for sweeping generalisations is slim. More
research needs to be done to understand how exactly natural history
knowledge was made to circulate globally, how it was stored in
metropolitan institutions and individual collections, and how it was
retrieved time and again to be redeployed in new contexts. I hope
I have demonstrated, however, that the study of the tools and

infrastructures of classical natural history's 'information economy' promises to advance our understanding of this crucial period in the history of the life sciences. It is well known that the irregular patterns of species distribution that we saw emerging, on a small scale, in Forster's *Catalogue*, and, on a larger scale, in Linnaeus's attempts to chart out the 'natural affinities' among plants, inspired Alexander von Humboldt, Augustin Pyramus de Candolle and Charles Lyell in the early nineteenth century to concede the possibility that species had been created independently of each other, at different times and places, and enjoyed differential success in the 'struggle for life', and that the same patterns also formed one of the chief explananda of Darwin's theory of evolution by natural selection. Tables, charts and maps of increasingly complex format played a crucial role in these explorations.[46] At the same time, however, naturalists used these tools to meticulously take account of their own achievements and those of others, hence articulating themselves as a recognised, and recognisable, community with new professional roles and regulated ways of distributing credit.

The infrastructure of 'labels' and 'containers' that was created by Linnaean paper tools, in other words, began to acquire a life of its own, both as a research object, revealing phenomena that could not have been revealed without it, and as a social instrument, articulating communities that would not have formed without it. The urge to document everything in writing, some anthropologists have claimed, is at the heart of the difference between 'the West' and 'the Rest'.[47] A cultural history of natural history that attends to this aspect of scientific practice, therefore, has the potential to reveal that natural history, its techniques, and the affects and aspirations associated with these, form a central element of modernity despite their enduring antiquarian image.

Further reading

Blair, A., 'Note taking as an art of transmission', *Critical Inquiry*, 31 (2004), pp. 85–107.

Daston, L., 'Taking note(s)', *Isis*, 95 (2004), pp. 443–8.

Delbourgo, J. and Müller-Wille, S., 'Introduction to focus section "Listmania"', *Isis*, 103 (2012), pp. 710–15.

Krämer, F., 'Ulisse Aldrovandi's *Pandechion Epistemonicon* and the use of paper technology in Renaissance natural history', *Early Science and Medicine*, 19 (2014), pp. 398–423.

McOuat, G. R., 'Cataloguing power: delineating "competent naturalists" and the meaning of species in the British Museum', *British Journal for the History of Science*, 34 (2001), pp. 1–28.

te Heesen, A., *The World in a Box: The Story of an Eighteenth-Century Picture Encyclopedia* (Chicago, 2002).

Thomas, J. M., 'The documentation of the British Museum's natural history collections, 1760–1836', *Archives of Natural History*, 39 (2012), pp. 111–25.

13   Image and nature

In 1731, the town physician and naturalist Christoph Jacob Trew in Nuremberg heard, through a former student of his, of a gardener in the nearby town of Regensburg. This gardener's name was Georg Dionysius Ehret, and he was apparently very talented in drawing plants and flowers. Eventually, he would become the most famous botanical draughtsman of his time. Trew was a great enthusiast of natural history images and devoted a large part of his money, time and attention to collecting and publishing them. He became interested in Ehret's drawings, and eventually promised to buy two or three pieces per week – provided these drawings met Trew's expectations as a naturalist. His student was to inform the gardener-turned-draughtsman that Trew attached 'great importance to everything being drawn true to nature'; since Trew wanted it 'not only for decoration but also for practical purposes'.[1] A little later, Trew transmitted more detailed instructions as to how he wanted the images to be done. Purely aesthetic questions, such as the design of the background, he left to Ehret's discretion. Trew underlined, however, that he wanted 'nature to be expressed as clearly as possible'; he requested that details of the fruit and seeds were added; and he wanted each species drawn on an individual sheet of paper – a necessary precondition if the images were to form part of a systematic collection. Plants decoratively intertwined, Trew added, were unacceptable, even though this was a popular method of composition in the eighteenth century and one which Ehret would later use in a more popular work with great skill and success (Figure 13.1).

This episode epitomises the main themes of this chapter. Images of nature were central elements of natural historical practice in Europe. Watercolour drawings formed part of naturalists' collections, and circulated among experts and connoisseurs alongside seeds, plant specimens and other natural historical objects (on these exchange networks, see also Müller-Wille, this volume, Chapter 12). In order to reach a broader audience they were published, usually in the form of hand-coloured copperplates. These plates were highly aesthetic pieces, eloquently praised for their beauty. Yet at the same time, they were expected to conform to scholarly conventions, as Trew's detailed

Figure 13.1 Two species intertwined, the glade mallow and the morning glory; drawn, etched and published by Georg Dionysius Ehret (Tab. VII, *Plantae et Papiliones Rariores*, 1748). Image from the Biodiversity Heritage Library. Digitised by National Agricultural Library (www.biodiversitylibrary.org).

letter of instruction made clear. Images were an integral part of taxonomic research and botanical education; they were used to identify unknown species and to facilitate the recognition of others. They contributed in important ways to the increasingly professionalised training of experts in pharmacy, forestry and agriculture; and they served as references in scholarly debates. This chapter explores what these images actually contained; which functions they served; and how they were made, in a complex collaboration between naturalists, artists and craftsmen.

The contents of botanical images

Eighteenth-century botanical illustrations were presented to the public in many formats and with widely divergent audiences in mind, ranging from comparatively modest manuals to splendid volumes of elephant folio size, from compendia of pharmaceutically useful species to monographs of certain plant families and multivolume floras that documented the vegetation of regions, countries or even empires. Most of these works, however, had in common a composition consisting of full-page illustrations in the form of copperplates, with little accompanying text.[2]

Naturalists found these images essential to their work. Already, at the end of the seventeenth century, John Ray thought that most people 'looked upon a history of plants without figures as a book of geography without maps'.[3] In a similar vein, the Swabian botanist Johann Simon Kerner declared in 1786 that 'accurate illustrations by an expert are instructive, help settle doubts and grant certainty'.[4] In his *Vegetable System* (1777), John Miller explicitly stressed 'the Insufficiency of the most accurate verbal Description for conveying to the Mind an adequate Image of the infinitely various objects in the vegetable Kingdom'.[5] Images were considered indispensable by these authors if one wanted to acquire sound knowledge of botany. The eminent Carl Linnaeus is known to have dismissed the use of images (although he conceded that they might be 'of great importance for boys and to those who have more brain-pan than brain'); but in the pertinent paragraph of his *Genera plantarum*, Linnaeus only dismissed the use of images 'for determining genera', in the way his French colleague and rival Joseph Pitton de Tournefort had attempted earlier. In his *Institutiones rei herbariae*, Tournefort had introduced each genus through the depiction of one of its species. This Linnaeus found inadequate for the discussion of genera.[6] When Linnaeus introduced new *species*, in contrast, he usually had his description accompanied by an illustration.[7]

In order to understand why these images were so highly valued, it is worth looking at an example, such as the image of the sweet vernal grass, a common European grass species known for the high nutritional quality of its hay (Figure 13.2). The illustration appeared in 1798 in the first comprehensive flora of the German countries, edited and published by the Nuremberg engraver Jakob Sturm, who was interested in botany. The finished work comprised thirty-six volumes and

(a)

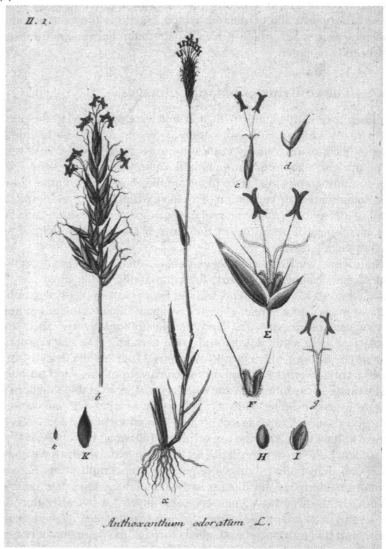

Figure 13.2 Sweet vernal grass (*Anthoxanthum odoratum* L.). (a) The version drawn and published by Jakob Sturm, *Deutschlands Flora in Abbildungen nach der Natur mit Beschreibungen* (Nuremberg, 1798–1855). (b) Johann Philipp Sandberger's partially copied version (Museum Wiesbaden, Germany). By permission of the SUB Göttingen and Museum Wiesbaden.

(b)

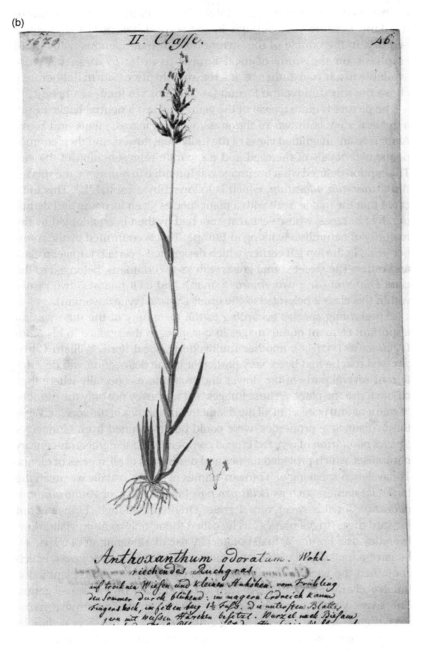

Figure 13.2 *(cont.)*

no less than 2,472 engravings in total. Multivolume, illustrated floras of this kind became very popular in the second half of the eighteenth century, in the course of the 'invention of the indigenous' which laid emphasis on the study of local floras.[8] In order to make his work available to a broad audience at a reasonable price, Sturm deliberately chose the tiny duodecimo format (13–15 cm or 5–6 inches in height).

The picture is quite typical of the genre. Against a neutral background, a general view is drawn of the grass, with its leaves, stems and roots. Alongside are magnified views of the individual flowers and their components, two details of the seed and a separate representation of the ear. The caption defined what the image was intended to represent: the species *Anthoxanthum odoratum*, which is followed by a capital 'L.'. This indicated that the image dealt with a plant species given its name and definition by Linnaeus, whose sexual system had by then been adopted by the majority of naturalists working in Europe. This is confirmed by the number 'II.2.', in the top left corner, which designated a certain Linnaean class and order. The sweet vernal grass, with its two stamens, belonged to the class *Diandria* (*di* = two; *andros* = man); and as it had also two pistils, within this class it belonged to the order *Digynia* (*gyne* = woman).

Representing species according to the taxonomy of the time was an important element of the images in question. In the preface to his *Flora Londinensis* (1777–98), another multivolume, local flora, William Curtis stressed that he had been 'very particular in the delineation and description of several parts of the flower and fruit, more especially where they characterise the plant'.[9] These images had to convey not only the number of stamens and pistils, but all the distinctive properties of the species. What these 'distinctive properties' were could be determined from Linnaeus's *Species plantarum* of 1753, the crucial compendium for eighteenth-century naturalists, which provided names and definitions of all species of plants. Today, when speaking of 'Linnean' names of plants, usually we mean the binomial names, such as *Bellis perennis* for the daisy, or *Anthoxanthum odoratum* for the sweet vernal grass. Originally, though, Linnaeus put forward these 'trivial names', as he called them, only as abbreviations for everyday use. For the actual, botanically useful, designation of a species, he gave a long descriptive name, the so-called *nomen specificum*, which contained the generic name plus a short phrase giving the plant's distinctive characters compared with other species in the same genus. The sweet vernal grass, for example, was given the descriptive name *Vernal grass with an ovoid-oblong spike and florets on short peduncles that are longer than the awn.*[10] All components of this definition are clearly depicted in Sturm's image: the spike is shown twice and its 'ovoid-oblong' shape is particularly obvious from the general view (detail *a*). (In fact, it is almost a caricature of an egg-shaped spike.) The magnified view of the inflorescence (detail *b*) shows the short stalks of the individual florets, while the proportion of the flower organs is best seen in details *E* and *g*: the floret is indeed shown as considerably longer than the two awns.

But the image also contained features of the species without taxonomic value. For example, it also shows the furry surface of the stigmata and the bifurcated x-shape of the anthers; and it demonstrates that the stigmata grow in the shape of a twisted moustache. The images served not only as taxonomical tools, but also as comprehensive descriptions of the species, including all its typical properties. Deciding whether or not properties were 'typical', however, was no trivial exercise – it required far-reaching botanical knowledge. The highest praise for images of this type was that they were drawn 'true to nature', just as Trew demanded of Ehret. This injunction is not only to be understood as the requirement that the image give the appearance of a plant as it was found in nature. It also implied that the image should represent the *inner nature* of the species. And this achievement, Daston and Galison have argued, required the judgement of an expert who had examined many exemplars. The image was then expected to present the distillation of all these observations.[11]

In this respect, botanical images differed fundamentally from the way species were represented in a *Herbarium vivum*, another form of preserving plants that had become essential to eighteenth-century naturalists. The dried and pressed plant specimens in a herbarium exemplified their species in a material sense, as pieces of nature, and they were prominently recommended by Linnaeus as foundational for botanical work. However, they did not usually display all the typical properties of the species in one exemplar – and in this sense, paradoxically, were not 'true to nature', in the botanists' sense. Furthermore, certain properties were lost in the preservation process. The renowned French draughtsman Pierre Joseph Redouté, for example, chose to publish his beautifully illustrated work on the *Liliaceae* because their specific features were particularly difficult to preserve.[12] Many had voluminous bulbs and large flowers, which were very hard to press and dry without losing their character. An additional advantage to printed illustrations was their longevity, praised, for example, by Kerner, who stressed this property of images in contrast to a herbarium, 'which is doomed to fall apart eventually'.[13]

Functions and uses of botanical images

Nevertheless, botanical images were drawn in such a way that they actually resembled herbarium sheets: a decontextualised specimen against a neutral background. Linnaeus had recommended in the *Philosophia botanica* that specimens of only one species should be mounted on a single herbarium sheet. Trew demanded the same for botanical images, as he made clear in his instructions for Ehret. Like herbarium sheets, these images were usually integrated into a collection system, they were exchanged and circulated through correspondence, and they were commented upon with regard to their beauty and botanical value. The images of exotic plants were also occasionally used by naturalists as a proxy for actual specimens to which they lacked access – even Linnaeus based some of his species definitions on images, because no material exemplar had been available to him.

Images were also used as observational evidence. This was particularly relevant in contributions to scholarly journals that reported the finding of a new species or an unusual variant. In such cases, the text would be accompanied by an image that confirmed the accuracy of the description. In 1746, for example, Trew wrote to his colleague Albrecht von Haller in Bern: '*HofRath* Schmiedel from Erlangen sent me a curious note, together with a meticulous drawing, which testifies that in his observations of the *Elatine* he saw the same metamorphosis of the flower that Dr. Linnaeus noticed in the Linaria'.[14] Trew wanted to publish this unusual observation in his scholarly journal and was obviously convinced that the drawing faithfully documented Schmiedel's finding. Others were more sceptical, and frequently societies would ask for the specimen itself to be supplied in addition to its drawing, particularly in the case of contributions from non-members. The correspondence between Linnaeus in Sweden and the Scottish naturalist David Skene serves as another example of this practice, albeit in the context of a private scholarly dispute.[15] In their conversation on certain properties of a group of zoophytes, both repeatedly referred to plates in the standard reference work on this group of organisms published by John Ellis. In the end, however, they were unable to resolve their disagreement by using the images, and Skene tried to convince Linnaeus by sending him alcohol-preserved preparations of the species instead. The evidential value of images obviously had its limits.

If we now redirect our view from the individual images to the botanical works in which many of them were published, yet other functions and uses come to the fore. The authors of many of the illustrated eighteenth-century works thought that their volumes would provide a thorough introduction to studying the plants of a specific region or of a particular economic value. Many of these works aimed at secular improvement in fields such as agriculture, forestry and pharmacy, and they presented the local flora of a region for purposes of utility, pleasure and religious contemplation. The aim of Curtis's *Flora Londinensis*, for example, was 'to facilitate a knowledge of the plants of our own country, and establish each species and variety on a firm basis'; to which Curtis added: 'This arduous task once accomplished, a way will be opened, and a foundation laid for numberless improvements in Medicine, Agriculture, &c.'[16] It is telling that these usually were written in the vernacular, in contrast to more scholarly treatises, which were still being published in Latin.

The more luxuriously illustrated works were also of value as objects of prestige. The demonstration of national power and pride was a major incentive for public or private patrons to underwrite the high costs incurred during the production of illustrated botanical works. Redouté's work on the *Liliaceae* was not only botanically useful, it was also intended to demonstrate the supreme quality and extraordinary beauty of French art, nature and culture, and it was funded to that end by the empire. Napoleon's home secretary ordered no fewer than eighty copies of the work, while the secretary of state Charles-Maurice de Talleyrand reserved a substantial number to be given as presents to foreign statesmen.[17] Another impressive example of a project of national importance was the monumental *Flora Batava*, produced under the supervision of the naturalist Jan Kops after Napoleon's conquest of the Netherlands. The work was intended to document the floral abundance of the whole 'Batavian Republic', as the Netherlands were called in this period, in twenty-eight volumes. The project was initiated in 1800, but it was only finished in 1934, under completely different circumstances. In the same vein, the lavish *Flora Danica* was published in seventeen volumes between 1766 and 1883. It was begun under the authorship of the botanist Georg Christian Oeder, then royal professor and director of the botanical garden at Copenhagen, and was meant to cover all territories of the Danish kingdom. The territories, however, changed considerably before the work was finished. The double monarchy of Denmark–Norway was abolished in 1814, so that Norwegian plants were left out after this date, and similar changes became necessary after 1864, when Schleswig-Holstein became part of Prussia. These floras, hence, quite articulately mirrored in their choice of species the political circumstances of the time.

Collaboration between botanists, draughtsmen, engravers

The production of these ambitious works required the collaboration of several protagonists with complementary expertise, all of whom contributed to the success or failure of the project. Linnaeus saw this clearly when he wrote in his *Philosophia botanica*: 'A draughtsman, an engraver & a botanist are equally necessary to produce a praiseworthy image; if one of these is at fault, the image will be flawed'. But he was obviously sceptical of this arrangement, for he continued: 'This is why those botanists who were not only good at their discipline but also skilled in drawing and engraving have left us the most outstanding images'.[18]

In writing these lines, Linnaeus was probably thinking of his colleague Johann Jacob Dillenius, who himself had both drawn and engraved the 324 plates for his *Hortus Elthamensis* of 1732.[19] Most botanists, however, did not find their drawing skills up to the task, and sought to collaborate with talented draughtsmen. Under favourable circumstances, these collaborations could produce outstanding results which none of the protagonists could have obtained on their own. Prominent examples were the working relationships between Tournefort and the botanical painter Claude Aubriet, at the Jardin du Roi in Paris, and the British analogue: Sir Joseph Banks and Franz Bauer, 'Botanick Painter to His Majesty' at Kew Gardens. Most botanists, however, were unable to secure a permanent position for a draughtsman and had to find other ways to work with skilled artists on their projects. One of the most successful naturalists in this respect was the aforementioned Christoph Jacob Trew. Starting in the 1730s, he systematically collaborated on several natural history projects with talented draughtsmen and engravers.[20] Nuremberg was an ideal place to do so. From the Middle Ages, it had been one of the most important trading towns on the route from Italy to northern Europe, and from Flanders to the East. It hosted an Academy of Fine Arts, and scholarly printers and publishers had a long history there: it was here that the first printed star charts were made by Albrecht Dürer in 1515. In the eighteenth century, Nuremberg came to be known, in part thanks to Trew's activities, as a centre of natural history illustration, whose artists worked all over Europe. As mentioned above, Jakob Sturm's flora of the German lands was also produced in Nuremberg.

One of the problems that Trew and others had to solve was helping the draughtsmen understand the botany behind the images and master the techniques involved in drawing details to the level of accuracy shown in Figure 13.2, such as dissecting the tiny flowers of the sweet vernal grass and identifying all their elements. Ehret received hands-on training in this respect, first from Trew, when Ehret was visiting him in Nuremberg, and then from Linnaeus himself, when the two of them collaborated on the *Hortus Cliffortianus* (1737), a book documenting the botanical wealth of the Amsterdam banker George Clifford. In later years, Ehret would examine the details of the plants he drew under the microscope; his contemporaries attributed his increasingly poor eyesight to this exhausting practice. Even then, however, the problem remained that fruit and flower appeared at different times of the year. Ehret solved this problem by using a collage-like style of drawing, with different parts of the picture being done in different seasons, as documented in Plate 7. It is known that Trew completed some of Ehret's drawings with details drawn by others – obviously, it was more important to include this information in the image than to preserve a consistent drawing style.

Other draughtsmen in Trew's circle received a similar education. In his capacity as a town physician, Trew gave public lectures in botany and anatomy that were attended not only by apothecaries, midwives and surgeons, but also by draughtsmen. Trew furthermore trained draughtsmen by having them copy outstanding images by other artists, borrowing the originals from botanist friends. This was standard teaching practice in eighteenth-century art academies and workshops.[21] Copying was considered to educate an artist's visual judgement, manual skills and good taste – in short, it taught the apprentices the common conventions of the genre.[22] A case in point is *A Botanical Drawing Book*, published in 1788 by James Sowerby, one of the most renowned botanical draughtsmen at the time, whose thriving family business in London influenced British natural history illustration for over a hundred years.[23] Interestingly, Sowerby's manual began not with aesthetic principles or drawing techniques, but with a short commentary on the principles of the Linnaean system and an introduction of the seven fructification organs, each with an accompanying plate. The author began with the stamens because, as he put it, they were the 'most simple and easy for Learners to imitate'.[24] The same principle was followed in the rest of the manual: it provided the apprentices with a set of standard examples that beginners were to copy and, thereby, learn the visual language of botany and achieve proficiency in their craft.[25]

Yet, despite all this training, naturalists still closely observed and directed artists' work. Supervising contracted publishers, draughtsmen and engravers, and correcting outlines and proof plates took up a large proportion of the time spent on the production of a natural history work. Where original drawings survive, they frequently show traces of botanists' corrections and comments. The same holds true for the next step in the production of an illustrated work: engraving the plates. Trew, like many of his colleagues, usually collaborated with a local workshop on this. However, for his masterpiece, the *Plantae selectae* (1750–73), a publication of one hundred of Ehret's most beautiful images, Trew decided to work with the engraver and publisher Johann Jacob Haid, who had his workshop in Augsburg, another renowned trading town in the south of Germany, as well as a centre of printing and publishing second only to Nuremberg. Trew would comment twice, with varying levels of criticism, on each test plate that Haid produced. Trew's main concerns were the typical appearance of the plant in its general view, and the accurate execution of the details of flower and fruit. But Trew also took issue if the hand-colouring of the plates – or the 'illumination' as it was known – differed from the original drawing. 'The illuminator must have done this at night, or in his sleep or while drunk, or must suffer from a considerable defect of the eyes', Trew sarcastically snapped of one test plate.[26] As Trew, Haid and others knew, much of the value of the final work depended on the colouring of the plates. However, the latter task was the responsibility of the publisher, who also had to cover the expenses for the work of engravers and printers. Frequently, this was already so expensive that the cost for having the plates coloured was reduced to a minimum, meaning that colouring was usually done by unskilled workers, women and occasionally children.[27]

Given this complex procedure, it is no wonder that the completion of an entire multivolume work often took several decades – unless lack of funding forced the protagonists to abandon the project earlier. In the case of the *Plantae selectae*, for example, Haid became more and more desperate either to terminate the project or to transfer it to another workshop, in the (unsuccessful) attempt to sever the difficult working relationship with Trew. In the end, both author and publisher passed away before the work was finished.

Sources for the botanical images

After this discussion of the full production process, one particular aspect deserves closer attention: the sources for botanical images. Several different types can be identified here. Curtis, for example, emphasised that the plates for his *Flora Londinensis* were 'drawn from living specimens most expressive of the general habit of appearance of the plant as it grows wild'.[28] Living plants were an important source of information, and presumably the large majority of draughtsmen and authors used them in abundance. When drawing an illustration for Watson, Ehret had ordered 'a Quantity of [specimens] placed before him on a table'; but the putrid smell of the plants had made Ehret sick, 'so that … I never after placed before him but a small piece at a time'.[29] Authors also frequently emphasised that they had taken care to have the plants depicted in their natural state. Sometimes plants were even specifically grown for this purpose, in flower beds to which no manure or other forms of fertiliser had ever been applied.

Dried plant material was also made use of, as numerous examples confirm. This was common practice when botanists and draughtsmen returned from field trips and expeditions. In a letter to Joseph Banks, for example, John Sibthorp apologised for not having sent some of the dried specimens that he had collected on his expedition into the eastern Mediterranean in the years 1786–7, since, as he wrote, 'my Painter [is] pleading the necessity of several specimens to compleat his Figures, having taken only rough sketches in the course of our Route'.[30] The splendid plates of the eventual publication, the ten-volume *Flora Graeca,* are in many cases strikingly similar to the dried plants in Sibthorp's herbarium. Yet the draughtsman, Ferdinand Bauer, did not use just one dried specimen for his drawings but rather combined the features of several: the general appearance was taken from one specimen, the shape of the leaves and the organisation of the roots from another; while the illustrated flowers combined different life stages, from the buds to full blossom and the development of the fruit.[31] The same procedure was used for the drawings of Oeder's *Flora Danica,* as the natural historian Johann Beckmann had the opportunity to observe on his journey through Scandinavia in 1765. The species were, Beckmann wrote, 'painted in *loco natali,* however not always the whole plant, but occasionally only the flowers, buds and a leaf, from which the painter afterwards painted the whole plant with the help of a dried specimen'.[32] The Englishman John Hill even devised a special technique, involving heat, water and paste, to revive the original appearance of the living plant in the dried specimens, as he explained in the preface to his *Exotic Botany* (London, 1759). The only problem was that the specimens were destroyed during this procedure, but they were sacrificed to a good cause, Hill assured his readers, and 'their remembrance will live in the designs'.[33]

Dried and living specimens were not the only source materials for plant illustrations. When, in 1740, a specimen of *Camellia mellia* flowered in Britain for the first time, Ehret was engaged to record this rare event in a drawing. In Ehret's sketchbook, next to the outline for this drawing, is a note that reads: 'The fruit is copied from a Chinese drawing which I received from Sir Hans Sloane in 1741'.[34] Earlier illustrations of species were also important sources of information for draughtsmen, and not only in cases such as this, where the Chinese drawing was perhaps the only available reference for the fruit of this plant.

This practice can be observed in more detail starting from the illustration of the sweet vernal grass by Jakob Sturm. In Figure 13.2, this image is shown side by side with a watercolour drawing by the German teacher and botanist Johann Philipp Sandberger, which was executed at the beginning of the nineteenth century and formed part of an extended collection of similar images.[35] Clearly, Sandberger was copying Sturm's earlier image. He took over the grass stalk of Sturm's main view in its entirety, but disregarded the peculiar egg-shaped ear, and reproduced the detail of the inflorescence in its place. Of the other details, Sandberger chose to depict only the individual floret with stamens and stigmas (detail *g*), and he painted the anthers pink, not red like Sturm's.

Copying links of this type between images of the same species were very common in the eighteenth century: botanists and draughtsmen made ample use of earlier representations when they composed their own views of the subject matter.[36] However, this should not lead us to assume that the later images were of lower quality, something which has been claimed of copying practices in earlier periods. Usually, only selected elements were copied by the draughtsmen, sometimes from different earlier versions that were specifically consulted for that purpose. Sandberger, with his almost complete takeover, is the exception rather than the rule. The copied elements were then integrated in a slightly modified form in the new image, and these modifications were not only a matter of style. Sandberger, for example, changed one significant detail of the main view of the plant: he fused the edges of the top leaf of the grass into a sheath, which is very typical for grasses of this family, but was not depicted by Sturm. Thus, Sandberger not only copied Sturm's image, he improved it by introducing an additional property of the species (the sheath). This is typical of copying practice at the time: the copied elements were carefully selected, meticulously scrutinised and corrected if necessary.

This use of earlier images was not mentioned in the text of the book. But the practice was so widespread that it is hardly convincing to think that the authors deliberately tried to conceal it. It rather seems that earlier images were consulted as a matter of course in preparing a new work, in the same way that earlier textual descriptions would be for writing a new description. Kerner, for example, wrote in the preface to his forestry compendium that he had made drawings 'which I always compared with nature and with descriptions by Linnaeus and other good botanical writers, which I constantly had beside me'.[37] There is no reason why Kerner should have ignored the visual elements of the sources he consulted. In some ways, this procedure resembled Linnaeus's concept of 'collating species', that is, it involved assembling the definitive properties of the species from a number of different sources.[38]

Conclusion

As has been shown in this chapter, the analysis of the content, use and production of botanical images provides insight into working practices that were not explicitly discussed at the time. Images were an integral part of botanical education and research. They were praised for their beauty, but had to comply with strict conventions of the genre in order to be useful, something ensured by complex collaboration between botanists, draughtsmen, publishers and engravers, all of whom contributed in crucial ways to the final result. The meticulous 'collation' of these images, which involved copying and integrating visual elements from earlier illustrations into one's own improved representation, underlines the knowledge, sophistication and skill that was required for drawing these images. The eventual illustrations were then either exchanged in correspondence networks or published as full-page copperplates in botanical works, which in the eighteenth century appeared in many formats and were directed at a variety of audiences.

These images embody many aspects that are frequently recognised as characteristic of eighteenth-century natural history: they were part of the extensive system of collection and exchange, which connected naturalists all over the world; they were sites where scholarly conventions, aesthetic preferences and issues of prestige merged; and they were documents attesting to the close collaboration between naturalists, artists and craftsmen that was so typical of early modern natural history. Epistemologically speaking, they were indispensable for the generation, communication and stabilisation of knowledge about nature. Images defined how nature ought to be seen. And it was the nature of these images, pleasant and useful, that made them so successful in this respect.

Further reading

Bleichmar, D., *Visible Empire: Botanical Expeditions and Visual Culture in the Hispanic Enlightenment* (Chicago, 2012).

Blunt, W. and Stearn, W. T., *The Art of Botanical Illustration* (London, 1994).

Charmantier, I., 'Carl Linnaeus and the visual representation of nature', *Historical Studies in the Natural Sciences*, 41 (2011), pp. 365–404.

Cooper, A., *Inventing the Indigenous: Local Knowledge and Natural History in Early Modern Europe* (Cambridge, 2007).

Daston, L. and Galison, P., *Objectivity* (New York, 2007).

Desmond, R., *Great Natural History Books and their Creators* (London, 2003).

Lack, H.-W., *The Bauers. Joseph, Franz & Ferdinand: Masters of Botanical Illustration* (London, 2015).

Nickelsen, K., *Draughtsmen, Botanists and Nature: The Construction of Eighteenth-Century Botanical Illustrations* (Dordrecht, 2006).

14   Botanical conquistadors

Eighteenth-century botanists produced many statements, textual and visual, celebrating their science and detailing their aspirations. Such statements often communicated two central messages: first, that botany was a global practice that concerned itself not only with nearby specimens but even more so with distant and rare ones; and, second, that botany would collect the world's plants and transport them to Europe for enjoyment, use and profit. The vision was thus both global and imperial. Numerous botanical publications from the period, for instance, contain engravings depicting Europe as a botanical monarch who receives floral tribute from other regions of the world (Figure 14.1).

What was the botany of empire in the long eighteenth century? Was all botany imperial at the time, or did something particular distinguish botany practised in imperial contexts from domestic botany, and if so what? How did specific geographies and imperial circumstances impact botany, and what is to be gained by examining the subject in a comparative framework?[1]

This chapter addresses the heyday of Spanish imperial botany, describing its aspirations, practices and results. I first introduce five expeditions that explored the flora of the vast Hispanic world between 1777 and 1816. I discuss the imperial motivations for funding these projects, the new institutional and political landscape in which they operated, and how, in the Hispanic world, the very practices that other European naturalists and nations considered innovative were viewed as promisingly retrograde. The travelling naturalists saw themselves as engaged in a botanical *Reconquista*, attempting to remedy imperial decline through a return to a period of imperial glory and expansion in the sixteenth century. The chapter then uses the Spanish expeditions to discuss two aspects that characterised European imperial botany more generally: the challenge of seeing and the challenge of distance. Those who engaged in imperial botany in this period – travelling naturalists and artists, the many locals who aided them in the field and the naturalists, administrators and commercial companies who awaited their findings and shipments back in Europe – faced

Figure 14.1 Frontispiece to C. Linnaeus, *Hortus Cliffortianus* (Amsterdam, 1737). Peter
H. Raven Library, Missouri Botanical Garden.

these two issues, and as a result imperial botany developed techniques to address them.

The promise of imperial botany

Between Charles III's accession to the Spanish throne in 1759 and the Napoleonic invasion of Spain in 1808, almost sixty scientific expeditions travelled through the vast Hispanic Empire. These expeditions addressed scientific, economic, administrative and political goals. Their tasks included investigating the viceroyalties' flora and fauna, exploring imperial frontiers, charting coastlines and producing maps, particularly of lesser known or contested areas, conducting astronomical observations and measurements, and reporting on the political and administrative state of the kingdoms.[2]

Amid this flurry of scientific activity, botanical expeditions held a privileged position. In 1777 a royal order launched the Royal Botanical Expedition to Chile and Peru (1777–88), led by Spanish naturalists Hipólito Ruiz and José Pavón. King Charles III pronounced the expedition:

advisable for my service and for the good of my vassals, not only to promote the progress of the physical sciences, but also to banish doubts and adulterations in matters of medicines, dyes, and other important arts; and to increase commerce; and in order that herbaria and collections of natural products be formed, describing and delineating the plants that are to be found in those fertile dominions of mine; [and] to enrich my natural history cabinet and court botanical garden.[3]

The expedition thus operated in three interrelated domains: taxonomic botany, economic botany and collecting. The first aspect, the 'progress of the physical sciences', refers to surveying and classifying specimens according to Linnaean taxonomy, a task accomplished through the accumulation and study of specimens, written descriptions and illustrations of American flora. The expedition also actively pursued economic botany, seeking to solve controversies regarding *naturalia* with medical and industrial uses and to identify valuable natural commodities. Specific goals included fostering the exploitation of cinchona, the only known reliable antimalarial medicine and a valuable Spanish commercial monopoly, exploring whether natural commodities traded by European competitors, such as coffee, tea, pepper, cinnamon or nutmeg, existed in the viceroyalties and identifying potential replacements for such products. Finally, collections of objects and illustrations would enrich two recently established Madrid institutions, the Royal Botanical Garden (f. 1755) and Royal Natural History Cabinet (f. 1771).

During the following twelve years, comparable orders authorised two other royal botanical expeditions, one to the New Kingdom of Granada (1783–1816) under the direction of José Celestino Mutis, the other to New Spain (1787–1803), led by Martín de Sessé and José Mariano Mociño. In addition, the naval expedition led by Alejandro Malaspina (1789–94) employed botanists who pursued these same goals, and the Spanish physician Juan de Cuéllar conducted botanical investigations in the Philippines (1786–1801). As a group, these five expeditions employed more than fifteen naturalists and about four times as many artists, who worked in a sustained fashion over a period of thirty years on a global mission to investigate the floras of Spain's vast overseas territories in the Americas and the Philippines.

As they explored imperial nature, Hispanic naturalists worked closely with members of the colonial administrative network and also capitalised on the availability of many other individuals who became engaged with their efforts. From town to town in the Americas and the Philippines, a wide range of local inhabitants collaborated with the travellers, including governors, treasury officials, administrators at all levels, physicians, surgeons, pharmacists, clergymen, young students, enthusiasts of natural history and labourers. And as the expeditions travelled throughout the empire, a complex institutional apparatus in Spain and the viceroyalties mobilised continuously for decades, from the early days of organising and funding the expeditions, identifying the appropriate personnel and supplying them with all the necessary equipment, artists and accoutrements, through many years of maintaining active correspondence, to welcoming both the travellers and the images and materials they sent or carried back with them so many years later. The expeditions did not function alone, but rather in concert with imperial and colonial institutions and networks that sustained them, both in the peninsula and throughout the viceroyalties. Expeditions, institutions and administrative networks came together as parts of a complex 'scientific colonial machine' for the exploration, rediscovery and reconquest of the Spanish Indies.[4]

In Spain and the viceroyalties alike, old and new institutions strove to further the useful pursuit of the sciences, technology and industry. Naval academies revamped mathematical and astronomical instruction, while army hospitals and pharmacies strengthened medical and surgical training. In Madrid, the Royal Botanical Garden and Natural History Cabinet – and to a lesser degree the San Fernando Royal Academy of Fine Arts (f. 1744) – worked especially closely with the expeditions. Parallel institutions emerged in the colonies, including botanical gardens in Lima (f. 1778), Mexico City (f. 1788), Guatemala City (f. 1796) and Havana (f. 1816), all of which had direct contact with the expeditions.[5]

Madrid's Royal Botanical Garden was the most active institution in the project to rediscover and reconquer nature in the Hispanic Empire. It directly helped to organise and staff the expeditions, trained many of its members, secured funding through courtly patrons and supervised the naturalists as they travelled. It also received many of the specimens, manuscripts and illustrations that the expeditions gathered or produced. The garden's directors and instructors carried out a major overhaul of Spanish botany, improving the garden's collections and reputation and training a new generation of botanists.

The other Madrid institution that was closely linked to the expeditions was the Royal Natural History Cabinet, whose collections grew from shipments sent by the expeditions and by contributors from all corners of the Spanish Empire. The Cabinet was housed in the same building as the San Fernando Royal Academy of Fine Arts, where many of the expeditions' artists trained. Thus, the objects collected on these voyages were exhibited on the floor above the academy where many of the young artists who participated in them had been trained. This cohabitation of the fine arts and natural history was not coincidental, but rather the result of a Spanish Enlightenment understanding of the practical applications of art. An inscription in Roman script above the entrance to the building makes that point to this day, proclaiming, 'King Charles III united nature and art under one roof for public utility'.[6]

In addition to working with the natural history expeditions, these new institutions also sought to benefit from the widespread imperial administrative network already in place, turning it into a system of collectors and informants. In 1776, Pedro Franco Dávila, director of the Royal Natural History Cabinet, published an *Instrucción* that was widely distributed throughout the peninsula and the viceroyalties. The document outlined in detail the appropriate manner in which to collect, cleanse, preserve, pack and properly archive minerals, animals and plants so that they could be transported to Madrid. The Cabinet received a huge number of contributions from throughout the viceroyalties and Spain, some of which it displayed and some which it exchanged with other European collections.[7]

Casimiro Gómez Ortega, the director of Madrid's botanical garden between 1771 and 1801, followed suit with his own *Instrucción* (1779) for transporting live plants from 'the most distant countries'. The document sought to enlist the eyes and hands of administrators and amateur naturalists throughout the Indies on behalf of the botanical garden, through written instructions and an engraving depicting the custom-built type of crate they should use for transporting live plants. Culled largely from earlier French published instructions, the Spanish publication included an original section that detailed the most desirable plants expected in Madrid from the Indies and provided their Latin and vernacular names, known location and properties. The wish list included cinchona, cinnamon, pepper, cloves and nutmeg – Asian natural commodities that, according to Gómez Ortega, could surely be located in the Spanish Americas.

This utilitarian bent characterised Spanish imperial botany. Gómez Ortega proposed that botany could replace mining in economic benefits, claiming that Spain 'examining its true interests, prefers to the laborious American gold and silver mines other fruits and natural products that are easier to acquire and no less useful in increasing prosperity and wealth'.[8] Plants were not only easier to harvest than minerals, they were also a renewable resource, and one that could ideally grow locally in Spain through transplantation, rather than necessitating import from overseas.[9] 'The vegetable riches of Spanish America', he explained, 'have over the mineral ones the advantage that they can be propagated and multiplied ad infinitum once they are possessed and naturalised' in the peninsula.[10] Botanical exploitation, he promised, would benefit both Spain and its viceroyalties:

It is useless to possess the most benign and fertile territories in the world, if we do not attempt to profit from the natural products that they grant us, extending knowledge and consumption of them within the country, and fostering their extraction through free trade. Without these measures, the most expansive territories become sterile deserts, as useless to their colonists as to the metropole.[11]

The naturalists in the Spanish expeditions shared this belief in the potential economic value and utility of colonial nature. And, like administrators and ministers, they linked new measures to the restoration of past glories. Looking back to the early days of the empire, eighteenth-century Spanish naturalists saw themselves as latter-day botanical conquistadors. In 1777, as the first of the botanical expeditions set out to explore the flora of Chile and Peru, Gómez Ortega suggested to the minister of the Indies that sending twelve naturalists and as many chemists or mineralogists to investigate American nature would yield 'a greater utility than a hundred thousand men fighting to add a province to the Spanish empire'.[12] Twenty years later, Antonio José Cavanilles, his successor at the Royal Botanical Garden, continued to present botanical exploration in the language of conquest, writing: 'Whether from the joy of Botany, or the conviction of the utility that it brings to States, each day new supporters enlisted under her flags. Many of them, as conquistadors of vegetal riches, went forth to reconnoitre new countries braving risks and hardship.'[13]

The associations between eighteenth-century imperial science and a glorious sixteenth-century Spanish imperial past were more than a metaphor. The Enlightenment botanical expeditions were not conceived as radically new ventures, but rather as continuing and extending the work of the humanist physician Francisco Hernández, who between 1570 and 1577 conducted the first European scientific expedition to the Americas. Hernández travelled to New Spain to gather information on New World medicinal practices and products, with a heavy emphasis on botany. In Mexico, Hernández consulted native healers, assembled collections, commissioned drawings of medicinal substances from native artists and drafted a manuscript that he hoped would provide a complete natural history of the Indies, doing for the New World what Pliny had done for the Old.[14] Eighteenth-century Spanish botanists saw themselves as following in Hernández's footsteps, especially after the discovery in the 1770s of a previously unknown copy of his manuscript at the Jesuit College in Madrid. Gómez Ortega used this as the basis for a new three-volume edition of Hernández's work that he published in Madrid in 1790, returning full-circle to the exploration programme undertaken 200 years earlier.[15]

Likewise, the role of imperial institutions in supporting scientific investigations and the appeal to colonial administrators for information were not Enlightenment novelties but rather extensions of long-standing Spanish imperial techniques.[16] In the sixteenth century, new Spanish institutions like the *Casa de la Contratación* (House of Trade, f. 1503) and the Council of the Indies (f. 1524) gathered information about the New World and stimulated natural history investigations and technological innovation in navigation, cartography and cosmography.[17] In 1577, administrators throughout New Spain received a printed fifty-point questionnaire requesting information about the history, natural history, mineral deposits, trade and navigation routes, and landscape of their regions. Versions of these questionnaires, known as the *relaciones geográficas*, made their way across the Atlantic again in 1603, 1743 and also in 1777 – the same year that the Royal Botanical Expedition set out for Chile and Peru, highlighting the continuity between early practices and Enlightenment imperial science.[18]

The eighteenth-century Spanish expeditions tend to be less familiar to Anglophone historians than other voyages from that period. However, this brief overview of their activities demonstrates just how actively the Hispanic world participated in the pursuit of imperial science in general, and imperial botany in particular. In some aspects, Spanish imperial botany is rather different from English or French projects: the Spanish expeditions explored territories that had formed part of the empire for more than two centuries, rather than new frontiers, and thus their goal was not discovery but rediscovery. They tended to last many more years than the voyages organised by other nations, and to draw on extensive and well-established viceregal networks, administrative and otherwise. Most significantly, they occurred at a time of imperial decline rather than expansion, and despite their purported promise to remedy the ailing empire failed to do so. The Spanish example suggests that imperial science could as easily fail as succeed in its political and economic goals, making it impossible to assume a direct relation between natural historical and imperial growth. However, in other aspects there are great similarities between the Spanish expeditions and other voyages, demonstrating a shared pan-European approach to imperial botany. These shared elements are as important to note as the differences, since they suggest that botanists could engage in imperial competition while at the same time collaborating in a shared project of a budding 'global science'. I now turn to two of the central challenges that European imperial botanists faced during the long eighteenth century.

The challenge of seeing

During roughly thirty years of sustained work, the Spanish botanical expeditions produced 12,000 botanical illustrations (see Plate 8). They employed more artists than naturalists, and yielded many more images than textual descriptions, specimen collections, taxonomic classifications or marketable natural commodities. Spanish naturalists went to enormous efforts to employ, train and supervise artists, and frequently addressed visual materials in their writings. And they shared this visual obsession with their European counterparts, who followed similar visual strategies and practices, even when their pictorial output often pales in comparison with the Hispanic missions. Clearly, images were of central importance to imperial botany. But why? The answer to this question is twofold. First, it involves the widespread practices of European natural history, which are addressed in this section; second, the difficulties that distance and mobility posed to imperial natural history, which are discussed in the next section.

Imperial botanists used, produced and circulated images because eighteenth-century European natural history – both in the Spanish Empire and elsewhere, at home or abroad – was a predominantly visual discipline (as discussed in Nickelsen's chapter in this volume). The work of natural history required carefully conducted and disciplined practices of observation and representation, so that multiple observers could look in similar ways and produce understandable findings. Naturalists needed to look together, and to see the same thing. For that reason, botanical eyes were considered specialised instruments, and carefully calibrated through a multimedia training that involved plants, texts and images, and which was designed to develop specialised ways of seeing. Educated as expert observers and collaborating closely with artists, naturalists worked within a visual culture based on standardised ways of viewing nature and on pictorial conventions guiding its depiction. They resorted to images and visual metaphors in research and communication, whether published or manuscript. This visual approach to botanical knowledge was premised on a 'visual epistemology', and involved both the training and the everyday work of naturalists.

Naturalists considered visual skill the defining trait of their practice and the basis of their method. Collecting and classifying, the twin obsessions of early modern natural history, were predicated on the ability of the trained eye to assess, possess and order. Becoming a naturalist implied gaining familiarity with a rigorously defined series of texts that imparted a specific methodology, one that involved observing and describing in highly structured ways, as well as using books to connect and compare observations. This was particularly true with the ascendancy of the Linnaean taxonomic system, introduced in the *Systema naturae* (1735; see Müller-Wille's chapter, this volume). Best known for outlining the sexual system of botanical classification, the *Systema* also proposed a methodology based on observation. By looking at a plant's flowering structure and answering a series of questions, the observer could classify any plant within one of the twenty-four classes Linnaeus proposed.

Linnaeus emphasised visual epistemology even more strongly the following year, when he hired the great botanical artist Georg Dionysius Ehret to create a pictorial table illustrating his system (Figure 14.2). This famous table includes twenty-four figures, one representing the distinguishing traits of each of the Linnaean botanical classes, which are characterised by the structure of the flower and seed.[19] By eclipsing the interrogation procedure, the table promoted Linnaean taxonomy as transparent and suggested that all it required was visual training. In order to classify a plant, it promised, the botanist needed simply to look at a flower, count its stamens and pistils, and note their structural arrangement. Who could not do something so simple?

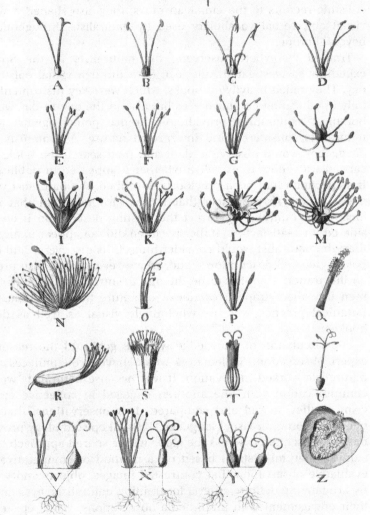

Figure 14.2 G. D. Ehret, *Clariss. Linnaei M. D. Methodus plantarum Sexualis in Sistemate Naturae descripta* (1736). Biblioteca del Real Jardín Botánico, CSIC, Madrid, Of LIN S-39

Ehret's table and the plates included in other works by Linnaeus soon appeared in practically every botanical textbook of the time – including Antonio Palau and Gómez Ortega's *Curso elemental de botánica* (1785; 2nd edn. 1795), used for instruction at the Royal Botanical Garden. The extensive reproduction of these illustrations contributed greatly to the popularity of Linnaean classification, making it appear simple, direct and seemingly foolproof. In addition to enlisting recruits to the Linnaean ranks, they also shaped a shared visual and verbal vocabulary used by naturalists throughout and beyond Europe.

Trained as expert observers, the naturalists in the Spanish expeditions investigated American flora through visual epistemology. They relied heavily on books, which were key instruments not only in educating botanical eyes but also in the day-to-day work of botanical exploration, providing reference points against which naturalists considered and interpreted nature. As botanists travelled, they would observe a plant and then search the books they carried with them to establish whether it appeared in publication. If a specimen had been previously described, the naturalist would make a note to the effect that it was also found in that other locality, or improve or correct the existing description if he considered it unsatisfactory. If the specimen did not appear in any text, then the naturalist would consider himself its discoverer and compose a detailed description – and, whenever possible, also prepare an illustration – hoping to be the one to introduce the new specimen into the European catalogue of nature through publication. Botanical practice was overwhelmingly visual as well as deeply bookish.

The circulation of knowledge was the goal and the reason for expert observation. Skilled eyes would have been unnecessary if naturalists worked in isolation. It was because naturalists were in communication with one another, engaged in collective empiricism as they shared and compared their observations, that they needed a common visual language.[20] Visual epistemology provided naturalists scattered across the globe with a shared approach to the study of natural history, based on a method of comparative and evaluative observation that contrasted images, objects and words. By articulating definite criteria for training naturalists' eyes and for their engagement with multimedia observations, visual epistemology turned multiple and far-flung observers into calibrated instruments that worked in concert. This synchrony was particularly crucial in imperial contexts. If taxonomic botany – the identification and classification of plants – required collective empiricism, the stakes were even higher for economic botany, which engaged

Figure 14.3 J. del Pozo (Malaspina expedition), self-portrait drawing a Patagonian woman (*c.*1790), pen and wash drawing, 18 × 24 cm. Location unknown. Reproduced from C. Sotos Serrano, *Los pintores de la expedición de Alejandro Malaspina* (Madrid, 1982), vol. II, figure 38.

naturalists as imperial agents. Standardised observation and representation allowed naturalists to communicate not only with one another but also with imperial and colonial administrators, and to make a case for how and why their expertise mattered.

Travelling artists were central to this process, as they translated naturalists' observations into the representations that would embody and transport both specimens and visual expertise. The Spanish naturalists worked extremely closely with their artists, constantly engaging with their work – literally looking over their shoulders as they drew, as suggested in a self-portrait by one of the expeditions' artists (Figure 14.3). The relationship between naturalists and artists was thus both close and hierarchical. Naturalists exercised great authority over their artists, supervising and

Figure 14.4 *José Celestino Mutis* (*c.*1800), oil on canvas, 124 × 92 cm. Attributed to Salvador Rizo. Archivo del Real Jardín Botánico, CSIC, Madrid.

directing artists' work and use of time, regulating their bodies by mandating where and when they should travel, and affecting their productivity by allocating work supplies.[21]

The way in which vision came to define the practices and the very persona of the naturalist can be seen in the portrait of one of the Spanish botanists, José Celestino Mutis, director of the Royal Botanical Expedition to the New Kingdom of Granada (1783–1816) (Figure 14.4).[22] The painting depicts Mutis in the very act of conducting a botanical observation. It invites the viewer into Mutis's study, allowing us to witness him at work. Mutis is shown sitting at a table, deeply engaged in the pursuit of his scholarly craft. His focused gaze fixes on the viewer with weary patience, as if we had just burst into his study of muted greys and browns and interrupted his silent labour. He has lifted his head but his body remains hunched over in concentration, eager to resume the examination of the flower he holds up. A branch of the same plant lies ready to be pressed between sheets of paper in order to become a specimen in a herbarium or collection of dried plants. Books scattered around the table will help Mutis corroborate his description and classification of the plant. The books outline the task at hand: if the plant that Mutis examines has already appeared in a publication, he will determine whether it has been assessed correctly or whether the entry needs emendation. Any discrepancy between published materials and the specimen that Mutis observes will provide a chance to contribute to the literature with a correction. Even better, if the plant does not appear in any of the existing sources on South American flora, Mutis could describe it in publication and in this way become the discoverer of a new species.

The naturalist's job, the portrait claims, is to observe. The magnifying lens that Mutis holds in his right hand serves as a symbol of the acute visual skills that characterise him as a botanist. A simple instrument, it suggests that the truly magnificent and sophisticated machinery at work is the naturalist's gaze. This is not simple *looking* but rather expert, disciplined, methodical *observing*. And expert looking is connected to other activities: collecting, as evidenced by the plant, books and herbarium, comparing and classifying, as signalled by the presence of books, and writing and drawing, as indicated by the pen in Mutis's right hand and the sheet of paper before him on the desk.

Mutis's portrait not only addresses the process of observation, but also hints at some of its goals and rewards. Like the magnifying lens, the flower that Mutis so attentively considers celebrates his work. It is carefully presented to the viewer, painted in a bright red that stands out vividly against the muted colours that dominate the portrait. This particular plant has a starring role in the canvas because it is a specimen of *Mutisia*, a new American genus named in his honour. By calling attention to his namesake plant, the portrait celebrates Mutis's talents as a botanical discoverer and relates them to his capacities as an observer.

Although Mutis is portrayed as a solitary figure, his work depended on close collaboration with plant collectors, artists, patrons, imperial administrators and other naturalists. Given the global interests of eighteenth-century natural history, visual epistemology was a collective process that involved bridging distances.

Trained observers voyaged to remote lands to gather observations and specimens, and in turn these stabilised incarnations transported distant nature so that it could be studied by observers who had not themselves travelled. At the core of the notion of observation lay an individualistic rhetoric of autopsy – the process of having experienced or witnessed for oneself, with one's own eyes. However, observation was often a collective endeavour, one that drew on the first-hand analysis of secondary materials in various media, or on the first-hand observations of others. As the portrait suggests, visual epistemology not only certified botanists' abilities and standardised botanical findings, it also served to make observations and specimens mobile and thus to overcome the challenge that distance posed to imperial botany.

The challenge of distance

The thorny issue of distance and place can be approached by juxtaposing Mutis's portrait with another painting produced by an artist in his workshop, which elucidates the long-distance circulation of people, specimens and images. The painting, shown in Figure 14.5, depicts the renowned Spanish botanist Antonio José Cavanilles, director of the Royal Botanical Garden of Madrid between 1801 and 1804, and Mutis's long-time correspondent and supporter.[23] Both portraits were produced at the turn of the nineteenth century in Bogotá, both emphasise the importance of visual epistemology in natural history, and both connect observational acuity to the honour of discovery. The paintings also allow us to reflect on the role of travelling images in allowing long distance observation. Cavanilles's portrait shows him in profile, from the waist up, sitting before a table. With his left hand, he points to a botanical illustration that he studies with unblinking attention. The image is clearly recognisable as one of the works produced by the New Granada expedition. Gazing attentively at the image, the botanist observes the various parts of the plant and immediately transforms his visual analysis into a textual taxonomic description, which he writes with a quill pen on a notebook that lies open on the table. Eye and hand work in coordination, image produces text. Set against a dark background and the lustrous velvety black of Cavanilles's priestly garments, the light-coloured pages pop brilliantly. The botanical illustration is as much a protagonist of this painting as the man rapt in its study. It serves to connect the naturalist in Spain and the artist in America, erasing the distance between them. Although the portrait is unsigned and undated, the artist inscribed himself into the painting in a clever way. The name visible at top of the image, *Rizoa*, points to the identity of both a South American

Figure 14.5 *Antonio José Cavanilles* (*c.*1800), oil on canvas, 86 × 66 cm. Attributed to Salvador Rizo. Museo Nacional, Bogotá, Colombia.

plant and a South American artist, Salvador Rizo, who was the expedition's lead artist and Mutis's second-in-command, and in that capacity directed the artistic workshop that produced illustrations exactly like the one Cavanilles is examining. Through this painting, Rizo thanked Cavanilles for naming this American genus after him.

Rizo's portrait of Cavanilles celebrates both artist and naturalist through a botanical identity, much as Mutis's portrait honoured him through the *Mutisia*. The replacement of a botanical specimen with an image is significant because it demonstrates how Mutis and Rizo expected naturalists in Europe to use the expedition's illustrations. A naturalist based in South America could observe multiple fresh specimens over the years and work with an artist to create an image that presented a composite result of all those observations. This would be impossible to achieve with a single dried specimen, which inevitably included accidental particularities. The picture, by comparison, incarnates not an individual American specimen but the many plants and observations that allowed the team of naturalist and artist to produce an idealised version of this type of plant. This painted composite specimen made it unnecessary for Cavanilles to travel, allowing him to sit at his desk in Europe and observe South American flora 'firsthand', using this rendition to classify and name it.

Conclusion

In the last decades of the eighteenth century, the Hispanic world buzzed with scientific activity. A flurry of expeditions criss-crossed the empire, surveying, collecting and documenting its natural history, as part of a process of reassessing and rediscovering kingdoms that – though long-held – remained in crucial ways half-known. In addition to these travellers, throughout the viceroyalties, local institutions, and a wide cast of characters that included both Europeans and Americans, participated in a concerted effort to identify useful and valuable natural products. In the Hispanic world, these attempts at imperial renovation through useful science were framed not only within the context of Enlightenment ideologies and policies, but also as the continuation of ventures that had taken place at a glorious moment of imperial expansion in the sixteenth century.

Imperial botany was central to these efforts. Natural history expeditions pursued three related goals, the first of which was economic botany. In a climate of fierce economic and political competition among European nations vying for primacy as commercial and imperial powers, botany appeared particularly well suited to fulfilling ambitions of controlling trade in useful and valuable natural commodities. Eighteenth-century European botanists – in Spain and elsewhere – promised administrators that they could identify desirable plants throughout the world and if necessary successfully transplant them to other locations, convincing their patrons of the need to fund expeditions that would exploit items such as coffee, tea, spices and medicinal plants. Casimiro Gómez Ortega, director of Madrid's Royal Botanical Garden, assured the Spanish imperial administration that its vast and fertile territories must surely hold valuable botanical commodities like cinnamon, pepper, tea and nutmeg, which would allow Spain to compete with the valuable Dutch trade in Eastern spices, British profits from tea and French successes with coffee transplantation. He followed with interest and concern activities at the Jardin du Roi in France and Kew Gardens in England, emulating their examples and seeking to outdo them. Meanwhile, Britain and France cast envious glances at the Spanish monopolies in cinchona and cochineal, two valuable American natural commodities that they attempted to locate in their colonial holdings or to steal away from the Spanish Americas. Spanish botanists, however, failed to deliver on their economic promises.

The expeditions proved much more successful at fulfilling their two other objectives, namely taxonomic botany and collecting. With the ascendancy of the Linnaean system of classification and the increasing access to non-European flora provided by imperial and commercial voyages, European botanists in the second half of the eighteenth century embarked on a global mission to survey and classify all the world's plants. Spanish naturalists had privileged access to the Spanish Americas, and as a result other European naturalists followed their work with anticipation and great interest. With their lengthy stays in the Americas, access to local resources and expertise and incomparable artistic workshops, the Spanish expeditions were uniquely poised to satisfy the European demand for depictions of New World flora.

Scientific voyaging was expensive, uncomfortable, exhausting and dangerous, when not altogether deadly. Cabinet naturalists who stayed behind in Europe, Mutis noted, had no idea of the 'unspeakable hardships' that voyagers faced:

Savants, in their cabinets or in schools, spend their days in great comfort, gathering the fruit of their diligence without moving. A traveler must spend a great part of each night ordering and describing what he gathered in the field during the day, and this after having suffered the conditions of that Season; the roughness and pitfalls presented by the ground he surveys, which tend to be greatly varied; the discomfort of insufferable insects that surround him everywhere; the frights and dangers of many poisonous and horrible animals that at every step scare him, terrify him about the austerity of a truly austere and boring life that through the heat, moors, and deserted places breaks down and wears out his body.[24]

For imperial botanists, images offered solutions to two important problems: the challenge of collective empiricism, that is of making sure that individuals looked at plants in the same way and saw the same thing, and the challenge of distance, allowing them to stabilise and mobilise specimens from far-off lands, allowing cabinet naturalists to observe them first hand at a convenient distance.

Further reading

Barrera-Osorio, A., *Experiencing Nature: The Spanish American Empire and the Early Scientific Revolution* (Austin, 2006).

Batsaki, Y., Burke Cahalane, S. and Tchikine, A. (eds.), *The Botany of Empire in the Long Eighteenth Century* (Dumbarton Oaks, 2017).

Bleichmar, D., *Visible Empire: Botanical Expeditions and Visual Culture in the Hispanic Enlightenment* (Chicago, 2012).

Cañizares-Esguerra, J., *Nature, Empire, and Nation: Explorations of the History of Science in the Iberian World* (Stanford, 2006).

Drayton, R., *Nature's Government: Science, Imperial Britain and the 'Improvement' of the World* (New Haven, 2000).

Lafuente, A., 'Enlightenment in an imperial context: local science in the late eighteenth-century Hispanic world', *Osiris*, 15 (2000), pp. 155–73.

Pimentel, J., 'The Iberian vision: science and empire in the framework of a universal monarchy, 1500–1800', *Osiris*, 15 (2000), pp. 17–30.

Schiebinger, L. and Swan, C. (eds.), *Colonial Botany: Science, Commerce and Politics in the Early Modern World* (Philadelphia, 2005).

15   Bird sellers and animal merchants

On Friday 22 March, 1776, after a day trip out to the Tower of London, Hester Thrale and her children left Moore's Carpet Manufactory in London's Moorfields, an establishment well known for its expensive and fine carpets, and went home. Thrale, a society patron of the arts, diarist and literary figure, noted in her journal how 'we drove home-wards, taking in our way Brooke's Menagerie, where I stopped to speak about my peafowl. Here Harry was happy again with a lion intended for a show who was remarkably tame, & a monkey so beautiful & gentle that I was as much pleased with him as the children'. This seemingly impromptu call at the premises of a London animal merchant speaks to the marketplace for exotic birds and animals in eighteenth-century London. Brooke's Menagerie was, to contemporaries, the most reputable and pre-eminent animal merchant in Georgian London, part of a much broader bird and animal trade in the city. The businesses of cage makers, bird sellers, animal merchants and menagerists were spread throughout the city, stocked by extensive trade networks that brought birds and animals, both familiar and rare, to the attention of a metropolitan public eager to consume exotic goods. This chapter principally traces the emergence of these sites as spaces for the spectacle, production and consumption of natural history in eighteenth-century London – but also touches upon the trade in exotica in the cities of Paris and Amsterdam.[1]

These three European cities, principal cities in maritime or colonial powers, swelled in the seventeenth and eighteenth centuries to become cultural and commercial goliaths. With a population of around 900,000 at the end of the eighteenth century, London was the biggest city in Europe. Shops, showrooms, galleries, museums, theatres and menageries constituted part of the visual spectacle and appeal of the city. The city was, too, the home of learned societies, publishing houses and instrument makers. The residences of the metropolitan elite hosted salons, literary or natural historical in nature, and so the city functioned as a hub of cultural production. Anne Home, a poet, hosted a literary salon, or 'conversation parties', in the drawing room of her Leicester Square home – guests included Haydn,

Horace Walpole and Elizabeth Montagu. Meanwhile, elsewhere in their home-cum-anatomy school-museum, her husband John Hunter, surgeon and anatomist, dissected bodies – both human and animal. Hunter was very much au fait with the exotic animals that made their way into the city for display, whether in the homes of friends and acquaintances, or the premises of bird and animal merchants. Indeed, eventually, a good number of these exotic birds and beasts ended up under his surgeon's knife. Another salon, though somewhat less cerebral, illustrates too this urban milieu of natural history, spectacle and sociability. Miss Poll, a parrot, was not only long-lived but also unusually gifted – she would sing songs and dance on her perch. Between the 1770s and her death in 1802, Half-Moon Street was 'filled with carriages and an admiring crowd' on the days of her morning levees. After her death, which was said to have 'thrown the West-end into condolence and confusion', this bird was dissected and preserved by the anatomist Joshua Brookes, the son of the bird and animal merchant Joshua Brookes. London, then, as a centre of consumption, cultural production and spectacle is a good place to study the place of bird and animal merchants in the history of natural history.[2]

'A class of men in London who are called animal merchants'

In the summer of 1805, Benjamin Silliman, a 25-year-old American, was in London to purchase the necessary books and equipment to take up his chair in geology at Yale College, as well as to attend lectures on the sciences. Silliman recorded his impressions of the city in letters to his brother. On 23 July he wrote:

Having occupied my leisure hours, of late, in perusing Buffon, Shaw, and other writers on zoology, I have been naturally led to visit the museums and collections of animals, which are found in such profusion in London. With these views I spent several hours before dinner in Pidcock's menagerie at the Exeter Change, and at the Leverian Museum. There are not many animals of importance which one may not see, at this time, in London; to mention only a few of those which I have examined today; – the lion and lioness, royal tiger of Bengal, panther, hyena, tiger cat, leopard, ourang-outang, elephant, rhinoceros, hippopotamus, great white bear of Greenland, the bison, elk or moose deer, the zebra, &c. Most of these were living.

In a later letter dated 15 August, he elaborated further on the spectacle of exotic animals in London:

There is a class of men in London who are called animal merchants. They keep, both for sale and exhibition, collections, more or less extensive, of living animals. Pidcock, whose menagerie I have already mentioned, is a dealer of this description, and this morning I visited another similar collection, Brooks', at the corner of Piccadilly and the Haymarket.[3]

This 'class of men' had a long history in London: animal merchants, some of whom were women, too, were part of a trade that brought exotic birds and animals into the capital in such abundance that even guidebooks to the city designed for foreigners remarked that the premises of the animal merchants were worthwhile points of interest. In the early 1800s, the premises of bird and animal traders were peppered throughout the city, lining the Strand and Piccadilly, but the roots of this trade extended well back into the seventeenth century. The trade in exotic birds and animals in premises dedicated to that purpose began with itinerant bird sellers, market stalls and canary packets – parcels of canaries imported from continental Europe – sold in taverns or coffee houses. Likewise, the public display of curious birds and beasts and the commerce in these animals has a long history, stretching back into the early modern and Renaissance periods.[4]

Collections of exotic animals, or indeed individual rare exotic specimens, have usually attracted the attention of historians within the context of princely collections. The royal menageries of European monarchs, in particular those of the Iberian peninsula, Italian lands and France, have long been of interest to historians. So too, singularly impressive animals, such as Pope Leo X's elephant 'Hanno', and his rhinoceros, famously depicted by Dürer based on a description of the animal, are well known in the history of natural history. In addition, fifteenth- and sixteenth-century portraits of the European aristocracy teem with parrots and monkeys; some are certainly depictions of actual animals kept by individuals and households, while others may have a more emblematical or representational function. Historians have traced the history of pet-keeping to the courtly and ecclesiastical cultures of medieval Europe; songbirds, squirrels, ferrets, dogs, cats and other native species were joined, in a smaller number, by parrots and monkeys. By the eighteenth century, 'pets' were a common feature of everyday life, and attracted both censure and approval; recent scholarship has shown how the presence of pets prompted contemplation on appropriate conduct, emotions and behaviours at a time of an explosion in consumer goods and luxuries. A fashion for natural history and a taste for the expensive exotic led to a vigorous trade in exotic birds and animals. A study of exotic animals in eighteenth-century Paris has drawn attention to a culture of exhibition and exotic pet-keeping that extended well beyond the Versailles menagerie and the estates of the aristocracy, and into the homes of the city's well-to-do merchants and professionals.[5]

The trade in exotic birds and animals is part of a broader history of the marketplace for natural history; it is also an integral part of the history of colonial expansion and trade. In the seventeenth century, the Dutch Republic, in particular the Dutch East India Company (Vereenigde Oost-Indische Compagnie or VOC), had access to a mercantile empire based on trading concessions and colonies; spices, textiles, porcelain, plants, bulbs and animals flowed into Amsterdam to be sold on the domestic and wider European markets. Still-life oil paintings by Dutch and Flemish artists bear witness to the floral marvels of empire, whilst portrayals of the homes and gardens of the affluent burghers depict parrots in domestic spaces or exotic waterfowl in canal gardens (see Plate 9). The permanent premises of the Dutch animal merchants emerged several decades before those in London. Dutch animal merchants sent their animals on the road to tour neighbouring countries where they were exhibited at fairs, taverns and coffee houses; 'Hansken' the elephant toured Europe between the 1630s and 1650s with Cornelis van Groenevelt (Figure 15.1).[6]

The album of Jan Velten, an inhabitant of Amsterdam, is filled with watercolours, gouaches, sketches and handbills. These document the birds and animals – both living and dead – that he had seen in Amsterdam in the years around 1700.[7] The city, at that time, could boast a far greater array of places to see and purchase exotic birds and animals than either contemporary London or Paris. The warehouses of the VOC frequently hosted animals to be seen or sold, and eventually specialist merchants emerged to deal in this trade. From 1675, Jan Westerhof operated the Menagerie van Blauw Jan on the Kloveniersburgwal, a canal lined with the residences and warehouses of merchants. In a sketch by Velten, the Blauw Jan is shown with an array of animals in cages or tethered on ropes, in addition to preserved specimens on shelves or hanging from the ceiling. An early eighteenth-century print of the business by Isaac de Moucheron depicts an elegant Baroque courtyard populated with a menagerie of birds and beasts (Figure 15.2). Bartel Verhagen established his menagerie, de Witte Oliphant, in 1681; his premises were two lean-to houses on the Botermarkt. The sign outside his shop read 'Here are sold for the benefit of the public, apes and monkeys, baboons, and parrots, and the like.' In addition to selling animals, Verhagen also toured them across Europe. He bought an elephant and toured with it through the German-speaking lands, stopping at Königsberg, Frankfurt and Danzig, and going on to Bologna, Paris and London during the 1680s and 1690s.

Exotic birds and animals were also brought into London by English merchants. East India Company ships arriving from the East Indies carried such novel cargo with them. On 24 July 1649, *Tuesday's Journal* brought the following news to the attention of Londoners: 'Some East

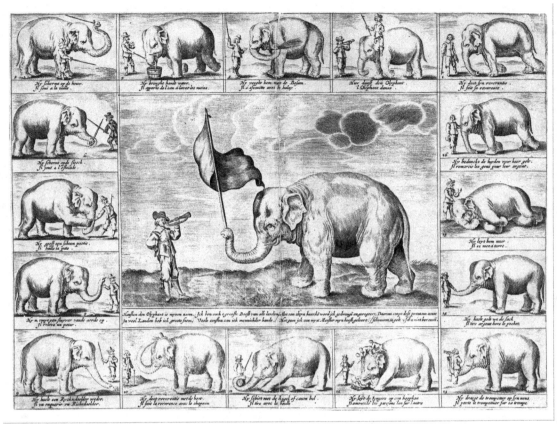

Figure 15.1 This handbill was printed in both French and Dutch. 'Hansken', a female Asian elephant, was first brought to Holland in 1637 and toured across Europe until her death in 1655 in Florence. Hansken's skeleton is preserved and on display at the Museo della Specola. *Die Oliphant Hansken* (1641), 29.5 x 38.3 cm. Rijksmuseum, Amsterdam.

Indies ships arrived and brought some very strange birds and beasts, such as were never seen in England. A great number of persons from the city of London and other parts flock daily to see the rarities that they have brought over'.[8] London coffee houses and taverns hosted travelling exhibitions of animals, in addition to auctioning or selling birds and animals on an ad hoc basis. London did not, at that time, have establishments on the scale of de Witte Oliphant and the Menagerie van Blauw Jan. A few decades later, however, around 1700, bird sellers were a well-established presence in London, indeed so well established that, in March 1704, David Randal, a bird seller

Figure 15.2 The Blauw Jan traded in exotic birds and animals as well as functioning as a tavern for over a century. I. de Moucheron, *Hof van Jan Westerhof te Amsterdam* (1687–1744), etching, 36.5 x 44.7 cm. Rijksmuseum, Amsterdam.

who traded from his house in Channel Row, Westminster, was prompted to place an advertisement for his business assuring the public that whilst 'a great many' traded in his name, persons of quality seeking birds would be 'kindly dealt with' if they came to him. Randal's stock included 'choice singing canary birds', 'fine talking parrots' and a 'sweet monkey that whistles like a bird'. Canaries were a staple of the bird trade in this period, and they were usually sold as 'imported packets' at coffee shops and taverns by bird sellers from the German-speaking lands; canaries were at that time commonly referred to as 'German birds'. The canaries, originally from the Spanish Canary Islands, had been bred selectively in central Europe for their singing voice and plumage – they were available in white, yellow and brown. In France, these itinerant German bird-sellers competed with the *oiseleurs*, the guild of bird trappers and sellers, and the latter attempted to protect their privileged status as sole purveyors of exotic birds and animals. In London, similarly, one of the early prominent retailers of canaries in London was 'the German' J. C. Meyer, with premises on St James's Street, whose advertisements asserted the superiority of all his products to those sold by others.[9]

Native songbirds had long been sold by bird catchers, but this was not a lucrative trade, regardless of the expertise and know-how involved in skillfully luring and trapping birds. Likewise, cage-makers did not stand to make much from their occupation. Rattan and wooden cages could be purchased inexpensively, and the market for high-end cages, by comparison, was small. Native songbirds could be bought for a shilling – imported canaries, in the early eighteenth century, could be purchased for ten to fifteen shillings, for a desirable bird. Money was to be made from imported birds and not, usually, from the cages that housed them. Thus, cage-makers usually sold birds on their premises or located their premises in the near vicinity of a bird seller. One such was Thomas Ward, who traded from the *Bell and Birdcage* on Wood Street. He had previously been a bird catcher, and had some 20 years of experience of working with birds. Three editions of his pamphlet *The Bird Fancier's Recreation* appeared between 1727 and 1740. It provided a crash course in the prudent selection of a bird and tips for avoiding the scams of unscrupulous bird sellers. Bird-fanciers' clubs gathered in coffee houses and taverns, selling and exchanging birds, socialising around a mutual interest in canaries. These men were not typically from the city's elite or middle ranks. Instead, they were labourers, journeymen and craftsmen.[10]

The emergence of specialist bird shops also fits in the long tradition of itinerants selling on the streets and in taverns, as part of a broader change in consumption patterns. From 1675 to 1725 there was a marked increase in the introduction of material goods into domestic spaces. Goods that had hitherto been out of the reach of ordinary Londoners, for example, became more commonplace. An increase in the ownership of relatively ordinary goods, like simple utensils and inexpensive china or clocks, was accompanied by the consumption of products imported at fairly modest prices. Canaries, for example, became cheaper and well within the reach of labourers. In the early eighteenth century, a 'shop' could be a wooden stall, the cellar of a house or the modified front of a house converted to trade purposes. Later in the eighteenth century, the purpose-built shop or showroom included fittings such as glaze-panelled windows with crossbars, decorative mouldings and display cabinets. Advertisements in newspapers placed by bird sellers in the 1750s and 1760s show that they – and new animal dealers – were dealing in a much broader range of species than earlier in the century. This was due, in greater part, to the territorial gains made during the Seven Years' War (1756–63) and the commensurate growth in East India Company shipping. Ships returning from overseas brought with them new species, many originating from newly occupied territories. Many different sorts of dealers in birds and animals proliferated in these decades. Some, like Edmond's Menagerie, offered more parrots and fewer canaries, alongside interesting novelty animals like chipmunks; cardinals, box turtles and chipmunks were clearly brought onto the market by the expansion of territories in North America. The Noah's Ark menagerie advertised a wolf, crocodile, several camels and a medley of parrots and other caged birds.[11]

Brookes of Holborn and other bird and animal merchants

From the 1750s the Brookes family became prominent as a family of bird sellers in London. Although the Brookes represented the 'top end' of their trade, they are a useful lens through which to view the animal trade in London. The lives of London's more prominent merchants and dealers can be traced through wills, insurance policies, trade cards and handbills. Much less is known about the itinerant traders or those who operated precarious and unprofitable businesses. Joshua Brookes, or Brook's, was the son of a poulterer, who had begun as a bird seller in the 1750s and 1760s, and later extended his business to become a bird and animal merchant – he was also known as 'Brookes of Holborn'. He traded from multiple sites, including his Original Menagerie on Gray's Inn Road, and later, from 1776, at premises on the New Road at Tottenham Court. A catalogue for customers listed some unusual animals like antelope, lions, monkeys and porcupines, although birds constituted the bulk of his business. A 1775 promotional handbill listed 160 different avian species, including cockatoos, flamingos and cassowaries. Brookes sometimes described himself as a 'zoologist', at other times, less grandiosely, as a 'bird merchant'. His professional identity was also based on a claim that he could convey his birds 'to any part of the world'. Brookes would travel annually to Holland to acquire birds, too, doing business with a Mr Echardt; whilst in Holland he must have also visited the Blauw Jan.

A 1779 trade card printed for his business announces that he will buy, exchange and sell 'all kinds of curious English and Foreign birds' at 'reasonable rates'. Joshua Brookes also periodically sold imported plants, seeds, shrubs, ferns, acorns and cones from the American colonies. These were acquired through the services of the

'King's Botanist', William Young, who was paid by Queen Charlotte, a patron of the arts with a keen interest in botany. In 1762, when Brookes first received a shipment from Young, he advertised for sale boxes of plants and seeds. This shipping of plants, birds and animals by Young continued until the American War of Independence (1775–83), after which Brookes did not advertise his boxes again until 1786. Little is known about Brookes's other agents, though they almost certainly must have existed; a large menagerie stock of curious birds and beasts could not have been sustained on the back of ad hoc exchanges or agreements to purchase birds and animals made with returning sailors or merchants. Brookes extended his business to occupy other premises, and also went into some form of partnership with a woman called Mary Cross.[12]

Brookes had previously gone into business with Cross's late husband, John, a proprietor of a menagerie in St James's until his death in 1776. The following year, Brookes took out a fire insurance policy on a property ('two houses laid into one') worth £300; the policy names Mary Cross as a 'dealer in live fowls and birds'. At least one other woman in Georgian London was closely involved in the same trade as a proprietor: Mary Wombwell, the owner of a menagerie on Piccadilly at the turn of the nineteenth century. The surnames of a handful of families (Brookes, Cross, Wombwell) regularly appear in relation to the bird and animal trade, indicating that it was continued by successive generations. Servants and keepers employed by these families would move between proprietors, and in some cases go on to open their own business. The wills of some of the animal merchants suggest that the trade could be quite comfortable, even lucrative. In 1803, when Joshua Brookes died, his wife Elizabeth was left with an annual income of around £200 derived from an annuity and income from estates. His daughters, too, were left with annuities of £50 – two of his sons received £200 as a lump sum. Joshua, another son, was given the leasehold on the piece of land that would later become his anatomy school. The youngest son, Paul, received £20. In this period a tradesman might need £40 or £50 a year to sustain a family, and household servants might only earn between £4 and £8 a year; Elizabeth Brookes, with her £200 a year, was a comfortable widow. She was not, however, rich; the gentry and aristocracy who bought her late husband's birds measured their annual incomes in thousands, and even tens of thousands, of pounds.[13]

The fire insurance records of another animal merchant, Gilbert Pidcock, give a deeper sense of the financial scale of the trade. Pidcock had toured animals in a travelling menagerie around the country since the 1770s, and by the late 1780s had put down roots in London. His premises were on the Strand at the Exeter 'Change;

a shopping arcade occupied by drapers, milliners and hosiers. Like any prudent proprietor, Pidcock knew the value of good fire insurance. He had good reason to; a fire in his menagerie had once killed a zebra that he had purchased for £300. Pidcock's 1803 insurance policy details at Sun Fire Insurance can be compared with those of average policy holders. Typically, the properties and possessions of tradespeople were insured at between £300 and £400. 'Gentlemen' were insured for upwards of £500 – with paintings and china insured separately from £30 to £50. Gilbert Pidcock's 'living birds and beasts & preserved specimens' alone were insured for £1,565, a sum so large that it required an itemised catalogue for the animals and birds to be lodged at the office of the insurer.[14]

Establishing oneself as an animal merchant or menagerist required both financial capital and expertise in acquiring, handling, rearing and feeding exotic birds and animals. Joshua Brookes sent his son to South America to acquire new stock and he returned in June 1784. Their collection of 'rare birds from the Brazils and the southern parts of America' impressed the royal family so much that they placed orders for some of this new stock. In 1803, Paul Brookes took his inheritance and travelled the world again to acquire the birds and animals with which to establish his own business. First, he went to Africa and Asia, and later on made a second journey to South America. Brookes advertised on a handbill that he had arrived back in London with a 'choice collection of curious quadrupeds' – this voyage to 'various parts of the globe' had, apparently, taken several years and had been for the purpose of 'collecting' and establishing a 'correspondence' that would allow him to obtain a 'regular supply of the most rare and interesting animals'.[15] The logistics of trapping, feeding and shipping live birds and animals across land and sea required both good credit and a good reputation; in short, London's animal merchants needed to know people who could get them animals or else have the ability to go collecting by themselves.

A few families seem to have dominated the animal trade in the last decades of the eighteenth century. Yet servants employed in these families' businesses occasionally became proprietors. In the case of John Bobey, a black man with vitiligo, his transition was not merely a transition in professional or financial status: it was a transition from exhibit to exhibitor, owned to owner. Bobey was born to slaves in Jamaica in 1774, and not into a family of animal merchants; however, his biography shows how servants equipped with knowledge garnered while working for animal merchants could eventually become proprietors themselves. In 1789, fifteen-year-old Bobey was bought as an indentured servant for one hundred guineas, and was eventually sold for fifty guineas to Gilbert Pidcock. Bobey was exhibited as a 'piebald' racial curiosity, in addition to being put to work in the menagerie. In time, he was released from his indenture and married an Englishwoman. Together they established a 'collection of monkeys, birds, and beasts' that required the 'keep of five horses and men'. Bobey's biographer supposed that by prudently managing their business the couple would amass a 'decent fortune'.[16]

Bird sellers and animal merchants as authorities

In France, the king endowed one of the *oiseleurs* with the title *oiseleur du roi*, or the king's bird seller. A gold-lettered sign above the shop of the *oiseleur du roi* on rue Saint Antoine in the late seventeenth century read 'governor, preceptor, and regent of the birds, parakeets, and monkeys of His Majesty'. The royal warrant of another *oiseleur*, Ange-Auguste Chateau and his son, stated in 1762 that they had been chosen on account of the knowledge they possessed about 'different kinds of birds and animals from foreign countries and about how to feed and conserve them'. Along with the prestige of a royal warrant came the not insignificant task of sourcing the rarest and most splendid birds and animals for his majesty's aviaries and menageries. Competition between the Parisian animal merchants was stiff – and the *oiseleurs* were often entangled in lawsuits, fistfights and unauthorised sales.[17]

London's bird sellers and animal merchants, too, vied for status and prestige. At least one, a merchant called Grainger, had a royal warrant. Grainger's premises in Gray's Inn Lane, Holborn, had the 'Prince's arms over the door' and Grainger styled himself 'Purveyor of Birds to His Royal Highness, George Prince of Wales'. Other sellers attempted to capitalise on Grainger's status, and his trade card took pains to mention that 'many impositions have been made on the public by using my name'. The handbills and trade cards of other Georgian bird sellers and animal merchants frequently advertised that they, too, supplied 'to the aristocracy and gentry' or 'their Majesties and the Royal Family'. The business of selling exotic birds and animals by its very nature brought the animal merchants into contact with a range of society figures, intellectuals and artists. In some instances, these connections could be long standing. Brookes is known to have provided Joseph Banks and the surgeon-anatomist John Hunter with interesting birds and animals. Each man entrusted Brookes with the acquisition of birds and animals of interest – both dead and alive. The animal merchant William Gough, of Gough's Menagerie on Holborn Hill, was also familiar to Hunter. Access to exotica provided metropolitan anatomists like Hunter and Brookes with the opportunity to amass large specimen collections that bolstered their credentials and authority as practitioners. In some cases, animal specimens played an important role in comparative anatomy. Publications of curious anatomical structures, and even straightforward anatomical accounts, were of value in establishing oneself as an authority. The exotic birds and animals present in London were, then, raw material for the production of anatomical expertise and knowledge (Figure 15.3).

In addition to providing advice and services to customers, animal merchants were also a significant resource for naturalists and anatomists eager to find out more about the origins, habits and behaviours of birds and animals. When the naturalists Eleazar Albin and George Edwards compiled their natural histories of birds and other animals between the 1720s and 1750s, they relied in part on the opportunity of drawing the birds and animals that could be seen in the taverns and coffee houses of London, the premises of bird sellers, and also the homes of private individuals (Figure 15.4). This was also the case in Paris. Georges-Louis Leclerc de Buffon's *Histoire naturelle* (1749–1804) contains many references to animals and birds, some seen by him, some described in accounts communicated to him. Likewise, the pages of popular English-language natural histories contain references to animals seen in menageries or at animal merchants. Useful information could be gathered from animal merchants and their keepers – though naturalists sometimes dismissed their words as useless or inaccurate. This was especially true in the case of animals that found their way onto the premises of animal merchants with little clue as to their provenance; indeed, some birds and animals were sold on multiple times, losing scant knowledge as to their origins along the way. In death, too, choice specimens of birds and animals could move from the premises of the animal merchants and into museum collections or private cabinets, via the taxidermist.

Merchants and menagerists also drew upon printed natural histories to position themselves as authorities. Indeed, that the seventeenth- and eighteenth-century materials produced to promote these commercial collections of animals often cited well-known descriptions of animals is reason for supposing that at least some of the animal merchants were readers of popular natural histories. Gilbert Pidcock deftly used natural historical descriptions of animals to add a level of respectability to his travelling menagerie.

Figure 15.3 This parrot was painted in watercolour and gouache by George Edwards. The parrot was owned by Lady Wager, the wife of an admiral, and a well-known collector and 'great admirer' of birds. In 1740, Wager gave the bird to Sir Hans Sloane. Her collection was described as 'a greater living collection of rare foreign birds than any other person in London'. G. Edwards, *Guinea Parrot* (1736), 27.3 x 22.2 cm. Yale Center for British Art, Paul Mellon Collection.

The material printed to promote the tour of his cassowary in 1779, for example, included descriptions of the six-foot bird that people would recognise from either Oliver Goldsmith's *History of the Earth and Animated Nature* (1774) or Samuel Ward's *A Modern System of Natural History* (1775). Visitors to menageries or the premises of animal merchants expected the proprietor to tell them something about the animals on display; sometimes, however, visitors were

Figure 15.4 Gilbert Pidcock purchased this rhinoceros for £700 and exhibited it both in London and around the country. This painting was commissioned by John Hunter. The rhinoceros died near Portsmouth in 1793; the skin was preserved and kept on display by Pidcock. G. Stubbs, *Rhinoceros* (1790), 70 x 92 cm. Royal College of Surgeons of London.

dismayed by indifferent keepers or tall tales. On other occasions something of note was imparted by a knowledgeable keeper. One Richard Heppenstall displayed some camels at the Talbot Inn on the Strand between December 1757 and May 1758, and dutifully imparted knowledge about the camels to the assembled crowds. An account of the camels was published in the *Gentleman's Monthly Intelligencer*, and despite Heppenstall's useful information pertaining to the diet and habits of his camels in captivity, the writer of the account scathingly opined: 'Heppenstall was very communicative, though some matters, that he says have fallen within his observation[,] are denied by the best writers'.[18]

The painted wooden hoardings or canvases created to promote a menagerie or booth could be equally misleading. Martin Lister, naturalist and physician, visited the St Germain Fair in Paris in 1699 and formed a dim view of one of the vendors:

I was surprised at the impudence of a booth, which put out the pictures of some Indian beasts with hard names; and of four that were painted, I found but two, and those were very ordinary ones, viz. a leopard and a raccoon. I asked the fellow, why he deceived the people, and whether he did not fear a cudgeling in the end?[19]

The patter of showmen did not always contain useful pearls of wisdom, nor were their animals always deserving of note; this notwithstanding, animal merchants and proprietors were clearly solicited for information and expected to provide some form of account as to the origins and habits of an animal or bird. Legal cases heard in the Old Bailey suggest another arena in which bird and animal merchants held expert authority. As valuable commodities, birds were frequently stolen or reported missing in eighteenth-century London, and some merchants were called upon to identify birds positively and recognise their stock or that of other merchants. In 1789, for example, a man called Thomas Andrews was indicted for the theft of a pair of gold China pheasants. The pheasants had been stolen from an aviary, and Andrews had walked into William Gough's premises on Holborn Hill with a pheasant stuffed in each of his coat pockets. Andrews asked Gough to buy the birds for thirty shillings, but, unluckily for Andrews, Gough recognised them as stolen. Similarly, in an earlier case in 1768, a dealer in Holborn called Mr James recognised another pair of gold pheasants as being stolen from an aviary. These cases hint at a network of informants as well as an expectation that bird merchants would honestly and competently identify stolen birds.[20]

Authority, then, could sometimes be contested, especially by practitioners of natural history from within the professions of medicine and anatomy; however, this notwithstanding, animal merchants and menagerists were also able to self-fashion as authorities and were, to a limited extent, recognised as such on the birds and animals they sought to sell or exhibit for show. The bird and animal merchants of the eighteenth century, as well as the keepers and servants they employed, emerge as important producers and holders of knowledge about exotic birds and animals; the premises of these merchants and menagerists were, too, important sites for spectacle and consumption. These speak of an extensive and far-reaching marketplace, catering for a fashionable, and sometimes lucrative, interest in natural history – birds and beasts on the market amongst a world of material goods.

Further reading

Grigson, C., *Menagerie: The History of Exotic Animals in England* (Cambridge, 2016).

Lothar, D. and Rieke-Müller, A., *Unterwegs mit wilden Tieren: Wandermenagerien zwischen Belehrung und Kommerz 1750–1850* (Marburg, 1999).

Plumb, C., *The Georgian Menagerie: Exotic Animals in Eighteenth-Century London* (London, 2015).

Robbins, L., *Elephant Slaves and Pampered Parrots: Exotic Animals in Eighteenth-Century Paris* (Baltimore, 2002).

Tague, I., *Animal Companions: Pets and Social Change in Eighteenth-Century Britain* (University Park, PA, 2015).

Velten, H., *Beastly London: A History of Animals in the City* (London, 2013).

16   Vegetable empire

In 1807 Alexander von Humboldt, the Prussian naturalist and explorer who became the pre-eminent natural philosopher of the early nineteenth century, published an *Essai sur la géographie des plantes*. The geography of plants, he claimed, was 'a science that up to now exists in name only', and his essay sought to describe the relationships between plant communities and their environments, on the basis of measurements and observations from his South American travels. Humboldt claimed that botanists had previously been 'concerned almost exclusively with the discovery of new species of plants, the study of their distinguishing characteristics, and the analogies that group them together into classes and families'. Instead of taxonomy, his was 'the science that concerns itself with plants in their local associations in the various climates' and also with 'the great problem of the migration of plants' in which 'the primary factor is man'.[1] However, many of those interested in the vegetable kingdom, and in natural history in particular, had long been concerned with where plants grew, under what conditions they thrived and, crucially, how they might be transported and grown successfully in other places. What was significant in the 40 years prior to Humboldt's *Essai* was the way in which these concerns became tied into the political economy and geopolitics of European empires made of colonial settlements, global networks of trade, naval power and scientific enterprise.

This culture of natural history has been called 'colonial botany', making a strong connection between power and the production of knowledge about plants and their uses in the encounters among European empires, colonists, Creoles, indigenous people and nature.[2] It is, however, important to differentiate these modes of encounter as they extended across centuries and over the globe. World regions differed in when and how far they afforded access to what is now called 'bioprospecting', the gathering and commodification of plant material.[3] For example, imperial China's long withholding of the mysteries of tea contrasts with the multiple exchanges of plants, including coffee and cocoa, around the Atlantic.[4] Trade, empire, settlement, warfare and exploration had long afforded the means for undertaking natural history – particularly after the mid seventeenth-century shift to realising value from vegetable productions (especially sugar in the Atlantic world and cotton around the Indian Ocean) rather than primarily from precious metals. However, it was only after the end of the Seven Years' War in 1763 that the geography of plants became a matter of concerted imperial interest.[5] By defeating France and Spain in a war that was fought in Asia, the Americas, West Africa and Europe, Britain assumed a new global role and oceanic reach, and the other European powers were forced towards re-evaluation and reform. One outcome was modes of imperial ambition within which natural history played an important part. Examining this further demonstrates that, in the late eighteenth century, the geography of plants was a more varied enterprise than can simply be contained by the term 'colonial botany'.

Voyages

After 1763, attempts were made to integrate natural history into state-sponsored voyages of exploration, particularly in the Pacific.[6] While the search for new lands and trade routes took primacy, the best-selling account of John Byron's 1764–6 *Voyage Round the World in His Majesty's Ship The Dolphin* was advertised as giving 'A faithful Account of the several PLACES, PEOPLE, PLANTS, ANIMALS, &c. seen on the VOYAGE.'[7] Similar attention to islands such as Tahiti, with abundant food supply and healthy climates, is evident in the accounts of the voyages led by Samuel Wallis and Philip Carteret between 1766 and 1768 (Figure 16.1). However, the first European Pacific voyages to carry naturalists were those of Louis Antoine de Bougainville, who, between 1766 and 1768, engaged Philibert Commerson and Jeanne Barré to document and collect plants; and James Cook, whose bark *Endeavour*, between 1768 and 1771, carried the Royal Society's sponsored naturalists Joseph Banks and Daniel Solander, the botanical artist Sydney Parkinson and the astronomer Charles Green. Banks, the son of a wealthy Lincolnshire landowner, had already travelled to botanise in Newfoundland and Labrador on the fishery protection vessel *Niger*, and he provided ample funds towards the *Endeavour* voyage. He returned to England from the Pacific in 1771, bearing accounts, drawings and dried specimens of over 1,400 previously unknown plants, becoming something of a natural historical sensation.[8]

Banks's example, and his own subsequent attempts to shape the imperial geography of plants, did much to cement the place of natural history on other European voyages. There were Cook's further Pacific voyages, which between 1772 and 1775 carried the naturalists Johann Reinhold Forster and Georg Forster, father and son, to New Zealand, and the Linnaean botanist Anders Sparrman to the Cape of Good Hope, and, in 1776–80, took William Anderson and William Ellis to the north Pacific. France sent out the La Pérouse expedition in 1785, which ended

in 1788 with the loss of both its ships. In the early 1790s, both the Spanish Malaspina expedition and the British Vancouver expedition explored the northern Pacific as part of more extensive voyages. Finally, the first years of the nineteenth century saw the competing Australian expeditions of Nicolas Baudin, in *Le Naturaliste* and *Le Géographe*, and Matthew Flinders, in *The Investigator*, from which the naturalist Robert Brown published nearly 1,200 descriptions of new species.

For the British, French and Spanish, these voyages were part of an understanding of the global geography of plants in relation to late eighteenth-century empire. What was new was that these powers now also wanted to reshape that geography for imperial ends by moving useful and profitable plants around.[9] In the British case, Banks – president of the Royal Society from 1778 to 1820, and in control of Kew Gardens from 1772 – played a crucial role. He provided a 'centre of calculation' for botanical knowledge, and was the prime instigator and facilitator of imperial botanical projects, even if the coherence of the 'visions of empire' involved and the power of Banks and others to achieve their ends via state support is easily overstated.[10] There were also practical difficulties. These years saw increasing experimentation

Figure 16.1 Tahiti. J. Hawkesworth, *An Account of the Voyages Undertaken by the Order of His Present Majesty for Making Discoveries in the Southern Hemisphere* (London, 1773). © British Library Board, 214.c.6.

with techniques to enable live plants (and their seeds) to survive lengthy ocean voyages, exposed to salt spray, excessive heat or cold, and too much or too little sunlight. For example, in the early 1770s John Ellis described various containers which would serve the purpose, although success in practice was less easily attained (Figure 16.2).[11]

Significantly, as well as being a Royal Society fellow, Ellis was the London agent for the Caribbean island of Dominica, ceded to Britain by France in 1763 along with Grenada, St Vincent and Tobago. His designs were published at the behest of West Indian planters, whose dependence on enslaved African labour working monocultural plantations, primarily of sugar cane (itself originally a Pacific plant that had crossed the Atlantic on Christopher Columbus's second voyage in 1493), produced a need for globally transplanted food crops such as 'the Mangostan and the Breadfruit' (Figure 16.3).[12]

As early as 1772, Valentine Morris, a sugar planter and later governor of St Vincent, sought Banks's assistance in introducing breadfruit to the Caribbean. In 1775 London's West India Committee offered a £100 prize to 'any Commander of an East India Ship, or other Person, [who] shall bring to England from any part of the World, a plant of the true Bread Fruit Tree, in a thriving Vegetation, properly certified to be of the best sort of that Fruit'.[13] The difficulties of getting the right plants, and having them survive the journey, are apparent in a 1787 letter from Stephen Fuller, Jamaica's agent in London, to the Home Secretary, Lord Sydney. Sydney, overseeing the new Australian penal colony, was, Fuller noted, 'endeavouring to get the Bread-Fruit into the Sugar Colonies, by the Ships returning from Botany Bay'. Fuller cited a series of botanical works to identify a breadfruit that 'grows in Batavia, & Amboina, and the Molucca Islands; and . . . there produces seeds or nuts, in great perfection, that are as good as Chestnuts, and very like them in the taste'. To simplify matters he suggested that Sydney 'should chuse to puzzle the Captains with no other name than that of the Bread-Fruit with the divided leaves', and that

I would recommend it to the Captains to get as much of the seed as they can, if in a state perfectly ripe; and to put some of it into pots immediately, and to leave some of the Pots at the Cape of Good Hope, and also some of the seed; the same at St Helena; the same at any of the Islands they may touch at on the Coast of Africa; conceiving that the only way to get it with certainty, is to do it by degrees.[14]

This was not, however, the eventual route that breadfruit took to the Caribbean. In 1788, with Banks behind the scheme, Captain William Bligh was engaged to carry thousands of live plants from Tahiti in the *Bounty*. However, the voyage was interrupted by a mutiny among the

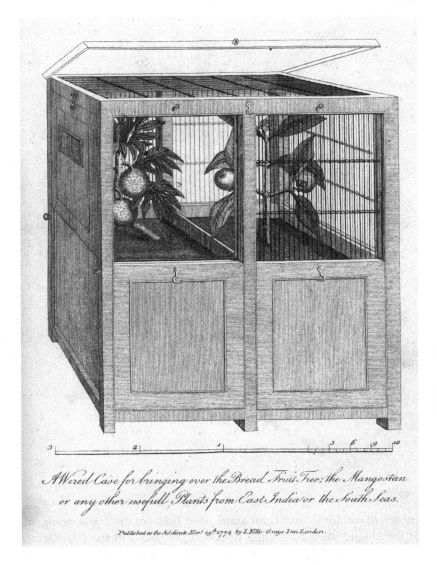

A Wired Case for bringing over the Bread Fruit Tree; the Mangostan or any other usefull Plants from East India or the South Seas.

Published as the Act directs Nov.ʳ 29ᵗʰ 1774 by I. Ellis Grays Inn London.

Figure 16.2 Oceanic plant transportation. J. Ellis, *A Description of the Mangostan and the Breadfruit . . . To which are added, Directions to Voyagers, for bringing over these and other Vegetable Productions, which would be extremely beneficial to the Inhabitants of our West India Islands* (London, 1775). © British Library Board, 34.e.21.

crew, and the breadfruit did not reach the islands until Bligh sailed again in 1793.[15]

Successfully changing the global geography of plants involved sending multiple specimens on multiple voyages, with expertise and instruction at either end and en route, along with the appropriate growing conditions. French botanists were also moving plants from east to west. In February 1782 two chests of plants were sent by Jean-Nicolas Céré, director of the botanical garden on the Isle de France

Figure 16.3 The Bread Fruit. J. Ellis, *A Description of the Mangostan and the Breadfruit . . . To which are added, Directions to Voyagers, for bringing over these and other Vegetable Productions, which would be extremely beneficial to the Inhabitants of our West India Islands* (London, 1775). © British Library Board, 34.e.21.

(now Mauritius), to Guillaume de Bellecombe, governor of St Domingue (now Haiti), the richest sugar island in the Caribbean, and the most deadly for the enslaved. One chest contained 700 mango seeds, along with the seeds of vacoua (a palm used for making hats and roofing huts). The other contained seeds and plants from places around the Indian Ocean and beyond: mangos from Goa; jackfruit, starfruit and bilimbi from the Malabar coast; ravensaras – a spice tree – from Madagascar; cinnamon from Ceylon; and the *arbre de Cythère* (or ambarella) from the Pacific islands.[16]

Yet they failed to arrive, so in August 1783 Céré dispatched another 'four enormous chests of trees', noting that it would be 'a most useful thing for our American colonies if these plants could arrive there alive'. Again, all were carefully identified, numbered and accompanied by detailed instructions on propagating the seeds (sent packed in dry earth), when and how to plant them out, and their potential uses. But another request for plants was received just a month later, and two chests were packed with cardamom, bamboo and a vetiver rootball, with instructions to divide it before planting in a humid spot. These came with Céré's assurance that he would, if necessary, send the whole 'Jardin du Roi' to St Domingue, although he may have been hinting that he thought he already had.[17]

Gardens

Rather than sending botanical gardens across the oceans, these voyages were increasingly intended to move plants between gardens, envisaged as crucial sites for developing knowledge about plants and how to grow them. This involved both the gardens established in Europe, especially those at Kew (1759), the Jardin du Roi in Paris (1635), which became the Muséum national d'Histoire naturelle (1793), and the Real Jardín Botánico in Madrid (1755), which were increasingly equipped for the acclimatisation of tropical plants, and those established in the colonies of each empire. The botanical garden on the Isle de France was laid out in 1735, and that at Port-au-Prince, St Domingue, in 1777. Others were established in Lima (1778), Mexico City (1788), Guatemala City (1796) and Havana (1816). In a new phase of global warfare between Britain and France, Céré's concern over the safe passage of the plants he sent around the Cape and across the Atlantic was well founded. In 1782, Britain's Admiral Rodney captured a ship carrying plants from the Isle de Bourbon (now Réunion) to St Domingue. Although the French sailors destroyed what they could, specimens of cinnamon from Ceylon, several varieties of mango, jack-fruit, Chinese hemp, the 'Otaheite Plumb', a moringa tree and 'Oriental ebony' were all diverted to the botanical garden established at Bath in eastern Jamaica in 1779, where they would grow alongside plants from Europe, Africa, Asia, the Pacific and elsewhere in the Americas. Rodney's gift was memorialised on an obelisk placed over the garden's waterworks. The message that imperial geopolitics made the garden grow could hardly be missed.[18]

Within the changing British Empire, botanical gardens emerged through a combination of specific local circumstances, the desirability of establishing particular long-distance linkages, and the variable ability of alliances of metropolitan advocates (particularly Banks and the Society of Arts) and local naturalists (usually Scottish surgeons) to persuade institutions such as the East India Company, the navy or Caribbean island assemblies dominated by sugar-planting slave-holders that their interests might be served by supporting the development of natural history. A garden was established on St Vincent in

1765 to acclimatise Asian plants in the Caribbean, as was the Jamaican garden. Alongside a local desire to provide drought-resistant food crops, such as the sago tree, the Calcutta garden was established in 1786 to provide plants for St Vincent, and the garden at St Helena (from 1790) was lobbied for from both East and West as a necessary waystation to enable plant transfers. Other Indian gardens of the late 1780s and early 1790s, at Madras and Samulcottah on the Coromandel coast and at Bombay, catered for particular political and environmental regions. Subsequent gardens at Penang (1800, but destroyed in 1805 and re-established in 1822), Ceylon (1810), Trinidad and Java (1817) and Sydney (1818), mapped the progress of empire.[19]

Yet the dreams of centrality of the state-sponsored botanical gardens were often unfulfilled in practice, and always required constant work. There were, of course, many gardens and many gardeners within this imperial geography of plants. In late eighteenth-century Jamaica, for example, alongside the Bath botanical garden, and often growing the same plants, were the extensive private gardens of wealthy gentlemen such as Matthew Wallen and Hinton East; the plantation gardens of horticulturalist slaveholders and overseers; and the provision grounds of the enslaved themselves. The latter have been called the 'botanical gardens of the dispossessed' and were, despite the claims of other gardeners, where vital food crops were nurtured, many with African origins.[20]

The place of the official botanical garden within this varied vegetable economy was not assured. In 1790, Thomas Dancer, the 'island botanist' in charge of the Bath garden, argued the case for 'the necessity or importance of a public garden' in a widely reported speech to the garden directors appointed by the Jamaican assembly. He understood the need to justify public funding, and did so by appealing to economic utility, competition with the French (who, he said, made exchanges of plants between their botanical gardens a matter of imperial policy), and the higher goals of civilisation, enlightenment and improvement at a time when the ethics of slave-grown sugar were being questioned as never before. Dancer's insistence on botany's public benefits also countered those who thought private interests could ensure Jamaica's botanical future. He acknowledged the important work of 'certain Gentlemen' – identifying elsewhere Wallen and East – 'for the zeal they have shown in enriching this Island with useful and ornamental plants', but informed his audience that 'I must take the liberty of suggesting, that the necessity of a Public Garden can never be superseded by their efforts as individuals – Private property is fluctuating, and the sentiments, taste, and persuits of different possessors, will seldom correspond.' He put it to 'the Public of Jamaica' that 'You will therefore concur with me in opinion, that a private

Garden can never be deemed a fit deposit of plants that are intended for the public use.'[21]

Alongside these fine words came the need to channel flows of plant material to the garden for public benefit. Rodney's imperial gift was exemplary here. More troublingly, in 1790 Dancer had to ask the Jamaican planter, politician and historian Edward Long, then in London, to ensure that plants sent by Banks were addressed directly to him, 'otherwise it is Chance if ever I get them'. This needed continual vigilance. As Dancer noted, 'Had not some of my Friends in the House of Assembly made powerful Exertions, the Seeds [the new governor Lord Effingham brought in 1789] ... would have been, in a very illiberal manner, engross'd wholly by Individuals.' When he heard of Bligh's first breadfruit voyage, Dancer hoped the ship would dock at Port Morant, close to Bath,

> otherwise I fear the Object in view will be defeated. The Gentlemen of the Country will be scrambling for all they can get (as they did for the Seeds brought out by Lord Effingham) & the Bot. Garden, the <u>Public Depository</u>, will be forgotten or come in only for a Remnant ... Duplicates of every species at least ought to be deposited in the Public Garden before any further distribution be made of them.[22]

Yet other gardeners, and the 'Public of Jamaica', at least as defined by its taxpayers, were not so sure. Hinton East's garden was certainly understood as more than a private endeavour. *Hortus Eastensis*, the published account of its contents, stressed its role – and that of gentlemen like Wallen – in bringing useful plants to Jamaica (Figure 16.4). Indeed, on his death in 1794, East's garden was acquired by the assembly along with 'thirty-nine Negroes belonging to it, many of whom are valuable gardeners', although for Dancer this only overstretched the constrained funds available for public botany.[23]

Elsewhere, Thomas Thistlewood – a slaveholding landowner on western Jamaica's plantation frontier, who started a garden in 1767 on his estate at Breadnut Island Pen – had heard in 1775 that a botanical garden was planned for Jamaica (under Wallen) on what he judged an unpromising spot at Bath. Thistlewood's accounts of his own garden, often planted with seeds from commercial nurserymen in England, demonstrate many earlier introductions than are recorded in *Hortus Eastensis*, and a pride in his claims to priority.[24] As well as engaging in extensive local exchanges of plant material, Thistlewood demonstrated a keen awareness of horticultural work as a conjoining of the local and the global. He complained about the growing conditions in his part of Jamaica – too hot, poor soil and extreme seasonality – and was a keen observer of how plants reacted to different localities. In 1776, he wrote to Long that 'In this Island is such a Variety of Soils & Situations of diff^t Temperature, that am apt to

[11]

Claſsis VIII.

O C T A N D R I A.

M O N O G Y N I A.

Tropæolum	*minus*	Indian Creſs	Peru	M. Wallen, Eſq. 1774
Melicocca	*bijuga*	Genip	South-America	
Ximenia	*inermis*	Smooth Ximenia	Eaſt-Indies	H. Eaſt, Eſq. 1784
Fuchsia	*triphylla*	Scarlet Fuchſia	Chili	H. Eaſt, Eſq.
Lawsonia	*inermis*	Smooth Lawſonia	Africa	Monſ. Nectoux*, 1789
	ſpinoſa	Prickly Lawſonia	Eaſt-Indies	H. Eaſt, Eſq. 1785
Vaccinium	*Arctoſtaphylos*	Mad^a. Whortle-Berry	Madeira	Tho. Hibbert, Eſq. 1787
Erica	*multiflora*	Many-flower'd Heath	S. of Europe	M. Wallen, Eſq. 1784
Oenothera	*pumila*	Dwarf Primroſe	N. America	M. Wallen, Eſq.
Nov. Gen.†		The Akee	Africa	Dr. Tho. Clarke, 1778

TRIGYNIA.

* Botaniſt to the French King at Hiſpaniola.

† This Plant was brought here in a Slave Ship from the Coaſt of Africa, and now grows very luxuriant, producing every Year large Quantities of fruit; ſeveral Gentlemen are encouraging the propagation of it. I do not know that it has hitherto been deſcribed; it's Characters are as follows:

Cal. Perianthium pentaphyllum inferum, foliolis ovatis acutis concavis, perſiſtentibus, villoſis.

Cor. Petala quinque oblongo-lanceolata, acuta, villoſa, ad baſin ſurſum flexa et receptaculo adpreſſa, calyce alterna et eo longiora.

Stam. Filamenta octo breviſſima, piloſa, ad baſin Germinis receptaculo glanduloſo inſerta. Antheræ oblongæ in orbem circa Germen diſpoſitæ et ejuſdem fere longitudinis.

Pist. Germen ſubovatum triquetrum piloſum. Stylus longitudine Germinis, cylindricus, piloſus. Stigma obtuſum.

Per. Capſula carnoſa, oblonga, utrinque obtuſa, trigona, trilocularis, trivalvis, apice dehiſcens.

Sem.

Figure 16.4 Plant provenance. *Hortus Eastensis: Or, A Catalogue of Exotic Plants in the Garden of Hinton East, Esq.* (Kingston, 1792). Published by permission of the Natural History Museum, London.

imagine, most of ye Fruits, both of the Torrid, & Temperate zones might be cultivated with success, and that the latter might be Naturalized by degrees.' He suggested, from his extensive reading, that a Madagascan grass, the dry rice of Cochin China, a large-leaved tree from Guiana and 'the Bread Fruit & other Productions of Otaheite' might all be 'Noble acquisitions'.[25] This practical horticultural understanding – the movement of species between places, and close attention to local conditions of production and reproduction – was made, here and in the official botanical garden, with and through the knowledge and skill of enslaved gardeners. Indeed, it was also evident in the botanical practices the enslaved pursued in their own gardens to stay alive under the deadly conditions of plantation slavery. When attention is paid to plants for food and medicine – guinea grass, pigeon peas, yams and plantains – Africa's botanical legacy in the Atlantic world comes into view, along with the work of Atlantic African gardeners.[26] Within this horticultural version of natural history as practice there are, therefore, many gardeners and a broad sense of the geography of plants. This culture of natural history might be defined as an understanding of plant characteristics in relation to the particular environments within which they thrive, and is concerned with working out what that means for moving them from one place to another.

The geography of plants

Debates over deforestation and climate from the mid eighteenth to the mid nineteenth centuries involved localised experiences on islands such as Mauritius, St Vincent and St Helena becoming networked together and globalised by connections between botanical gardens, and then generalised across imperial territories, particularly in southern Africa and India.[27] What grew where, and why, became the subject of a remarkable 'number and variety of theories relating to the geography of animals and plants' produced in first half of the nineteenth century, both before and after William and Joseph Hooker's extraordinary efforts to strike a new bargain between the British imperial state and natural history centred on Kew and its global plant collection.[28]

Humboldt's 1807 *Essai* shared, and in many ways shaped, these new concerns with mapping the global geography of plants, but also demonstrates the continuing complexity of the relationship between natural history and empire into the early nineteenth century. It was the first substantial publication resulting from his travels in the Americas between 1799 and 1804, and sought to redefine natural history within what he called 'a general physics' (*physique générale*) or 'a global physics' (*physique du monde*). This shifted attention from classification and taxonomy to the intersection of a multiplicity of natural processes at a range of scales, from the microscopic to the cosmic.[29]

'Global physics' reconceived the horticulturalist's question – what grows where and why? – on a grand scale, and as combining science, human history and aesthetics. 'This science', Humboldt argued, is 'as vast as its object, [and] paints with a broad brush the immense space occupied by plants, from the regions of perpetual snows to the bottom of the ocean, and into the very interior of the earth, where there subsist in obscure caves some cryptogams that are as little known as the insects feeding on them'. The geography of plants would also assist in 'shedding light on the prehistory of our planet' using fossil plant distributions – 'tropical bamboos buried in the ice-covered lands in the north' – to understand the implications of the cooling of the Earth's crust. It would also illuminate the span of human history, one that was certainly colonial. As Humboldt put it, 'In the European colony of the two Indies, one small cultivated plot may contain coffee from Arabia, sugar cane from China, indigo from Africa, and a host of other plants belonging to both hemispheres.' This micro-geography would then recall 'to one's imagination a series of events that caused the human race to spread over the entire surface of the earth and to appropriate all its productions'. Inquiries extended, however, way beyond utilitarian concerns to 'the character of vegetation, and the variety of effects it causes in the soul of the observer'.[30] Here, Humboldt differentiated the *botaniste nomenclateur* – 'our miserable archivists of nature' – from the *botaniste physicien*.[31] The former elucidated the specific structures of plants, the latter's imagination was excited by 'the large picture, the ensemble', nature as a landscape.[32]

Characteristically, this was to be realised through precise measurement in the field; careful calibration to determine means, variations and errors; and the visualisation of the results in striking diagrammatic forms. In their ascent of Chimborazo, then thought to be the world's highest peak, Humboldt and the botanist Aimé Bonpland collected thousands of tropical plant species, but, more importantly, quantitatively identified where and how they grew:

Since we were also carrying out at the same time astronomical observations as well as geodetic and barometric measurements, our manuscripts contain materials that can determine exactly the position and elevation of these plants. We can show the breadth of the latitudinal zone occupied by those plants, their maximum and minimum elevation, the nature of the soil in which they grow, and the temperature of the plants' native soil.[33]

This information was presented on a 'tableau' which illustrated the complex relationships between physical processes demonstrated by the variation of vegetation by elevation (Figure 16.5). Humboldt insisted that 'he would much rather know the exact geographical and elevational limits of an already known species than discover fifteen new'.[34]

Central to the tableau was an image of the geography of plants. This was partly a picture of a section of the Andes, albeit with a significantly shortened horizontal axis to fit it on the page. From bottom to top it showed the mountain landscape from beneath land and sea, through variegated vegetation and low clouds, up to the permanent snow line, the peaks (with smoke rising from Cotopaxi's cone), the high clouds and the gradually intensifying sky. It was also what Humboldt called a 'botanical map', where his observations

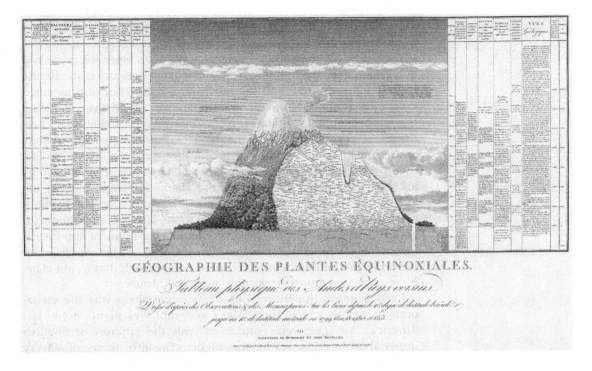

GÉOGRAPHIE DES PLANTES ÉQUINOXIALES.

Tableau physique des Andes et Pays voisins

were displayed by writing the names of plant species or genera onto the mountain at the appropriate elevation, often at a slant to indicate an elevational range. This had its methodological limits. Only a relatively small number of plants could fit on, and the 'new genera ... whose names we are not yet sure about', had not been included. Other limits were epistemological. To ensure that 'the notions that one should have about the situation of these plants [were] from a perspective that is more general and more worthy of physics', Humboldt identified botanical regions in larger letters ('as one does for provinces on ordinary maps'), such as the 'Région des Palmiers' (sea level to 990 m) or the 'Région de Chinchona' (700 m to 2,900 m).[35] Identifying these areas allowed a more synthetic discussion of plant assemblages and initial comparisons between tropical and temperate environments.

Yet, for Humboldt, 'global physics' could not and should not end with plants. He argued that the tableau 'could also help us understand the totality of our knowledge about everything that varies with the altitudes rising above sea level', a set of general physical laws. So, on either side of the central image, framed and ordered by vertical scales against which the data could be read off, he presented 'many numbers resulting from the large quantity of research conducted in various

Figure 16.5 Humboldt's *'Tableau physique'*. A. von Humboldt and A. Bonpland, *Essai sur la géographie des plantes* (Paris, 1807). David Rumsey Historical Map Collection, Stanford University, reproduced under a Creative Commons License.

branches of general physics', including many more of his own observations. These included numerical measures of refraction; the distance at which mountains are visible from the sea; the decrease in gravity, measured by pendulum; the intensity of the sky's blueness, measured by a cyanometer; the humidity, pressure and temperature of the air; the boiling point of water; and the intensity of light. They also included more descriptive information: comparative elevations, including European mountain summits; plant cultivation (noting that 'Few African slaves' are found in the zone of European wheat, quinoa, maize and potatoes, compared to the lower zone of sugar, indigo, cocoa, cotton and bananas where 'African slaves [were] introduced by the civilized peoples of Europe'); the chemical composition of the atmosphere; the limit of perpetual snow by latitude; types of animals; and underlying geology.[36] In explaining these, albeit briefly, Humboldt began to set out a science based on precise measurement and instrumentation, the interrelation of many processes at a range of scales, and the ordered equilibrium of the cosmos: an approach that would characterise his vast future production in natural history.

Early nineteenth-century European imperialism was the environment within which Humboldt's geography of plants grew. His American voyages were conducted with the support of Spanish imperial authorities, yet he was highly critical of them, of slavery and of the treatment of indigenous people.[37] His methods deployed all the power of Enlightenment rationality, but combined with a Romantic sensibility which damned the modes of cultivation 'where civilization perfected itself' and 'which makes the European cultures seem so monotonous and hopelessly dull'. Yet he also argued that near the equator 'man is too weak to tame a vegetation that hides the ground from view and leaves only the ocean and rivers to be free'.[38] His engagement with the *Naturphilosophie* of Schelling and Goethe, which posited an organic dynamism as the soul of nature, set him against the narrow utility of much imperial science. Yet Humboldt's envisioning of environments via both accurate survey and rich narration opened them up to European colonisation.[39] He has been seen as drawing on older Creole versions of Andean nature, as well as presenting something radically new.[40] He has also been seen as articulating a different version of voyaging and empire to that promoted by Banks, yet simultaneously as connected to Banksian natural history via the influence and example of Georg Forster.[41] How central Humboldt's notions of 'what grows where and why' were to natural history in the British Empire depends on where, when and how you look. The evidence from his vast scholarly production and its complex legacy lies neither on one side nor the other. What is clear, however, is that such ways of thinking emerged from the long and complex imperial engagement with the global geography of plants and its horticultural vision.

Vegetable empire

It has been argued that 'the study of animal and plant geography in nineteenth-century Britain was one of the most obviously imperial sciences in an age of increasing imperialism'. In particular, William Hooker's Kew Gardens, which collected 'natural objects gathered during various expeditions, and those sent back to Britain from the nation's far-flung outposts', is seen to have represented 'the whole culture of imperial enterprise'.[42] Instead of focusing on botanical knowledge via plant collection and taxonomy, or tracing the development of the later term 'biogeography', this chapter has examined the broad notion of the geography of plants by working forward from the end of the Seven Years' War, rather than looking back from the 1860s. Considering how this geography was understood and acted upon in the late eighteenth century offers a way to decentre a unified 'colonial botany' by acknowledging the variety of forms of knowledge and practice associated with growing and moving plants, and the many different relationships with empire that they forged.

Examining how the question of what grows where and why was understood and addressed brings a range of engagements with plants under consideration, and encourages a focus on matters such as the material practicalities of plant transportation in an age of oceanic sail and imperial geopolitics, as well as issues of collection, classification and taxonomy. This is a horticultural view of nature, and one that multiplies the sorts of gardeners who were involved in the processes of 'naturalisation'. It means situating the particular histories of imperial botanical gardens, and those who ran them, within a broader historical geography of horticultural practice – one that must include enslaved women and men, for example, as well those who claimed ownership of them – which shaped engagements with an increasingly global geography of plants, and within which state-sponsored botany had difficulty staking its claim. Finally, alongside the undoubted transformations in early nineteenth-century natural history, there were also continuities with this broadly horticultural understanding of the geography of plants. Humboldt's attempt to create a 'complete historical geography of the earth' could both underpin and undermine imperial projects of different sorts as he tried to define 'the natural divisions of the vegetable empire', as they existed 'not in the greenhouses and books of botany but in Nature itself'.[43]

Further reading

Carney, J. A. and Rosomoff, R. N., *In the Shadow of Slavery: Africa's Botanical Legacy in the Atlantic World* (Berkeley, 2009).

Dettelbach, M., 'Humboldtian science', in N. Jardine, J. A. Secord and E. C. Spary (eds.), *Cultures of Natural History* (Cambridge, 1996), pp. 287–304.

Drayton, R., *Nature's Government: Science, Imperial Britain and the 'Improvement' of the World* (New Haven, 2000).

Grove, R. H., *Green Imperialism: Colonial Expansion, Tropical Island Edens and the Origins of Environmentalism, 1600–1860* (Cambridge, 1995).

Humboldt, A. von and Bonpland, A., *Essay on the Geography of Plants* (1807; Chicago, 2009).

Miller, D. P. and Reill, P. H. (eds.), *Visions of Empire: Voyages, Botany and Representations of Nature* (Cambridge, 1996).

Ogborn, M., 'Talking plants: botany and speech in eighteenth-century Jamaica', *History of Science*, 51 (2013), pp. 251–82.

Schiebinger, L. and Swan, C. (eds.), *Colonial Botany: Science, Commerce and Politics in the Early Modern World* (Philadelphia, 2005).

Plate 1 A drawing of a parrot from Brazil. Ink and watercolour, cut out and pasted onto paper. Because he had not seen the bird himself, Gessner described this parrot from Brazil from the picture he had, as follows: 'its neck, wings and back are entirely green, though the colour on the back tends to blue, as does the tail, whose largest feathers are red. The longer feathers in the wings are black. In the green feathers of the same wings are some spots or dots that are red. The head is grey. The belly is of a yellow colour' (*Historia animalium*, vol. III, p. 691). This matches the original drawing that has survived in Felix Platter's album in Basel, which was not reproduced in *Historia animalium*. MS K,I,1, 56r, Universitätsbibliothek Basel.

Psittacus uarius ruber.

Gesn. p. 88.

Plate 2 The drawing of a parrot that Gessner named '*Erythroxanthum*' (from the Greek for 'red–yellow'), because of its plumage. Ink and watercolour, cut out and pasted onto paper. MS K,I,1, 52r, Universitätsbibliothek Basel.

Plate 3 One of the earliest known drawings of the American tomato, in the image collection of the Venetian naturalist Pietro Antonio Michiel, *c.*1540–60. Biblioteca Nazionale Marciana, Venice. MSS Marciano It.II, 26 (= 4860), fo. 80r.

Plate 4 Drawing of the Mediterranean '*mullus barbatus*' (red mullet) by an anonymous painter. It was sent from Rome to Conrad Gessner, who did not use it as model for his printed woodcut illustrations. Special Collections of the University of Amsterdam, MS III C 22, fo. 88r.

Plate 5 A portal onto Mexica ritual life and historical record. The entry on the '*vitzitzili*' or 'hummingbird' depicting the animal in the various stages of its life. *Florentine Codex*, Book 11, MS Mediceo Palatino 220, Biblioteca Mediceo Laurenziana, Florence, fo. 24r. Reproduced with permission of MiBACT. Further reproduction by any means is prohibited.

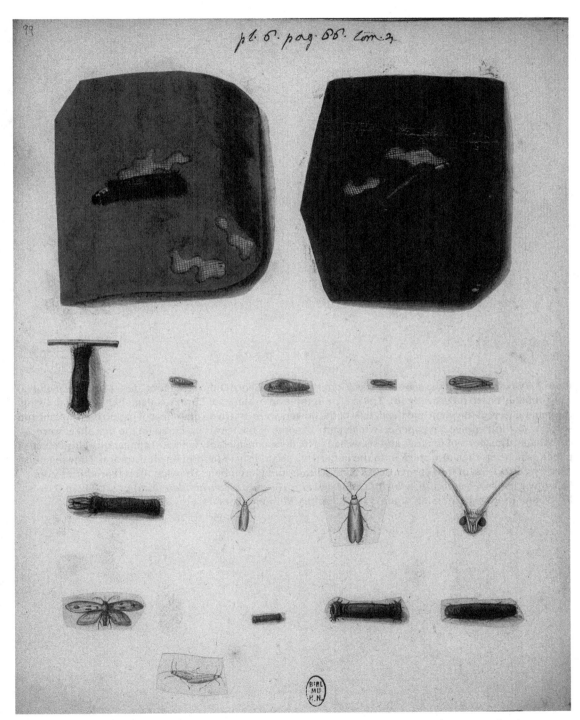

Plate 6 Tinea larvae on woollen cloth. C. Aubriet, drawing for R.-A. Ferchault de Réaumur, *Mémoires pour servir à l'histoire des insectes* (Paris, 1734–42), vol. III, plate 6. MNHN (Paris), Direction générale déléguée des collections, Bibliothèque centrale.

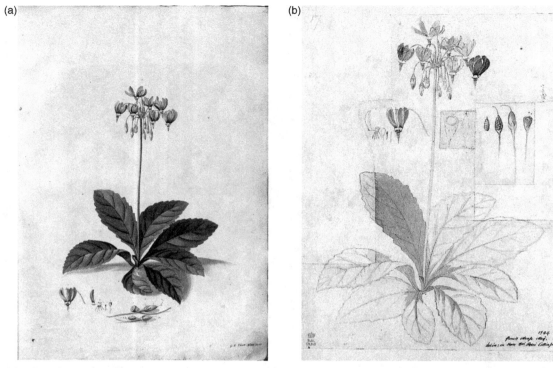

Plate 7 Two drawings of the shooting star (*Götterblume*) by Georg Dionysius Ehret, dated 1744. Drawing (a) was probably based on drawing (b). The sketch in (b) is an example of Ehret's collage-like style of drawing: the general view of the plant and the details of the flowers were drawn on one sheet of paper, then cut out and stuck onto another piece of paper on which parts of some of the leaves executed on the first sheet were completed. The views of the fruit and seeds had to be done considerably later and appear on a third piece of paper, which was then also stuck onto the underlying second sheet to form one illustration. Drawing (a) is used by permission of the Germanisches Nationalmuseum Nuremberg. Drawing (b) is reproduced from H. Ludwig, *Nürnberger naturgeschichtliche Malerei im 17. und 18. Jahrhundert* (Marburg, 1998), p. 216, XXXV, and used by permission of the Natural History Museum, London.

Plate 8 '*Heliconia*', tempera on paper, 54 x 38 cm. J. M. Carbonell (Royal Botanical Expedition to the New Kingdom of Granada, 1783–1816). Archivo del Real Jardín Botánico, CSIC, Madrid, Div. III, lám. 610.

Plate 9 This oil painting depicts a woman feeding a parrot in the kitchen of a wealthy burgher's home. *Interior with a Woman Feeding a Parrot*, known as '*The Parrot Cage*', Jan Havicksz Steen, *c.*1660–70, oil on canvas on panel, 50 x 40 cm. Rijksmuseum, Amsterdam.

Plate 10 Volumes from the xylothek made in Nürnberg, Germany in the 1800s. (a) Each volume is constructed using the wood of the tree it describes and the spine is covered with the corresponding bark and associated mosses and lichens. (b) Inside there are dried leaves, flowers, seeds, a piece of the root, cut branches and a compartment for the botanical description of the tree from which the wooden book was made. Courtesy of the SLU University Library, Alnarp, Sweden. Photographs by Mikael Risedal.

Plate 11 Loose seaweed specimens in a folder labelled '*Delessaria*', placed opposite the descriptions of the species in this genus in William Harvey, *A Manual of the British Algae* (London, 1849), between pp. 54 and 55. Author's collection.

Plate 12 Elephant-gate, Berlin Zoo, 1904. Postcard. Author's collection.

Plate 13 Concert salon, Berlin Zoo, 1904. Postcard. Author's collection.

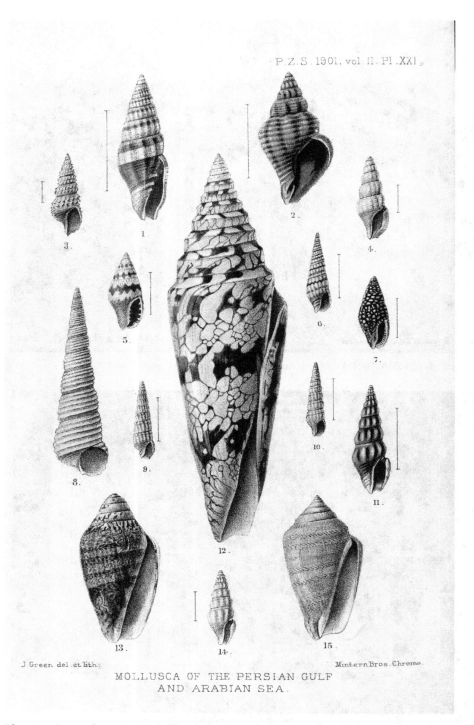

Plate 14 *Conus clytospira* (Melvill and Standen, 1899) in *Proceedings of the Zoological Society of London*, 1901, plate 21. National Museums Scotland.

Plate 15 *Colonial and Indian Exhibition Daily Programme* (London, 1886). Author's collection.

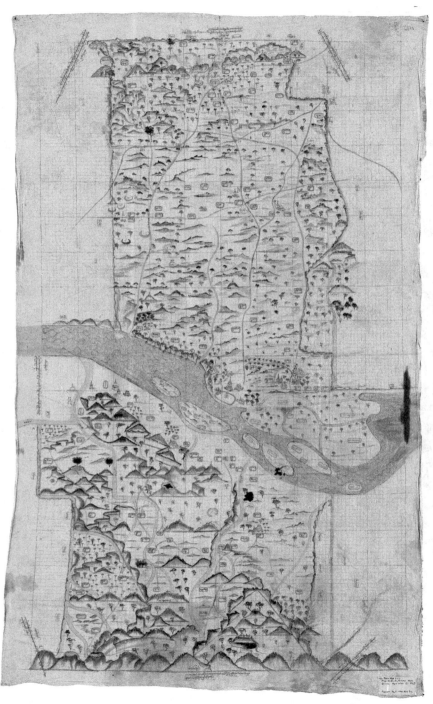

Plate 16 This large textile map measures 300 x 176 cm and was painted by a Burmese artist around 1860, showing mountains, waterways and Buddhist temples. 'Map of Sa-lay township, Upper Burma: present-day Sale, on the Irrawaddy river, just south of Chauk in the Bagan Region', University Library, Cambridge. By kind permission of the Syndics of Cambridge University Library, Maps MS. Plans, R.c.2.

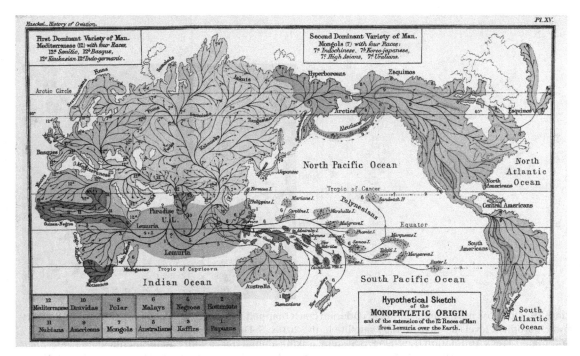

Plate 17 Hypothetical map assuming a single origin for the distribution of humans from the sunken continent of Lemuria. E. Haeckel, *The History of Creation* (London, 1876), vol. II, frontispiece pl. XV; translated by E. Ray Lankester from *Natürliche Schöpfungsgeschichte* (2nd edn, 1870). Reproduced by kind permission of the Syndics of Cambridge University Library.

Plate 18 Specimens of Zoological Collection, MASMO 4524, National Collection of Birds, Ciudad Universitaria, Mexico City. Photo by Miguel Ángel Sicilia Manzo. Banco de Imágenes CONABIO, all rights reserved.

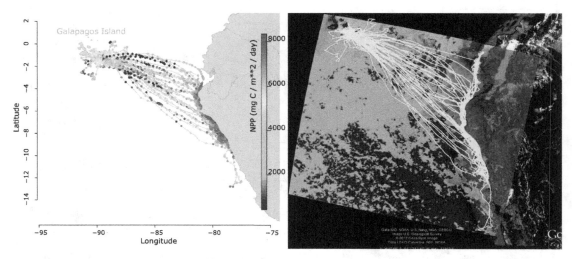

Plate 19 Albatross movements off the Ecuadorian coast mapped in relation to ocean net primary productivity and chlorophyll levels as measured by satellite in the 2010s. S. Dodge et al., 'The Environmental-Data Automated Track Annotation (Env-DATA) System: linking animal tracks with environmental data', *Movement Ecology*, 1 (2013), DOI: 10.1186/2051–3933-1–3, figure 5. Dodge et al.; licensee BioMed Central Ltd, 2013.

Plate 20 David Attenborough tussles with mountain gorillas on location in Rwanda during filming for the BBC series *Life on Earth* (1979). His entirely ad-libbed speech on the connection between humans and other primates demonstrated his consummate skill as a narrator. © John Sparks/naturepl.com.

III Publics and empires

17   Containers and collections

When Michel Foucault analysed how ways of seeing were structured by the framework of eighteenth-century natural history, he located change from an earlier age of curiosity not in the *things* observed but in the *spaces* in which they were observed:

The documents of this new history are ... unencumbered spaces in which things are juxtaposed: herbariums, collections, gardens; the locus of this history is a non-temporal rectangle in which ... [objects] present themselves one beside another ... grouped according to their common features, and thus already virtually analysed, and bearers of nothing but their own individual names.[1]

This decontextualisation was necessary for classification and for the production of order from disorder. Natural history, as characterised by Mary Louise Pratt, 'extracts all things of the world and redeploys them in a new knowledge formation whose value lies precisely in its difference from the chaotic original'.[2]

In this chapter I will suggest that the 'unencumbered spaces' so necessary for the new way of seeing the natural world as described by Foucault were, in fact, containers, and that notions of containment and the work involved in containing things provides us with a more fruitful way to investigate the practices that comprised natural history in this period. Foucault's generalisation reflects collections once they have been scientifically arranged; that is, the point at which the intellectual labour and observational skill involved in their compilation and organisation appear to have been erased. In contrast, by drawing on studies of material culture and recent histories of scientific observation, I shall argue that containers, considered as concrete things, allow us glimpses into the processes whereby collectors not only learned to observe and to order their thoughts, but also reveal some of the ways in which novices could be drawn into natural history and become useful contributors. I will illustrate this by focusing on British botany, as it remained a largely private activity during the first half of the nineteenth century, with independent participants spread across the country. As a consequence, it provides much evidence that containers and the notion of containment were essential to the progress of science.

Vegetable boxes

Historians have begun to discuss the mental dispositions of collectors and the relation of collections to educational and mercantile practices in the eighteenth and nineteenth centuries. The ordering of objects and information in containers was an important part of the construction of individual identity.[3] Indeed, the mind itself had long been likened to a container: John Locke saw it as a cabinet and Joseph Addison as a drawer of medals, but it had most commonly been described as a storehouse characterised by its contents. These depictions of the mind as container-like show how mental life was embedded in its material context; explanations of mental activity were modelled on the observable features of collections of things, while collections of objects reflected a set of mental habits and a way of ordering the world.[4] Just as the drawers and shelves of a cabinet served to contain specimens and other objects, so the categorised pages of a commonplace book or the columns and compartments of custom-made notebooks for naturalists were the material forms for the organisation, retrieval and comparison of ideas and observations. The eighteenth-century naturalist Gilbert White of Selborne jotted down his lifetime's observations of the natural world in *The Naturalist's Journal*, a notebook consisting of printed forms designed for the systematic recording of a variety of natural phenomena over a year. He continued to purchase the *Journal* annually even though after the first couple of years his observations frequently overflowed the boundaries of the allocated spaces (Figure 17.1). When, in 1787, White did not have a copy of the published *Journal,* he ruled out columns on similarly sized sheets of paper to correspond with the pages in the printed version, and used these to contain his observations.[5]

Analyses of collecting as a cultural and behavioural phenomenon encompassing both the social and the psychological are particularly pertinent in the context of British botany with its reliance on private collections well into the nineteenth century. When the Norfolk botanist Dawson Turner travelled to France in 1814 following the cessation of the Napoleonic Wars, he was surprised to discover that 'all the Botanists & all the men of science in France are collected in Paris ... & everything public'.[6] French naturalists were contained in their institutions just as much as their specimens. Turner conceded that large public collections in which objects were gathered together in one place possessed the

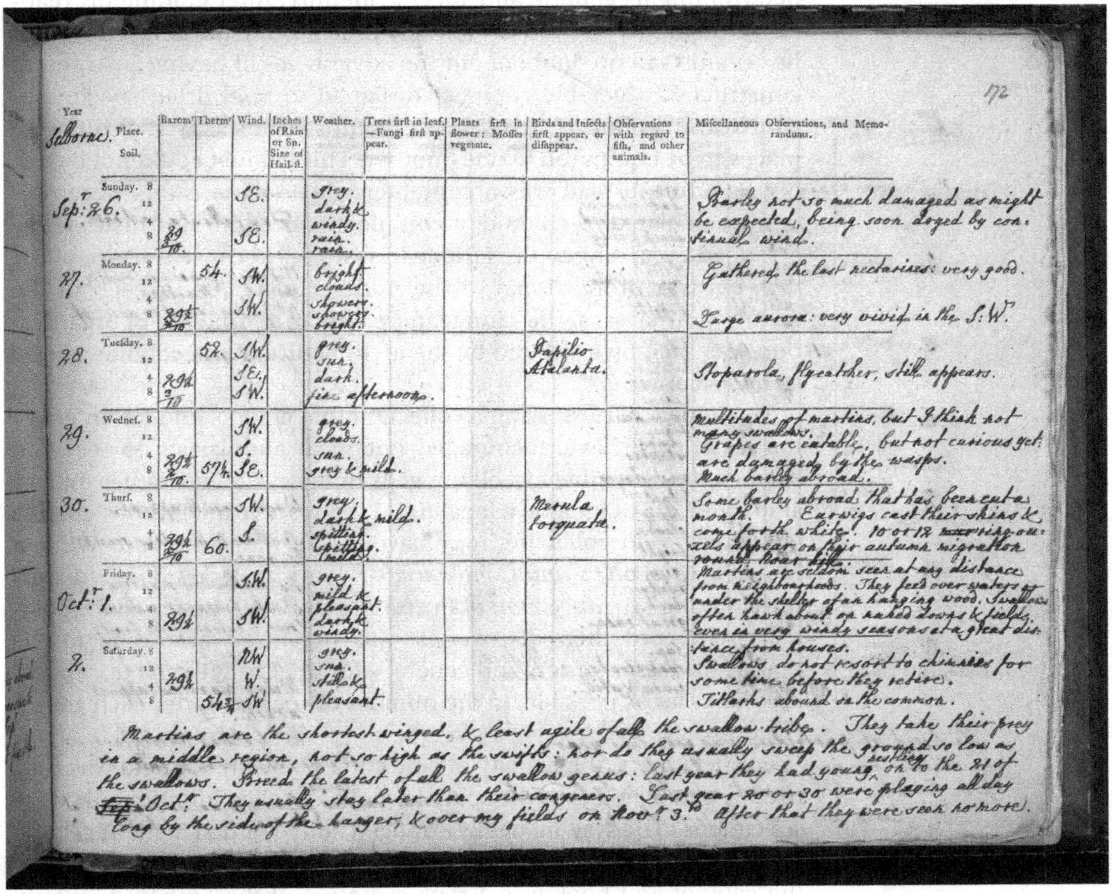

Year Selborne. Place. Soil.	Barom.	Therm.	Wind.	Inches of Rain or Sn. Size of Hail-st.	Weather.	Trees first in leaf. —Fungi first appear.	Plants first in flower: Mosses vegetate.	Birds and Insects first appear, or disappear.	Observations with regard to fish, and other animals.	Miscellaneous Observations, and Memorandums.
Sepr: 26. Sunday.	29 3/10		SE. SE.		grey. dark & windy. rain. rain.					Barley not so much damaged as might be expected, being soon dryed by continual wind.
27. Monday.	29½	54.	SW. SW.		bright. clouds. showers. showery. bright.					Gathered the last nectarines: very good. Large aurora: very vivid in the S.W.
28. Tuesday.	29½ 3/10	52.	SW. SE. SW.		grey. sun. dark. fine afternoon.			Papilio Atalanta.		Stoparola, flycatcher, still appears.
29. Wednes.	29½	57½.	SW. S. SE.		grey. clouds. sun. grey & mild.					Multitudes of martins, but I think not many swallows. Grapes are eatable, but not curious yet: are damaged by the wasps. Much barley abroad.
30. Thurs.	29½	60.	SW. S.		grey. dark & mild. spitting. spitting. mild.			Merula torquata.		Some barley abroad that has been cut a month. Earwigs cast their skins & come forth white. 10 or 12 ~~ring-ou~~ zels appear on their autumn migration round Noar hill.
Octr: 1. Friday.	29½		SW. SW.		grey. mild & pleasant. dark & windy.					Martins are seldom seen at any distance from neighbourhoods. They feed over waters or under the shelter of an hanging wood. Swallows often hawk about on naked downs & fields, even in very windy seasons at a great distance from houses.
2. Saturday.	29½	54¾	NW. W. SW.		grey. sun. still & pleasant.					Swallows do not resort to chimnies for some time before they retire. Titlarks abound on the common.

Martins are the shortest winged, & least agile of all the swallow-tribe. They take their prey in a middle region, not so high as the swifts: nor do they usually sweep the ground so low as the swallows. Breed the latest of all the swallow genus: last year they had young sitting on to the 21 of ~~Sep~~ Octr. They usually stay later than their congeners. Last year 20 or 30 were playing all day long by the side of the hanger, & over my fields on Novr. 3. After that they were seen no more.

Figure 17.1 Gilbert White's records in *The Naturalist's Journal* from 26 September to 2 October 1773. © The British Library Board (Add. MS 31846, fo. 172r).

advantage of allowing a scholar to 'compare as well as observe'. Nonetheless, when considering public French collections in relation to private English ones, he concluded that while there was 'more splendor' in France there was 'more enjoyment' in England.[7]

Botany had become popular in Britain in the 1770s when the Birmingham physician William Withering published *A Botanical Arrangement of All the Vegetables Naturally Growing in Great Britain* and introduced the British public to the 'System of the Celebrated Linnaeus'.[8] The Linnaean system of classification provided an easy way of acquiring and arranging information about plants, based as it was on the number, shape, position and proportion of a flower's sexual parts – its stamens and pistils. These variations allowed Linnaeus to divide plants into twenty-four classes; knowing which class a plant belonged to was the starting point for determining its generic and specific identity. After guiding his readers through this method of plant identification, Withering left would-be botanists in no doubt about the advantages of having a carpenter construct a 'vegetable cabinet', designed to hold dried specimens, which would allow plants from different seasons as well as different places to be compared to one another. This cabinet could not be any old box, but instead was a container whose dimensions were precisely calculated to contain 'a compleat collection of British plants' arranged according to the Linnaean system.[9] Because the size of the drawers was proportional to the number of plants in each of the twenty-four classes, the Linnaean system as a principle of organisation was thereby exhibited by the architecture of the cabinet as well as by its contents.[10]

The appeal of possessing a collection that could stand for an entire flora may well have encouraged botanical enthusiasts to seek out specimens from others to fill the gaps in their cabinets rather than to study the flora of their own localities. Even works explicitly designed to encourage field collecting may have led to the same end. William Mavor's *The Lady's and Gentleman's Botanical Pocket Book adapted to Withering's Arrangement of British Plants* (1800), designed for botanists to record their discoveries, was organised according to the Linnaean classes, orders and genera, with 'spaces left to fill up, apportioned, as far as possible, to the number of species under each genus' (Figure 17.2).[11]

For beginners, the activity of collecting plants in the field was thus usually connected with the spatial order of a collection. Not only were specimens to be systematically arranged and recorded in the vegetable cabinet and in the collecting book respectively, the very structure of the cabinet and book was a representation of the Linnaean system, even when the objects and records contained were not visible. Such order, no matter how artificial, was considered essential for comprehending nature at a glance, by allowing the viewer to understand not only an individual object but also its relation to others by means of the container. Although the physical confines of Withering's vegetable cabinet and Mavor's collecting book rapidly became too small to contain representative collections as knowledge of even local floras increased, they serve as a striking reminder that a collection, as Susan Stewart points out, 'is not constructed by its elements; rather it comes to exist by means of its principle of organization'.[12] Nature is taken to be nothing more or less than the group of decontextualised natural history objects which is articulated by the classification system to hand; through this 'fiction', Stewart makes clear, 'it is the Linnaean system which articulates the identities of plants . . . and not the other way round'.[13]

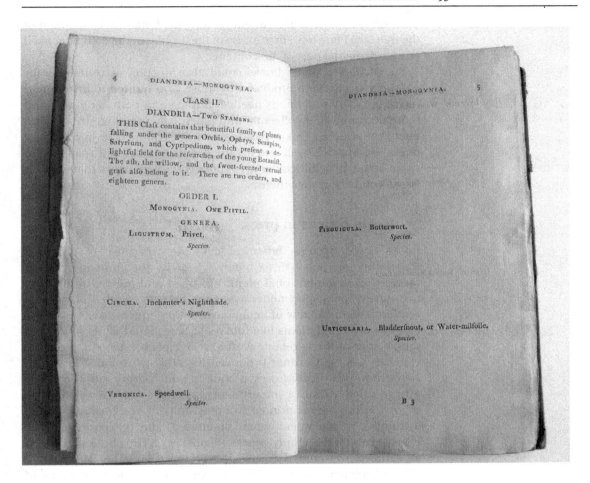

In its aim of promoting the study of indigenous botany through an understanding of the Linnaean system, Mavor's book resembles a box because it relied less on words than space. Filling this box with the Linnaean names of found plants, however, meant adding words, thus making it more book-like. This slippage between books and boxes is more characteristic of botanical practice than is generally recognised. The most literal overlap between boxes and books occurred when specimens that defied easy containment were used to construct boxes that looked like books. The resulting wooden libraries, or xylotheks, were particularly popular in Germany but can also be found in some other European countries. They consist of books made out of particular types of wood, whose spines are covered with the corresponding bark and associated mosses and lichens (see Plate 10a). Once opened, these wooden books reveal samples of dried leaves, flowers, roots,

Figure 17.2 An example of the blank spaces left for recording the plants observed when botanising in the field, organised according to the Linnaean system. From William Mavor's *The Lady's and Gentleman's Botanical Pocket Book adapted to Withering's Arrangement of British Plants* (London, 1800). Author's collection.

etc., with a special compartment holding the botanical description of the tree (see Plate 10b). In the case of these objects, the container itself is a fundamental part of the contents.

But published paper books could also function as containers for specimens, and were produced by those who wanted to inspire and discipline the visual habits needed to transform both neophytes and more accomplished collectors into potentially useful contributors to botanical science. This aspiration was not confined to introductory books like Mavor's: well into the nineteenth century, it was applied to some of the most difficult areas of botany through books that were, in effect, cabinets of specimens.

Specimens in press

Books could become botanical containers because of a spatial transformation peculiar to botany; namely, the dramatic loss of form as three-dimensional plants are dried and pressed to become two-dimensional specimens. It was the flatness of dried plants that rendered them capable of being incorporated into books. And in the case of very small plants like mosses, a book could allow you to put this section of the vegetable world in your pocket. In 1836, the plant collector George Gardner produced his *Musci Britannici, or Pocket Herbarium of British Mosses,* a book in which every other leaf 'has a page ruled off into compartments, suited to the size of the species ... & the names printed in lithography'.[14] The book contained a number of specimens when purchased, and the possessor was expected to fill the empty spaces (Figure 17.3). Thus books, as much as cabinets, through the provision of space and the flattening of plant specimens, possessed the potential to act both as works of reference and as records of their owner's field experience. This is underlined by the fact that other than the title page, Gardner's book has no explanatory text: it is simply a paper cabinet. The visual acuity required to detect tiny mosses in the field was honed by placing their pressed forms in the framework of a classificatory system, which was itself quite literally represented in the labelled blank spaces of a book.

The observational skills required by this activity, and the resulting suites of dried specimens, were essential for the study of groups of plants like mosses and seaweeds, whose classification was difficult. Filled copies of Gardner's *Pocket Herbarium* are testimony to their owners' visual acuity and existing knowledge. In order to develop the observational skills necessary to distinguish between different moss species in the first place, beginners could turn to introductory books that also came equipped with actual specimens. A notable example is a modest guide to the study of mosses

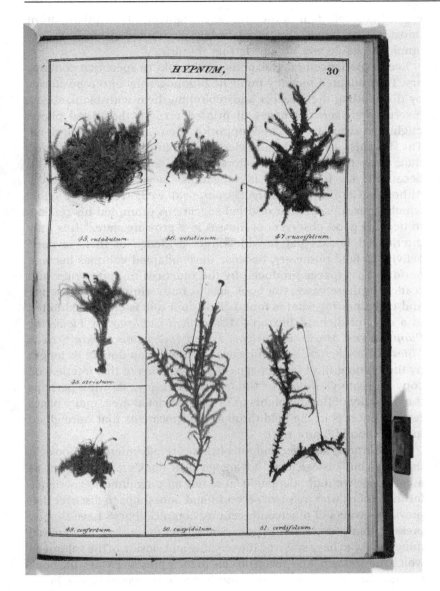

Figure 17.3 Specimens of different species of *Hypnum* in George Gardner, *Musci Britannici, or Pocket Herbarium of British Mosses* (Glasgow, 1836), p. 30. By kind permission of the Master and Fellows of Selwyn College, Cambridge.

published in 1846. This work, *Twenty Lessons on British Mosses*, by the Dundee umbrella maker William Gardiner, was aimed at young novices. Gardiner's 'little' publication devoted to 'minute' plants rejected engravings in favour of 'real specimens' because 'the eye can more readily recognise a plant in the growing state by this means, than by the most careful delineations of the pencil'.[15] His *Twenty Lessons* included a brief introduction to mosses, an explanation of their structure, and eighteen descriptions from 'most of the larger and more widely-distributed genera' as 'examples to guide you in your farther inquiries'.[16]

Even experienced moss experts used books as specimen containers. They adapted recently published monographs into repositories by dismantling these works and rebinding them with blank sheets inserted between the pages of printed text, so that dried plants might be securely attached opposite their printed descriptions. The botanist and Yorkshire clergyman James Dalton advocated that all serious moss collectors should follow this procedure because it 'facilitated investigation beyond conception'.[17] Although Gardiner's *Twenty Lessons* sold well, we do not know whether his presentation of dried specimens prompted his readers to become good observers of mosses in a growing state.[18] It is only interleaved books that provide direct evidence of their owners' activity as field observers, because these adapted volumes become performative spaces produced by the practices of both writer and reader. In these cases, the book acts as both stimulus to collecting and container for what is found. This dual role is clearly exhibited in a copy of Richard Buxton's *A Botanical Guide to the Flowering Plants, Ferns, Mosses and Algae, Found Indigenous within Sixteen Miles of Manchester* (1849), increased to more than double its length by the binding-in of blank pages. While sections of this interleaved copy of Buxton's *Botanical Guide* remain empty, showing perhaps the difficulty of finding plants in even a restricted area, many pages register success in the field through the specimens that were glued into the book (Figure 17.4).

Book owners, however, did not always carefully interleave books in this way, but instead took advantage of a book's ability to act as a container without adaptation. It is not an uncommon experience for those of us who frequent second-hand bookshops to discover that previous owners of nineteenth-century botanical books have slipped pressed plants between the pages. Sometimes these are poor battered things, lucky to have survived the opening and closing of the volume in which they remain, and giving no clue to their presence when the book is shut. In other cases, loose specimens can overwhelm a book, increasing its girth to such an extent that the covers bulge and no longer fulfil their protective function. These records of practice are unlikely to be encountered through systematic research, as they are only rarely found in libraries or museums.[19] Rather, such specimens and their host texts are serendipitous discoveries, because the context of their use and preservation is that of private ownership. They are thus accidental remnants of past practice, lucky survivors that prove to be rich sources of evidence. One example is a copy of William Harvey's

A *Manual of the British Algae* (London, 1849), bursting at the seams with specimens. Far from signalling a lack of serious study or indicating a slovenly attitude to collections, the book suggests that the collector used Harvey's descriptive work as an ordering device, in effect as a cabinet or box. The specimens were slipped in between the pages in labelled paper folders, and although they are loose within the book, care was apparently taken over their preparation and positioning in relation to the text, for they are placed between the pages on which the genus to which they belong was described (see Plate 11).

Figure 17.4 Specimens glued into an interleaved copy of Richard Buxton's *A Botanical Guide to the Flowering Plants, Ferns, Mosses and Algae, found Indigenous within Sixteen Miles of Manchester* (London, 1849). Author's collection.

Thinking outside the box

These examples of books used as containers akin to botanical cabinets or boxes – that is, as things whose contents and structure reflect a classificatory system – provide ways to understand the shock experienced by a Victorian gentleman on encountering a working-man's collection stored in books. After the Warrington botanist William Wilson had received some rare and well-prepared specimens from the handloom weaver John Martin, he considered the man '*addicted* to neatness'.[20] However, when Wilson later visited Martin's cottage, he was shocked to discover that Martin's plant specimens were 'rather carelessly mixed in the leaves of a copy of Withering & in other Books, which are not so clean as I expected', and he was puzzled that there were few outward signs of '*order & arrangement*' when Martin's mind seemed to be 'very well regulated' and he was 'an original & patient thinker'.[21] The clue to Wilson's distress lies less in the use of a book as a container, and more in his references to neatness, care, arrangement and cleanliness – all essential attributes to a social class that ranked order among its highest attributes. The perceived careless mixing meant that Martin's collection was not articulated and rendered scientific by the classification system of the book. Despite Martin's keen ability to identify plants according to the Linnaean system, the outward disorder of his specimens caused Wilson to revise his view of the inward order of Martin's mind.

It is therefore ironic that one of the neatest examples of a book as specimen container was created by a collector using a Manchester flora produced by one of Martin's botanical confrères, the destitute Lancashire shoemaker Richard Buxton, who declared in his autobiographical introduction that it was his custom 'to observe living plants on the spots where they grew, and not to make great collections of dried specimens'.[22] Although Buxton first learned the Linnaean classification from Withering's book, and was well aware that many people 'attempted to learn botany from dried specimens', he himself 'could never look upon a dried specimen with much satisfaction': to obtain 'a true idea of a plant', he held, 'let me see it alive and flourishing in the place where it grows, surrounded by all the conditions necessary for its growth'.[23] Collections can thus become a way of distinguishing between artisanal and genteel ways of seeing. For middle-class collectors, whose observation was shaped by classificatory systems, collections became the *basis* of their knowledge of nature; for artisans, whose knowledge of plants usually preceded their understanding of classification, collections were viewed as the *product and result* of their knowledge of nature. Botanical containers and their contents are therefore 'things that talk' in the sense of imparting their meaning through their particularity and context.[24]

Although the specimens trapped in privately owned books were not necessarily destined for circulation, exchange or display, they symbolised the potential of their collectors to contribute to botany. Specimens gave a wide range of men and women, from artisans to the aristocracy, the capacity to act in the largely private world of British botany; dried plants were not only material objects of exchange, but also signs of the attainment of the requisite observational skills to participate in the science. They linked the individual to the whole, and tied private collecting to a notion of public good. In 1800, Mavor ensured that the users of his collecting book were aware that 'the united labours and observations' of botanical observers would, in time, probably result in 'a more perfect work on indigenous botany ... than by any other method hitherto attempted'.[25] We do not know how many of the collectors of the specimens placed in books went on to contribute to the researches of expert botanists. But regardless of whether these individuals' enthusiasm for botany fizzled out, or their visual acuity came to be evaluated in terms of the scientific usefulness of their collections, ultimately it is the evidence their books provide of active engagement with botany that is significant. Moreover, botanical guidebooks and audience responses encompassing specimens cut across class and gender lines, thus indicating that textual spaces that acted as containers played an important role in creating a wide and diverse community of botanical observers.

Botanical housekeeping

The network of collectors scattered across the country was of immense importance in providing specimens to serious students of nature for whom a collection came to be seen as an essential attribute. William Hooker, for example, although already celebrated as field collector, marked his decision to dedicate himself to botany in 1806 by aspiring to fill a four-foot-square specimen cabinet.[26] By the time he moved to Kew in 1841 as director of the botanic gardens, his collection of dried plants, which remained his private property until his death, was so vast that it occupied seven rooms of his house.[27] While Hooker had little trouble in attracting contributions in the form of specimens for his taxonomic work, 'what I much want', he admitted three years later, 'is a competent person to assist in keeping my Herbarium in order'.[28] By the following year, things were so out of hand that his son, Joseph Dalton Hooker, recently returned from a voyage to the Antarctic regions, was 'bewildered' by the amount of work needed to maintain order in his father's herbarium and to get even the general collection into a state adequate to allow it to be used for reference; the 'rooms are so lumbered', he fretted, 'that the cabinets cannot be got to without great trouble'.[29]

The fact that even large collections like Hooker's were private meant that not only the arrangement of collections, but the very work of natural history, was seen to be dependent upon a bourgeois domestic economy.[30] In order to have a useful collection, that is, one from which scientific generalisations could be made, the preservation and ordering of specimens was essential. In 1831, therefore, the botanist George Arnott Walker-Arnott gave up his summer trip to the Scottish Highlands when he found that he was 'so much in arrears' with keeping his herbarium in order that he had '10 or 15 thousand specimens to intercalate'; earlier, the Edinburgh botanist Robert Kaye Greville, who ultimately possessed a herbarium of 30,000 specimens, grumbled that he had 'so much on my hands in the house I have comparatively rare opportunities of searching for plants myself'.[31] Even modest collections required much time and dedication. The Welsh surgeon and botanist John Roberts, who wished to make a complete 'local Herbarium', confessed that 'I have often spent a day in botanising [&] collected several good & rare plants but in consequence of my professional avocations allowed them to get mouldy & have thrown them away to *my great regret*.'[32] Expert moss classifier William Wilson complained about having to set aside his work on botanical systematics because of his 'time being so very much engrossed with manipulation, arranging and stowing away specimens'.[33]

By 1856, a domestic advice manual included both information regarding household chores and the preservation of natural history collections.[34] This was pertinent not only for those who collected as a hobby, but also for botanists engaged in scientific study. The association of botanical *science* with domesticity was made clear by James Edward Winterbottom. In 1852, together with Captain Richard Strachey of the Tibetan Boundary Commission, he greatly enjoyed collecting mosses in the Himalayas, but claimed 'I am no botanist' because he disliked the 'sorting and arrangement of dried specimens'. In his opinion the 'dry department' was 'better suited to married & settled gentlemen, than to those who are not blessed with an inducement to remain at home'.[35] Botanical science was increasingly tied to the need for dried specimens, and scientific study could not be separated from the domestic labour of maintaining a collection. Through this work, in Winterbottom's view, the scientific botanist was contained within the cabinet.

Inside the cabinet

The relation between collectors and their collections was an important aspect of the practice and work of early nineteenth-century naturalists. As Susan Stewart has pointed out, the 'ultimate term in the series that marks the collection is the "self", the articulation of the collector's own "identity"'.[36] The articulation of identity through a collection was made strikingly explicit by William Wilson, the Warrington botanist who was so shocked by the disorder of Martin's specimens. A devout Congregationalist of modest independent means, Wilson abandoned a legal career because of nervous debility. He underwent much soul-searching before deciding upon a botanical vocation in the 1820s. William Hooker soon came to rely on him for information: 'I ask you, because I know you study plants, more perhaps, than any Botanist in the field, & have thus the advantage over us who are obliged to work more in the closet.'[37] By this time, however, Wilson was declaring his intention to study the natural system of classification and was 'already sensible of the value of a well stored herbarium for such a purpose'.[38] Initially unwilling to place a specimen in his collection unless he had seen it growing, it was not until 1831 that Wilson could state 'I am willing now to know a plant in a dried state without first going to gather it myself.'[39] Wilson ultimately dedicated himself to the study of mosses and soon proved 'something more than a mere discriminator'.[40] In 1846, Hooker invited Wilson to prepare a third edition of the *Muscologia Britannica*, a task which Wilson completed only after years of arranging both his own moss collection as well as that of Hooker, which was shipped from Kew to Wilson's house in Warrington.[41] When the leading continental bryologist Wilhelm Schimper visited Wilson, he was impressed by the 'neatness & order & extent' of Wilson's collections.[42]

Events of the night of 26 October 1847 gave the fastidious and highly strung Wilson cause to demonstrate the way in which his identity was utterly bound up with these collections. The carelessness of builders repairing the roof of Wilson's house resulted in a leak during a rainstorm; the following morning his agitation was extreme as he explained his plight:

I am just now in a state of great disorder and perplexity; assailed in a most tender point, as a Naturalist, whose nervous system naturally extends and ramifies throughout his collections and materials for study – My 'sanctum' of books & hoarded parcels &c was threatened late last night with innundation ... We have got through the night better than I anticipated; though I have had little rest, and am now in no condition to study ... I am crowded up with displaced and damped bundles and books, as well as fatigued & feverish with my midnight visit to the stolid plumber who left us so much exposed ... I hope my health will not again suffer from what I underwent last night; but I had to risk it in several ways, in order to avoid other evils – A naturalist becomes vulnerable to an extent which more compact and self-contained people do not readily comprehend . . .[43]

Here we have Wilson describing his almost literal *incorporation* in his collection; his own body has lost its boundaries and he his sense of containment. Although an independent gentleman, Wilson regarded himself as having a 'professional reputation' in botany.[44] Neatness, order and domesticity were essential attributes of the well-regulated naturalist, who maintained a good stock of duplicate specimens to ensure a credit-worthy place in the circuits of specimen exchange, and continually worked to enlarge and refine collections according to his or her interests. The loss of control over a collection entailed a loss of identity as a naturalist and even threatened bodily dissolution, both of the individual and potentially of the natural history community. In Wilson's view, without a well-ordered cabinet, a botanist ceased to exist.

Conclusion

In this chapter, I have examined a few of the containing things that botanists used to define their world. These ranged from wooden boxes with multiple drawers, to books that could hold the pressed specimens of a herbarium, to entire rooms and buildings dedicated to collections. As we have seen in the case of William Wilson, the water that leaked through the roof of the 'sanctum' containing his moss specimens threatened the very basis, 'the tender point', of his personal identity. More than that, books, boxes and other containers were taken to signal the moral qualities of the botanist, and thus his or her qualification for inclusion in a broader community of naturalists.

A collection, whether arranged in a box or a room, despite the disparity in size, is still contained, and the same word – cabinet – can apply to both. The meaning of a collection is determined by the boundaries of its container, be that a box, a cupboard, or a series of shelves or drawers. However, when the container changes from a pocket herbarium collected by oneself to the vastness of a room containing specimens sent in by collectors around the country, if not the world, the difference of scale is not just one of size but of lived experience. Artisan botanists and other local collectors could exist without a collection of specimens because they could easily collect the plants from their locality again, but the botanist whose work depended on a vast private collection of specimens made up of plants contributed by many botanists could not replace such a collection readily if it were lost or destroyed.

Foucault characterised the space of the collection as 'non-temporal' and 'unencumbered': a geometrical abstraction. I have looked at early nineteenth-century botany to argue the opposite. The space of the collection is always a concrete thing, whether it be made of paper, wood or bricks. To understand collections of any kind, I would suggest that we need to think about their actual material containers, those time-bound, context-laden objects that define space in the practice of every-day life. The role played by these containers is often taken for granted, but they are as important as the natural objects within a collection for the way that natural history itself is structured. They maintain nothing less than what we believe to be the order and stability of the natural world.

Further reading

Bellanca, M. E., *Daybooks of Discovery: Nature Diaries in Britain, 1770–1870* (Charlottesville, VA, 2007).

Bleichmar, D., 'Seeing the world in a room: looking at exotica in early modern collections', in D. Bleichmar and P. C. Mancall (eds.), *Collecting Across Cultures: Material Exchanges in the Early Modern World* (Philadelphia, 2011), pp. 15–30.

Brusius, M. and Singh, K. (eds.), *Museum Storage and Meaning: Tales from the Crypt* (Abingdon, 2018).

McCook, S., '"Squares of tropic summer": the Wardian case, Victorian horti-culture, and the logistics of global plant transfers, 1770–1910', in P. Manning and D. Rood (eds.), *Global Scientific Practice in an Age of Revolutions, 1750–1850* (Pittsburgh, 2016), pp. 199–215.

Müller-Wille, S., 'Linnaeus' herbarium cabinet: a piece of furniture and its function', *Endeavour*, 30 (2006), pp. 60–4.

Opitz, D. L., Bergwik, S. and van Tiggelen, B. (eds.), *Domesticity in the Making of Modern Science* (Basingstoke, 2016).

Rose, E. D., 'Natural history collections and the book: Hans Sloane's *A Voyage to Jamaica* (1707-1725) and his Jamaican plants', *Journal of the History of Collections*, 30:1 (2018), pp. 15–33.

Silver, S., *The Mind Is a Collection: Case Studies in Eighteenth-Century Thought* (Philadelphia, 2015).

te Heesen, A., 'Boxes in nature', *Studies in History and Philosophy of Science*, 31 (2000), pp. 381–403.

te Heesen, A., *The World in a Box: The Story of an Eighteenth-Century Picture Encyclopedia*, trans. A. M. Hentschel (Chicago, 2002).

Yeo, R., *Notebooks, English Virtuosi and Early Modern Science* (Chicago, 2014).

18 Natural history and the scientific voyage

Resolution, Discovery, Adventure, La Recherche, L'Espérance: the names alone of the ships that travelled to the Pacific in the late eighteenth and early nineteenth centuries convey the ambition and drama of scientific voyages.[1] Their significance for natural history seems self-evident. Such voyages brought back crates and crates of collections; their discoveries built both the authority of key institutions, like the Muséum national d'Histoire naturelle in Paris, and the reputation of individual naturalists, such as Sir Joseph Banks in London. Historians of science have turned to these voyages to trace many features of natural history: from debates over taxonomy to theories of biogeographical distribution, from scientific illustration to the global reach of political and commercial networks. The patterns and collective influence of later voyages have attracted much less scrutiny. Yet the scientific voyage, as both a state and a private enterprise, continued to be a remarkable feature of the nineteenth century and well beyond. There is much to gain from shifting our focus from the voyages of discovery to a consideration of the voyage as a routine form of modern scientific enterprise. What can the character of scientific voyages tell us about natural history, and the changing forms of scientific knowledge in the nineteenth century?

Most importantly, looking beyond the voyages of discovery reminds us that natural history cannot be extracted from the other scientific work of the voyage in the way that a crate of specimens is extracted from the ship's hold. Voyages had always pursued many scientific aims beyond collecting – astronomical measurements, magnetism surveys, ethnography – and this mixed character was increasingly important once the initial novelty of the natural history expedition had passed. The natural history of a voyage, then, was part of a wider set of practices.[2] With the dramatic expansion of global maritime trade in this period, the most significant of those practices during the voyage itself was hydrography: charting straits and coastlines, identifying navigation hazards and studying currents to find efficient routes. Chart-making was the inevitable background to the collection of geological, ornithological or botanical specimens, and to the accumulation of ethnographical, linguistic and historical records. Exploring the relationship of hydrography to natural historical practice allows us to consider the 'routine' of voyages in two senses of the word – 'ordinary practice' on the one hand and 'regulation' on the other. In many ways, the logic of chart-making underwrote the natural history of the nineteenth-century voyage.

Voyages: variety or system?

The number and variety of expeditions in this era is easily underestimated. Ships sailed with bundles of motives: to chart coasts and trade routes; to respond to curiosity about other peoples and landscapes; to supply authority to consuls, traders and whalers; to pursue economic botany projects; to Christianise Pacific islanders; to promote national prestige and to engage in international rivalry. Even if we limit the count to the state-sponsored voyages of the first half of the nineteenth century, France sent a dozen notable expeditions to the Pacific in the 1820s, 1830s and 1840s; the British Admiralty sponsored voyages to South America, to Africa, to Alaska and to China, as well as Australasia and the Antarctic. The Russian voyages to the Pacific of Otto von Kotzebue in 1815–18 and 1823–6 deeply impressed contemporaries. The United States sent out a major exploring expedition to the Pacific from 1838 to 1842. Beyond all these state enterprises, moreover, lies the category of 'unofficial' or private expeditions. French investors in Brest were behind the voyage of the *Vénus* (1836–9): seeking to explore the possibilities of the China trade and Pacific whaling, they made scientific observations and collections part of their projects.[3] Privately funded Arctic voyages set out from Britain in the 1840s to search for the missing expedition of Sir John Franklin; these too pursued scientific goals as a matter of course.[4] Once underway, the ship's scientific activities were pursued by a variety of figures: sometimes exclusively naval officers, as in the French tradition; often by the ship's surgeon, who combined the post of naturalist with his medical duties; and at other times by 'supernumeraries' who joined the ships for the purpose of scientific work through the good will of the navy or at the wish of the captain.

Faced with this array, the historiography of these voyages has tended to split rather than to lump, often treating voyages in their singularity. Notably, we learn about voyages because of the reputation of the naturalist or captain – Charles Darwin or Thomas Henry Huxley, say, or Jules Dumont D'Urville.[5] Yet historians have also considered the important series of French voyages of the 1820s and 1830s, and the expeditions promoted by Sir Francis Beaufort, hydrographer of the British Admiralty from 1829 to 1855.[6] There are histories of the polar voyages, Pacific voyages, or voyages to the northwest coast of North America.[7] Nevertheless, scientific voyages in this period have common features that defy these compartments of biography, geography or national traditions. After the Napoleonic era, scientific voyages shared in the expansion of global maritime trade, especially to South America and the Pacific, but also to south and east Asia. Ships from different national expeditions often crossed paths repeatedly, stopping at the same ports. Officers extensively consulted each other's charts and publications, carrying the works of past expeditions in the ship's library. Contemporaries saw this common ground, responding with ideas for coordination and system. The French government, for instance, experimented with a short-lived training school for naturalists during the Restoration, hosted by the Muséum national d'Histoire naturelle.[8] In 1849, the British Admiralty published a *Manual of Scientific Enquiry Prepared for the Use of Officers in Her Majesty's Navy and Travellers in General* (1849), a volume that envisioned observers around the globe acting whenever and however opportunity arose. The voyaging tradition was a well-understood network of 'intelligence and enterprise'.[9]

One important shared feature of scientific voyages is the ambiguity that comes from their varied motives. The voyage of the *Bonite* commanded by Auguste-Nicolas Vaillant from 1836 to 1837 offers a good example. Was this a scientific or a diplomatic voyage? Since a major goal of the voyage to the Pacific was the installation of French consular agents in South America and some of the Pacific islands, the government's official instructions to Vaillant noted that 'the voyage does not have a scientific purpose'. But clearly this did not mean that the voyage did not undertake considerable scientific work. Vaillant's instructions continued, 'you will however find frequent occasions to render useful services to the various branches of science'; that is, Vaillant was to correct details in hydrographic charts, to make observations on meteorology, electricity and magnetism, and to 'add a few new specimens to the [national] collections' in Paris. Vaillant carried a hydrographer, a physician–naturalist and a botanist among his officers and sailed with detailed notes from leading members of the Academy of Sciences on how to take physical observations, and on collecting techniques in botany, geology and zoology.[10]

Perhaps not surprisingly, these mixtures could lead to tensions between naval and civilian participants. The most notorious example is the acrimonious voyage of Captain Baudin in the ships *Géographe* and *Naturaliste* to Australia (1800–4), where naval officers and naturalists vied for authority during and after the voyage. The French navy apparently avoided carrying civilian scientific men in subsequent voyages as a consequence.[11] Lesser degrees of friction were commonplace. In 1836, Richard King, naturalist on a voyage funded by the Arctic Land Committee, noted with relief that collecting work could begin with the departure of Captain George Back, who, 'like Sir John

Franklin', treated natural history as 'an excuse for omission of duty'.[12] Evidently, there could be something to these suspicions: a clerk on the HMS *Adventure* in Tierra del Fuego in 1828 recorded that the seamen teased a young lieutenant by pretending to see an extraordinary black snake slither under the roots of a tree – the plot to capture it 'for a specimen' turned into rowdy fun, with the tree as a bonfire.[13]

This last incident hints at another source of tension, one that directly affected the practices of natural history. This source of tension concerned the value of specimens, their ownership, and the question of when during a voyage ownership was decided. Officially, collections made by naval officers were public property. Yet, clearly, this was a complicated issue. On the same South American voyage that witnessed the black snake incident described previously, Captain Phillip Parker King ran into several difficulties with the Admiralty regarding his collections. He had to justify sending skins and specimens directly to an ornithologist at home, and he had to remind the Treasury not to apply duties to these crates. Revealingly, he had also to defend his instructions to his officers:

I soon found that by exacting every specimen from the officers that a lukewarmness might be caused which would very much interfere with the interests and increase of the collection. I therefore caused it to be understood that I required one or two specimens of every novelty that was found beyond which I should take no more, but at the same time without presuming to permit them to make any use of the residue foreign to the wishes and determination of their Lordships. This was followed by everyone paying the greatest attention to collecting and I feel convinced has been the cause of my possessing what may be considered a very large number of very interesting things.[14]

This exchange tells us much about the networks at sea and on land that developed to transfer, catalogue and publish materials from a voyage. But it is of interest here chiefly as a barometer to both the conventions and the negotiations of the voyage. Scientific work was embedded within naval hierarchies, to be sure, but also flowed within private networks of advisors and colleagues, and in step with a lucrative market for natural history collections.

The most routine and familiar result of the voyage was not the production of collections, however, but the production of multiple and multiply authored texts that flowed from a scientific expedition, often for years and even decades following the actual return of a vessel to harbour. Authorised publications included hydrographic charts, and the pilot guides written to accompany them: texts that listed details of harbours, hazards and local resources such as water, wood or other supplies.[15] But a voyage's standard publications went far beyond those designed explicitly for navigators. A general narrative summarising the voyage and its results was the usual practice.[16] In addition, the state contributed (sometimes reluctantly, sometimes generously) to the expense of printing and illustrating other catalogues and descriptions of specimens and findings. Weighing down the shelves of libraries all over the world, the ranked spines of printed volumes generated by scientific voyage eponymously invoke the ship itself: they are the printed 'voyages'. These volumes are remarkable sources for the conventions and expectations of the voyage. Considered as a genre, they can illuminate our understanding of natural history by illustrating the distinctive model of knowledge that the voyage represented. To trace this model, we need to investigate a double-sided question: the character of print 'voyages', and the scientific practice of hydrography to which they were linked. To do so, the next section focuses on a particular voyage, perhaps the most famous of the century.

Voyages as repetition, synthesis and uncertainty

In 1836, Captain Robert Fitzroy was returning from the second of two surveying expeditions in South America. The first, from 1827 to 1829, had been led by Captain Philip Parker King in HMS *Adventure* with HMS *Beagle* along as the secondary vessel, and the second featured Fitzroy in command of the *Beagle* from 1831 to 1836. Both captains exemplified the scientific sailor of the nineteenth century: they were keenly interested in geology, botany, ornithology and ethnography as well as being expert hydrographers.[17] As Fitzroy approached England, he wrote to his sister describing the tasks ahead:

I must sit down quietly in London – to work at the materials collected in this voyage[.] I am obliged to turn writer (I will not say Author) as a point of duty – on subjects connected with Hydrography – and have a number of Spanish works to translate – which together with charts and other matters will keep me occupied during two or three years.[18]

As writer-not-author, he presents these literary activities as 'connected' to his charts. What exactly did this claim mean?

Fitzroy's assertion that, arriving home, the captain's work shifted to *writing* points us first of all to the general narrative of the voyage for which he was editor. Besides the *Journal of Researches* (the volume written by Darwin, and better known now as *Voyage of the Beagle*), the *Narrative of the Surveying Voyages of His Majesty's Ships* Adventure *and* Beagle (1839) included two thick volumes on the first and second expeditions written by King and Fitzroy respectively, and a further, even larger volume of appendices related to the second expedition produced by Fitzroy. Darwin dismissed King's volume as a heavy 'pudding': it 'abounds', he thought, 'with Natural History of a very trashy nature'.[19] Fitzroy himself described the *Narrative* as 'unavoidably of a rambling and very mixed character' and his own volume of appendices as 'disorderly'.[20]

The variety of materials and the multiplication of voices did indeed have a 'mixed character'. But in this sense they represent precisely the connection Fitzroy had alluded to in his letter to his sister. The form of Fitzroy's *Narrative* and the work of hydrography mirrored each other. The captains' accounts were chronological; they shifted points of view and broke off to summarise side expeditions or a lieutenant's log. They were repetitive, recording another cove, another island, another round of angles taken or lead-lines dropped overboard. Fitzroy's 'disorderly' appendices are especially revealing as an illustration of the many related goals and interests of a scientific voyage. In contrast to the chronological approach elsewhere, they contained texts produced before, during and after the voyage itself. They introduced more voices and more lists: on instruments, indigenous vocabulary, phrenology, meteorology, ornithology and translations of historical records. Unlike Darwin's own *Journal*, Fitzroy's text underlined the voyage as a collective endeavour, subject to the hierarchies of the navy, the genealogy of past exploration and many bodies of specialist expertise.

The literary features of the *Narrative* presented knowledge as a process leading back and out to a web of past and future surveys and other texts. It was not so much a summary or an encyclopedia as an active library, a collection to be added to and rearranged.[21] As a text, then, it signalled a mode of observation showing how particulars are scaled up to wider and wider scales, and so translated into global knowledge. That is, the *Narrative* itself reflects in striking ways the typical labour of hydrography. What did this labour involve? Entering uncharted waters, the hydrographer established an observation base, after which officers and crew would then head out with notebooks, sextants, chronometer, compass and sounding line to survey the areas near this secured observation point (Figure 18.1).

Further land measurements would be later combined with the 'seawork' to produce a chart. But charting was much more than measurement. A hydrographer turned to many other sorts of evidence for comparison and verification. The goal was to combine his own observations with all other impressions and records that he came across. This aspect of hydrography, then, like the *Narrative*, involved repetition and miscellany: hydrographers literally retraced past voyages, combing through older charts and texts and consulting all available local documents as they went.[22] Pilot guides expanded upon the chart. These 'pilots' added descriptions of local circumstances that helped

Figure 18.1 *Sailor Sounding from Merchantman.* Nineteenth-century sketch by Gordon Grant. The navigational practice of sounding with a lead-weighted line to measure depth and to bring up distinguishing sedimentary traces of the ocean floor inspired many of the sampling techniques of natural history at sea. National Oceanic and Atmospheric Administration/ Department of Commerce, Central Library: Historic Coast and Geodetic Survey (C&GS) Collection, theb2124.

navigators interpret the chart and assess its reliability. Similarly, the general voyage narratives were bound up with the work of the chart. They elaborated on its dynamic process of composition, and reminded the reader of its diverse threads.

This work of compilation and integration encouraged a distinctive and broad attitude to historical evidence. Surveys called attention to repeated changes in an observer's points of view both geometrically, in the building up of small-scale triangulations, sketches and soundings, and historically, by revisiting established points of longitude, reconciling older descriptions, and modernising measurements. This awareness of uncertainty associated with diverse historical records was a critical part of hydrography's logic, and translated to questions of nature and evidence more generally.

The best example in the *Adventure* and *Beagle* voyages is both captains' interest in the Telhuelche, inhabitants of the southern Pampas (Figure 18.2).[23] In their voyages, both King and Fitzroy were preoccupied with sorting out the sensational question of these reputed Patagonian 'giants', and providing a modern (re)measurement. King begins his account of the native population with measurements of some unusually large trees, including one enormous specimen, 21 feet in circumference, in a bay that that the explorer Captain John Byron of HMS *Wager* had visited some 60 years earlier. Confirming Byron in this detail, King then turns from timber to men, describing the sight from *Adventure*'s deck of three or four natives standing with their horses on a hill. The men 'loomed very large', King wrote, and they 'should certainly have thought them giants' if this was their first experience with the population. Instead, he was confident that this view was 'an optical delusion', a mirage caused by the haze in the moist air and the perspective of ship deck to headland.[24] By combining trees and men, King arrived at conclusions that both justified the reliability and explained the unreliability of past observations. The Telhuelche, in fact, were exactly like *vigia*: the uncertain reports of shoals or rocks that the hydrographer must investigate, to either fix on the charts or dismiss as error.

The discussion of the Telhuelche, then, conveys the ambition of complete and definitive knowledge, but also a sense of what is or could be unreliable about measurement, description and historical record. In a striking passage in his volume of the *Narrative*, Fitzroy interrupted his description of the skin colour of these natives to remind the reader that his point of view has moved across continents and oceans: 'The colour of these aborigines is extremely like that of the Devonshire breed of cattle. From the window of a room in which I am sitting, I see some oxen of that breed passing through the outskirts of a wood, and the partial glimpses caught of them remind me strongly of the South American red men.'[25] His knowledge is authoritative, *autopsia*, but still a matter of 'partial glimpses' – the sailor's and surveyor's acknowledgement that a limited and fleeting perspective was normal, and inescapable. Shoals shift, coral grows, earthquakes cause harbour fronts to rise and currents to change course. This meant the hydrographic task was above all a picture of a dynamic world, in which scientific work is a matter not of completeness, but repeated, ongoing observation.[26]

This appreciation for the sensibilities of a travelling observer emerges interestingly in Darwin's chapter for the Admiralty *Manual of Scientific Enquiry* mentioned earlier. Darwin's formal subject for this volume was geology, but his remarks slid readily into other departments of knowledge,

Figure 18.2 'Patagonian' by T. Landseer. A portrait of the South American Telhuelche natives with the details of bulky clothing, position on horseback, and landscape that King referred to in his re-evaluation of their reputation as a gigantic race. P. P. King, *Narrative of the Surveying Voyages of His Majesty's Ships* Adventure *and* Beagle, *vol. I: Proceedings of the First Expedition 1826–1830* (London, Colburn, 1839), frontspiece.

and into general advice about observation, record keeping and speculation. A collector should make 'copious' notes 'before it is too late (a grief to
which every sea voyager is particularly liable)', he wrote, and he exhorted
the reader to use the dredge, to collect floating logs and seaweed to look for
seeds and minute marine life, to dissect sea birds, and to observe all kinds
of organic and inorganic distribution. 'By collecting . . . by acquiring the
habit of patiently seeking the cause of everything which meets his eyes,
and by comparing it with all he has himself seen or read of, he will . . . in
a short time become a good geologist.'[27] This description of a good scientific traveller can of course be interpreted biographically – that is, it can be
read as an illustration of Darwin's omnivorous natural historical interests.
Yet, directed to an audience of scientific officers, and recalling Darwin's
long exposure to the work of the *Beagle* voyage, it seems to be equally
modelled on the habits of a hydrographer.[28] Darwin's instructions suggested that in terms of observation, there might be little distinction
between the various subjects of investigation in a scientific voyage. Print
voyages and hydrographical charts both mustered the authority of multiple particulars into a larger picture of natural order, in combination with
an acute historical sensibility about existing authorities. These parallels
between literary and hydrographic practices help to summarise the power
of natural history description for contemporaries. The renowned French
naturalist Isidore Geoffroy Saint-Hilaire began his contribution to the
publications of the *Vénus* voyage, a French circumnavigation of 1836 to
1839, by outlining his procedures. His description of how to produce
a catalogue of mammals evoked the hydrographer's language of perspectives, tracks and comparisons, explaining that

In order for me to carry out my task in a manner worthy of the importance of the
expedition and more fruitful for science, I thought it my duty to follow again
here the plan that I have followed in several similar publications . . . I place [the
specimens] alongside those of other similar species [leurs congénères], either
already known or recently discovered and still unpublished; I adopt the comparative perspective, studying each by comparison with the other.[29]

Geoffroy Saint-Hilaire's description of the naturalist's method illustrates the affinity of chart and catalogue. The affinity further suggests
the close relationship between two types of collections that super-
ficially look distinct: the collection of measurements on the one
hand, and the collection of specimens on the other.[30]

Legacies

The nineteenth-century scientific voyages, both from their number
and their debt to the pattern of earlier Enlightenment voyages, seem
to be predictable and routine enterprises. They represent an obvious
ambition for geographically extensive and comprehensive natural

knowledge. The mixed motives of the state in deploying its ships, or the relationships between naturalists and navy, point to the complex social and economic dimensions of scientific projects. Similarly, the opportunism of a voyage in which a scientific traveller seized on all available phenomena matches well with our sense of natural history as a fundamentally eclectic pursuit. The epistemological virtue attached to miscellany that we see in the specimens, the catalogues, and the voyage narratives, express this sense as well. Yet the tradition of voyaging in this period takes on different dimensions when we examine the nature and influence of its most characteristic scientific exercise, hydrography. Certainly hydrography, like other forms of mapping, displayed the authority of precision measurement. Yet it did not excise what was local, contingent and fluid. One of the most telling remarks about scientific voyages comes from the *Vénus* circumnavigation of 1836 to 1839. As the captain Dupetit-Thouars quipped, he was involved in an enterprise of 'undiscovery'.[31] Voyages as *undiscovery* confirmed a vision of order – a hydrographer had to correct and refine older charts. But it also emphasised the perpetually unfinished nature of such scientific work – for the work of undiscovery has no self-evident conclusion. It encoded the instability of knowledge, and the continuous task of inquiry, as a consequence both of uncertain records and of natural changes in the environment. In that sense, the knowledge produced by a voyage, including the hydrographic chart, was more similar to the natural history catalogue than it seems. However ambitious its theoretical system, however exhaustive the collection it recorded, the catalogue was by its nature incomplete, subject to additions or revisions. Lists, in short, 'are not stable objects of contemplation'.[32] The voyage, then, whether encountered as charts or as narrative, afforded natural history another model for this key practice of description and comparison. It combined the task of measurement with a rich historical and comparative sensibility to arrive at an appreciation of local particularity and flux in the natural and social world.[33]

How much these traditions of the voyage influenced scientific practices towards the end of the century is an intriguing question. Certainly scientific voyages continued. Spain dispatched a Scientific Commission to the Pacific from 1862 to 1865. Germany sent out the *Gazelle* from 1874 to 1876 under the leadership of the director of the Natural History Museum of Bern, T. R. Studer, a marine biologist and ornithologist. Both France and the United States sent expeditions to the southernmost oceans in the 1880s. Italy commissioned the circumnavigation by the *Vettor Pisani* from 1882 to 1885, its officers first receiving extensive training in collecting practices by Anton Dohrn and the Naples Zoological Station. Later in the nineteenth century, too, wealthy yacht owners joined in enthusiastically, and their activities built links between field naturalist societies on shore and the growing interest in deep-sea biology.[34] By the 1890s, Scandinavian scientists began to push for a coordinated sequence of voyages in the North Sea. These proposals culminated in the formation of the International Council for the Exploration of the Seas (ICES) in 1902.

Are these later nineteenth-century voyages to be seen as projects that continued to insist on the survey-like vision of scientific work, with the voyage as the home of the generalist scientific observer? This characterisation cuts across the development of science into tighter specialist communities, institutions and forms of training. The most well-known scientific voyage of the later nineteenth century, the *Challenger* expedition of 1872 to 1876, poses the historical problem of discipline formation well. Designed to investigate questions about life in the depths of the sea, the *Challenger* voyage is often identified as the foundation of the big science of modern oceanography and a break with the older tradition of scientific voyages.[35] The *Challenger* expedition is not, in short, usually described as either natural historical or hydrographical. Yet the *Challenger* also gives us the culminating example of how nineteenth-century natural history had become interwoven with hydrographic practice (Figure 18.3). Many *Challenger* arrangements, during and after the voyage, made clear how natural history and survey work were still conceived together. Even the workspace of the ship made the parallel endeavours visible (Figure 18.4). The *Challenger*'s captain and the lead naturalist Charles Wyville Thomson shared a work cabin, while a natural history workroom on the port side of the deck was matched by a chart room for the magnetic, meteorological and hydrographic observation on the starboard side: the larger equipment was on the shared deck space.[36] These relationships continued after the voyage, with the staff-commander and chief navigating officer, Thomas Henry Tizard, working as the naval counterpart to Thomson in the *Challenger*'s post-voyage Edinburgh office to co-author a general narrative. As Thomson planned the scientific reports, it seems fitting that he referred to a projected volume of initial results as 'a running outline', like the rough survey of a shore by a ship in passage.[37]

Legacy of an older tradition, or a signal of new directions and disciplines? In either case, the significance of the *Challenger* is inseparable from the voyage-in-print. The *Challenger* set a new bar for the monumental nature of such publications. In 1882, the task of the Scientific Report was taken over by John Murray, who became general editor to the seventy-six authors and fifty volumes (1880–95), and co-author of a general narrative (1885).[38] The *Challenger*'s publishing enterprise was quickly matched by

Figure 18.3 'Botanising'. The expedition of HMS *Challenger* in 1872–6, often seen as the foundation of modern oceanographic deep-sea research, followed the classic model of nineteenth-century scientific voyages in pursuing a miscellany of scientific activities. Woodcut from T. H. Tizard et al., *Narrative of the Cruise of HMS* Challenger *with a General Account of the Scientific Results of the Expedition* (London, 1885), p. 695.

a similarly extensive set of materials that flowed from the Kiel Commission's Plankton expedition of 1889, whose volumes appeared for nearly two decades (1892 to 1913). The long textual afterlife of both expeditions confirmed the importance of marine biology as a subject, and created new communities of specialists within it. By the end of the century, then, printed voyages had developed from the general narrative and catalogues into something more recognisably like a combined disciplinary journal and work of reference.[39] The International Council for the Exploration of the Seas, the institution designed to set the course for marine research in the new century, included from the beginning not only its arrangements for controlling its own ships (thus ending constant negotiations with the navy) but also arrangements for

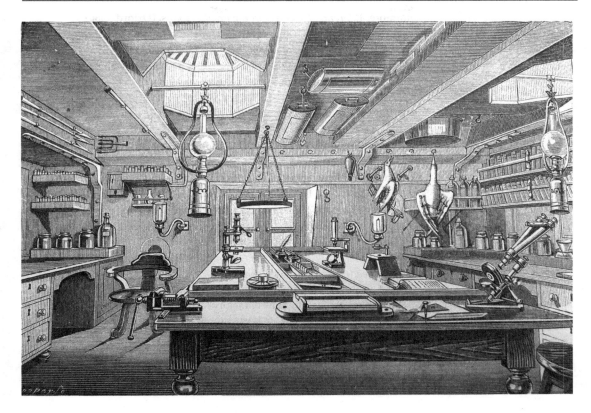

a comprehensive set of data exchanges and research publications.[40] The system of divided labour, of prolonged publication of reports appearing for years following an expedition, of the setting of agendas and questions from past expeditions built on the older genre's practices of collective authorship and variety. The 'disorderly' narrative may be the most enduring legacy of the voyage, a signal that science was a serial, collaborative and permanent enterprise.

Figure 18.4 'The zoological laboratory of HMS *Challenger*', T. H. Tizard et al., *Narrative of the Cruise of HMS* Challenger *with a General Account of the Scientific Results of the Expedition* (London, 1885), p. 6.

Further reading

Burnett, D. G., 'Hydrographic discipline among the navigators: charting an empire of commerce and science in the nineteenth century Pacific', in J. Ackerman (ed.), *The Imperial Map* (Chicago, 2009), pp. 185–259.

Daniell, C. and C. E. Harrison, 'Precedence and posterity: patterns of publishing from French scientific expeditions to the Pacific (1785–1840)', *Australian Journal of French Studies*, 50 (2013), pp. 361–79.

Driver, F., 'Distance and disturbance: travel, exploration and knowledge in the nineteenth century', *Transactions of the Royal Historical Society*, 14 (2004), pp. 73–92.

Dunmore, J., *French Explorers in the Pacific*, 2 vols. (Oxford, 1969).

Keighren, I., Withers, C. and Bell, W., *Travels into Print: Exploration, Writing and Publishing with John Murray 1773–1859* (Chicago, 2015).

Millar, S. L., 'Sampling the South Seas: collecting and interrogating scientific specimens on mid-nineteenth century voyages of Pacific exploration', in D. A. Finnegan and J. J. Wright (eds.), *Spaces of Global Knowledge: Exhibition, Encounter and Exchange in an Age of Empire* (London, 2015), pp. 99–117.

Miller, D. P. and P. H. Reill (eds.), *Visions of Empire: Voyages, Botany and Representations of Nature* (Cambridge, 2011).

Nyhart, L. K., 'Voyages and the scientific expedition report, 1800–1940', in R. D. Apple, G. J. Downey and S. L. Vaughn (eds.), *Science in Print: Essays on the History of Science and the Culture of Print* (Madison, 2012), pp. 65–86.

Pratt, M. L., *Imperial Eyes: Travel Writing and Transculturation*, 2nd edn (New York, 2008).

Rozwadowski, H., *Fathoming the Ocean: The Discovery and Exploration of the Deep Sea* (Cambridge, MA, 2005).

Sponsel, A., 'An amphibious being: how maritime surveying re-shaped Darwin's approach to natural history', *Isis*, 107 (2016), pp. 254–81.

19   Humboldt's exploration
at a distance

Alexander von Humboldt, viewed by contemporaries as the most significant naturalist of the nineteenth century, is best known as an independent explorer. Born into the Prussian aristocracy in 1769, together with his brother Wilhelm he received an exquisite education from private teachers; later he attended the universities of Frankfurt and Göttingen, the Academy of Commerce in Hamburg and the Freiberg School of Mines. For 5 years from 1799 to 1804 he explored the Spanish colonial territories in America, which at the time were divided into the viceroyalties of New Spain (including much of what is now the southwestern United States), New Granada, Peru and the island of Cuba. The goal of this expedition was to study all aspects of natural history in these regions. After his return Humboldt lived in Paris, the scholarly centre of Europe in those years, and dedicated his time to the elaboration and publication of his results. In 1829, after having been called back to Berlin by the Prussian king, he undertook a six-month expedition to Russia, from St Petersburg to the Chinese frontier. Humboldt connected diverse fields of knowledge, seeing natural history as a universal science, and played an outstanding role in the production, circulation and representation of both textual and visual information. He became one of the most important intellectuals of his era and a key figure in the globalisation of science, both within Europe and in the Atlantic context (Figure 19.1).[1]

As this chapter suggests, Humboldt's impact on the production of knowledge extended to the scientific reconnaissance of territories that he had never visited. Through his extensive transdisciplinary and transnational networks, Humboldt was able to participate in numerous scientific projects in various ways. These networks were the foundation for the collaborative work he developed over his entire life, and the basis for his global approach to all fields of knowledge. Today we can call his methodology *Humboldtian science*, a term introduced in scholarship in 1978 and widely discussed since then.[2]

The key concept for Humboldt was a holistic understanding of nature: he envisaged the Earth as an inseparable organic whole, all parts of which were interdependent. Detecting these interdependencies between the different phenomena and understanding their structure was much more important for Humboldt than making a contribution to

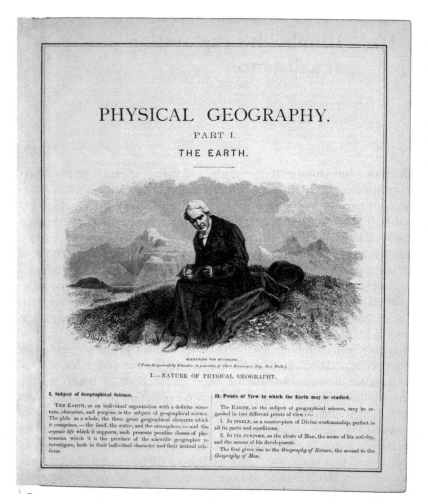

Figure 19.1 Humboldt transposed in old age to the mountains of the New World he explored as a young man, from a painting of 1859 by Julius Schrader. A. Guyot, *Physical Geography* (New York, 1873), p. 1.

a specific field of knowledge. He was a great synthesiser, who in his works connected and elaborated partial and isolated advances of science established by others. It was through this interrelation of facts and data that he was able to create innovative knowledge. He applied his holistic concept not only to different academic disciplines, but also in a geographical sense, exploring the ties between particular regions. His methodology therefore required intense collaboration across national and disciplinary borders, and he relied upon persons willing to provide him with data on geographic regions or specific fields of knowledge where he had little or no information at his disposal.

The contribution of Humboldt to the exploration of the American West demonstrates what a naturalist in his position was able to achieve

Figure 19.2 'Captains Lewis & Clark Holding a Council with the Indians', P. Gass, *A Journal of the Voyages and Travels of a Corps of Discovery* (Philadelphia, 1810), facing p. 26. Courtesy, American Antiquarian Society.

at a distance and by working through collaborative networks. In June of 1799, after undertaking the preparations for his American expedition in Spain,[3] he expressed interest in adding California – then a northern province of New Spain – to his itinerary through the Spanish possessions in the New World.[4] However, the travel route underwent several changes, and although he stayed a year in New Spain, in the end Humboldt neither travelled north to California, nor did he set foot in the region today known as the American Southwest. His visit to the United States in spring of 1804, on his way back to Europe, was limited to a six-week stay on the East Coast, between Philadelphia and Washington.

Nevertheless, an interesting coincidence would direct future scholarly interests. In 1803 the United States acquired the Louisiana Territory from France, the large region west of the Mississippi, and precisely at the moment when Humboldt was finishing his own voyage, President Thomas Jefferson initated a systematic government-sponsored exploration of the West. The first corps to start was the Meriwether Lewis and William Clark expedition (1804–6), which had departed from the Missouri River only two weeks before Humboldt's arrival in the United States, aimed to explore the lands west towards the Pacific (Figure 19.2). In October of 1804 George Hunter and William Dunbar ventured to travel along the Washita and the Red River, in August 1805 Zebulion Pike ascended the Mississippi to find its headwaters and the next year, in May, Jefferson sent the Red River Expedition, led by Thomas Freeman and Peter Custis, to explore the southern part of the Southwest towards Santa Fe. Possibly, if he had not been pressed to return to Europe at that moment, Humboldt might have modified his itinerary once again by participating in this challenging enterprise.

From comments in correspondence, it does appear that Humboldt would have preferred to remain longer in the United States. Nevertheless, after an absence of almost five years, he felt the need to end his travels and start working on the publication of his findings. He seemed confident of a quick return, however, aware of the government's interest in exploring the western regions. In one letter he expressed his wish to see the United States again, once the way from Missouri to the Pacific Ocean was open, since the Great Lakes and the area from Pittsburgh to the Rocky Mountains offered, in his view, a vast field for geological exploration.[5] He envisioned travelling north as far as Mount Saint Elias in Alaska and the Russian possessions. On another occasion he mentioned that 'the country that extended to the west of the mountains offered a wide field to conquer for the sciences'.[6]

Despite these plans, once back in Europe Humboldt was overtaken by his ambitious editorial tasks, and, much to his regret, he was unable to return to the United States in order to participate directly in the scientific exploration of the West. Yet in many ways he did have a considerable impact on this enterprise. During decades of correspondence with naturalists, travellers, artists, diplomats, politicians, publishers and writers on both sides of the Atlantic, Humboldt discussed the latest studies, evaluated or promoted the works of specific explorers, provided his expertise on particular questions, and established contacts between his correspondents. Through his contact with those responsible for planning and leading expeditions, he remained well informed about the latest projects and their outcomes. He read new publications, sent to him mostly by the American authors themselves, but also via institutional and diplomatic connections. His publications incorporated an in-depth and comparative analysis of these works, and he used his close contacts with publishers to make them known in the Old World, acquiring and circulating a remarkable amount of information. Although Humboldt never crossed the Atlantic again after 1804, for the following half-century and more he was probably the best-informed person in Europe on the American West.

The fascination of the West

Why was the exploration of the American West such a fascinating project for Humboldt? Here we need to distinguish between his intellectual engagement in the early voyages of exploration undertaken by different European nations between the sixteenth and eighteenth centuries, which he studied closely, and Western expansion as a politico-scientific project directed by the US government during the nineteenth century. While the first approach formed part of Humboldt's historical interest in expeditions directed by the Spanish government in particular, in the second case he was able to observe a nation-building process, in which to an extent he was implicated. The fact that Humboldt was enthusiastic about the future of the United States, as a place where the values of Enlightenment were put into practice, gives it an additional dimension. Western exploration combined several aspects of his scholarly interests in bringing to light a rich and varied natural world, with unique geological formations, mineral resources and climatological conditions. Humboldt remained committed to the romantic idea of travelling through unknown territories in remote areas, far from European civilisation, and to his vision of connecting this world through the new transportation and communication technologies, with a view to the potential future impact of the United States on international politics and commerce.

It is thus unsurprising that, while in Washington, Humboldt's meeting with Jefferson, advocate of Western expansionism, was very fruitful. The president sought reliable and up-to-date information regarding the Spanish possessions in America, in connection with his plan for the territorial expansion of the new nation, but also with his concern to possess strategic knowledge about the neighbouring country. Given these circumstances, Humboldt's visit, shortly after the Louisiana Purchase and just at the beginning of the US exploration of the West, was perfectly timed for a personal encounter with the president, who doubled as president of the American Philosophical Society, the nation's most prestigious learned society. The Prussian's knowledge of western North America, drawn from Spanish archives in Madrid as well as Mexico City, was crucial to Jefferson in his pursuit of more precise information concerning the acquired territory and its disputed borders.[7] In translating parts of his early work on New Spain, *Tablas geográfico-políticas* ('Geographical-political Tables'),[8] into French, Humboldt added a two-page summary specifically on the border region of the Louisiana territory, something which had not been included in the original text, written for the viceroy of New Spain, José de Iturrigaray. These documents contained valuable statistical data on the different provinces, descriptions of their population, climate, agriculture and commerce, as well as a proposal for the geological exploitation of the new territory.[9]

From early Spanish voyages to American westward expansion

Impressed by Humboldt's formal training in the Freiberg School of Mines and his experience working in the mines of Bayreuth and the Fichtel Mountains, the Spanish crown hoped that this expertise could be used to exploit their possessions in America. Therefore, the Prussian was granted unprecedented permission to explore these territories and to access confidential unpublished documents in the Spanish archives. At the beginning of 1799, while in Madrid preparing his expedition, Humboldt studied the early Spanish voyages at the Hydrographical Office and the libraries of other institutions. He was fascinated by the different phases of the scientific reconnaissance of the American West by generations of travellers at different moments and within distinct political contexts. The expedition of Alejandro Malaspina and José de Bustamente (1789–94) particularly caught his attention, and while in Madrid he was fortunate to meet some of the expedition members.

Later, in Mexico City from October to December 1803, Humboldt consulted all available material in the colonial archives on exploration voyages to the southwestern part of North America, today belonging to Mexico and the United States. His notes detail the documents he consulted there: records of the Malaspina expedition and its exploration of the west coast; the manuscript diary of Juan Francisco de la Bodega y Quadra's travels to Nootka; other maps of the northwest coast; and documents on Juan Pérez's exploration of the coast north of Monterey.[10] Much of this material was in a rough state, not for publication. These circumstances made his findings in the archives relevant and his analysis based on these sources unique.

Humboldt discussed these early voyages in *Views of Nature*,[11] published in Germany in 1808; his *Examen critique* (1836–9); and his final synthetic work, *Cosmos* (1845–62).[12] Most of the information was included in his *Political Essay on the Kingdom of New Spain* (1808–11), regarded as the first comprehensive statistical and economical analysis of this region.

In Philadelphia and Washington, the scholarly and the political centres of the young nation, Humboldt's interest in the West took a new direction. He was intrigued by the government-promoted exploration of the newly acquired Louisiana Territory and over the next decades tried to keep up with its progress. The scientific reconnaissance of the American West occupied a considerable part of his communication with the United States; his correspondents included explorers like John C. Frémont and Amiel W. Whipple, naturalists like Louis Agassiz and John Torrey, and German explorers like Balduin Möllhausen.[13]

With Humboldt's international fame growing constantly, many American scholars, explorers, naturalists, writers and artists sought contact. Members of the US Coast Survey like Alexander D. Bache, as well as officers of the Corps of Topographical Engineers, the branch of the Army tasked with western exploration, knew about Humboldt's methods and applied them.[14] Some enthusiastically discussed their research questions with him, keeping him informed about their publications and projects, and asking to visit him in Paris, and later, in Berlin, or for introductions to third parties. Numerous others did not establish a personal contact, but expressed how he had influenced their understanding of the American West, its landscape, its biogeographical implications and its artistic representation.

Humboldt was also well connected within political and diplomatic circles. He knew most of the American representatives in Germany, as well as Prussian diplomats in the United States. With several of them he became close friends, and these officials took personal responsibility for sending the latest news on the United States. These sources, such as Johann Gottfried Flügel, the American diplomat in Leipzig from 1839 to 1855, who also worked as an agent for the Smithsonian Institution in Washington, founded in 1846 'for the increase and diffusion of knowledge', and Friedrich Freiherr von Gerolt, the Prussian envoy in Washington since 1844, were thus also important conduits of intelligence.[15]

There were many reasons why Humboldt's correspondents provided information. Some sought to disseminate their work on an international level, others had an interest in being connected to European institutions or in publishing their work in the Old World, or they required Humboldt's personal expertise on specific topics. The military engineer and surveyor Amiel Weeks Whipple, for instance, sent living cacti, knowing that Humboldt had collected similar species in the plains of Mexico, as well as volcanic rock samples, which Humboldt had examined by a geologist before replying with a detailed analysis.[16] The German geographer Johann Georg Kohl contacted Humboldt in 1856 on a very different matter. He had been asked by the US government to prepare a critical edition of all important cartographic material in different archives on the history of the discovery and first settlement of North America. Knowing of Humboldt's expertise as cartographer, he sought advice for this important task.[17] Finally, there were also those who had personal or professional interests in transmitting information regarding the progress of their new nation to Europe through Humboldt. John Buchanan Floyd, Secretary of War, mentioned that he would continue sending information regarding western exploration, so that Humboldt could judge the progress of civilisation in those vast regions, to which – Floyd added – Humboldt more than any other, through his voyages and publications, had attracted the attention of the American people.[18]

In consequence, Humboldt's library became one of the best German collections on the American West, containing travel narratives, studies on American geography, geology, climate, cartography and Native American languages, proceedings of scholarly societies and reports from various institutions.[19] Humboldt thus remained knowledgeable about the latest developments and expedition projects, from the Lewis and Clark expedition through the Pacific Railroad Surveys from 1853 to 1855, which were organised to find the best route for the transcontinental railroad across the continent. In his publications, Humboldt continually referred to these works, commented on their results, and established comparisons with other regions he had personally explored. No specific publication was dedicated to western exploration, but there were discussions in several of his works – in his essays on New Spain and Cuba, his *Views of Nature*, *Researches Concerning the Institutions and Monuments*, *Asie Centrale* and *Cosmos*.

Mapping, climate and communication

Given his professional training in mining, Humboldt developed a strong interest in gold production in different parts of the world.[20] After his Russian expedition of 1829, where he studied the geology of the Urals, he was eager to obtain data for comparative analysis. He began an intensive dialogue with American correspondents who could provide data on North American gold production. His questions show the links between his interest in gold production and other fields such as economics, politics, geography and geology. He forwarded the responses to several of his correspondents in Germany and elaborated it in publications such as *Asie Centrale*, where he described the discovery of gold in Virginia, North and South Carolina and Georgia between 1824 and 1836.[21] Comparing the mountain ranges of the Urals with the Alleghenies in regard to their gold production he predicted where else in the United States mineral deposits might be discovered.[22] Not surprisingly, the discovery of gold in California in 1848 reignited Humboldt's interest. He analysed the mineralogical conditions, established comparative studies with other regions and warned on several occasions of the possible consequences of the Californian gold fever.[23]

The geographical and geological features of the West also attracted Humboldt's attention. In *Asie Centrale*, we find not only numerous analogies and differences between mountain ranges in Asia, America and Europe, but also detailed discussions of different chains and the geological particularities that distinguish them. His *Views of Nature* provided general observations on the configuration of North America, describing distinct regions in the West: the Pacific alpine mountains of

California; the dry and sparsely inhabited high plains; and to the east, between the Alleghenies and the Rocky Mountains, the water-rich, fertile and densely populated Mississippi basin.[24] The Great Basin between the Rocky Mountains and the Sierra Nevada fascinated him, as did the extension of the Great Salt Lake in Utah, depicted as Lake Timpagonos on a large map of Mexico he drew in 1804.[25] Rivers, in their role as both connecting links and dividing boundaries, caught his attention: he described how the Mississippi divided the United States into two large regions – an eastern region with rapid advances in terms of culture and civilisation, and a wild and uninhabited western part – and he discussed the possibilities offered by the Missouri River for transcontinental communication with the Pacific Ocean.

A key figure in the development of modern cartography and geography – both fundamental disciplines for the emerging United States – Humboldt also contributed to the cartography of the American West. His famous map of New Spain, *Carte Générale du Royaume de la Nouvelle Espagne* ('General Map of the Kingdom of New Spain', Figure 19.3), collated the geographical findings of Spanish explorers, relying mostly on Bernardo de Miera y Pacheco's map of the 1776 expedition conducted by Franciscan priests. Before his departure in 1804, he lent a draft of that map to the Secretary of State's office in Washington and thus served as a connection between Spanish cartographic knowledge and early US territorial imperatives. Given his first-hand access to unpublished material in the Spanish archives on both hemispheres, the map was unusually well documented for its time, and was copied by several cartographers. In addition, it marked an important step towards the application of modern methods to American cartography with its special attention to scale. It thus remained the standard map of the area for the next decades. Other results of Humboldt's geographic research were published in Heinrich Berghaus's *Physikalischer Atlas* ('Physical Atlas'), which includes both an ethnographic map and a map of the mountain chains of North America.[26]

Humboldt also elaborated comparative tables with geographical positions of numerous locations, determined by astronomical observations.[27] His interest was not in geographical features in isolation, but rather in the interconnection between topographic characteristics and other aspects of the natural history of these places, such as vegetation, human settlement or mineral resources, in order to elaborate his larger view on the social conditions and the progress of culture.

The American West offered in addition a vast terrain for climatological studies. In his *Views of Nature* Humboldt sought to 'put together a single picture of the diverse causes of America's humidity and decreased heat'.[28] He searched for explanations for the climatic difference between the area east versus west of the Alleghenies.

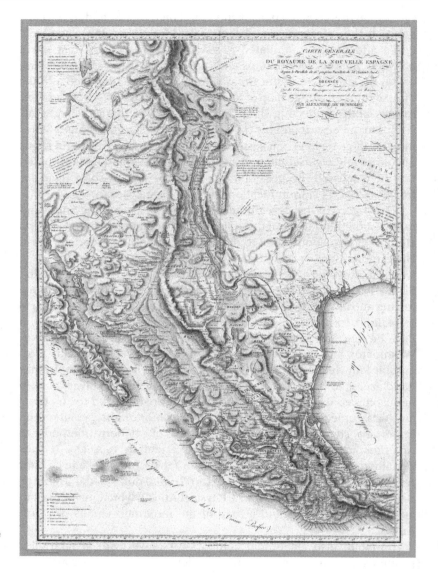

Figure 19.3 Alexander von Humboldt's 'Carte générale du Royaume de la Nouvelle Espagne', *Atlas de la Nouvelle-Espagne* (Paris, 1809). David Rumsey Historical Map Collection, Stanford University, reproduced under a Creative Commons License.

Drawing upon Samuel Forry's work on climatology,[29] he studied the distribution of heat in three regions of the United States: the Atlantic states east of the Alleghenies; the western states in the broad basin between these mountain ranges and the Rocky Mountains; and the plains between the Rocky Mountains and the coastal range of New California.[30] As part of this work, he was also interested in explaining the higher temperature of the Pacific coast as compared to the Atlantic coast.[31] In 1843, Humboldt's analysis of the differences in mean annual temperatures in littoral areas of the east and west coasts of the United

States and Europe made him one of the first proponents of the healthy climate in California, just as this region was coming to political attention.[32]

Humboldt sought to incorporate the American West into a global network of communication, envisioning both commercial and intellectual exchange. He understood the potential benefits of this interconnection in terms of the progress of civilisation; therefore, new communication and transportation technologies like steamboats, electric telegraphy and railroads caught his attention. In particular, the idea of establishing a canal between the Atlantic and the Pacific remained a topic of profound interest for over 50 years, which led him to draw a map with nine different options for its course, discussing the advantages of each.[33] Early on, he predicted that such a connection would shorten navigation times, creating new markets for European and American traders and positioning the United States as a player in global affairs. Through such a connection, he argued, the wealth of America could provide economic prosperity for the Old World, and European knowledge and technology could bring similar advantages to people from the new continent.

Clearly, Humboldt's contribution to the scientific exploration of the West was never merely descriptive. He typically addressed questions in analytical terms, establishing comparisons between different people, periods of time and regions. These comparisons sometimes relied on material or observations from other areas such as New Spain, Cuba or Russia; but at other times he analysed and evaluated data provided by others.

Humboldt's interest in western exploration was connected with his lifelong concerns in understanding nature. The reconnaissance of this region was necessary to complete the global comparisons that would produce world knowledge. Humboldt adopted a holistic and integral approach, and did not consider the West merely as an interesting laboratory of particular strata or vegetation. In analysing the scientific and political implications of the West, he always kept in mind its larger significance for the growth of the nation through future settlement and the improvement of living conditions.

Humboldt's impact on western exploration

Humboldt inspired a generation of travellers on both sides of the Atlantic to visit the American West (Figure 19.4). Through his intensive and extensive connections with the highest scholarly and political circles in the United States, he was in a position to support numerous Germans interested in exploring the large territory. Humboldt's influential support allowed the explorer Balduin Möllhausen to accompany Amiel W. Whipple in his expedition from the Mississippi to the Pacific

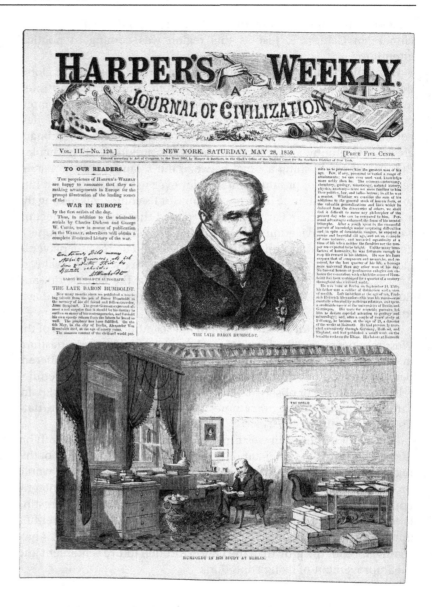

Figure 19.4 Audiences in the United States were fascinated by the details of Humboldt's working life. 'The Late Baron Humboldt' with a wood engraving of 'Humboldt in his Study at Berlin', *Harper's Weekly* (28 May 1859), p. 337.

Coast in 1853–4. A few years later, when Möllhausen wanted to participate in another expedition, Humboldt again used his name and fame strategically to reach out to the highest political levels: in 1857, he contacted Jefferson Davis, the Secretary of War,[34] and shortly afterwards his successor John B. Floyd, presenting the skills and interests of his protégé in the most flattering words. Other travellers lucky enough to count on Humboldt's active support were the palaeontologist and geologist Carl Ferdinand Römer and the botanist Heinrich Karl Beyrich. Conversely, the work of many explorers or scholars, including Frémont, Wislizenus, Emory, Nicollet, Agassiz and Catlin, was circulated in Europe by the famous Prussian. In some cases, Humboldt made a special effort to express his esteem publicly and honour their scientific achievements. In 1850, Frémont received the *Preußische Große Goldene Medaille für Wissenschaft* ('Prussian Great Golden Medal for Science') from the Prussian King Friedrich Wilhelm IV, for his outstanding exploration of the territory between St Louis and the Pacific, at Humboldt's instigation.[35] In his position as first chancellor of Prussia's highest order of merit for sciences and arts, Humboldt recommended Agassiz and Frémont for this distinction, which both received in 1860, a year after his death.[36]

With his unparalleled knowledge of Western exploration, Humboldt provided first a connection between the Spanish and the US exploration of the West and later a significant conduit of information between the German-speaking world and the young American republic. In many cases he was the first to be informed or to receive reports concerning new publications, projects or exploration journeys. Once Humboldt had reviewed these letters, maps and publications, he forwarded them to those who might find them of interest. Humboldt maintained an elaborate and well-organised correspondence with the geographer Heinrich Berghaus, leading to plans for joint editorial projects, such as the idea, suggested by Humboldt in 1850, of collecting all the reports on the exploration of the American West, translating them into German, and publishing them in a single volume.[37] Unfortunately, negotiations with the publisher were unsuccessful, and they were unable to carry out this ambitious scheme. Nevertheless, whenever Humboldt was able to publish material on the westward expansion, he took the opportunity, as with some of the letters he received, such as Albert Gallatin's account of North American metallurgy and geography,[38] his study of the Native American tribes of North America,[39] or Möllhausen's expedition to the Rio Colorado.[40]

A sign of the admiration for Humboldt and a recognition of his pioneering contribution to the exploration of the American continent is that more places were named after him there than in any other area of the world (Figure 19.5). Humboldt's name was given to towns, streets and counties across the nation, in California, Arizona, Nevada, Colorado, Iowa, Kansas, Michigan, Minnesota, Ohio, South Dakota, Tennessee, Texas, Pennsylvania and Wisconsin. Several geographical features carry his name, including a lake, river, bay, mountain, forests and parks. When the territory of Nevada became a state, the name Humboldt was one of the options along with Washoe and Esmeralda, although it did not pass a vote at the Constitutional Convention in 1864.[41]

GOING TO HUMBOLDT.

Figure 19.5 'Going to
Humboldt', Mark Twain,
Roughing It (New York,
1872), p. 199. Courtesy,
American Antiquarian
Society.

Different motives stood behind this spatial appropriation of
Humboldt's name. German immigrants could establish a connection
to their famous countryman, allowing them to identify simultaneously
as Americans and Germans. For others it was simply a way of marking
the significance of science in American life. In other cases it constituted a statement against slavery, as when Free State immigrants to
Kansas expressed their abolitionist politics by calling their town
Humboldt.[42] Frémont too showed his deep admiration by naming
places after his hero. In 1848, he added Humboldt's name to a river
previously known as Mary or Ogden River, as well as to a chain of
mountains on his map of Oregon and Upper California.[43]

In 1858, Floyd sent Humboldt an album of nine maps, showing the
places named after him, with the words: 'The name of Humboldt is not
only a household word throughout our immense country, from the
shores of the Atlantic to the Waters of the Pacific, but we have honored
ourselves by its use in many parts of our territory, so that posterity will
find it everywhere linked with the names of Washington, Jefferson and
Franklin.'[44] For Humboldt it was doubtless flattering that his name was
associated with the Founding Fathers – a connection he himself established on several occasions. Nevertheless, this territorial commemo-
ration also meant his political appropriation in the context of the
Manifest Destiny, the belief that settlers from the United States were
destined to occupy the entire North American continent. Humboldt's
interest in the reconnaissance of the West cannot be disconnected

from its political and social implications for European settlement and the growth of the new nation. Both the cartographic material he produced and the scientific exploration of the continent he promoted contributed to these processes. However, in spite of his moral approbation for the expansion of a society based on enlightened values, Humboldt also observed the imperial growth of the country with suspicion and occasionally expressed critical comments about the expansion of slavery or the invasion of Mexican territory. Yet this was another aspect of the ways in which Humboldt and his science were instrumentalised for geopolitical interests beyond his own concerns. It also illustrates his delicate position between science and politics. While in works such as *Cosmos* his own focus was on creating global science, gathering knowledge on natural history of all regions and understanding the American West in this context, these initiatives were used to defend Manifest Destiny and American exceptionalism.

What if Humboldt had been able to explore the American West as he had initially envisioned? Particularly in geology this might have led to significant findings in an area rich in natural resources. Yet his overall contribution to its scientific reconnaissance might have been less impressive if he had focused on one geographical area or scholarly field. Being only able to participate in this project from a distance, through different types of networks and forms of collaboration, opened the way for an involvement that turned out to be much broader and more far-reaching in its effects. Thanks to these connections, his physical presence and personal involvement in the actual exploration process were unnecessary, and he could play a major role in shaping this process. This was possible through his wealth of knowledge, the importance of his name, his influential connections, and, last but not least, his specific way of interconnecting and combining Spanish, Prussian, British and French traditions with emerging American science.

Further reading

Brescius, M. von, 'Connecting the New World: nets, mobility and progress in the age of Alexander von Humboldt', *Humboldt im Netz*, XIII:25 (2012), pp. 11–33, available online at www.uni-potsdam.de/romanistik/hin/hin25/brescius.htm.

Crampton, G. C., 'Humboldt's Utah, 1811', *Utah Historical Quarterly*, 26 (July 1958), pp. 269–81.

Goetzmann, W. H., *New Lands, New Men: America and the Second Great Age of Discovery* (New York, 1986).

Lange, A. J., 'The Humboldt connection', *We Proceeded On*, 16:4 (November 1990), pp. 4–12.

Mathewson, K., 'Alexander von Humboldt's image and influence in North American geography, 1804–2004', *Geographical Review*, 96:3 (2006), pp. 416–38.

Rebok, S., *Jefferson and Humboldt: A Transatlantic Friendship of the Enlightenment* (Charlottesville, 2014).

Sachs, A., *The Humboldt Current: Nineteenth-Century Exploration and the Roots of American Environmentalism* (New York, 2006).

Sherwood, R., *The Cartography of Alexander von Humboldt: Images of the Enlightenment in America* (Saarbrücken, 2008).

Walls, L. D., *The Passage to Cosmos: Alexander von Humboldt and the Shaping of America* (Chicago, 2009).

Wulf, A., *The Invention of Nature: Alexander von Humboldt's New World* (New York, 2015).

20   Publics and practices

Natural history has a weirdly bifurcated image. On the one hand, it is associated with classification: putting creatures into their taxonomic places and establishing their phylogenetic relationships. Widespread images presenting endless rows of nearly identical butterflies, birdskins, or more recently, printouts of DNA sequences, mark a highly technical and, for outsiders, dull-seeming set of scientific practices. On the other hand, natural history as the study of living organisms in their natural habitats and behaviours has a lively existence today. We see it in ubiquitous films and television shows that portray birds migrating, accompanied by invisible orchestras; animals stalking and devouring one another to hushed-voice commentary; and plants exploding into time-lapsed blooms.[1] Local and online communities, open to all interested participants, are devoted to observing and identifying plants and animals 'in nature' (for example, the British Naturalists' Association, or crowd-sourcing sites like iNaturalist.org and www.zooniverse.org), or to promoting natural history as a form of knowledge imbued with aesthetic and nature-protectionist dimensions (for example, naturalhistorynetwork.org and the related natural historiesproject.org). How can a historical perspective shed light on this odd dichotomy?

Historians have recognised that this tension goes back a long way, at least to the eighteenth-century rivalry between the Swedish naturalist Carl Linnaeus and Georges-Louis Leclerc de Buffon, his Parisian rival. Linnaeus became known as the champion of systematic classification, especially for his binomial nomenclature (think *Homo sapiens*) and his controversial yet highly successful sexual system of plants. Buffon, by contrast, sought to create a natural history that offered the full exposition of living animals in their anatomy, physiology, habits and moral lessons for humans.[2]

The classificatory agenda, much enhanced by comparative anatomy, came to stand at the centre of specialist and professional natural history in the nineteenth century. The further development of a Buffon-style focus on the living organism has received less historical scrutiny, perhaps because this approach was often aimed at and taken up by a wider public. Over the nineteenth and early twentieth

335

centuries, this approach expanded to embrace new attention to animals' relations to other living things and their surroundings. Well before animal behaviour and ecology were accepted as significant subdisciplines of academic biology, their study was a prominent part of a natural history championed by writers and institution-builders who conceived of their audience as a general one, literate but not always highly educated.

Natural history reformers in the nineteenth century sought simultaneously to enliven natural history and to make it accessible to a broad public. They did so by bringing living animals to the general urban public through the zoo movement, by redirecting the great research-oriented natural history museums to a broader audience and opening new ones intended primarily for public education, by writing about nature in books and magazines, and by encouraging people to engage nature directly, out in the field or even at home. Through all of these means, popular natural history emerged as a recognisable leisure-time pursuit, sometimes in close contact with professional biologists, but very often not. This chapter argues that the rise of institutions like the zoo and the public natural history museum, together with the development of organised hobbyist natural history pursuits such as bird-watching and aquarium-keeping, came to form a *cultural system* of public-oriented natural history, focused on living organisms.[3]

In developing this argument, this chapter concentrates on the German setting. Broadly speaking, the same sorts of public natural history institutions emerged around the same time across Europe, the Americas and some European colonies in the century before World War I. But the character and patterns displayed by these settings were not identical, and what we know about them is uneven. The history of public natural history in the German-speaking lands was inflected by the absence of a single leading centre of activity (as Paris was for France); the presence of a well-developed educational system extending from the primary grades, for both sexes, through over two dozen research universities and technical colleges (for men only) and a tradition of voluntary organisations stretching back to the late eighteenth century. Together, these features produced a robust, well-educated cadre of secondary-school teachers and civic institution-builders who became significant drivers in the development of the institutions of public natural history. However, in sharp contrast to the United States, the lack of educational opportunities for German girls and women rendered them almost entirely invisible as participants in producing and furthering natural history knowledge until the early twentieth century.

German natural history was also shaped by certain geographical facts. Most critically, 'Germany' did not exist as a single country until 1870–1. Before then, the political landscape was divided into many separate states, with their own capitals and university towns. The German cultivation of public natural history took place in towns and cities that supplied a critical mass of institution-builders, where nature was still accessible not far beyond town limits. As elsewhere in Europe, this 'nature' almost never meant primeval wilderness but rather referred to un-built landscapes populated by plants and animals, along with trees that had often been planted by humans. This was the nature for which older nature lovers yearned and which they sought to preserve, and it was love for this nature – or its remnants – that they sought to instil in a wider public.

Natural history's publics

Before delineating in more detail the development of this cultural system, we need to address a basic question. What do we mean by 'public natural history' and 'the public(s) for natural history'? Both *public* and *natural history* have long histories associated with tortuous theoretical and historiographic discussions. Consider the subtle differences among three common phrases: *the public good* concerns the general population; *my literary public* points to a specific readership; *out in public* opposes *in private*. Then there is Jürgen Habermas's much-discussed bourgeois 'public sphere', a space of open cultural and political expression that – somewhat confusingly on first acquaintance – is part of the private realm, which in Habermas's schema encompasses all that which is separate from the state.[4]

Natural history has often been more open than other areas of science to non-specialists, and thus 'the public' in its first sense here. Yet a main theme of the history of nineteenth-century science is the definitive emergence of science as a paid profession, with a clear social role for the researcher, and the subjects encompassed by natural history have been among their areas of expertise. The botanist, zoologist, mineralogist, geologist, biogeographer, and later the marine biologist and ecologist all might be naturalists, but they are now not 'amateur' in any way. Further complicating matters is a common image of popular science as merely a simplification of scientists' knowledge for the public. This model has been discredited. Any scientific writing conveys not only information but also values; moreover, 'the public' is not a uniform, passive entity but comprises different active agents who select and contextualise the information presented.[5] Yet the emergence of popular writing and civic institutions of natural history indicates a phenomenon increasingly distinct from how specialists communicated with one another. So what's a historian to do?

I propose that three different but overlapping meanings of 'the public' will help capture the breadth of 'public natural history'.

The first sense views the public as a *market* for natural history, including its books and magazines, and its private institutions – most nineteenth-century zoos, aquariums and 'artistic taxidermy' museums. This consuming public had to have money to buy these products and time to visit these venues; these two kinds of costs stratified public access to natural history into many economic layers.

The second meaning of the public shifts our perspective from a marketplace to a *civic* standpoint. Here the public is again a target audience, but not one primarily organised around market forces. Rather, this public is the projected recipient of civic organisations and sometimes governmental ones – typically through non-profit, entertaining education. The main exemplar of this sense of the public is the projected visitor to the civic natural history museum. Here, too, social entry has tended to be regulated through timing and price: in the early nineteenth century, free opening hours of large public museums were often confined to a few weekday hours per week, but this typically expanded over the latter part of the century to increase visitor access.

The third meaning of the public views it in the Habermasian sense of civil society, focusing on the natural history activities undertaken through voluntary organisations and clubs – activities that might be considered 'doing' natural history as opposed to consuming it. In the German-speaking lands, these activities expanded and changed dramatically over the nineteenth century, exactly in the direction of hobbyist activities working with living organisms.

The rest of this chapter traces these three overlapping senses of the publics for natural history and their practices through German examples from the long nineteenth century, between the beginning of the French Revolution in 1789 and the start of World War I in 1914. It treats the institutions and activities of natural history as an evolving system that developed and differentiated over this period, beginning with the emergence of voluntary associations for the natural sciences in the early nineteenth century (sense three of the public) and continuing through the zoo movement and the rise of the popular periodical (sense one) at mid century to the reform of natural history museums that opened them to a mass public at the end of the century (sense two). Across all of these, influential promoters expressed a commitment simultaneously to reform natural history by focusing on living animals and plants, and to spread enthusiasm for nature to a broader public. By the early twentieth century, a new iteration of the nature-oriented voluntary association (sense three) had emerged that reflected both these reforms. It is best represented through the history of the aquarium hobby, which will be treated in some detail in the section 'Natural history as a hobby'.

Public natural history as an evolving system

Early modern princes, pastors, teachers, apothecaries, landed gentry, businessmen and others built personal collections containing *naturalia*, but only gradually developed formal organisations to support these interests. Echoing the seventeenth-century founding of scientific academies and societies elsewhere in Europe, German-speakers founded local societies for investigating nature – enough to qualify as a visible social phenomenon in the eighteenth century, part of the broader emergence of Habermas's bourgeois public sphere (public sense three). Their numbers grew steadily between the 1770s and World War I, with a particularly intense period of establishing new associations from the 1840s through the 1870s.[6] Such societies tended to become the centres of avocational study of the natural sciences, though sometimes, especially in non-university towns, they took on more academic features like courses. Typically, members met monthly (or more often) to present their work, show new experiments and specimens, and enjoy the company of like-minded colleagues. These societies were often explicitly open to people of different economic, religious and political backgrounds.[7] Some might organise hikes and collecting excursions into the countryside. Nearly all such societies produced a journal for contributions and proceedings by members.

In a pattern well established by the 1840s, some societies housed collections of the *naturalia* owned and classified by their members. Initially the society typically rented a space, but eventually, perhaps a few decades on, it would contract with the town or state government or crown to turn the collection of collections into a public good, amalgamating it with an existing governmental collection or founding a new one. In this pattern, citizens exhibited local patriotism by donating their time and energy to developing their collections, and the city, state, or crown government would eventually repay this patriotism by graciously taking them over when they grew beyond individuals' means. This was one way the civic museums of the nineteenth century formed.

Other roots of civic museums could be found in older princely collections or those of schools and universities. The Berlin natural history museum combined all of these: it was simultaneously the repository for the Prussian royal collections, the teaching and research collection for the University of Berlin, and the home of the leading local natural science society. It occupied the increasingly crowded second floor of the main university building until its removal to an expansive new campus near the university's medical and agricultural buildings in 1887. In 1889 – much to the dismay of rival museums – it was also named the official natural history repository for Germany's African and Pacific colonies, acquired since 1884 (and relinquished after World War I).[8] Although it was exceptional in its diversity of functions, similar amalgamations occurred as the public museum stabilised as a cultural institution.[9]

By the 1880s, natural history museums frequently formed a standard feature of local civic natural history scenes. As these institutions became more established, separate owner-curated collections were merged into a more anonymous institutional form, with volunteers giving way to hired curators. Increasingly, these were not avocational experts in a particular group of organisms, but science-educated PhDs who brought with them higher expectations for the collections' conservation and knowledge of the geographical provenance of individual specimens.[10]

The image of the museum as a space for a narrowly professional enterprise of classification often masked a more flexible reality, but it made for an easy target. Already in the 1850s and 1860s, reformers sought to broaden the reach of natural history, to take it out of the exclusive hands of the experts (especially museum scientists), and place it in the hands of 'the people'. A leading spokesman for this view, the museum taxidermist, zoo enthusiast, nature protectionist and sometime popular writer Philipp Leopold Martin, argued that professionals' preoccupation with classification forced their attention away from its proper object, the living animal, to focus instead on minute differences among dead carcasses. This focus made museums excruciating places to visit. 'Even with the very first [exhibit] cases', he wrote, 'we can hardly suppress the feeling of an unending boredom, and it requires an unusual disposition to sustain patience and interest all the way to the last case'.[11] 'Practical' natural history, instead, would concentrate on the living animal, and practical naturalists would care for wild creatures in nature and in captivity.

The German zoo boom of the 1860s and 1870s benefitted considerably from this movement. As with the development of associations and museums, it participated in a broader trend. The classic public zoo was at the Jardin des Plantes in Paris, transformed through the French Revolution from a royal menagerie into a site of pleasurable viewing and strolling open to the public. Thereafter, zoos were opened in London (1828), Amsterdam (1840), Antwerp (1843), Berlin (1844), Brussels (1851) and Ghent (1851). As zoos burgeoned elsewhere across Europe, they also popped up across Germany: between 1858 and 1881, over a dozen German cities founded zoos (usually supported by a local zoological society – another instance of a voluntary association).[12]

Unlike museums, zoos had relatively few pretensions to science, though some did partake in a broader acclimatisation movement that sought to import exotic animals into Europe for their beauty, exotic qualities or economic uses.[13] Local fauna, too, were kept and bred at the zoo. These were not the modern zoological gardens of natural habitats and physiologically appropriate props, but parks containing 'collections' of caged and penned living animals, housed in increasingly fanciful architecture to evoke a sense of exotic places.[14] Zoos were built by the wealthy urban bourgeoisie to display civic pride and power – a spacious, public new answer to the older royal menageries, and a more genteel and animal-friendly form than the tiny ill-kept cages of the travelling menagerie.[15] 'It is not princes, not scholars, not pedagogues, not education ministers who founded the zoological gardens in Dresden, Cologne, in Hamburg, Amsterdam, Antwerp, Rotterdam, Brussels', a leading German zoo-booster wrote in 1862. 'It is instead the educated majority of the citizens of the cities, who have been driven by what we almost might call an unconscious urge toward the observation of living nature.'[16] Yet at many zoos, the company of living humans was more compelling than that of the animals. One writer reminisced about the Berlin zoo of the 1870s and 1880s, 'It was the gathering place of Berlin's elegant world. A posh restaurant with a large concert hall and a broad terrace in front of it was the meeting place of the upper crust', while another noted in passing that for zoo-going Berliners, 'zoology ... was unimportant'.[17] (See Plates 12 and 13.)

Nevertheless, the zoo movement was an important element in public natural history, through its insistence that the German public should be interested in the living animals of the world, and not just their preserved counterparts in museums. Natural history writers stoked this interest, filling the pages of the new general family magazines of the second half of the nineteenth century – magazines ranging from ones with a largely regional coverage such as *Über Land und Meer* ('Over Land and Sea') to the blockbuster bourgeois magazine *Die Gartenlaube* ('The Garden Arbor') – as well as popular magazines specific to science. Living animals, in zoos and beyond, were popular subjects. For example, both the pastor–ornithologist Christian Ludwig Brehm and his son, the much-loved natural history writer Alfred E. Brehm, wrote numerous illustrated articles about living creatures in *Die Gartenlaube* starting in 1856. By the 1870s, Alfred Brehm's articles served to advertise the Berlin Aquarium, a commercial institution he founded in 1869 that included birds and beasts as well as aquatic creatures.[18] Popular natural history writing thus amplified the effects of the zoo movement and the new attention to living animals it expressed.

Some of the same people involved in the zoo movement would also champion breathing new life into natural history museums themselves via more lively exhibits. The taxidermist Martin, who ranted against what he called 'the terrorism of system' in museums, made 'artistic' mounts of animal groups for private customers while he worked at the Stuttgart natural history museum – itself a pioneer in lifelike display. He became a zoo enthusiast in the 1860s, though he never won the positions he coveted at the major new civic zoos and had to be content with an affiliation with a private zoo run by a Stuttgart innkeeper (where he built a 'Museum of the Primeval World' as an added attraction). Martin's sometime scientific collaborator in Stuttgart, Gustav Jaeger, was a university-educated zoologist who ran an 'aquarium-salon' and then a small zoo in Vienna for a few years in the 1860s. Jaeger later worked as an adjunct teacher in various institutions while writing popular science works and editing a popular science magazine.[19] Such reformers were essential in spreading the gospel of 'living' natural history across institutions; they justify our viewing zoos, museums and popular writing as parts of a common system of public natural history.

Perhaps the most influential of these reformers was Karl Möbius, a schoolteacher and natural history activist who helped found the Hamburg zoo and aquarium in the early 1860s before gaining a professorship at the University of Kiel in 1868, where he built a new natural history museum with cutting-edge architecture and display organisation. By the end of his remarkable career, he would oversee the Berlin Natural History Museum's transformation into a model of what came to be called 'the New Museum Idea'.

For the first two-thirds of the century or so, natural history museums had primarily been intended as places for those conducting research on and with the collections – mostly a classificatory enterprise. But beginning in the 1860s, and accelerating through 1900 and after, reformers like Möbius introduced a new approach that took root across the globe. The New Museum Idea radically broadened the museum's intended public, while typically changing its physical organisation from a single unified collection to a form called 'dual arrangement'. This separated the research collection – the vast majority of specimens – from the displays made available to the public.[20]

The unhooking of the organisation of displays from that of the research collection did two things: it clearly marked a separation between professional research and casual museum visiting, and it facilitated a wave of exhibitionary novelty designed to bring in the masses. While an epitome of the systematic collection still filled the majority of the public space, the highlights now featured novel displays. At Berlin's natural history museum, exhibits informed visitors about insect life cycles, mutual dependencies of plants and animals and agricultural pests; allowed the visitor to compare bones and musculature across types; and invited observation of the tunnelling and building habits of living ants between glass walls. Collectively, such exhibits made the museum into a kind of three-dimensional textbook of natural history, aimed at teaching the general public the most up-to-date, essential knowledge of living nature. Here the public was projected as a consumer of expert knowledge, building on assumptions about the desire for lifelong learning.[21]

New exhibits also spoke to the German public's patriotism. While atrium-filling whale and dinosaur skeletons awed through sheer size, the first new star attractions were native German fauna – wolves, deer, raptors, etc. – mounted in spectacular 'artistic' taxidermy groups and dioramas depicting them in characteristic activities.[22] Pioneered in the 1870s, these became increasingly *de rigueur* at museums both large and small after about 1900, often as central features of new 'homeland' galleries.[23] In the 1900s and 1910s, the same techniques yielded spectacular new groups of African mammals, Arctic bird-nesting grounds, and Antarctic life at a few leading museums, inviting visitors to partake in imperial Germany's self-image as a player on the global stage.[24]

Although the public was typically treated to more spectacle at the private, commercial zoo or aquarium than at the civic museum, these institutions were knit into a common system of public natural history in multiple ways: by the circulation of practical naturalists who moved between them, carrying with them their reformist ideas of enlivening representations of nature; by the illustrated books and magazines (to which some practical naturalists contributed) that further spread descriptions and images of living animals; and by the positioning of the public itself as having the job of consuming these views of nature. Yet even as living nature became a more familiar object of this indirect sort of consumption, many nature-lovers also interacted with nature more directly. This is the subject of the next section, which will eventually return us to the world of voluntary associations and close the loop of the evolving system of public natural history.

Natural history as a hobby

Many of the same people targeted as the consumer public of natural history institutions also engaged more directly with living non-agricultural animals. While collecting and classifying remained a mainstay of serious naturalists, hobbyists increasingly found other ways into natural history – again, often expressed through clubs and other voluntary organisations. By the end of the century, nature and hiking clubs such as the *Naturfreunde* ('Friends of Nature') and the *Wandervögel* ('Birds of Passage') invited budding naturalists to enjoy the outdoors in the company of others.[25] Natural history was especially popular among schoolteachers. The Teacher's Union for Nature Study (*Lehrer-Verein für Naturkunde*), a subscription society established in 1887 by Karl Gottlob Lutz, reported 1,500 members by its third year in 1890, 11,000 members by 1899 and over 38,000 by 1915.[26] Cleverly organised around regional chapters that met regularly on a frequent basis and then joined forces for an annual general meeting, this association also offered an illustrated magazine that provided a publishing outlet for naturalist–teachers, and undertook large-scale projects like the republication of the famous illustrated *Deutschlands Flora* ('Germany's Flora') by Jakob Sturm (orig. 1796).[27]

The populist movement of practical natural history in late nineteenth-century Germany entailed a further fascinating but usually overlooked feature: the promotion of keeping native wild animals. Because this took its earliest organised form in the aquarium–terrarium hobby, this section will focus on how people did living natural history using the aquarium.

The aquarium rode into Germany on the coat-tails of the English craze, which started in the late 1840s. Although scientists and nature-lovers in Germany and elsewhere had long kept aquatic animals and plants in glass jars for close study, middle-class German readers were introduced to the 'balanced' saltwater aquarium in the pages of the family magazine *Die Gartenlaube* in 1854 and 1855, as a 'new scientific luxury article' decorating the living rooms of British high society.[28] In 1856, the natural history populariser Emil Adolf Rossmässler, writing in the same magazine, suggested the *freshwater* aquarium as a more appropriate living decoration for most Germans, who lived inland. He followed this up in 1857 with a small book titled *Das Süßwasser-Aquarium* ('The Freshwater Aquarium') – the first such book in German.

In these initial presentations, the 'salon' aquarium appeared as a household ornament, a beautiful and instructive conversation piece (Figure 20.1). Rossmässler proclaimed the pious hope that his book would spread 'scientific knowledge into the living-rooms of the wealthy'.[29] The science here mainly involved a basic understanding of what was needed to keep the aquarium's inhabitants alive. This included the chemical exchange between animals and plants that differentiated the balanced aquarium from the goldfish bowl, and appeals to the balance of nature were frequent in literature on the aquarium. But salon aquarists were not especially interested in exactly reproducing a real, particular community. This goal did come to exist, however, among aquarists connected to the effort to refocus natural history on living creatures in their natural conditions. This shifted attention away from the domestic emphasis on ornament and toward filling the aquarium with native plants, fish, amphibians, insects and molluscs that hobbyists collected themselves locally (in what aquarium books often called 'the hunt') and that paired well with an emerging interest in the 1870s and 1880s in reconstructing natural communities.[30]

What turned the aquarium hobby into a more public form of natural history was the introduction of tropical ornamental fish, largely from

Figure 20.1 Wood-
engraving by Gottfried
Kühn (1824–98) from
E. A. Rossmässler, 'Der See
im Glase' (The Lake in
a Glass), *Die Gartenlaube*,
1856, pp. 252–6, p. 253,
commons.wikimedia.org.

Asia and South America. Beginning in the 1870s and stabilising by the
mid 1890s, the import industry changed the character of the hobby
substantially. Imported fish were expensive to buy and to keep in their
heated aquariums. Not only did this limit who could afford to have
these fancy fish, it also put a premium on breeding, and it encouraged
special attention to fish of unusual colours, such as paradise fish, or
especially interesting behaviour, such as the climbing perch (introduced to Germany from the Indian Ocean in the early 1890s) or the live-
bearing guppy (introduced from Venezuela in 1908). The import men-
tality also encouraged aquarists to pay less attention to non-fish aquatic organisms and natural communities, and to put together into one
tank fish that would never meet in nature. Finally, the new value placed

on breeding reinforced interests in developing special varieties such as the flowing 'veiltailed' goldfish, or the telescope fish with its gigantic, buggy eyes – the fish equivalent of pigeon or orchid breeding.[31]

In conjunction with the increase in imported ornamentals, the hobby took a new turn, with the rise of aquarium and terrarium associations following the 1889 founding of the Berlin hobbyist society, Triton. By the outbreak of World War I, Germany had become the world's centre for the hobby, with a well-organised network of over 280 local clubs; numerous businesses devoted to supplying fish and other aquarium creatures, tanks and supplies; and a well-developed publishing market that included two nationwide journals and dozens of books advising aquarium keepers.[32]

From an introduction that initially connected aquaristics to domestic interior decoration, the visible aspects of the hobby soon moulded into the masculine forms of the German public sphere. The clubs that provided the market for the businesses and periodicals followed the forms of earlier natural history societies: members typically met at a restaurant, often weekly, to exchange tips on fish care and technological improvements, listen to each other give speeches, show off new acquisitions, and sometimes share the offspring of successful breeding events. These clubs might also organise group hikes to collect new material; pool resources to maintain a library; and, as the hobby quickly grew more robust, sponsor exhibitions for other hobbyists and for a broader public.[33] In developing these activities, aquarists drew on the long-standing public-sphere form of the general natural science society, turning it to the more populist goals of practical natural history, and focused on the peculiarities of caring for fish, reptiles, amphibians and their combination with plants, snails, and insects. With the institutional infrastructure afforded by the association, the avocational keeping of aquarium fish became perhaps the most prominent hands-on natural history activity to come out of the movement that sought to revivify natural history while returning it to 'the people'.

Conclusion

In the aquarium hobby, a new manifestation of the association (sense three of the public) appeared that was very much tied into the public as consumer (sense one) – but this was now the consumer of speciality hobby magazines and aquarium technologies. Although more loosely linked to the reform of civic museums, the aquarium hobby partook of the same spirit of returning life to natural history that earlier museum reformers sought as they reached toward a broader public (sense two). In this way, the new, populist approach to natural history striven for by activists such as Philipp Leopold Martin and Karl Möbius beginning in the 1850s and 1860s became instantiated in a robust system of institutions and practices by 1914.

The story certainly does not end there. The great era of museum dioramas was yet to come. The move to study animals in their life habits and environments would blossom academically into the discipline of ecology in the first half of the twentieth century. Concerns about the effects of habitat loss and species decline, already evident among some practical naturalists in the late nineteenth century, would burgeon in the conservation movement of the early twentieth century, which in turn would morph into the environmental movement of the 1960s and 1970s. Some members of the public would continue to read natural history literature. Families would still raise fish in tanks at home. The general public would continue going to natural history museums, zoos and public aquariums through many changes in their design and practices of caring for animals. Yet today the system of public natural history that developed over the nineteenth century is in decline (as is its professional, classificatory side), and the nature that urban dwellers now encounter is most often commercially or digitally mediated. In a world dominated by virtual experiences, the direct contact with living nature that the German practical naturalists of the nineteenth century worked so hard to bring to their public once again threatens to elude us. What might replace it? How will we live without it?

Further reading

Brunner, B., *The Ocean at Home: An Illustrated History of the Aquarium* (Princeton, 2005).

Daum, A. W., 'Science, politics, and religion: Humboldtian thinking and the transformations of civil society in Germany, 1830–1870', *Osiris*, 17 (2002), pp. 107–40.

Daum, A. W., 'Varieties of popular science and the transformations of public knowledge: some historical reflections', *Isis*, 100 (2009), 319–32.

Kisling, V. N., Jr (ed.), *Zoo and Aquarium History: Ancient Animal Collections to Zoological Gardens* (New York, 2001).

Nyhart, L. K., *Modern Nature: The Rise of the Biological Perspective in Germany* (Chicago, 2009).

Phillips, D., *Acolytes of Nature: Defining Natural Science in Germany, 1770–1850* (Chicago, 2012).

Rader, K. A. and Cain, V. E. M., *Life on Display: Revolutionizing U.S. Museums of Science and Natural History in the Twentieth Century* (Chicago, 2014).

21   Museum nature

As you enter the 'Animal World' gallery of the National Museum of Scotland in Edinburgh, a massive elephant confronts you, standing on a large podium surrounded by two still more massive blue whale jaw bones (Figure 21.1). Looking around, other large beasts are arranged in silent stampede. A *Tyrannosaurus rex* skeleton cast looms beside, while a shoal of sharks, dolphins, whales and other giant sea life swims past on wires above. The displays may be inanimate, but the gallery teems with families during the weekends and with tourists during the week. This and other institutions that collect, preserve and display parts of the natural world – often termed museums – have played a vibrant role in Western culture for some time. For although the dramatic National Museum of Scotland arrangement has been in place for only a few years, the elephant was similarly popular when it was first displayed over a century ago. Its origins, composition and afterlife over the decades tell us much about the role of museums in the history of natural history, and the place of natural history in museums.

The elephant will therefore be our first guide in the following pages on a brief journey to understand the material culture of natural history. But as iconic as they are, such large vertebrates on display make up only a tiny fraction of the numbers of specimens in museum collections. We therefore continue our exploration below using a humble shellfish, one of the tens of millions of museum specimens held in storage. Natural history is a particular kind of science – a way of knowing, collecting, classifying and describing the natural world – which has come to rely on such vast collections.[1] The subject considered in this volume, natural history, not only comprises words, texts and images, but also *things*. This particular material culture, I argue, coheres in *museum nature*: the construction of a version of the natural world using the bounteous physical resources of classifying institutions in major urban centres.

From their long history, much of this chapter will concentrate on museum nature in a period when these practices of collecting, production and consumption were at the apex of their quantity and credibility:

348

the years around 1900. To understand late Victorian natural history museums, however, we need to know how they came to be.

Figure 21.1 Specimen NMS. Z.1907.216 on display in the National Museum of Scotland natural history gallery in 2011. National Museums Scotland.

Museum histories

In the early nineteenth century, efforts to catalogue the world moved to an institutional basis as associations, universities and governments took over the princely cabinets and personal collections detailed earlier in this volume (see, for example, Chapter 12 by Müller-Wille and Chapter 16 by Ogborn). Hans Sloane's cabinet was unusually early in forming the nucleus of the British Museum in 1753; dozens of provincial learned societies later followed suit and absorbed personal collections into their new institutions.[2]

The physical and administrative structures were therefore in place by the end of the nineteenth century when a perfect storm of social, economic and macro-political shifts meant that museums came to be valued and supplied as never before. The quantitative peak in collecting in the disciplines we would now recognise as archaeology, anthropology, botany, mineralogy and palaeontology was in part a symptom of the imperial project. From the 'Scramble for Africa' of the 1880s to the zenith of European imperialism between the World Wars – the 'payday of Empire' – natural and material culture flowed from the far reaches of the world as never before or since. No less than the ethnographic collecting that has attracted more historical attention, by collecting natural objects from the colonies, Europeans strengthened their imperial infrastructure in the decades around 1900.[3] In major centres, curators arranged the colonial other for the benefit of the metropolitan self.

Museums were not the sole showcases for the products of empire, however. The 1851 Great Exhibition in London was only the first of the international expositions designed to promote the industries, people and phenomena of the major powers and their dominions; they reached their zenith in this period. The expositions – of which taxidermy was a common feature – were part of the same culture of display as permanent collections, and often stimulated new museums, including for example the Field Museum, based on the 1893 World's Columbian Exposition in Chicago.

Whether directly from colonial collecting, from expositions, or both, across Europe and the Americas the influx of objects was housed not only in new museums but also in a series of buildings that expanded existing institutions. In South Kensington, the natural history collections of the British Museum opened in a new 'cathedral of science' in 1881; in Edinburgh, the original conception for the imposing building that would house the elephant we have already met was finally completed in 1888; the Muséum national d'Histoire naturelle in Paris boasted a grand new zoology gallery in 1889; in the same year, the Imperial Natural History Museum in Vienna encapsulated European dominance of the natural world in its new premises in 1889; and the Manchester Museum (Figure 21.2), which we will encounter below, opened an imposing new building in 1890.

The following decades were an exciting time for those who visited these buildings, as well as those who studied and worked in them. At the vanguard was Sir William Henry Flower, Director of the British Museum (Natural History), a surgeon-turned-zoologist who, like his predecessor, the palaeontologist Sir Richard Owen, had begun his museum career at the Royal College of Surgeons of England before taking up the leading national position. Flower advocated the 'new museum idea' by which selected specimens were deployed for public educational displays, but the majority of these vast collections were to 'be used only for consultation and reference by those who are able to read and appreciate their contents'.[4]

Kept apart, the bulk of the collection was put to use in this period by an entangled assortment of practitioners: morphologists comparing multiple specimens; palaeontologists generating evolutionary journeys of long-extinct species; proto-ecologists studying the relations

Figure 21.2 Architect
Alfred Waterhouse's 1882
conception of the
Manchester Museum on
Oxford Road, opened in
1890. Whitworth Art
Gallery, University of
Manchester.

of the flora and fauna gathered from a particular place; and taxonomists feverishly classifying new species (at the rate of a thousand a year
by this time).[5] Counter-intuitive as it may be, those we would now term
environmentalists began to collect and study specimens in museums
to understand the connections between human behaviour and the
natural world – as we will see in the career of the hunter responsible
for the death of the elephant now at the National Museum of Scotland.

   This frenzy of collecting activity was well known to those who worked
in museums in the decades that followed, who inherited custody of this
mass of material and their voluminous accompanying records. For
much of the century that followed, historical perspectives on the collections were provided by these curators and by historians of natural
history.[6] But more recently they have been joined by a herd of scholars
from other disciplines. Architectural historians, for example, have found
museums a rich source – as temples of science and taxonomies in
stone.[7] To understand the construction of museum nature, I want to
draw on two perspectives from the growing literature on museum
history: one broadly speaking anthropological, the other geographical.

Anthropology as a discipline emerged from museums in the very period in question. In recent years anthropologists have not only curated and used collections but have also taken museums and their practices as their focus of study.[8] Anthropological (and archaeological) perspectives on objects can also be profitably applied to natural history museums. Just as the ceramics in the neighbouring exhibition required work and had myriad meanings, so too the elephant in the natural history gallery is material culture. Specimens gathered meanings through associations with people they encountered on their way to collections, thus linking the history of museums to broader scientific and civic cultures. The specimen has a biography, a path through time and space that can be retold.[9]

In such a retelling, the specimen's trajectory illustrates the extensive web through which people, information and material passed to, from and between museums. Here a geographical eye is helpful, focusing on space and place, on the flow of things, people and ideas. Such an approach helps us to understand how these networks were formed and how within museums the world can be 'understood as a mosaic of cultures and natures ... represented through the groupings of objects'.[10] Museums in general intended to present the world in microcosm; natural history collections specifically brought together elements of the natural world and recreated it in a case. 'Museum nature' is this collection of things from outside reconstructed inside. Both of the examples of museum nature encountered in the following sections – elephant and mollusc – are zoological, both collected in or near Africa just after the turn of the nineteenth century and both brought to museums in Britain. Many of these reflections may be applied to botanical and geological objects, however, and to things from (and taken to) other places and other times.

Material nature: the elephant in the room

The Edinburgh elephant features in Figure 21.3.[11] Even if it doesn't show the elephant in situ in the museum, this 1912 postcard speaks volumes about museum nature in this period. It was produced by the recently renamed Royal Scottish Museum, successor to the 'Industrial Museum of Scotland' and the 'Edinburgh Museum of Science and Art', which had absorbed the natural history collections of the University of Edinburgh next door.[12] Such changing names, scopes and ownerships are common in the history of museums as collections grew, changing hands and shifting collecting foci. But at their heart remained specimens like the elephant, which caused quite a stir upon its arrival in 1907:

Natural History Series, No. 1
AFRICAN ELEPHANT
THE ROYAL SCOTTISH MUSEUM, EDINBURGH

Figure 21.3 Royal Scottish Museum postcard, Natural History Series no. 1, *African Elephant*, 1912. National Museums Scotland.

Thousands of visitors are being attracted to the Museum in Chambers Street to see the giant new elephant now on view, which has the distinction of being the finest stuffed elephant in the world. This forest king is three inches taller than Jumbo [actually it is likely to have been a foot smaller – under 10 feet] ... He was killed by Major Powell-Cotton in the Lado Enclave [now South Sudan] in December 1904, and is classed as one of the finest examples ever obtained. The Major shot the monster at night while drinking [the elephant rather than Powell-Cotton, presumably], and followed the spoor of blood until the next day.[13]

Such behemoths catch the eye – in the postcard and in the gallery – and here its scale is emphasised by the attendant and calf beside it. The latter was an unrelated West African Forest elephant with an equally fascinating afterlife. Brought to London from Sierra Leone by one 'Dr Randall' in the early nineteenth century it was part of the collection that William Bullock sold in 1819.[14] Bullock was a showman, collector and naturalist, who established the 'Egyptian Hall' and within it charged visitors a shilling to visit his novel 'Pantherion', an early example of animals displayed in a naturalistic setting. A century later and 400 miles north, the calf was still set in a recreated nature disguising its urban context. It was depicted next to the 'magnificent adult male of the *African Elephant*'; a vast chunk of mobile landscape ripped asunder and transplanted to the Edwardian city as part of a culture of the spectacular.[15] But unlike in its isolated state on the postcard, in the gallery the mount was part of a wider display strategy.

Bringing the world to Scotland, the Royal Scottish Museum was intended to be a universal collection, like some larger civic museums and many national institutions such as the Smithsonian and the British Museum (at least before the natural history collections departed). Other galleries at the Royal Scottish Museum presented the human world, whereas that which the elephant joined offered a survey of the natural world. Following the taxonomic practice established by Linnaeus, it was assigned a binomial classification, in this case *Elephas africanus* (Blumenbach, 1797).[16] In this period, however, curators were no longer content to arrange their exhibit specimens solely by taxonomy in serried, encyclopaedic ranks. There had long been a tradition of imaginative settings, often involving violent encounters between predators and prey – none quite so dramatic as those displayed by William Bullock – but a range of other approaches was now applied.[17] From the later nineteenth century, species were arranged alongside others from their region in geographical groups, or with related species in a biological group.

The most dramatic and controversial of the new techniques, as Nyhart discusses in Chapter 20 in this volume, was the habitat diorama. This presented mounted animals in lifelike poses against a large naturalistic painted background, often foreshortened and sometimes curved at the edges to merge with carefully constructed scenery that used plants and rocks from the locale. The Royal Scottish Museum's birds were arranged in this way at this time by the appropriately named ornithologist William Eagle Clarke, Keeper of Natural History.

There were many invisible and intangible factors at play in the stunning visuals of the habitat diorama: for example, the advances in plate glass manufacture that enabled joiners to generate the 'fourth wall' of the cases, which were often huge.[18] Woven into the displays were changing attitudes to the environment which elsewhere at this time gave rise to the creation of nature reserves.[19]

Habitat dioramas were only as convincing as the animals at their centre, and this period was the acme of lifelike reconstructions of megafauna. The Royal Scottish Museum bought the elephant from the premier taxidermy firm Rowland Ward in London, and the Museum's director considered it 'a triumph of the taxidermist's art'.[20] Ironically, such verisimilitude is only based on a fragment of the original animal, the hide, which was arranged over a mannequin, and parts of the face and extremities were often painted or modelled. Their lifelike countenance masked the hybridity of these 'remnant models', surface without substance.[21] Many kinds of models were to be found in natural history museums, from plaster casts to stunning glass flowers, but none exemplified the institution like taxidermy mounts. They were all were hybrid creations of fur and straw, feathers and plaster, but this bull elephant especially so, for it was actually a composite of two beasts: the tusks came from an even better-endowed elephant shot four months later on the same expedition.

Both elephants were victims of a rampant hunting culture that peaked in this period, evidenced by rack upon rack of trophy heads that then adorned country houses, inns and museums. There was always an element of hunting in museum acquisition, whether in the literal or metaphorical sense. Hunters shot to collect (if only the head, horns or hide). The two constituent animals that formed the display elephant were among sixteen shot by Percy Powell-Cotton and his party on their 1904–7 African hunting expedition. Like the calm, commanding museum attendant in the postcard, hunting showed how man – heroic, masculine, European man – dominated the natural world. Powell-Cotton was far from alone on this expedition, exploiting the well-oiled indigenous networks of hunters, trackers and ivory merchants.[22] Tellingly, however, it is his name that remains connected with the specimen, not the Africans who hunted with him. Likewise, in the postcard, a lone European man is depicted – his uniform echoing martial prowess, exerting dominion over wild African nature. Yet Powell-Cotton, while providing for other collections and establishing a museum in the grounds of his home in Kent, was also a dedicated film-maker and conservationist; embodying the Edwardian duality we now find surprising – collecting the world to preserve it.

Collecting does not stop at the moment of triumph. Powell-Cotton may have tracked and shot the elephant, but many others were also involved. The great beast was dismembered, packed and transported.

Once in Britain, preparing the elephant would have been a complex, expensive and time-consuming process. A central armature, sculptural in elegance and artistry, was carefully wrapped in straw and wood strands until the precise shape was achieved with a high degree of anatomical accuracy.[23] Not all mounts were so successful: the Horniman Museum in London has an 'over-stuffed' walrus first displayed at the Colonial Exhibition in 1886. Not that taxidermy involved stuffing, of course: rather, after sealing the mannequin with plaster, papier-mâché or clay, the hide was painstakingly layered around it. Even then the mount was not complete, as the eyes, mouth and other tricky outliers remained to be improvised, painted and modelled.

Furthermore, after taxidermy the elephant and other mounts needed considerable care and attention to maintain and preserve them.[24] Likewise, plants needed drying, pressing and arranging; even minerals, seemingly so unfettered by human hand, could require considerable effort to collect, section, polish and store.[25] Thinking about the elephant and other museum specimens anthropologically, as the products of considerable *work*, we appreciate that they are material culture. We can look at them not only for what they tell us about the natural world, but what they can tell us about human activity: the conceptual, affective and physical traces of work help us to read them as historical documents.

Museum objects that survive (and the textual and photographic traces of those that do not) therefore allow the historian to understand the human elements of the construction of museum nature with all the strategy, effort and sheer serendipity involved. Curators, conservators, backdrop artists and taxidermists set out to recreate the natural world within the confines of the museum gallery. And like other collections, what they constructed was sanitised and idealised: the idyllic backdrop of Figure 21.3; the peaceful, uncontaminated environmental balance in the habitat diorama; or else nature red in tooth and claw as carnivores (and even herbivores such as gorillas) snarled and snapped at their prey or at the visitor.[26] Taxidermists tried to cover their tracks, to delete evidence of artifice by stitching up bullet holes or substituting different tusks, as they did with this elephant. Museum nature reflects its contingent historical origins: with an anthropological eye we can see natural history displays as culture crafted from nature. Apparently effortlessly natural, they were (like the birds of paradise specimens discussed in Lawrence, Chapter 6, this volume) actually painstakingly artificial.

Networked nature: the mollusc's movement

The elephant is massive and on display; as such, it is statistically unusual for a museum specimen. Gradually over the centuries, expanding natural history museums have exhibited smaller and smaller proportions of their collections, until today some 99 per cent are in stores. Early modern collectors tended to display everything they collected; Victorian curators developed museum furniture that had drawers below or above the displays for related and duplicate material; from the late nineteenth century, as we have seen, William Flower and his peers began to gather sizeable research collections deep in dedicated storerooms away from the exhibition halls. Skeletons were stored in racks; soft tissues stored in alcohol in jars; plants in large boxes; but the vast majority of zoology specimens, from bird skins to beetles, were in drawered cabinets. In principle, stored collections were vital archives of the natural environment, but we may also think of them as the material basis of a collective societal memory: in repose, to be woken by use, conservation (or insect attack).[27] Accessible only to researchers and staff, they appear markedly different from their peers on display. They were visibly extracted from their environment without the conceits of fabricated context; obscured by wood, not revealed by glass; and they were very clearly dead.[28]

The Manchester Museum was one such late Victorian institution prone to storing. Emerging from the cabinet of a private collector that had been absorbed and expanded by a learned society, from 1888 the collections were housed in a grand neo-Gothic building – designed by Alfred Waterhouse, fresh from his success with the British Museum (Natural History) – as part of Owens College, predecessor to the University of Manchester (see Figure 21.2). It was one of a gaggle of collections absorbed into public ownership in the later nineteenth century as museums became part of the cultural make-up of most European, American and colonial towns. And like many other university, civic and national museums that expanded at this time, an increasing proportion of its footprint was devoted to storing specimens that were not suitable for display, especially in the basement and tower. These included duplicates or very similar items; and at the other end of the value spectrum, *type* specimens, those from which a new species was first described, which became the standard for further description.

Among the stores in Manchester the shell shown in Figure 21.4 remains to this day. Brown and white, 119 mm long, it has a spiral at one end and a curious dark stain down its side. With its associated records, this humble mollusc exoskeleton can tell us as much about museum culture as the mighty elephant that starred in the previous section, especially about the geographies of the connections between the museum and the outside world and of the networks that filled these stores in this period.

The shell is known as *Conus clytospira* (Melvill and Standen, 1899), commemorating two men and the year in which they used it to

Figure 21.4 *Conus clytospira* (Melvill and Standen, 1899), Manchester Museum, University of Manchester.

describe a new species – or so they thought. Just months afterwards they found that a smaller example of the same species had been named *Conus milneedwardsi* (Jousseaume, 1894) by the French doctor–naturalist Félix Pierre Jousseaume after Henri Milne Edwards, director of the Muséum national d'Histoire naturelle. By the laws of zoological nomenclature so painstakingly established in the previous decades and enshrined in the International Commission on Zoological Nomenclature from 1895 – which established type specimens as taxonomic icons, and cemented the authority of collections – *milneedwardsi* became the formal name of the species, and *clytospira* the junior synonym.[29] (Furthermore, the shell was still later identified as 'Le Drap d'Or Pyramidal' that had been known to some eighteenth-century French collectors.)

Messrs Melvill and Standen did not know this when they excitedly unpacked this shell in 1899. James Cosmo Melvill was the son and grandson of senior colonial officials who had settled in Manchester. He had established himself as an East India merchant, but was a keen naturalist in his leisure time. He amassed vast collections of molluscs, insects and plants by purchase and via an extensive web of collectors that encompassed travelling and expatriate Europeans as well as permanent inhabitants of the regions from which he collected. He was an active member of the Conchological Society of Great Britain and Ireland. So, too, was Robert Standen, a curator at the Manchester Museum, which stored the Society's collection in return for key specimens. A former teacher who had studied zoology at Owens College, Standen was a molluscan expert. Together he and Melvill studied and described shells that came to Manchester from around the world.

Because of its conical shape they ascribed this example to the genus *Conus* and the apparently new species *clytospira* – 'illustrious spire'. Thus, as *Conus clytospira* (Melvill and Standen, 1899) the shell became indelibly associated with two particular people in its trajectory, with a brief moment in its biography. For neither Melvill nor Standen had themselves collected it. Rather, it had been sent by Frederick W. Townsend, once of Manchester, first officer aboard the cable dredging steam ship *Patrick Stewart*. In September 1899 the *Patrick Stewart* was in the Arabian Sea, replacing a section of the Eastern Telegraph Company cable. Townsend saw this striking shell on the cable and managed to grab it. (A larger example slipped through his fingers, it being difficult to catch all the material that fell off during cable inspection.) He sent it to Melvill, like other shells he collected from the Persian Gulf, the Gulf of Oman and the Arabian Sea from 1890 onwards.[30] The gutta-percha stain that remains visible is testament to its telegraphic connection.

Along acquisition routes parallel to that which brought *Conus clytospira* to Melvill and Standen came shells from the Falkland Islands, the Antarctic, Madras, Rhodesia, the Torres Straits and from the Loyalty Islands, which had passed through the hands of locals, expatriates and travellers.[31] The routes then intersected in Manchester, where they were brought together, worked on and selected. The specimens then either remained, or continued on their journey to other private collections and museums – Townsend had also sent another of the species, which Melvill duly dispatched to the British Museum (Natural History). Similar and overlapping systems are evident for plants, animals, rocks and fossils; fuelled by indigenous traders, by maritime personnel like Townsend, and by those travelling the empire for military and missionary reasons.

The journey from cable to cabinet may be considered 'the coming into being of the specimen'.[32] And this process was repeated thousands, even millions, of times – especially during this period as objects poured into collections. Where possible, museums were proactive in this respect. Some curators were able to undertake fieldwork; others curators cultivated patrons, asked for donations and purchased en masse when they could afford it. Some of the larger institutions were able to sponsor expeditions, while others sent circulars to colonial officers and agents with precise desiderata.[33] The economy of acquisition was widespread and at times frantic, with complex systems of value and patronage at play. Professionals like Standen were few and far between, and they relied on amateurs like Melvill, or else those like Townsend who were salaried in other sectors, to fuel their collecting. Melvill and Standen were careful to credit Townsend in the publications relating to the shell; these and other museum periodicals were widely distributed, forming textual and visual layers over the network of material exchange (see Plate 14). Each of the human links in these chains thereby left a documentary trace of their involvement, recoverable in the publications, object record and/or institutional archive of the museum.

Just as Townsend sent multiple shellfish to Manchester, so other routes within the network were well worn. Museums were connected to other sites of natural history: to botanical and zoological gardens (the death of a zoo animal could be the birth of a museum specimen); to the taxidermy shop, auction house and other commercial sites; to universities and their laboratories; and, especially, to 'the field', whether the local meadow or a far-flung jungle. The museum was therefore part of the global circulation of knowledge and materials that is the subject of this volume; to adapt James Secord's helpful phrase, it was integral to 'knowledge in transit'.[34] Secord advocates breaking down the purported boundaries between the practice of science (how scientists understand nature) and popular understandings of science (how the rest of us understand nature), which is especially apposite in natural history, then as now both a vocation and a pastime. We should think geographically about the circulation of knowledge, about the sciences as a series of 'communicative actions'. This circulation involves not only scientific ideas but is also (and especially effectively) channelled through specimens – *material* knowledge in transit. We can then understand museum nature with a combination of anthropological attention to material culture and a geographical approach to place and networks.

Reflections

The journey of the mollusc and the materiality of the elephant both illustrate how specimens were brought into the museum via local and global networks and used to construct (a version of) nature among the urban sprawls of northern Britain. Each was a bearer of meaning via its complex journey to Edinburgh and Manchester. Once in their respective collections each became part of museum nature: a construction of the natural world generated by collecting, preserving and displaying inside a selection of things from outside – animals, plants, fossils and rocks.

Museum nature is not a static, single entity; rather, collections are dynamic and mutable, adapting and adapted by taxonomies, collecting practices and display (or storage) strategies. Museum nature is generated and regenerated with each generation. Museums may be considered engines of difference, connected machines that take from a mass of potential objects and classify, separate and distinguish them, helping us to make sense of the world. If so, in the late nineteenth and early twentieth centuries natural history museums were engines at full steam, fuelled by an unprecedented volume of acquisition.

These activities bequeathed stacks of records, which reveal the geographies of the specimens that comprise museum nature. We can recreate the journeys they travelled and the afterlives of these animals and plants. So too we can learn a great deal from the materiality of the specimen – the bullet hole in the hide, the gutta-percha mark on the shell. But most of all one can learn about *people*, and the relationships between them. Specimens were propelled upon their journeys by Powell-Cotton, Townsend and countless others involved in collecting these things – which included, we should remember, expedition bearers and unnamed seamen upon whom they relied. Museum nature was then physically constructed by Eagle Clarke, Melvill, Standen and their peers, from taxidermists to charwomen to researchers in the stores. All too absent from the historical record, however, are the predecessors of those who now teem around the elephant: the museum visitors. For all its careful construction, viewers gave their own meanings to museum nature: they too are reflected in the glass of the cabinet.[35]

Further reading

Alberti, S. J. M. M., *Nature and Culture: Objects, Disciplines and the Manchester Museum* (Manchester, 2009).

Alberti, S. J. M. M. (ed.), *The Afterlives of Animals: A Museum Menagerie* (Charlottesville, 2011).

Kohler, R. E., *All Creatures: Naturalists, Collectors and Biodiversity, 1850–1950* (Princeton, 2006).

Livingstone, D. N., *Putting Science in its Place: Geographies of Scientific Knowledge* (Chicago, 2003).

MacGregor, A., *Curiosity and Enlightenment: Collectors and Collections from the Sixteenth to the Nineteenth Century* (New Haven, 2007).

Poliquin, R., *The Breathless Zoo: Taxidermy and the Cultures of Longing* (University Park, PA, 2012).

Rader, K. A. and Cain, V. E. M., *Life on Display: Revolutionizing U.S. Museums of Science and Natural History in the Twentieth Century* (Chicago, 2014).

Thorsen, L. E., Rader, K. A. and Dodd, A. (eds.), *Animals on Display: The Creaturely in Museums, Zoos, and Natural History* (University Park, PA, 2013).

Wonders, K., *Habitat Dioramas: Illusions of Wilderness in Museums of Natural History* (Uppsala, 1993).

Yanni, C., *Nature's Museums: Victorian Science and the Architecture of Display* (London, 1999).

22   Peopling natural history

Naturalists have long pondered over people. What makes us human? Do we belong to a universal family or are we divided into numerous species? How have we come into being? When did we appear in Earth's history? By the mid nineteenth century, European scholars interested in these questions tried to redefine the physical, social and cultural criteria used to classify humans. European understandings of humanity's origins and place within the natural world had been deeply informed by the biblical account of the Creation. God created Adam and Eve, who were expelled from Eden to beget humankind, and everyone now living is descended from them through Noah by virtue of having survived the Great Deluge. Thus, humans have been customarily viewed as a single family with a custodial role in nature.

Drawing primarily on British and American examples, this chapter explores the making of humans into natural historical subjects. It argues that natural history, performance and anthropology were linked through a continual exchange between world fairs and museums and the emerging apparatus of disciplinary publications, institutions and professional practitioners. Recognising the importance of displayed peoples for natural historical research does not entail that the shows were 'human zoos'. This catchy designation is commonly used but misleading. Displayed peoples were always interpreted within multiple contexts such as broader debates on enslavement, human development, moral philosophy, imperial politics and natural history. Likewise, displayed peoples appeared in museums, theatres, art galleries, public gardens, world fairs, museums and zoos. Historians need to attend precisely to this geographic diversity. The 'peopling' of natural history is worth recapturing because of its lasting significance for theories of race, for the emergence of anthropology as a discipline and for histories of the human sciences more broadly.

The riddle of our ancestors

In Europe, nineteenth-century debates on humanity's past were dominated by discussions of antiquity, evolution and descent. Up until the early nineteenth century, humans were usually considered to be late arrivals on the Earth. Their history was conflated with that of literate peoples and dated using biblical chronologies.[1] The brief human past was radically revised between the 1820s and 1860s. Claims to have found ancient human remains from the 1820s onwards were consistently met with scepticism and extreme caution and both stone tools and fossils were incorporated into established understandings of human history.[2] For example, in 1833 Philippe-Charles Schmerling found two human skulls in a cave in Liege, Belgium. He claimed that the remains were ancient fossils. His medical education, scholarly publications and insistence that the he had personally excavated undisturbed cave sites lent weight to his case. Nonetheless, his critics countered that the bones were aged but not ancient. In 1856 a partial skull, thigh bones and other fragments were discovered above the banks of the Neander River, near Düsseldorf, Germany (later interpreted as the remains of an unknown human; Figure 22.1). As further finds accumulated in the 1840s and early 1850s, murmurs of antiquity continued to fascinate geologists, ethnologists, philologists and historians. The religious consequences of rejecting a biblically dated human past were profound and considered carefully for decades. Notions of human antiquity also depended upon accepting that Earth had a considerably longer history than previously imagined. Thus, fossil finds alone did not ensure the rewriting of the human past; rather, achieving a consensus on human antiquity depended upon interpreting ancient remains within newly established visions of 'deep time'.[3]

The discovery of an undisturbed cave in Brixham proved significant for debates on human antiquity.[4] The cave's untouched floor was peeled back under careful geological supervision to yield heaps of animal bones and seven flint tools. Previous sites were not pristine or were excavated by men who, despite their expertise, lacked formal affiliations with elite societies that might have invested them with sufficient authority to validate claims. The finds prompted renewed investigations of other sites in Europe and a new consensus quickly emerged that humans had a 'prehistory'. The immediate controversy over human evolution is well known within the European context and, increasingly, a global one, but the importance of antiquity is often neglected.[5] Yet the vast new timescales of human development made claims regarding human evolution substantially more plausible. Charles Darwin began the *Descent of Man* (1871), his first extended discussion of human evolution, by expressing his debt to the fact that the 'high antiquity of man has recently been demonstrated ... this is the indispensable basis for understanding his origin. I shall, therefore, take this conclusion for granted.'[6]

Simultaneously, naturalists deliberated on the issue of human descent. The predominant position always held that humans were a single species, but marginalised dissenters remained. In Europe, as early as

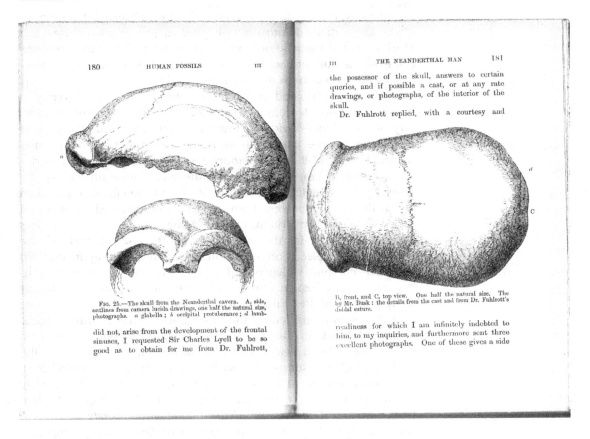

180   HUMAN FOSSILS   III

III   THE NEANDERTHAL MAN   181

the possessor of the skull, answers to certain queries, and if possible a cast, or at any rate drawings, or photographs, of the interior of the skull.

Dr. Fuhlrott replied, with a courtesy and

FIG. 25.—The skull from the Neanderthal cavern. A, side, outlines from camera lucida drawings, one half the natural size, photographs. *a* glabella ; *b* occipital protuberance ; *d* lamb-

B, front, and C, top view. One half the natural size. The by Mr. Busk : the details from the cast and from Dr. Fuhlrott's doidal suture.

did not, arise from the development of the frontal sinuses, I requested Sir Charles Lyell to be so good as to obtain for me from Dr. Fuhlrott,

readiness for which I am infinitely indebted to him, to my inquiries, and furthermore sent three excellent photographs. One of these gives a side

1655, Isaac de la Peyrère's *Prae-Adamitae* suggested that a human race had existed before Adam's creation and had been the progenitors of a different species.[7] Even those rejecting la Peyrère's daring proposition insisted that racial variation was both sufficient and necessary to cleave humans apart. For instance, the French physiologist Marie Jean-Pierre Flourens argued that white skin consisted of 'three distinct laminae or membranes – the *derm*, and two *epiderms*', whilst black skin had an additional 'two layers, the external of which is the seat of *pigmentum* or colouring matter of the Negroes'.[8] Charles Darwin believed that the 'most weighty' argument in favour of unified descent stemmed from the lack of consensus on this very issue:

Man has been studied more carefully than any other organic being, and yet there is the greatest possible diversity amongst capable judges whether he should be classed as a single species or race, or as two (Virey), as three (Jacquinot), as four (Kant), five (Blumenbach), six (Buffon), seven (Hunter), eight (Agassiz), eleven (Pickering), fifteen (Bory de Saint-Vincent), sixteen (Des Moulins), twenty-two (Morton), sixty (Crawfurd), or as sixty-three, according to Burke.[9]

Figure 22.1 Neanderthal skull fragments, from T. H. Huxley, *Man's Place in Nature* (London, 1894), pp. 180–1.

Significantly, Darwin amalgamated authors who insisted that humans ought to be divided into multiple species (for example Virey and Morton) with those who proposed a lower order difference into races (for example Blumenbach). Darwin's own explanations of racial variation were rooted in his theory of sexual selection.[10]

Between the 1840s and early 1870s, scholars interested in debates on human antiquity, descent and variation transformed the natural history of race. Most obviously, interest in humans fissured into new disciplines. In the late eighteenth and early nineteenth centuries, natural history, philology, anatomy, theology and history all contributed to studies of the human past. By the later nineteenth century, the sciences of archaeology and anthropology had been firmly established across Europe and become home to studies of fossilised humans and racial variation.[11] In Britain, the Aborigines' Protection Society was founded in 1837 in the aftermath of the 1836 report of the Parliamentary Select Committee on Aborigines. The report detailed extensive colonial violence and argued for British imperialism to become a more benign form of custodianship. In 1843, a breakaway faction founded the Ethnological Society of London and in 1863 a further splintering led to the emergence of the Anthropological Society of London. After years of fractious, albeit often overstated, rivalry, the latter two societies amalgamated to form the Anthropological Institute of Great Britain and Ireland (founded 1871). Their publications included the *Journal of the Ethnological Society of London* (1848–56, new series 1869–70), the *Ethnological Journal* (1848–66), and *Transactions of the Ethnological Society of London* (1861–9). These institutions and their periodicals brought together formative discussions on anthropology's methodologies, intellectual scope and practitioners with significant repercussions for the natural history of humans.

Up until the early to mid nineteenth century, European scholars interested in the natural history of humans were often natural historians, physicians, philologists and colonial officials steeped in humanistic research methods; this began changing with newer practitioners trained in zoology, anatomy and medicine in the late nineteenth and early twentieth centuries. Early ethnologists frequently drew on the writings of natural historians, travellers, missionaries, philologists and colonial officers for detailed information on peoples they may never have encountered personally. In the later nineteenth century, new practitioners increasingly distanced themselves from their forebears with attempts to establish professional positions and freshly demarcate what counted as anthropologically useful knowledge. Alfred Cort Haddon was appointed to a lectureship in ethnology at the University of Cambridge in 1900. Three years later his presidential address to the Anthropological Institute of Great Britain and Ireland eschewed early ethnologists and sought newly 'trained observers and fresh investigations in the field'.[12] Ten years later, his colleague on the 1898 expedition to the Torres Strait and future doctor to shell-shocked soldiers of World War I, William H. R. Rivers continued to argue for the importance of 'intensive work'.[13] Dismissing earlier generations was a highly partisan strategy for garnering prestige and securing funding for emerging professionals. Subsequently, histories of anthropology are often narrowly focused on the 'rise' of fieldwork.[14] Although useful in highlighting the increasing importance placed on in situ, extended observations (as was the case with many other naturalists and collectors in the field), this emphasis has led to a neglect of the substantial research that continued to be done in European and American metropolises throughout the late nineteenth and early twentieth centuries. In contrast, it is worth revisiting how anthropologists conducted research closer to home.

The natural history of performance

Foreign people have been imported from their homelands to be exhibited abroad for centuries.[15] On returning from the New World, Columbus presented two Arawaks to the Spanish court, where they soon died.[16] Shakespeare is said to have been inspired to create the pitiful Caliban for *The Tempest* after having seen an exhibition of Native Americans.[17] Missionaries have often displayed converts as evidence of their success. For instance, in 1804 the London Missionary Society displayed three Khoekhoe converts, John and Martha van Rooy and Martha Arendse, alongside their missionary Mr Kircherer in local congregations.[18] Early displays featured individuals or small groups who had been brought to foreign metropolises by voyagers, missionaries and entrepreneurs. In the nineteenth century, displayed peoples were imported to order and exhibited in ever larger groups, sometimes featuring dozens of performers in the world fairs of Britain, continental Europe, the United States, South Africa, India, Australia and Japan. Just as such shows became more frequent and commercially successful, managers and anthropologists expended considerable labour in making performers into specimens.

Sara Baartman was the first and most famous person to be exhibited as an exemplar of ethnic alterity in nineteenth-century London, due to claims that she had unusually large buttocks and long labia.[19] Born in the 1770s on the South African frontier, she eventually found employment in Cape Town as a maid. By 1808, she was being exhibited at Cape Town's hospital by her employer Hendrick Cesars. He was approached by the surgeon Alexander Dunlop who wanted to display Baartman in England. All three set sail in April 1810. In London's Piccadilly, customers paid two shillings to poke and prod her body whilst she wore a dress so tight that her bosom and behind were 'as visible as if the said female were naked'.[20] The abolitionist Zachary Macaulay was so convinced that Baartman had been enslaved he initiated a case against Dunlop at the court of King's Bench. Dunlop argued that Baartman had signed a contract and so consented. Baartman was interviewed and provided with a Dutch translation of the contract. By November, the case had been dismissed and patrons kept flocking to the show. By 1814, Baartman was being exhibited in Paris under new management. She was examined by the men of science at the *Jardin des plantes*, where she appeared naked before them but refused to allow an intrusive examination of her labia. Baartman died in late December 1815. The leading French comparative anatomist Georges Cuvier obtained her corpse, dissected it, made a full body cast, preserved her brain and pudenda and removed her skeleton. His report of the procedure is laced with voyeuristic triumph and veers between describing Baartman as an intelligent woman able to speak several languages and someone he is convinced exemplifies the lowest form of humanity.[21] Many people followed in Baartman's footsteps from their homelands onto stages in strange lands and, all too often, into natural history collections. They were exhibited in commercial exhibitions at precisely the same time as scholars fought to revise the meaning of 'race'.

In the 1840s, managers began to advertise such shows as explicitly relevant to scientific debates. Posters, promotional pamphlets, newspaper reviews and playbills all proclaimed that, for a fee, one could see foreign peoples who had 'Just Arrived' and were 'THE FIRST OF THEIR TRIBE EVER SEEN IN EUROPE' or 'Two New Races of People, the First of either Race ever discovered'.[22] Reviewing the 1847 exhibition of San or 'Bosjesman' (Figure 22.2), the *Theatrical Journal* claimed that, since they were a 'diminished, and diminishing, species', 'the naturalist as well as the philosopher must look on them with interest'.[23] Similar claims continued into the later nineteenth century. In 1884, the Great Farini exhibited a group of San as the 'Earthmen'. The *Era* felt that 'Everyone should see these Earthmen; they constitute an exhibition without any repulsive features, and interesting alike to the ethnologist and the general public.'[24] Advertising tried to conjure excitement and urgency by emphasising how recently performers had arrived, how soon they would leave and their novelty. The commercial viability of displaying foreign peoples may seem obvious. Yet, a city like London was home to significant immigrant populations, and commentators frequently claimed that walking its streets was enough to witness global human diversity. By showcasing what made performers unique and different from the resident immigrant populations in bigger metropolises, promotional materials created a clientele willing to pay

Figure 22.2 'The
Bosjesman, at the
Egyptian Hall, Piccadilly',
Illustrated London News,
(12 June 1847), p. 381.
Author's collection.

to see ethnic difference.[25] Likewise, promotional material that emphasised the value of the shows for rational reflection and recreation
forged strong associations between the shows and the natural history
of race.

Many scholars interested in human difference took up opportunities
to examine displayed peoples. In the early 1840s, George Catlin exhibited groups of Native Americans who were examined by phrenologists
such as Mr Bally. The science of phrenology divided the mind into
numerous faculties, each corresponding to a moral or intellectual trait,
such as 'pride' or 'affection', that resided in specific physical locations
of the brain.[26] Devotees believed that the shape of the skull directly
reflected the maturity of each faculty and could be mapped to reveal
a person's character. Bally made plaster casts of the Anishinaabe

performers' heads.[27] Likewise, a Parisian phrenologist begged for an audience with the Bakhoje. Jim was the first performer to be examined. Afterwards, Jim reclaimed the favour and scrutinised the doctor's head.[28] In 1847, the Edinburgh anatomist Robert Knox began his season of lectures on race accompanied by a group of San at Exeter Hall. After his lecture, the group went among the crowd demonstrating their weapons and accepting gifts from Knox's patrons.[29] The lecture was almost certainly the basis for his chapter on the 'Dark Races of Man' in the notorious *Races of Men* (1850).[30] In 1853, the children Maximo and Bartola, exhibited as the 'Aztec Lilliputians', were proclaimed to be the last remnants of the ancient Aztecs rescued from the lost city of Iximaya (Figure 22.3). They were exhibited alongside Flora and Martinus, who were marketed as 'the Earthmen Children'. All four drew the attention of doctors and ethnologists. Contemporaries agreed that Maximo and Bartola were singular children but were sceptical of the fantastical claims regarding their origins and were more interested in their development. Both children almost certainly had severe mental impairments. They were examined by Richard Cull, secretary of the Ethnological Society of London, and Richard Owen, palaeontologist and comparative anatomist. They co-wrote a paper in the *Journal of the Ethnological Society of London*. The men were keen to denounce the pair's authenticity and establish whether the pair were in fact 'idiots'.[31] Martinus died shortly after appearing with Maximo, and Bartola and Flora's deaths followed in 1864. Flora was dissected by William H. Flower, the conservator of the Museum of the Royal College of Surgeons, and James Murie, a prosector for the Zoological Society of London, and the two men compared her to Baartman throughout.[32] As late as 1883, a group of Krenak were being promoted as an 'Anthropological Exhibition' in Piccadilly Hall.[33] Thus public lectures, private viewings, dissections and articles became the technologies that materially incorporated performers' bodies into scientific practice.

One of the most important uses of performers' bodies for scientific research was made by Robert Gordon Latham, a philologist and ethnologist, who created a museum dedicated to the natural history of race within the new Crystal Palace at Sydenham.[34] Built in the wake of the 1851 Great Exhibition, the new palace was twice as large as the original, and showcased historical development from antediluvian monsters to modern industrialisation. Latham curated the ethnological material for the 'court' devoted to natural history, alongside Edward Forbes, a Professor of Botany at King's College, London. The court featured tableaux of flora, fauna and model peoples from the Americas, Africa, China, India, Australia and the Pacific Islands. The displays for Africa included a group of Zulus using sorcery to locate a lost article, and a San family looking out over the horizon.

Figure 22.3 Poster advertising Maximo and Bartola's wedding breakfast. Wellcome Library, London.

The models were made as lifelike as possible, painted in flesh hues, with hair and individual facial features. Crucially, many were plaster casts from living subjects, including the Zulus, Flora and Martinus, who were exhibited in London in 1853. Latham hoped the court would serve as a small-scale museum of natural history in which humans were thoroughly embedded.

On the whole, press reports were positive. The 'ethnological collection is nearly perfect', noted the *Lady's Newspaper*, and the 'life-like appearance of these figures is remarkable ... It will instruct both the well-read man and woman, and the young child. Nothing, however, but a visit to this department can convey an adequate idea of its excellence and value.'[35] Latham's handbook to the court defined ethnology as the science of 'different varieties of the human species', and provided readers with detailed explanations of the tableaux and an account of debates on the cause of human skin pigmentation (Figure 22.4).[36] Given that over 1.3 million people visited Sydenham in the first year alone, the guidebook is likely to have been one of the most widely read ethnological works of the 1850s. In 1866, a fire reduced the models to ashes. The *Anthropological Review* regretted the loss and hoped that a new collection, based on the Gallery of Anthropology in Paris, could be founded.[37] Latham's court of natural history made performers' bodies into publicly available specimens that were encountered by an extraordinary number of people. Moreover, the museum was open during the precise period in which ethnologists and anthropologists were debating the remit of their interests in humans.

Anthropologists' visions

In the latter half of the nineteenth century, world fairs provided particularly important opportunities for research. In 1886, the Colonial and Indian Exhibition hosted one of the largest cohorts of anthropologists at a British exhibition (see Plate 15). Based in South Kensington, the exhibition explicitly celebrated the British Empire, especially India, as a source of goods, colonial labour and imperial pride. Eighty-nine living 'natives' worked as shopkeepers, artisans and servants throughout the event. Several meetings of the Anthropological Institute were held at the exhibition. Anthropologists heard lectures on artefacts and peoples before being led around the site for closer inspections.[38] A significant number of articles drawing on research conducted at the exhibition appeared in the Institute's Journal in 1887 and 1888. In the 1889 *expositions universelles* in Paris, numerous 'native villages' were constructed in the shadow of the Eiffel tower (Figure 22.5).[39] In 1893, the World's Columbian Exhibition in Chicago hosted an international congress on anthropology. The human material was primarily curated by Frederick Ward Putnam, a Harvard Professor and Curator of Archaeology and Ethnology, who had originally campaigned for its inclusion.[40] William J. McGee, the first president of the American Anthropological Association, curated all the villages of displayed peoples for the 1904 Louisiana Purchase Exposition. As well as displaying 3,000 people, a Field School was established at this exhibition and was run by Frederick Starr, the first lecturer in anthropology at the University of Chicago, who spent three weeks giving lectures, supervising independent research and holding practical demonstrations.[41] The course was officially accredited by the University of Chicago, and earned the institution an award in

Figure 22.4 Cover of Robert Gordon Latham and Edward Forbes, *A Hand Book to the Courts of Natural History Described* (London, 1854). Reproduced by kind permission of the Syndics of the Cambridge University Library.

recognition of the value of the Field School from the exposition's organisers. World fairs not only staged race for the public, they also provided ideal, albeit short-lived, opportunities for scientific research and collaboration.

Despite anthropologists' best efforts, they were often left frustrated that more could not be done. Francis Galton, the eugenicist and president of the Anthropological Institute, identified three problems

1. Indian Café.   2. Dome of the Palais des Arts Libéraux.   5. A Corner of the Tunisian Section.
4. A Doorway in the Cochin China Pavilion.   3. An Annamite Carriage.   6. A Shady Corner.   7. At the Foot of the Eiffel Tower.

SKETCHES AT THE PARIS EXHIBITION.

Figure 22.5 'Sketches at the Paris Exhibition', *Illustrated London News* (10 April 1889), p. 187. Author's collection.

with conducting research at the Colonial and Indian Exhibition, despite its being a 'great event of anthropological interest'. The 'chief difficulty' arose from the time constraints on conducting research at an event that only lasted a few months. Galton also considered the displays insufficient for a comprehensive anthropological survey of the empire's subjects. For instance, the Canadian displays focused almost entirely on settlers, with the material devoted to the 'whole of the Red Indian' occupying 'no more horizontal space than would be afforded by a moderately-sized dinner table'.[42] The frustrations blossomed into discussions on founding an Imperial Institute. Galton was keen and proposed that a 'an Ethnological Museum of the races in the British Dominions' ought to feature prominently.[43] The campaign failed. British anthropologists had to be content with the British Museum's reorganised Ethnographical Gallery, which followed the establishment of the Natural History Museum in London (1881) and Oxford's Pitt Rivers Museum (1884).

American colleagues shared similar hopes and challenges. The World's Fair Congress of Anthropology, held in Chicago in 1893, was assembled too hurriedly for it to 'rise fully to the dignity expected of an international congress' with fewer international participants than originally envisioned.[44] Nonetheless, it featured figures such as Harvard's Frederic Ward Putnam, Franz Boas, later known as the pioneer of relativistic American anthropology, Otis T. Mason, curator of anthropology at the National Museum, and D. G. Brinton, as the president of the Congress. The Congress dedicated many sessions to considering the research value of the artefacts brought together by the fair. As early as 1890, Putnam had dreamed of a 'great anthropologic museum' to house the collections he had amassed, and so boost the professional standing of American anthropology. At the Congress, his colleagues sensed the 'opportunity of a century'.[45] Unlike Galton's dashed hopes, Chicago's anthropological displays found a permanent home when the Field Columbian Museum opened in 1894. Initially, Putnam's vision was only partially realised. The city's donors wanted the museum to be a permanent tribute to the entire fair. The original museum contained substantial sections devoted to themes such as the industrial arts, transport and agriculture that fell outside the boundaries of natural history. By 1910, the Museum had become an institution devoted to natural history and anthropology, after significant restructuring of the collections. In 1905, the new name of the Field Museum of Natural History was adopted.[46] To this day, visitors can visit dinosaur skeletons, stuffed animals and ethnographic artefacts all under one roof. The continued inclusion of indigenous art and artefacts in the museum is a jarring testament to the racialised hierarchies that the fairs helped create and that remain with us.

Conclusion

Scientific research on displayed peoples has been consistently overlooked or misrepresented in contemporary histories. Many have touched upon scientific interest in displayed peoples within the broader context of world fairs and the history of anthropology; nonetheless, such work is dismissed as pseudoscience, even in otherwise impressive accounts of anthropologists curating world fairs, conducting research on site and using the fairs as training grounds.[47] These analyses perpetuate the damaging assumption that science cannot create, be entrenched in, and informed by, racism. Yet, modern historians have also misread the claims of some nineteenth-century practitioners. For example, in 1885 the president of the Ethnological Society of London, John Connolly, declared that 'specimens showing the progress made in arts or in science among rude people and in remote regions, and even the natives of such regions ... have been merely regarded as objects of curiosity or of unfruitful wonder ...'[48] By 1900, even in a new era of institutionalised anthropology, W. H. R. Rivers was frustrated that 'At present, little or nothing is done to utilise the anthropological material which is thus brought to our doors, although in other countries, and especially in Germany, much useful work has been done.'[49] These laments were never neutral claims. The polemics were intended to rally colleagues into making even more systematic investigations of displayed peoples than the investigations we know took place across Europe, America and in Britain. Rivers, for instance, noted with admiration the concerted efforts to investigate displayed peoples under the auspices of the Berlin Anthropological Society and its head, Rudolf Virchow.[50] Connolly, Rivers and their peers sought to bolster the institutional and intellectual standing of their discipline by campaigning to become the *sole* arbiters of racial authenticity. In doing so, they deliberately overlooked the shows' importance for broader public engagement with the natural history of race. Yet, even without desired exclusivity, anthropologists expended considerable labour in making use of exhibitions.

Displayed peoples underpinned anthropology's foundations in numerous ways. First, the shows provided both the lay and the learned with opportunities to participate in debates on the nature of being human. Lay engagement has been particularly misunderstood as vulgar misunderstanding or uncritical consumption of managers' promotional claims. Yet patrons were frequently provided with significant resources, often drawn from travel literature, to inform their visits, and consistently encouraged to consider performers as exemplars of human development.[51] Second, exhibitions often provided possibilities for formal scientific research. Whether examining the clothing, weapons or other artefacts imported for the shows or examining performers' bodies, scholars sought to use the shows for their research. Such opportunities were particularly important for first-hand research on physical anthropology and language. Third, much of this research was subsequently published in prestigious journals such as the *Lancet*, *Journal of the Anthropological Institute of Great Britain and Ireland* or *American Anthropologist*. Thus, performers were inscribed into broader experimental and publishing practices integral to modern scientific research. Fourth, the shows and world fairs provided both the impetus for establishing permanent museums of anthropology and the nuclei of their founding collections. Even when unsuccessful, calls for such museums were important indications of the value anthropologists placed on material collections. When fruitful, as with the Field Museum, such institutions had a profound and lasting impact on the tendency to view some human cultures as of natural historical interest. Finally, pleas to use the shows for research and to found museums were often deeply rooted in concerns about educating current and future cohorts of practitioners. Hailing from diverse backgrounds in medicine, zoology and anatomy, early anthropologists were acutely aware of a lack of systematic and shared means of enskilling the next generation and, through their campaigns, sought to sustain the long-term viability of their new discipline. By informing the intellectual debates on human variation and methodology and playing a role in research and training, displayed peoples helped shape anthropological research.

The history of natural history needs to be peopled. Naturalists' interest in humans and their pasts shaped broader debates on who was considered human and how differences between peoples were racialised. It is easy to forget that humans have been the agents and subjects of natural history, partly because we are so accustomed to the consequences of disciplinary fissures between natural history and anthropology. Yet recapturing the collection, display, dissection and inscription of humans into natural history texts provides a way of materially tracing how hierarchies of human worth were created. Meanwhile, who counted as human was of utmost concern to naturalists in the nineteenth century; as we have seen, displayed peoples were often compared to animals and placed on the lower spectrum of human development. Yet, such comparisons underscored that the 'line of demarcation between man and the lower order of animals' was both real and 'very slight indeed'.[52] This profound interest in performers' bodies was well known. Tellingly, an exhibited Zulu was once said to have recalled being taken to see the 'doctoring houses' where cadavers were 'cut up and dried'. He remembered that when 'we were at the door we saw dead men standing up as if they were alive, so we feared to go in'. When asked why the English cut up their dead, he recalled:

I heard that the doctors were the people who liked dead men, and that if the graves were not taken care of their people stole the bodies for them; we were also told that the man of our party who died at Berlin was only buried because we were there, and that he was afterwards taken out and cut up, to see if he was made inside like the white people.[53]

The Zulu's recollections were reported by a missionary. Whether they are the words of a traveller who survived or the missionary, they confirm that, dead or alive, people mattered to naturalists.

Further reading

Bethencourt, F., *Racisms: From the Crusades to the Twentieth Century* (Princeton, 2013).

Conklin, A. L., *In the Museum of Man: Race, Anthropology and Empire in France, 1850–1950* (Ithaca, 2013).

Crais, C. and Scully, P., *Sara Baartman and the Hottentot Venus: A Biography and a Ghost Story* (Princeton, 2008).

Hoffenberg, P. H., *An Empire on Display: English, Indian and Australian Exhibitions from the Crystal Palace to the Great War* (Berkeley and Los Angeles, 2001).

Manias, C., *Race, Science and the Nation: Reconstructing the Ancient Past in Britain, France and Germany* (London, 2013).

O'Connor, A., *Finding Time for the Old Stone Age: A History of Palaeolithic Archaeology and Quaternary Geology in Britain, 1860–1960* (Oxford, 2007).

Parezo, N. J. and Fowler, D. D., *Anthropology Goes to the Fair: The 1904 Louisiana Purchase Exposition* (Lincoln, 2008).

Qureshi, S., *Peoples on Parade: Exhibitions, Empire and Anthropology in Nineteenth Century Britain* (Chicago, 2011).

Zimmerman, A., *Anthropology and Anti-Humanism in Imperial Germany* (Chicago, 2001).

23   The oils of empire

This is a history of natural history from the banks of one of the greatest rivers of the world – the Irrawaddy, which flows lengthwise through present-day Myanmar into the Bay of Bengal. The story that follows is not of this river itself, but rather of a range of extractive relations which allowed kings and colonists to meet and compete on this great waterway. The geological extractions that flowed up and down the Irrawaddy, and through multiple agents, have the potential to change the way we think of a critical question facing historians: how did the West come to diverge from the rest? And what does it mean to understand that divergence by taking seriously the natural knowledge of the wider world?

Divergence continues to be a heated topic of world history today. Through the decades, commercialisation, industrialisation, the 'Scientific Revolution', the causal relation of Protestantism and capitalism, the definition of property rights, the control of population and orientalist cultural knowledge have all been held up as holding the key to the rise of Europe. Fifteen years ago, Kenneth Pomeranz added his influential thesis to the mix.[1] He urged that Europe and East Asia were undergoing similar transformations in the seventeenth and eighteenth centuries, with respect to labour, land and productivity. To explain his 'great divergence', he pointed to expansion in the New World and the flow of resources into Europe. He also pressed the case for the impact of a fossil fuel, coal. For Pomeranz, some of Europe's largest coal deposits were held in Britain, and their extraction benefitted from easy access and skilled craftsmen and came about in a society that had experienced a shortage of firewood. Though East Asia had significant coal deposits, knowledge of how to extract coal was wiped out due to invasion and catastrophe from the twelfth through to the fourteenth centuries. As for access: '... the Chinese situation – in which coal deposits were far further removed from the Yangzi Delta than they were from say, the Paris basin – throws England's good fortune into still sharper relief'. Critical too, in Pomeranz's telling, was the role of the steam engine: its success relied on its use in coalfields.

Responding from a footing in the historiography of South Asia rather than East Asia, Prasannan Parthasarathi's recent engagement with the origins of divergence is different once again.[2] The substitution of coal for wood in Britain still plays a role in Parthasarathi's explanation; indeed the way in which Britain advanced ahead of South Asia relied in turn on how Indian states did not need to devise a substitute for wood, since the Indian subcontinent was heavily forested. Yet Parthasarathi also insists on the importance of cotton: the competitive challenge generated by the global spread of Indian cloth culminated in new spinning technologies in the West. Borrowing from historians of science, Parthasarathi asserts that 'the Indian subcontinent in the seventeenth and eighteenth centuries was home to complex articulations between knowledge and skill, or mind and hand, and the economic and political order'. Science, in this model, did not originate solely in Europe. However, he adds that since energy was not a problem in India, techniques connected with coal mining, iron working or the steam engine could easily and successfully travel there, to be appropriated by technicians and knowledgeable elites.

In these histories of divergence, a teleology of energy is assumed, from wood to coal and from coal to petroleum. Accordingly, one important commentator, J. R. McNeill, writes that 'the adoption of fossil fuels made us modern'. He describes an 'age of coal', lasting from *c.*1750 to *c.*1950, followed by an 'age of oil' which has run from *c.*1950 to the present. 'Each of these marked a great shift in the human condition, in the sense that each encouraged greater complexity in human society and the human relationship with the Earth.'[3] Such an argument has gained further mileage as a result of debates around the Anthropocene, with its proposal that humans have become a geological force. Scholars who follow the suggestion of the Anthropocene inevitably stress increasing ecological convergence across the globe in the modern era, tied to patterns of urbanisation, population growth, migration, agriculture, deforestation and extinction; these are taken as part and parcel of how humanity has become a geological force.

Regardless of whether one adopts divergence or convergence, one thing is clear: the story of the natural history on the Irrawaddy does not fit into these narratives about transitions in energy capacities or processes of industrial globalisation in the wake of empires. Knowledge of what could be dug from the Earth and how to use it in order to energise political projects, quite literally, flowed in both directions. This flow was unexpected and sticky. In other words, there is no straight line from wood to coal and then oil in British colonial engagements with Burma or in Burmese practices connected to fossil fuels.

Natural histories of empire

In retelling this story, the first thing to note is that Britain went to war with the Konbaung dynasty of Burma three times in the nineteenth century. As these wars proceeded, Burma was sliced and annexed.

The first war (1824-6) saw the taking of Manipur, Arakan and Tennasserim, dramatically reducing Burma's access to the Bay of Bengal. The second (1852-3) saw the fall of Pegu to the British, while the third (1885) saw the whole of Burma become a province of British India. As these wars proceeded, petroleum became a more significant cause – the agitations of Western oil corporations based in Rangoon fed into the third war.

The first war illustrates that there was a tussle over wood, oil, fire and coal, all at the same time. In this tussle knowledge of nature was prized by both the Burmese and the British. On the Irrawaddy, colonial natural historians, who benefitted from the context of war, puzzled over 'Earth oil' and how it was used by the Burmese; they surveyed and classified the woods of Burma, intrigued at the technologies of boat making demonstrated by the Burmese. Britons at war in 1824-6 also experimented in using a steam engine which drew the curiosity of the Burmese, who went on to adopt steam ships. Even after the third war, when corporatised geology gained access to Burma's oilfields, trad-itional hand-dug wells and their hereditary technicians sat alongside the drilling rigs of Western experts.

This means that instead of straitjacket histories of convergence and divergence – or indeed, of modernisation and the Anthropocene – one is left with the unexpected transitions in a specific place as Europeans learn about petroleum from Asians and Asians locate knowledge of geology alongside their own techniques of digging minerals out of the Earth. Extrapolating outwards, this is also an attempt to illustrate what a global history of natural history might look like.[4] Global forces did work on the Irrawaddy, spreading armies and navies equipped with scientific knowledge; yet these forces did not create uniformity every-where, nor did the globalisation of scientific understanding follow a singular track or teleology, or a path without friction and resistance. Indeed, in what follows, it is vital to keep in mind that natural historical work in Burma undertaken by the British not only had London in view, but also, and as its prime audience, the governing elites in Calcutta. Histories of centre and periphery do not sufficiently account for this issue. A further end point of the story is the racially segregated nature of colonial labour on the oilfields, indicative of difference rather than exchange.

Approaching the history of natural history in this vein, as this chapter does, makes the history of natural history of significance to a wide array of political, economic and social programmes in world history, and collapses the distinction between natural history and empire, avoiding the assumption that these are discrete entities linked by a causal relationship. It is also vital to keep in view that non-European polities had the capacity to practise knowledge like the Europeans did. The value of starting from the First Anglo-Burmese

War is that Britain had still not overtaken Burma by way of technical capacity, for instance in military terms. From the perspective of the region, understandings of nature were tied up with the resurgence of Theravada Buddhist kingdoms across Southeast Asia on the eve of the advent of British colonialism.[5]

The puzzle of oil

To contextualise this attempt to pluralise and complicate the standard account of the transition from wood to coal and then to oil, it is worth paying heed to late eighteenth-century and early nineteenth-century understandings of petroleum. In this period, European and American natural historians struggled to situate petroleum in relation to coal and wood, or indeed to other types of vegetable and animal oil. Terminological confusion lasted into the mid nineteenth century: 'petroleum' was used interchangeably with naphtha, bitumen, tar, pitch and asphalt. The difficulty of classifying petroleum issued from two problems of comparison and standardisation: first, how to relate the products of chemical distillation to those generated by the Earth; and, second, how to create an equivalence between samples of hydrocarbons from various parts of the globe. As late as 1865, the American chemist Henri Erni provided his readers with a table comparing natural and artificial gases, asphalts, oils and empyreumatic liquids.[6] For example, the 'illuminating gas' derived from coal was said to be equivalent to the 'sacred fire of the Brahmins', 'issuing here and there from crevices of the rocks' and the 'thin or light oil of coal tar, containing benzole' was equivalent to 'naphtha ... oozing out of the earth in Italy and Persia'. The idea was that the Earth acted analogously to a chemist.[7]

Into this confused web entered samples of 'Rangoon oil'. The collection of specimens of 'Earth oil' in Burma started early. In 1795, Francis Buchanan wrote of how one of his companions took a sample 'from the bottom' of a 'deep well' – '[i]t is a sand mixed with mica, like the common soil of the place tinged green and cohering by means of the Bitumen or Earth oil'.[8] Calcutta was a scientific sorting house for specimens of the oil. After the First Anglo-Burmese War, specimens went from the secretary of the government of India to the Royal Society of Edinburgh, where they were chemically analysed alongside 'the *black varnish* used in different parts of Hindostan and the Burmese territories, with specimens of the juices of which these varnishes are said to be compounded', as well as 'specimens of *naphtha* from Persia', 'specimens of *wood-oil*' and 'specimens of crude caoutchouc'. Robert Christison, the toxicologist and physician, who gave a paper on this set of samples, wrote in 1835 that Rangoon petroleum was the most interesting of the lot. He reported its consistency at rising temperatures, its 'crystalline principle' when compared with coal, and how it responded to chemical solvents and distillation: '[f]rom the trials I have made, I consider that the Rangoon petroleum, when distilled on the large scale, will yield nearly a third of its volume of this colourless naphtha'.[9] By the time of the Second Anglo-Burmese War, the chemist and astronomer Warren de la Rue, together with the chemist Hugo Müller, reported on 'several tons of Rangoon tar, which was carefully collected at the source, and transmitted to Europe in well-secured vessels'. De la Rue was able to distil 'paraffin' from what he called the 'tar', and went on to patent the process.[10] Price's Patent Candle Company secured the right to work under this patent, and became the first foreign company to send out agents to oilfields in Burma.[11]

Natural history of war

Leaving the scientific debates of the West, it is time to journey to the Irrawaddy to witness the First Anglo-Burmese War (1824–6) as a conflict of wood, steam, fire and petroleum. This war cost the British £5 million and 15,000 lives.[12] It was a post-Napoleonic conflict which allowed British scientific and natural historical data gathering to travel smoothly to the frontiers of British India, alongside an army and navy eager for war, and merchants keen on free trade. British natural historical practices were entangled with war not only by synchronicity, but also by aim, reach and style. At the heart of the complex of wartime natural history was the question of how to make British forces move on the Irrawaddy.

Rare letters written by a Buddhist monk, Kyeegan Shingyi or 'The Elderly Novice of the village of Kyeegan Lake', indicate the trouble that was brewing for the kingdom of Ava in upper Burma. 'The Elderly Novice' wrote on behalf of a young man, 'Lotus Leaf', who had travelled down the Irrawaddy River to Rangoon.[13] For 'Lotus Leaf', Rangoon was a city to which people came from across the oceans: 'all sorts of sailors, strangers and aliens in habit and custom, and belonging to many races all of which I cannot name'. He continued: 'They are hairy people with moustaches, side-whiskers, beards and shaggy legs. Energetic and alert, they hustle and bustle from place to place, round and round and up and down, in and out and to and fro, winding and curving, to all nooks and corners, east and north and west and south.' It was the acquisitiveness and inquisitiveness of these 'hairy people' that led to war. The war was eventually won by the British. They advanced by sea, bringing a fleet into Rangoon, which, in the end, sailed up the Irrawaddy to surprise the Burmese. A contemporary colonial explanation of the tortured course of the war was the paucity of colonial knowledge of terrain, climate and people: 'defective intelligence – scanty supplies – a tropical climate, whose deluging rains were fatal to the European constitution – a most intricate and desolate country – a hostile population – and an active and harassing enemy'.[14]

This acknowledged ignorance led into British reliance on existing knowledge and a tussle that ensued over the deployment and possession of information and techniques. One aspect of this tussle over knowledge pertains to the use of fire. The firepower and scorched-earth techniques used by the British found their match in the Burmese use of petroleum to attack British ships on the Irrawaddy. In the influential report of a Scottish military officer, John James Snodgrass, British forces were surprised by the brilliant illumination caused by 'tremendous fire-rafts floating down the river towards Rangoon':

The fire-rafts were, upon examination, found to be ingeniously contrived, and formidably constructed, made wholly of bamboos firmly wrought together, between every two or three rows of which a line of earthen jars of considerable size, filled with petroleum, or earth-oil and cotton, were secured; other inflammable ingredients were also distributed in different parts of the raft, and the almost unextinguishable fierceness of the flames proceeding from them can scarcely be imagined.[15]

The private journal of Frederick Marryat (1792–1848), who became the chief naval officer in Rangoon, bears out how fire was responded to by fire, on the part of the British:

1824	
June 30	Fire rafts came down
July 1st	Mr. Fredk Brown. Mid[shipman]: died
	Fire rafts sent down
3rd	Burnt Dalla [town burnt by the British].[16]

The stand-off around fire, however, was not only connected to its use as a means of direct warfare. The British utilised a coal-fired steam vessel, the *Diana*, for the first time in conflict in the First Anglo-Burmese War.[17] This steamship provided an undoubted advantage, for instance in the launching of rockets, which, according to British reportage, amazed the Burmese for their rapid succession, 'dazzling light', 'fatal aim' and 'ominous hissing'.[18] In the words of one Englishman, who came aboard the *Diana* after being kept captive by the king of Burma for the 2 years of the war, the ship was a 'work of Enchantment', for the terror it induced among local people. They approached it like some 'infernal beast called into existence by superior art in sorcery'.[19] The king allegedly proclaimed a wish to own a steam vessel and when discussing the vessel, '[s]pecimens of Bengal coal were shown to him; and he and his courtiers immediately observed, that there was an abundance of the mineral in the country'.[20]

Yet the *Diana* counted as a symbolic use of the steam engine. For according to John Crawfurd, who became civil commissioner of Rangoon after negotiating a commercial agreement with Ava at the end of the war, and who later became president of the Ethnological Society of London, the *Diana* took 30 days to reach Ava from Rangoon. A Burmese warboat, travelling both night and day, took 4 days 'in the freshes' and 10 days 'in the season of the rains.' The steam vessel struggled on the return journey back to Rangoon on the Irrawaddy, given the drop in the level of the river. At one point, the assistance of 300 Burmese was required to drag it off a sandbank.[21] The steam engine then did not mark a decided change in British military or naval capacity in colonial theatres in this period, especially in a place where there was a prior understanding of petroleum. By 1844, the Burmese had bought their first steamship. Intriguingly, they used *mi* or 'fire' to refer to the steam, so calling a steamship a 'fire ship' in Burmese.[22] Knowledge of extractions from the Earth – and indeed, how to cope with the river – flowed in both ways along the Irrawaddy. Burmese traditions of fire stretched to encompass the steamship.

The natural history of colonial expansion is easily traced in the practices demonstrated by naturalists who worked alongside the soldiers and naval men of the First Anglo-Burmese War. At the site of the petroleum wells, close to where the *Diana* ran aground, an 'immense' collection of fossils in 'perfect state' was amassed, particularly by Crawfurd.[23] The geologist William Buckland presented an account of the seven large chests 'full of fossil wood and fossil bones' brought to London by Crawfurd and presented to the Geological Society of London. The numbers of bones were given as follows:

Number of bones

Mastodon	150
Rhinoceros	10
Hippopotamus	2
Tapir	1
Hog	1
Ox, deer and antelope	20
Gavial and alligator	50
Emys	20
Trionyx	10

Buckland praised their remarkable condition, which arose from their 'penetration with hydrate of iron'. He compared the district where these remains were found to the 'valleys of our European rivers', and argued that the geological structure of 300 miles of the course of the Irrawaddy up to Ava presented 'a repetition of the geological structure of Europe'. He claimed that the immediate district close to the petroleum wells displayed both alluvial as well as diluvial deposits.[24] Also of interest to wartime observers were Burmese sapphire and ruby mines, from which collections were made despite royal sovereignty over them: 'The King makes claim to every ruby or sapphire beyond a hundred ticals value, but the claim is not easy to enforce.'[25] Crawfurd took home with him a ruby weighing 3,630 grains or about 970 carats.[26]

Yet the most extensive natural historical survey to be conducted immediately after the First Anglo-Burmese War was Nathaniel Wallich's study of the forests of Burma. This bears out that wood continued to be of great interest to the British Empire, even at the moment when coal and petroleum were cast against each other on the Irrawaddy. Wallich was the long-serving Danish superintendent of the Calcutta botanical gardens.[27] Formerly a prisoner of war in India in the Napoleonic wars, he had successfully integrated himself within Calcutta's European elite. After arriving in Burma in 1826, Wallich collected an extraordinary number of wood specimens from the forests: 'sixteen thousand, of which five hundred and upwards are new and undescribed'.[28] The hunt for wood, in Wallich's explanation, was driven by the military and naval needs of the British in India. This was an economic botany in gestation, still seeking to tap the sources of Burmese knowledge of wood.

One of the abiding themes of wartime commentary was the ingenuity of the Burmese warboats. The use of 'magnificent teak, first roughly shaped, and then expanded by means of fire' for the construction of warboats was noted in one published account. The author, Captain Thomas Abercrombie Trant, ranked these warboats above British vessels for their rapidity, which arose from their lightness and their small surface above the water, combined with the 'uniform pulling of the oar falling in cadence with the songs of the boatmen'. He gave his readers three pages of musical notation for Burmese warboat songs.[29] Warboats carried about fifty or sixty rowers and were said to 'sail in fleets' and to mount 'impetuous' attacks (Figure 23.1).[30] The orientalised fascination with Burmese techniques of manufacture and war-making fed into Wallich's documentary on wood. He wrote of the 'Theyngan tree': 'It grows in vast abundance near the sea shore, where it constitutes one of the principal features of the forest.' The largest of the Burmese boats, he noted, was made of

'one single excavated trunk' of this tree.[31] Crawfurd noted its scientific notation: *Hopea odorata*.[32]

Yet the celebrated find of Wallich's tour of trees was not the 'Theyngan tree'; it was the 'Thoke tree', now the 'Pride of Burma'. Wallich's concern with woods of use to the imperial machine was twinned with a personal project of ingratiation with colonial elites, which depended on the discovery of ornamental beauty. His three-volume *Plantae Asiaticae Rariores* (1830) began with the 'Thoke tree', which he named *Amherstia nobilis*, after the wife and daughter of the Governor-General of India. Wallich wrote of how he found the tree difficult to locate until he entered 'a sort of monastery' 27 miles from Martaban, where the tree's flowers were used for worship (Figure 23.2): 'There were two individuals of this tree here; the largest, about forty foot high, with a girth, at three feet above base, of six feet ... They were profusely ornamented with pendulous racemes of large vermillion-colored blossoms, forming superb objects, unparalleled in the Flora of the East Indies.'[33] That this 'superb' delight was collected in the yard of a Buddhist temple is significant: the war saw British forces scouring the landscape for small cells or chambers in

Figure 23.1 'One of the Birman Gilt War Boats Captured by Capt. Chads, R. N. in his successful expedition against Tanthabeen Stockade', painted by T. Stothard, from an original sketch by Captain Marryat, in Joseph Moore, *Birman Empire* (London, 1825–6), part 2, no. IV, with permission of SP Lohia Foundation.

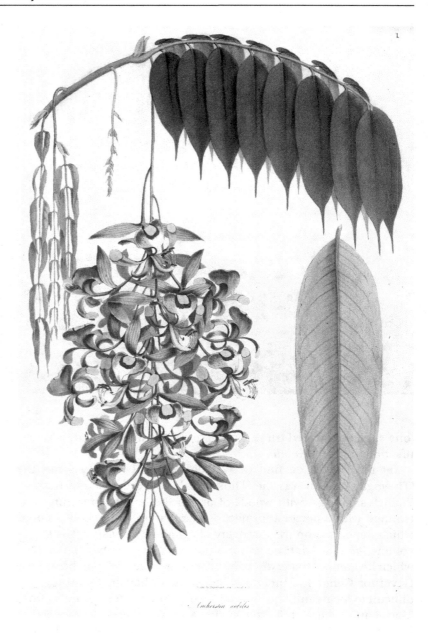

Amherstia nobilis

Figure 23.2 '*Amherstia nobilis*', in N. Wallich, *Plantae Asiaticae Rariores* (London, 1830), plate 1.

temple yards enclosing silver- or gold-covered images. The plunder of these cells was compared at the time to a form of 'sapping and mining' as it was driven by the perceived value of metals. What better context for understanding Wallich's natural historical plunder?[34]

As a result of this war, a dramatic scene of scientific inquiry was thus opened at the peripheries of the British Indian empire. War, driven by free-trade commercialism and an ethos of plunder, framed the natural historical exploration of Burma, as also the specific scientific questions asked of India's frontier. The violence of the moment was intertwined with the pursuit of information about wood and coal and petroleum, but there was no linear sequence across these sources of energy. War resulted in the rapid acquisition of partial knowledge rather than full knowledge (or to deploy the last metaphor, a type of 'mining'), and the knowledge possessed by the Burmese was often a match for that held by the British. This resulted in a two-way flow of knowledge, which places natural history at the heart of the imperial war machine. Natural history was plundered alongside Buddhist artefacts.

The fields of oil

It is time to move forward to the period after the Third Anglo-Burmese War, which saw the arrival of corporatised geology in Burma and the expansion of the oilfields close to the Irrawaddy. A good place to begin is an iconic photograph (Figure 23.3). Taken by the commercial photographers Watts and Skeen, it appeared in the British Indian geologist Edwin Pascoe's encyclopaedic account of the oilfields of Burma and India, which were part of the Geological Survey of India, published in 1912.[35] Pascoe's survey is seen to have coincided with the coming of age of the modern oil industry in Burma.[36] Indeed, he wrote of the oilfields of Yenangyaung (the same site investigated by Crawfurd and Wallich) that 'the native industry is rapidly becoming a thing of the past'.[37] Yet a history of the dramatic rise of Western oil interests in Burma – or their coming of age – must be seriously qualified with an attention to long continuities. These continuities stretch across the precolonial and the colonial periods, from eighteenth century travellers' accounts of landscape, soil and oil to the intensive work undertaken by geologists like Pascoe.

'Yenangyaung' translates as 'creek of stinking water'. The earliest English observer of these wells, George Baker, wrote of 'about 200 families who are chiefly employed in getting Earth-oil out of Pitts'.[38] The hereditary right to make oil was granted by the kings of Ava to families, who formed a corporation called *twinzayo*. The history of these oilfields is taken to stretch back some centuries, though there is scanty evidence to back up this claim.[39] With the expansion of the trade in the hands of foreigners, the rights of the *twinzayo* were hemmed in. The king declared a royal monopoly over the trade itself in 1852, after the Second Anglo-Burmese War, nationalising the fields in order to recoup the kingdom's finances after war. The arrival of various foreign corporations, such as Burmah Oil, which began drilling in

Figure 23.3 'Yenangyaung – Native Well-Digger in Diving Dress (The man on his right is holding the mirror)', in E. H. Pascoe, 'The oil-fields of Burma', *Memoirs of the Geological Survey of India*, 40 (1912), plate 18. Reproduced by kind permission of the Syndics of Cambridge University Library.

Yenangyaung in 1888, restricted the status of the hereditary traders of oil at Yenangyaung; yet the traditional hand-dug wells were not easily obliterated.[40]

The Burmese used oil for a variety of purposes in the eighteenth century, and there was a relatively far-reaching trade in the commodity. This precolonial trade and understanding of petroleum must therefore be placed alongside modern colonial uses of oil. Oil was used for sealing boats, for preserving palm-leaf manuscripts, for medicinal purposes, for illumination, for warding off insects and for lubricating the wheels of bullock carts.[41] In 1795, the diplomat Michael Symes wrote that oil was supplied to 'the whole [Burmese] empire, and many parts of India', and his naturalist companion commented on 'great piles of Jars [filled with oil] . . . on the Beach', at the riverside close to the oilfields.[42] In 1797, another diplomat, Hiram Cox (1760–99), observed that petroleum was used by the Burmese as 'a lotion in cutaneous eruptions, and as an embrocation in bruises and rheumatic affections'.[43] '[T]he white ant', Crawfurd wrote, 'will not approach it.' He estimated that two-thirds was used for illumination, and that the consumption of oil 'is universal wherever there is water-carriage to convey it; that is, in all the country watered by the Irawadi'.[44]

Part of the photograph that Pascoe published was later reproduced on the 45 kyat banknote issued by the 'Union of Burma Bank' (Figure 23.4). Here, two of the men in Pascoe's photograph are juxtaposed with drilling rigs at the new company wells. One of them wears a 'diving helmet', while another has a mirror to hand. The helmet and mirror technique was an innovation of the mid 1890s which enabled deeper wells to be dug by hand. The helmet was made out of the 'ubiquitous kerosine-oil can'; it had an air-line to a pump, and was tied down by a piece of cloth at its base, which was soaked with water. '[W]hen the air is turned on [the cloth] blows out like a balloon, and the air can be heard fizzing out through the meshes of the wet cloth.'[45] Pascoe wrote that this change meant that light could be 'flung down' by a mirror; and air could be pumped down by 'two coolies who turn the handles of an air pump'.[46] For the historian of natural history, the ancestry of this technique challenges a narrative of the rupture caused by corporatised colonial oil drilling in the late nineteenth century: a pearl diver from the Mergui archipelago happened to be at the oilfields at Yenangyaung, and his enterprise in reaching greater depths was then emulated by the locals.[47]

Within the colonial imagination itself is a chain of association, indicating how the modern petroleum industry's visual archive was built on earlier predecessors. The oil wells were a hotspot of colonial photography and visualisation. One of the earliest photographers of the British Empire, Linnaeus Tripe, who accompanied a diplomatic mission to the court of Ava in 1855, after the Second Anglo-Burmese War, produced some photographs at Yenangyaung. One of them is of a large tamarind tree; another of a 'Chatty Manufactory', showing a deserted site where pots were made for the transportation of oil. These sit alongside a long series of photographs of Buddhist pagodas, some in ruin.[48] With him was Colesworthy Grant, who painted water-colours, including one of boats at the shoreline of Yenangyaung carrying oil down the river, stowed in bulk 'like salt or coals'; two of the landscape meant to show its 'wild, broken, arid and barren nature'; and one titled 'The Oil Wells: Ye-an-nan-gyoung' (Figure 23.5). Grant wrote of the oilfields:

The road which led to them was marked by nature's wildest and strangest aspect. High hills, split as it were by some mighty convulsion of nature, resembling a stony avalanche – or one of the gigantic 'rock cuttings' for the railway in Europe, margined its early part. The path then continued to wind between a succession of hills and high broken banks, and over a sandy rugged

Figure 23.4 Bank note issued in 1987 by the Union of Burma Bank, worth 45 kyats. Private collection.

soil, dotted here and there with rocky substances and innumerable specimens of petrified wood.[49]

This watercolour is panoramic but thinly populated: there is only one oil well in focus, others merge with the landscape; a cart with 'chatty' (petroleum storage jars) sits alone in the landscape, and the rolling hills are studded with the barest vegetation. The line work on the watercolour accentuates the supposed barrenness of the scene.

The fascination with the oilfields certainly included a commercial and statistical concern with the number of wells and their depth. But it also included a wider set of scientific concerns which benefitted from the visual fascination that Yenangyaung generated. In keeping with Grant's aesthetic, observers from the end of the eighteenth century to the start of the twentieth were concerned with the lay of the land and its vegetation. Symes wrote that the country was 'cheerless and sterile'; '[t]he hills, or rather hillocks, were covered with gravel, and yielded no other vegetation than a few stunted bushes.'[50] The naturalist Buchanan, accompanying him, wrote: '[n]othing can be more irregular and broken that the low sandy hills in the neighbourhood of this place … Its temples are small and ruinous.'[51] A recurrent point of reference was the petrified wood which dotted the landscape. Buckland reported that trunks of fossil wood between 15 and 20 feet long and 5 feet in circumference had been found by Crawfurd.[52]

In the late nineteenth century, the most important geological work undertaken at Yenangyaung was conducted by the German–Australian Fritz Noetling who travelled far and wide in the empire, writing on the

Figure 23.5 *The Oil Wells: Ye-an-nan-gyoung,* watercolour no. 18, C. Grant, 'Album of 106 drawings of landscapes and portraits of Burmese and Europeans . . .', British Library, WD540, IOR. By permission of the British Library.

evolutionary history and material culture of Aboriginal Tasmanians, and undertaking geological work in Burma, Baluchistan and the Punjab. Noetling's study of Burmese geology extended to coal, Burmite, jade and other minerals.[53] His mammoth survey of the oilfields was still characterised by this romantic and picturesque gaze – a landscape of barrenness, soaked in the remains of the past. One of his photographs shows 'Native boats loading oil at Yenangyaung'; although more crowded than Grant's watercolour of the same scene from decades earlier, it perpetuates Grant's iconography (Figure 23.6). Noetling explained that the region had originally been one plateau, eaten into by the action of water: '[s]ome parts of the country, especially the ferruginous, conglomeratic beds, resisted in a more energetic action the surface water, which by washing away the surrounding softer strata modelled the harder ones to a kind of ridge'. Noetling announced that the oil-bearing strata were chiefly of the Tertiary formation, 'no later age than Miocene'.[54] The strata were said to form an anticline 'on the centre or top of which' the oilfields were situated.[55]

Figure 23.6 'Native boats loading oil at Yenangyaung'. The photo-etching faces the table of contents in F. Noetling, *Report on the Petroleum Industry in Upper Burma* (Rangoon, 1892), British Library, General Reference Collection I.S. BU.47/6. By permission of the British Library.

The density of information collected by Noetling, ranging from the marine fossils of the region to legends of the origins of the oilfields, and from its physical geography and geology to the social organisation of the *twinzayo* and how wells were dug, is quite astounding. But here too, it is striking to see how this scientist's statistical study of wells relied on Burmese information. In one of his tabulations, a 'key' to find a well, he placed the 'Burmese number' of the well next to what he called a 'consecutive number'; in his 'record of the wells', he gave the locality, depth in feet, 'stated depth in attaungs, at 20 inches'; the difference between the stated and actual depth; the yield; how often oil was drawn up and the stated age of the well.[56] Of course, this classification of knowledge was parochial, and meant to substitute for what he gleaned from his informants with what he took to be a scientific understanding. Yet what emerges from this survey is that the landscape of Yenangyaung was not empty: it was covered with numerous Burmese wells. For the modern oil industry to take these fields, it was critical for the sources of Burmese oil to be mapped in excruciating detail.[57] If this did not occur, there were problems. Pascoe wrote for instance of a 'congestion of inflammable derricks' giving rise to fires. 'The village of East Twingon was at one time entirely wiped out.' Less fatally, but critically for capitalism, if wells were dug too close to one another, their yields affected one other and the flow was unsatisfactory.[58]

The coexistence of Burmese and colonial wells and orders of knowledge might be contextualised in the racialised regimes of labour present on the Burmese oilfields. According to one author, the Burmese showed a 'natural aptitude for handling the cattle of the country', allowing them to be employed 'in bullock-car transport, which is responsible for the conveyance of timber, drilling tools, engineer and workshop details', while Indian labourers brought to the wells worked pumping wells, as boiler men and greasers in charge of engines, on river steamers and digging the earth. This made the Burmese oilfields a place where 'each race and *jat* [people] ha[d] its own well defined niche of labour'.[59] If this describes the segregation of labour on the colonial wells, by this time, the same style of differentiation also operated between the colonial and the colonised, between Western corporatised wells and Burmese hand-dug wells. While knowledge flowed between agents, and down the decades, different traditions and modes of work were not flattened into one, rather colonial extraction and reliance gave rise to ethnographic classification.

Conclusion

In order to understand the significance of this view of the history of natural history from the Irrawaddy, it is worth recording the myth of the origin of the petroleum industry. In his book of 1863, Norman Tate gave a history of petroleum stretching from Nineveh and Babylon to Persia, and he added a reference to Burma: 'The petroleum springs of Rangoon, on the banks of the Irawaddi, in the Burman Empire, are said to have been known and worked for ages and are at present some of the most powerful and copious springs yet discovered.'[60] Yet despite this long history, Tate exalted one event from another hemisphere to the position of originating moment for the modern petroleum industry: August 1859, when a certain Colonel Edwin Drake struck oil in Venango County, Pennsylvania. It is this account which has endured since the mid nineteenth century.

Looking at the origins of modern petroleum from the perspective of the Irrawaddy forces us to revise this narrative, by showing the productive interface between Asians and Britons. The Irrawaddy troubles a neat and linear sequence of energy-producing natural commodities and their scientific study that moves from wood to coal and from coal to oil. The histories of carbon capitalism and high-energy economies have had many twists and turns. All of these are important for what they tell the historian about how natural historical knowledge operates. Conversely, the history of natural history is relevant to explaining a key motor in the making of the modern world.

To accept this historical geography of petroleum is also to reject straitjacket histories of divergence or the Anthropocene. Natural historical knowledge sits at the heart of empire. It is an assemblage which often crosses locales and seeks to act globally; yet the crevices of this assemblage are apparent here, in the difficulties of terminological equivalence, mobility and cross-cultural collaboration and work. Natural historical knowledge is sticky – or perhaps one could call it oily; it can move in multiple directions, in jumbled sequences and can even move against empire, while also being part of the mechanics of colonial war and capitalism. Its stickiness should not preclude attention to its essential place in the military machine in Burma. In the First Anglo-Burmese War, wood, coal and oil were all subject to use, study and comparison. In the history of Yenangyaung, Burmese understandings of oil and digging for oil survived for a surprisingly long time alongside corporatised colonial geology. To understand the oiliness of natural historical knowledge requires a widening of the traditional geographies used by historians and the inclusion of the knowledge of the wider world.

Also of note here are the long-term continuities, from the eighteenth century to the early twentieth century – in forms of description and visualisation, including texts, watercolours and photographs produced by naturalists, diplomats, military men and colonists. The long-term continuities also pertain to the assembly of knowledge across the barriers between precolonial and colonial, Asian and Western. These continuities highlight how energy-intensive empires require the assimilation of information pertaining to natural resources from every possible source and place. The British Empire reached its peak before World War I, and since Burma was one of the only places containing oil which was under British rule prior to the war, the British and their foreign aides prised open the Burmese oilfields to feed their interests.[61] Yet before this happened the wells of the *twin-zayo* had to be known in every way. The interlacing of precolonial and colonial did not occur in a power vacuum; plunder, capitalist accumulation and racial differentiation were part of the results of this encounter of natural historical knowledge.

In Cambridge University Library lies a beautiful, large-sized and little-understood series of hand-woven and painted cotton maps of Burma, presented to it by a colonial inspector of schools in Burma and probably drawn by Burmese artists around 1860. One of these maps (see Plate 16) depicts the Sa-lay township (now Sale) in Upper Burma on the Irrawaddy river. Sale lies north and very close to the town of Yenangyaung. It is a perfect source with which to mount a retrospect on this chapter and to ponder how the scholarly field of the history of natural history might move in the future.

The challenge now for the historian of natural history is to move deeper still into Burmese sources on the Earth, its trees, rivers and mountains. Indeed, the kings of Burma were interested in botany.[62] The fabric maps in Cambridge indicate the richness of Burmese knowledge of nature; Burma had a long-standing state-sanctioned tradition of cartography which intersected with the unfolding wars with the British. On the black and red map of 'Sa-lay township' is evidence of attention to mountains of various kinds – there are ranges bordering on waterways, mountains showing volcanic action, some with flat tops and black crevices and forested hills too. It is super-imposed on a grid and includes pen drawings of Buddhist temples. Does this then provide evidence of the take-up of European methods of mapping or the evolution of Burmese cartographic tradition? The history of natural history on the Irrawaddy and of Yenangyaung would look different again if one were to begin with this fabric map. How can it be contextualised within Southeast Asian understandings of the Earth? This chapter has decentred the Euro-American history of natural history. Yet further revision starting with sources like this map awaits.

Further reading

Arnold, D., *The Tropics and the Travelling Gaze: India, Landscape and Science, 1800–1856* (Delhi, 2005).

Damodaran, V., Winterbottom, A. and Lester, A. (eds.), *The East India Company and the Natural World* (Basingstoke, 2015).

Driver, F. and Martins, L. (eds.), *Tropical Visions in an Age of Empire* (Chicago, 2005).

Gomez, P., *The Experiential Caribbean: Creating Knowledge and Healing in the Early Modern Atlantic* (Chapel Hill, 2017).

Grove, R., *Green Imperialism: Colonial Expansion, Tropical Island Edens and the Origins of Environmentalism, 1600–1860* (Cambridge, 1995).

Lambert, D., *Mastering the Niger: James MacQueen's African Geography and the Struggle over Atlantic Slavery* (Chicago, 2013).

Raj, K., *Relocating Modern Science: Circulation and the Construction of Knowledge in South Asia and Europe* (Delhi, 2006).

Rood, D. and Manning, P. (eds.), *Global Scientific Practice in an Age of Revolutions, 1750–1850* (Pittsburgh, 2016).

Safier, N., *Measuring the New World: Enlightenment Science and South America* (Chicago, 2012).

Schaffer, S. et al. (eds.), *The Brokered World: Go-Betweens and Global Intelligence, 1770–1820* (Sagamore Beach, MA, 2009).

Sivasundaram, S., *Islanded: Britain, Sri Lanka and the Bounds of an Indian Ocean Colony* (Chicago, 2013).

Winterbottom, A., *Hybrid Knowledge in the Early East India Company World* (Basingstoke, 2016).

IV Connecting and conserving

24   Global geology and the tectonics of empire

What can it mean to be 'global'? From the late nineteenth century onwards into the twentieth, new answers to this question emerged through the natural historical sciences and especially geology. This era witnessed unprecedented national rivalries, racial stereotyping and two world wars; but it was also an age of internationalism. Linked by capitalist relations of production and exchange, the world could be imagined as a whole, bound together through a single future, present and past. Geology, based on local studies and defined as the evolutionary science of the Earth, emerged as a key discipline for comprehending the foundations of this interconnected vision. The new approach went beyond classical mathematical astronomy and geography to consider the internal structure of the planet and its history. With innumerable regional accounts being scaled up to delineate a complex but unified globe, geology became the ultimate cosmopolitan science.

Rather than outlining the details of individual geological theories and seeing them as steps towards the triumph of plate tectonics in the 1960s, this chapter aims to understand what led such ambitious syntheses to be proposed in the first place. Scientific conceptions of the Earth were shaped by changes in practice, purpose and criteria of explanation, which in turn were embedded in debates about free trade, empire and colonial domination. The result is an essay in geopolitics.

The summit view

The age of European empires witnessed an outpouring of attempts to explain the history and structure of the Earth. The central inspirational leader of this movement was the naturalist Alexander von Humboldt (see the chapters by Ogborn and Rebok in this volume). During his wide-ranging travels, Humboldt had applied the perspective of his experiences in the Prussian mining districts towards an integrated vision of the terrestrial globe. It was epitomised by the view from the mountain summit, in which all phenomena were brought together through the aesthetically engaged consciousness of a single observer, controlled by reason and instrumental precision (see Figure 19.1). Humboldt's vision was as much spatial as temporal, focusing on the geographical display of measured data. The distribution of animals, plants, winds, ocean currents, magnetism, strata, volcanoes and mountain chains could all be determined by instruments and charted on maps. Among those who adopted a similar approach was the English naturalist Charles Darwin during his circumnavigating travels on HMS *Beagle*. Writing with the same speculative energy that would later fuel his species work, Darwin developed a theory involving the subsidence and uplift of the Earth's crust, which explained everything from the formation of coral reefs to the distribution of the remains of extinct creatures. As he scribbled in a notebook in 1836, 'geology of whole world will turn out simple'.[1]

The *Beagle* voyage, through surveying and mapping, aimed to extend free trade and a national sphere of influence. Geology – with its implications for discovering resources and avoiding dangers from undersea reefs and other hazards – was especially significant in this enterprise. As part of the repurposing of knowledge for empire that took place in Europe from the end of the eighteenth century, geologists had elaborated a standard sequence of fossil remains that could identify strata of similar age anywhere on Earth. Geology was forged as a spatial science in which the so-called 'discovery of time' was secondary. The focus was on ordering the materials of the terrestrial globe for mineral exploitation and extraction. By the end of the nineteenth century geological syntheses also relied upon results from oceanographic voyages, telegraph networks and seismological stations.

These activities vouchsafed the place of geology in furthering the progress of reason and civilisation. Classifying, surveying, collecting and comparing became ways of creating a single perspective. The significance of these activities became manifest in a wave of museum building from the 1850s onwards: Paris, New York, Berlin, Melbourne, Manchester, Rio de Janeiro and many other cities. Researchers could compare local specimens with their counterparts from other countries, while casual visitors were overwhelmed by the visual spectacle of an ordered past. The Museum of Practical Geology in London, opened in 1851, laid bare the history of the Earth in tens of thousands of individual specimens, all visible in a single glance (Figure 24.1). The casual visitor gained the impression of seeing more, and more deeply, than Humboldt had from the high peaks of the Andes.

Grand theories of the Earth tended to be the province of writers (almost always men) in Europe or the United States, where libraries, collections and other resources were relatively accessible. Positioned at the leading edge of the movement towards empire, natural history museums before World War I became centres for syntheses,

Figure 24.1 The fossil record at a glance in the Museum of Practical Geology in London. Photo 640481. Reproduced by permission of the British Geological Survey, CP17/023.

particularly by their directors, who were often domineering figures of authority. Constructing an overarching system became something that was also expected of professors in the universities of central Europe. Unified points of view were important not only for attracting students but also for cultivating broad readerships: this was a golden age for commercial science writing.

The most significant syntheses of the final decades of the nineteenth century were forged in the German-speaking lands of central Europe, notably by the Austrian geologist and politician, Eduard Suess, who is best known today for introducing terms such as 'Gondwana' (for an ancient supercontinent) and 'biosphere' (for the dwelling place of life on Earth). Suess wrote in Vienna, the capital of the Habsburg Empire, where the boundaries between Asia and Europe, East and West, were unstable and uncertain. Here the model of empire was land- rather than sea-based, and grounded in balancing the interests of multiethnic territories rather than in controlling distant colonies.

Suess worked by scaling up from local experience. Initially employed at the Austrian Geological Survey on invertebrate fossils, he was then assigned the mapping of a high alpine district in the Dachstein that sparked his long-standing interest in mountain structure, as expressed in *Die Entstehung der Alpen* ('The Origin of the Alps') in 1875. Suess's massive, multivolume *Das Antlitz der Erde* ('The Face of the Earth', 1885–1909) extended these conclusions to a planetary scale, showing how the gradual collapse of the crust of the cooling Earth could explain all the diverse phenomenon of the globe. The mountain ranges that trended across the continents were produced by the buckling and crumpling of the contracting crust. In the high Alps these contractions were so extreme that older strata had sometimes been thrust on top of more recent deposits.

Suess's approach to theorising exemplifies the place of geology within fin-de-siècle liberal imperialism. Active in politics, during the revolutionary uprisings of 1848 he was the youngest member of the governing committee until the old order was restored. Like many liberals frustrated by defeat, Suess saw the sciences as the most effective long-term route towards reform. From 1856, his position as 'extraordinary professor' in the University of Vienna provided a platform for civic engagement. In the 1850s, the ancient walls around Vienna were being replaced by a circular road and grand buildings that would showcase the power of the Habsburg Court. Suess pointed out that to be a modern city, its cholera-ridden water had to be replaced with a pure supply from the Alps, and he effectively became the plumber of liberal Vienna. Elected to parliament, Suess advocated free trade, disarmament and other liberal causes. He encouraged international scientific cooperation and the pursuit of knowledge beyond European borders. In a tract on silver and the international currency question, Suess underlined the connection between geology and history as a balanced equilibrium between different human races. 'Europe', he wrote, 'has been conscious of the leadership; that is a proud memory, but to-day Europe is obliged more and more to allow other continents to enter into equilibrium. They are honestly struggling onward to be the peers of Europe, and their claims must be recognized.'[2]

Underlying *The Face of the Earth* was a confidence in the power of the human mind, embodied in the creation myths of all cultures and manifested in science, to grasp the essentials of Earth history. In the first chapter, Suess offered a sympathetic consideration of Mesopotamian accounts of the deluge, thereby setting himself as the latest in a sequence of tellers of epic tales. Employing research from across the world in dozens of languages, the work was a statement of ecumenical philosophical principles. Seeing the completed volumes,

Suess's servant wondered how the myriad books, pamphlets and papers fetched from the library over many decades had been condensed into such a small compass.[3] It opened by inviting readers to envision the planet Earth as seen from outer space, to make connections not only between continents and oceans, but also between different peoples and intellectual traditions. Drawing on orientalist notions of the seer and visionary from India and China, this was the ultimate extension of the Humboldtian summit view of an interconnected terrestrial landscape.

The lost epics of the Earth

The geologist, on Suess's telling, was akin to Rama, one of the greatest heroes of the Hindu epics. In an often-quoted passage, Suess wrote that 'As Rama looks out upon the Ocean, its limits mingling and uniting with heaven on the horizon, and as he ponders whether a path might not be built into the Immeasurable, so we look over the Ocean of time, but nowhere do we see signs of a shore.'[4] Through synthesis, geology was one of the ways in which Western geologists combined Eastern epic and empirical science to recast the very idea of what it meant to see the history of the world as 'global'. This was a vision in which recondite details of national and colonial surveys could shed light on the most profound problems of planetary history. The search for fabulous lands of grandeur and significance, now abandoned and drowned, extended the imperial frontier into the deep past, at a time when the purpose of the empire was changing and the era of geographical discovery seemed to be ending.

The recovery of 'lost worlds' gained depth and resonance from references to what Europeans saw as the epic mysteries of the Orient. As the novelist E. M. Forster wrote in *A Passage to India* (1924):

The Ganges, though flowing from the foot of Vishnu and through Siva's hair, is not an ancient stream. Geology, looking further than religion, knows of a time when neither the river nor the Himalayas that nourished it existed, and an ocean flowed over the holy places of Hindustan. The mountains rose, their debris silted up the ocean, the gods took their seats on them and contrived the river, and the India we call immemorial came into being. But India is really far older.[5]

Such perspectives marked a regular traffic between natural historical, philosophical and literary perspectives, with participants from different traditions able to see their work as contributing to a common understanding. When Joseph Conrad wrote of the antiquity of Africa in *The Heart of Darkness* (1899) and Arthur Conan Doyle described *The Lost World* (1912), they took readers on a journey shared with contemporary geologists.

As explicit acts of cross-cultural exchange, the grand syntheses looked for inspiration to the ancient creation stories of China, India and the Middle East; but they were also intended to destroy and replace them with explanations drawn from modern scientific surveys. Sunken land-bridges and drowned continents, as explanations of the distribution of past life, proliferated in the decades of high imperialism around 1900. Melchior Neumayr, Suess's student and Professor of Palaeontology at Vienna, developed in his influential *Erdgeschichte* (1880) ideas about vanished continents, including a Brazilian–Ethiopian landmass extending across the South Atlantic. One of the most comprehensive over-views, with superb maps of the former distribution of life, was provided by the German schoolteacher Theodor Arldt's *Entwicklung der Kontinente und ihrer Lebewelt* ('Evolutionary Development of the Continents and their Life Worlds', 1907). In Britain and Germany, the history of life was one of evolutionary conquest rather than the shifting balances imagined by Suess. As the Royal Geographical Society's *Atlas of Zoogeography* (1911) put it, 'The natural tendency of any species which is successful in the ever-waging struggle for supremacy is to gradually spread over a wider and wider area.'[6]

Recovering lost worlds, far from being confined to Europe and the United States, became important in anti-colonial resistance move-ments and modern nation-building. Fossil remains, like human arte-facts, became part of cultural heritage; and access to local materials could provide a route towards gaining international attention. In the independent republic of Argentina, the palaeontologist Florentino Ameghino (with his brother Carlos) collected thousands of fossils in the rich deposits of Patagonia. Born in humble circumstances, he travelled to Paris in the late 1870s and established a position within European science. Upon his return, Ameghino made the La Plata Museum (founded in 1888) a centre for understanding the history of vertebrate life. Not only were the South American fossils vital to identifying the Tertiary (now Paleogene and Neogene) period as 'the Age of Mammals', but Ameghino's view that they appeared in the strata record contemporary with dinosaurs became the basis for claim-ing that the key vertebrate groups had originated in the Pampas. Humans, for example, had spread from Argentina in three successive waves: primitive hominids in the Eocene; protohumans at the begin-ning of the Pliocene; and ancestors of modern humans, who during the Pliocene had migrated northwards in the Americas and thence to Eurasia (Figure 24.2).[7] Ameghino's speculations were controversial but widely discussed.

In the Indian subcontinent, recovering the Earth's deep past also had a potent role in creating a sense of national identity. At the begin-ning of the twentieth century, intellectuals in the Tamil-speaking

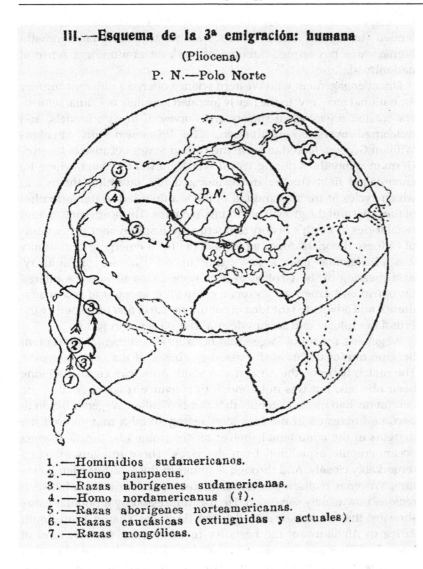

III.—Esquema de la 3ª emigración: humana

(Pliocena)

P. N.—Polo Norte

1.—Hominidios sudamericanos.
2.—Homo pampaeus.
3.—Razas aborígenes sudamericanas.
4.—Homo nordamericanus (?).
5.—Razas aborígenes norteamericanas.
6.—Razas caucásicas (extinguidas y actuales).
7.—Razas mongólicas.

Figure 24.2 Florentino Ameghino's third phase of the spread of humans from South America; from J. Ingenieros, *Las doctrinas de Ameghino: la tierra, la vida y el hombre* (Buenos Aires, 1919), p. 173.

region developed the notion of a drowned continent of Lemuria, the original homeland of the Tamils from which all humans had derived. Lemuria had a clear pedigree in European science, having been proposed by the English naturalist Philip Sclater and advocated by the arch-Darwinian Ernst Haeckel in Germany (see Plate 17). It had been taken up in the *Manual* of the colonial Geological Survey of India, and given relevance to the origin of humans in administrative publications for the Madras presidency. Such writings, repurposed in Tamil tradition (and in Forster's *Passage to India*), gave sanction to 'lost worlds' for dispossessed colonial subjects.[8] Western geologists, however, tended to ignore these efforts, which the historian Sumathi Ramaswamy has termed 'labours of loss', for grounding a sense of nationhood.

Direct engagement with Western science offered a different strategy for national recovery. In the newly founded Republic of China, reformers created a geological society and survey of foreign models, and welcomed overseas practitioners. The Wisconsin-born Amadeus William Grabau, forced to leave the United States because of his pro-German sympathies during World War I, moved to China, where he encouraged field studies and elaborated his 'pulsation theory', in which cycles in the rise and fall of the sea floor led to the formation of the crust and the growth of mountain chains. These attempts to find regularities in Earth's history and structure inspired later generations of Chinese theorists. Among the leaders of this emerging community was Li Siguang, later a key figure in the Chinese oil industry. At a meeting of the Geological Society of China in 1926, he offered European and American visitors a compelling account of recent syntheses, and advocated the idea of continental drift first proposed by the Prussian explorer and geophysicist Alfred Wegener in 1912.[9]

Wegener's theory of large-scale horizontal continental movement became one of the most discussed syntheses of the interwar years. The match between the African and South American coasts had long been obvious, but was not enough to convince researchers that the continents had moved laterally thousands of miles. Wegener began to become interested in the idea after seeing an atlas that showed the margins of the continental shelves as determined by the *Challenger* oceanographic expedition from the 1870s. These margins matched remarkably closely. And through readings in the post-Suessian literature, Wegener realised the extent of the fossil connections between regions now widely separated. He was impressed by a popular essay showing that a characteristic ice-age flora was shared from South Africa to Afghanistan, the basis for reconstructing the lost land of Lemuria.[10] However, Wegener opposed as 'a fantasy' any explanation involving sunken continents, and instead explained the faunal continuities through drifting continents, 'pulling apart from one another via some huge fault/fracture'.[11]

Wegener summed up his ideas in a short book, published in 1915 as *Die Entstehung der Kontinente und Ozeane* ('The Origin of Continents and Oceans'). The book could be brief because it recast evidence marshalled by Suess, Arldt, Neumayr and many others. Their works provided data 'from geology', which Wegener viewed from an outsider's perspective. Although he had collected specimens and made observations, Wegener's principal experience was as a geophysicist and polar explorer, notably in an expedition to Greenland supported by the Danish government to maintain its claim over what it considered an internal colony. Wegener's travels in this alien landscape encouraged him to take bold views. This is evident in images added to the third edition of 1922, which are schematic diagrams rather than geological maps (Figure 24.3). As a geophysicist, Wegener looked for proof of his theory not to local strata surveys of the kind that had motivated most of the grand syntheses, but to mathematical understandings of the dynamics of the Earth as a rotating body, and to new technologies such as radio, which he hoped would enable measurements of the ongoing progress of drift.

By the 1920s, Wegener's proposal was one of a host of competing syntheses, most of which had their origins before the war. Suess's contraction theory remained widely discussed, although the discovery of radioactivity challenged its basis in a cooling Earth. Ideas about planetesimal accretion, radioactivity-induced thermal cycles, and vertical uplift through isostatic compensation were also popular. Among the most pervasive theories was that of geosynclines, based on the case of the Appalachians in the eastern United States, in which sediments filled a linear trough and ultimately rose to become a mountain range. The idiosyncratic 'world-ice theory', proposed in 1913 by the Austrian engineer Hanns Hörbiger, posited ice as the source of all cosmic processes, with the impacts of a succession of icy moons producing the deposits of successive geological eras. Taken up by the National Socialists and a vast popular audience, the theory looked to Nordic legends for inspiration, contrasting their cleansing purity with the supposedly 'Jewish' origins of orthodox science. Hörbiger's mystical racism and anti-Newtonian physics, however, had no attraction for practising geologists.[12]

The dangers of such unbridled pluralism encouraged attempts to tie synthetic theories back to local studies. With the Germans easily caricatured as spinning out cosmic visions of the *Weltall*,[13] there was a need for Wegener's supporters to scale down his often sketchy hypothesis. Speaking at the International Geological Congress in Brussels in 1922, the French-speaking Swiss polymath Emile Argand brought drift theory more in line with contemporary work on particular regions. Borrowing techniques from developmental biology, and employing his extraordinary visual memory, Argand traced the deformations of the crust back to their original forms through a technique he termed 'embryotectonics'. The main feature of his lecture was a map of the trend lines of mountain chains, demonstrating the power of Wegener's synthesis.

Argand's performance aroused great interest, not least from Weng Wenhao, a vice-president of the Congress and leader of the Chinese delegation. Weng, who like many of his generation had studied in Europe, intervened at the Brussels meeting to note that the Chinese

Fig. 1.

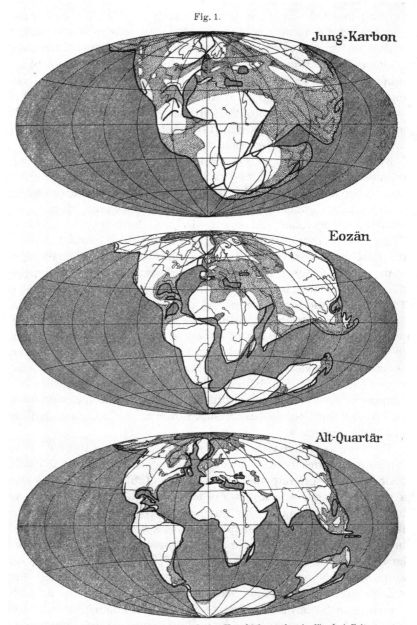

Jung-Karbon

Eozän

Alt-Quartär

Figure 24.3 Reconstruction of drifting continents at different stages of Earth's history, from A. Wegener, *Entstehung der Kontinente und Ozeane* (Braunschweig, 1922), p. 4. Reproduced by kind permission of the Syndics of Cambridge University Library, MF.75.74.

Rekonstruktionen der Erdkarte nach der Verschiebungstheorie für drei Zeiten.
Schraffiert: Tiefsee; punktiert: Flachsee; heutige Konturen und Flüsse nur zum Erkennen.
Gradnetz willkürlich (das heutige von Afrika).

had much to contribute to global geology, and that ancient records of earthquakes needed to be taken seriously rather than treated as legends. In founding the Chinese Geological Survey, Weng would later argue that understanding the Earth was vital to recovering the patrimony of the Chinese nation.[14]

Imperial and counter-imperial geological syntheses in different countries were motivated by radically different agendas, but those who proposed them were united by the aspiration to participate in a quintessentially modern international debate. Science gained authority by being seen as the common heritage of all humanity. In this view, ancient creation narratives and modern geological syntheses emerged from the same impulse. As Argand told the Brussels congress:

The waves pass and as in the old dreams of Asia they all together tell the evanescence of the universe ... Day after day, through infinite time, the scenery has changed in imperceptible features. Let us smile at the illusion of eternity that appears in these things, and while so many temporary aspects fade away, let us listen to the ancient hymn, the spectacular song of the seas, that has saluted so many chains rising to the light.[15]

A gifted linguist, Argand spent his later years as Professor of Geology in Neuchatel immersed in what his later biographers rather sceptically called 'eastern legends'. For many of his contemporaries, the distance between legend and speculation was tantalisingly close.

Post-war science and geopoetry

From the 1930s, the great burst of syntheses gradually came to an end. The status of theorising changed as geologists – especially in the emerging scientific powerhouse of the United States – focused on prospecting for oil, gas and (slightly later) uranium. As a profession, geology emerged during the first half of the twentieth century through career paths provided by petroleum companies such as Standard Oil, Royal Dutch Shell and the Anglo-Persian Oil Company, and great mining enterprises such as Homestake and Anaconda. After the failure of early hopes that drift might prove useful in locating new resources, most geologists (especially in America) turned instead to geosyncline theory, a framework that could be applied on a regional basis, and without reference to lost worlds or horizontal movements contested by geophysicists.

Attitudes towards speculation tended to harden. Driven by the insatiable quest for resources, geology became increasingly prosaic, shedding much of the philosophical sophistication characteristic of the traditional centres of central Europe. Bailey Willis, an American survey geologist who worked in places as diverse as China and

Argentina, condemned Wegener's treatise in 1944 as 'a fairy tale' of the kind that Germans were all too good at composing.[16] Geologists elsewhere found this frustrating, as in the case of Alexander du Toit in South Africa, whose book-length advocacy of continental drift in the late 1930s failed to make much impact. 'The Americans are about the toughest isolationists in existence', a colleague told him in a letter, 'geologically as well as politically'.[17]

A typical practitioner of the new order was Emerson McMillin Butterworth. Born in Ohio, Butterworth was educated at the University of California and went to work as a petroleum geologist in 1917. After extensive travels in the United States, he was sent to Alaska, Madagascar and the Dutch East Indies, returning in 1930 to set up oil concessions in Batavia. It is there, at the hill station of Buitenzorg, that Butterworth acquired Suess's multivolume work.[18] His purchase epitomises contemporary networks of international circulation: an American in a Dutch colony buying a French translation of a German work written in Vienna and published in Prague. As an oil man, however, Butterworth was probably more interested in the summaries of strata distribution and structure in Suess's volumes, than in the synthetic framework that held these together. Even academics such as the Dutch geophysicist Johannes Herman Frederik Umbgrove, in proposing his own dynamical theory in 1942, claimed to believe that such hypotheses were 'a necessary evil': 'We may expect to find a similar "geopoetical" aspect in many a geological treatise . . . However, authors should always keep their theories strictly separated from descriptions and conclusions of a more rigorously documented kind.'[19]

Particularly in the United States, World War II deepened the trend towards utility, as the flowering of imperial geology in the decades around 1900 was forgotten in a fresh phase of instrument-based enquiry into the oceans, atmosphere and planetary interior. The war had produced new technologies for surveillance, including monitoring submarines and tracking missiles and aircraft. Advances in computing and bureaucratic management offered the promise of comprehending vast amounts of data. Perspectives associated with natural history – palaeontology, taxonomy, biological diversity – were pushed to the margins. The imperial gaze from the summit was replaced by measured data points interpreted from the perspective of spherical trigonometry. This was a model of science derived from mathematical astronomy and naval exploration, and one that discouraged – though did not entirely rule out – the participation of women, who were encouraged to stay in the office as computers or in museums as preparators and taxonomists. As one aspiring physical oceanographer had been told in the late 1930s, 'My dear Mrs. Robinson, this is a man's field.'[20]

The quantifying approach came into the ascendant in the post-war era as understanding the physical environment of potential conflict became a key element in foreign policy. Enterprises such as the International Geophysical Year (1957–8), notable for collaboration between the Soviet Union and the United States, were inevitably linked to national security.[21] Yet even at the height of the Cold War, by no means all research was linked to military concerns. Books such as the structural geologist Hans Cloos's lyrical autobiography *Gespräch mit der Erde* (1947), translated into English in 1953 as 'Conversation with the Earth', were valued as romantic invocations of a vanished age. Syntheses were attempted, but they remained marginal. S. Warren Carey, who worked as an oil geologist before the war and then became professor at the University of Tasmania, proposed an expanding Earth in 1956 to explain the distribution of the continents. Most practitioners, however, considered Carey's theory as intellectually isolated as Tasmania was geographically remote. There were other attempts to revive drift, but even after studies of remnant magnetism in the mid 1950s suggested that the continents had moved, most geophysicists argued that such theories had to be rejected without more precise evidence and causal mechanisms.

With so much novel data pouring in, many felt that geological speculation was premature: for undergraduate lectures, textbook appendices, student societies and pub talk, not for refereed articles. One widely canvassed idea (later called 'sea-floor spreading'), in which basaltic magma rises from the ocean floor and then moves away from the mid-oceanic ridges, had been proposed by the British geologist Arthur Holmes in 1928 and was presented through his textbook of 1944. Tellingly, though, it appeared there as an appendix. Related ideas were revived by the Princeton geologist Harry Hess, who had served in the war and reached the rank of rear admiral. Yet the work, though widely circulated, was not published in a major journal, but was distributed first as a preprint and from 1962 as an offprint from a festschrift on petrology. Recalling Umbgrove's reference to 'geopoetical', Hess called his article 'an essay in geopoetry' to underline its speculative character.[22] If this involved a certain ambivalence, in the American context it was also an act of bravura, an effort to reclaim the high ground of theory. Privately, Hess argued that even fellow Princeton resident Immanuel Velikovsky deserved a fair hearing, if only to refute his theories of cosmic cataclysms and 'worlds in collision'.[23]

Practitioners of the post-war geosciences were often those whom the molecular biologist James Watson characterised as 'accomplished young men, exposed to sharp, non-emotional thinking and the need to aim for the best'.[24] The push towards rigour was thus part of the confidence of a generation hardened by war that pursued the 'Earth sciences' and dismissed 'geology' as old hat. One of the young men trained in this bracing masculine environment (though he had not served in the military) was Bruce Heezen at Columbia University's Lamont Geological Observatory. When his associate, the cartographer Marie Tharp, pointed out in the 1950s that the crustal rifts they had been tracing looked like strong evidence for continental drift, Heezen dismissed such speculation as 'girl talk' (Figure 24.4).[25]

For the lateral movement of continents to be accepted, it would have to be under a dramatically different guise than that of Wegenerian drift. That is what occurred with the development of plate tectonics in the 1960s. The continents, with their localised geologies of the kind understood by Humboldt, Darwin and Suess, were to be superseded in significance by underlying tectonic plates. Cold War surveillance science discounted the kind of natural historical evidence that had traditionally underpinned global syntheses, but provided evidence of a different kind, deemed more useful to the new breed of Earth scientists and to military strategists. In that sense, the joking comment of an American geophysicist was spot on: 'the plate tectonics revolution . . . was a product of the Cold War'.[26]

The mapping of the ocean floors conducted by Heezen and Tharp, for example, was crucial to locating enemy submarines. The Lamont Observatory, established by the charismatic autocrat Maurice Ewing in 1947, received over 90 per cent of its funding from military contracts. Development of the precision depth recorder and other technologies revealed features such as the abyssal plains and the rift valley along the Mid-Atlantic Ridge. Navy-sponsored ships also carried equipment to measure residual magnetism and its direction. The magnetometer had been deployed in wartime to detect submarines from the air, but began to be used to detect magnetic variations in the basalt that made up the seabed. It soon became clear that these variations were not random, but occurred in parallel bands of normal and reversed magnetism, reflecting the alternation of the magnetic pole from north to south and back again.[27]

The speculative geopoetry of sea-floor spreading and the meticulous measurements of the magnetometers came together when geophysicists realised that these magnetic bands matched across both sides of the mid-oceanic ridge off the coast of Vancouver. This could be taken as evidence that ocean floor was forming at the ridge and then being conveyed away in opposite directions. Here was precisely measured evidence for drift; yet it was not the continents that were moving, but the underlying plates on which the continents travelled. It took, however, the match between a particular magnetic profile (Eltanin-19) and evidence from radiometric dating of one reversal episode to convince most geophysicists (and ultimately, geologists as well) of the reality of what quickly became known as plate tectonics. That a theory of such scope

Figure 24.4 Marie Tharp and Bruce Heezen pose before their map of the Mid-Atlantic Ridge. By kind permission of Lamont–Doherty Earth Observatory and Debbie Bartholotta of the Estate of Marie Tharp.

could be accepted only after minute evidential precision is testimony to the constricting role that exact measurement had come to assume.

The ends of the Earth

The switch to plate tectonics was experienced by many as a bolt-from-the-blue conversion, so that there has been no more appreciative audience for Thomas Kuhn's *Structure of Scientific Revolutions* (1962) than among English-speaking geologists. With its stress on dramatic paradigm shifts, *Structure* seemed to offer a template for understanding the transformation from a consensus based on continental fixity to one involving large-scale horizontal displacements. On this view geology had been stuck in an outdated paradigm, or was even (to use a term Kuhn later abandoned) pre-paradigmatic.[28]

At a single stroke stability appeared to have been replaced by movement; geology's natural history side (clearly gendered feminine) had been superseded by physical analysis and demonstrative logic. As the Canadian geophysicist J. Tuzo Wilson tendentiously claimed in 1971,

'The acceptance of continental drift has transformed the earth sciences from a group of rather unimaginative studies based upon pedestrian interpretations of natural phenomena into a unified science that is exciting and dynamic.'[29] Geologists constructed a disciplinary myth, which made Wegener into a prescient hero. The deeper philosophical, theological and literary resonances that were necessarily part of any theory of the Earth were downplayed, local perspectives were denigrated, and the cross-cultural engagement characteristic of Humboldt and his successors was ignored. The limitations of this view should be evident from this chapter, in which continuities are at least as striking as any revolution. And if there was a revolution, it was a post-war shift in tools, methods, gender relations and disciplinary orientations, which prized instrumentally measured data and mathematical analysis. As we have seen, most supporters of this new, astronomical model for the Earth sciences initially dismissed drifting continents as idle speculation.

Geology is always embedded in geopolitics: the replacement of one theory by another has always been part of larger transformations in contexts, practices and aims. The making of geology was bound up with the rise of the European colonial powers on the world stage, through mapping, mining and surveying. It contributed to international trade, military conquest and the exploitation of resources. The most significant synthesis of the past 200 years, Suess's *Face of the Earth*, elaborated a perspective centred on the Vienna Ringstrasse and the Alpine peaks of the late Habsburg Empire. Continental drift theory was originally an idiosyncratic contribution to theoretical debates grounded in the age of high imperialism. During the Cold War, the situation was transformed by military patronage and the rise of American science, but the basic questions from earlier centuries and other cultural traditions were only temporarily obscured. This became clear in the 1970s and 1980s as geologists recognised that the history of life, the Earth and its climate were intimately connected. As a science pre-eminent in defining what it means to be 'global', geology offers the opportunity to forge narratives of the past, binding local history and epic into the evolution of the planet.

Further reading

Agar, J., *Science in the Twentieth Century and Beyond* (Cambridge, 2012).

Bowler, P. J., *Life's Splendid Drama: Evolutionary Biology and the Reconstruction of Life's Ancestry, 1860–1940* (Chicago, 1996).

Coen, D. R., *The Earthquake Observers: Disaster Science from Lisbon to Richter* (Chicago, 2013).

Greene, M. T., *Geology in the Nineteenth Century: Changing Views of a Changing World* (Ithaca, 1982).

Lightman, B., McOuat, G. and Stewart, L. (eds.), *The Circulation of Knowledge between Britain, India and China: The Early-Modern World to the Twentieth Century* (Leiden and Boston, 2013).

Oreskes, N., *The Rejection of Continental Drift: Theory and Method in American Earth Science* (New York and Oxford, 1999).

Oreskes, N., *Science on a Mission: American Oceanography from the Cold War to Climate Change* (Chicago, forthcoming).

Ramaswamy, S., *The Lost Land of Lemuria: Fabulous Geographies, Catastrophic Histories* (Berkeley and Los Angeles, 2004).

Rosenberg, E. S., *A World Connecting: 1870–1945* (Cambridge, MA, 2012).

Rudwick, M. J. S., *Earth's Deep History: How It was Discovered and Why It Matters* (Chicago, 2014).

Shen, G. Y., *Unearthing the Nation: Modern Geology and Nationalism in Republican China* (Chicago, 2014).

Sivasundaram, S. (ed.), 'Global histories of science', *Isis*, 101(2010), pp. 95–158.

Wu, S. X., *Empires of Coal: Fueling China's Entry into the Modern World Order* (New York, 2015).

Zakariyan, N., *A Final Story: Science, Myth, and Beginnings* (Chicago, 2017).

25   Zoological gardens

Menageries and zoological gardens have always been multifunctional institutions. As representative locations for the cultural history of animal–human relations, they have served different publics, often several at once, and have also been places for doing science with animals. Their history is well researched from the former point of view, but their role as places for science is far less studied. This chapter begins by outlining the emergence of modern zoological gardens as living natural history museums and sites of urban popular culture during the nineteenth century. The next section discusses the radical innovation introduced by Carl Hagenbeck around 1900, asking whether and how the open, apparently cageless design that he patented in 1896 can be regarded as science-based, and what kind of science actually took place in his 'animal park'. The third section very briefly surveys continuities and changes in zoos' cultural function, along with the emergence of two new zoo-related disciplines, ethology and zoo biology, during the twentieth century. The final section analyses two responses by zoos to the challenges from the environmental and animal rights movements since the last third of the twentieth century: the addition of immersion exhibits, or 'environments', to existing 'zoogeographical' displays, and a shift in the mission and research priorities of zoos toward conservation and species preservation.

The emergence of the modern zoo

The modern zoo is generally said to have begun in 1794 with the addition of a zoological department to the former Royal Botanical Garden (*Jardin des Plantes*) in Paris, after it was appropriated, and the Royal Menagerie at Versailles abolished, by the Revolutionary government. However, this facility remained under state control, and its scientific staff, including such luminaries as Georges and Frédéric Cuvier, Jean-Baptiste de Lamarck and Etienne Geoffroy Saint-Hilaire, were civil servants. The founding of the Zoological Society of London in 1828 marked the beginning of the zoological garden as part of emerging bourgeois culture. Sir Stamford Raffles and his collaborators plainly saw the zoo they founded as an expression of British imperial domination; the designation 'zoological' for both the society and its park signalled the scientific interests of the membership, distinguishing it from other imperial elites. Whereas the zoo in Regent's Park, London and Natura Artis Magistra (called Artis) in Amsterdam, founded in 1838, were owned by private associations, the Zoological Garden in Berlin, initiated by university zoologist Martin Liechtenstein and founded in 1844, was established as a joint stock company, with the Prussian king as chief shareholder and donator of the property – a segment of the Tiergarten, a hunting preserve then located outside the city walls – and leading nobles and Bürger as additional stockholders. New foundations followed in British cities and Belgium, in the German states after 1850, and in other European and colonial cities thereafter. In the United States, the movement continued with the founding of zoological parks in Philadelphia, Cincinnati and other cities in the 1870s, the National Zoological Park in Washington, DC in 1889, and the New York Zoological Park (known as the Bronx Zoo) in 1899.

The publicly stated aim of these zoological gardens was to serve public education and recreation, in that order. The London zoo and Artis Amsterdam were open at first to members only; recreation there evidently implied taking refuge from the urban masses. Over time, however, such ambitions proved impossible to sustain, due to the high cost of acquiring animals and maintaining the facilities. Membership of the Zoological Society of London became relatively easy to obtain, and members of the wider public were eventually permitted to visit these and other 'bourgeois' zoos for nominal fees. By the turn of the twentieth century, the Berlin zoo was one of the expanding metropolis's most popular attractions, with over 100,000 visitors on 'cheap Sundays' in the summertime.[1]

Despite this shift to mass entertainment, nineteenth-century zoos continued to articulate an educational mission, which was reflected in their design. They were often constructed as living natural history museums set in artificial landscapes, with animals' cages placed in fanciful 'houses' built in styles intended to serve as reminiscences of their exotic origins (Figure 25.1). Animals were often displayed on elevated 'stages' to make them more visible, in a manner analogous to displays of consumer goods in early department stores; in Berlin, the same architects designed both kinds of display.[2] The educational mission was enacted at first with rudimentary labels, identifying animals by their Latin and vernacular names. Later, more systematic presentation schemes were adopted, with taxonomically related species housed near one another, and official guidebooks provided to enable visitors to follow prescribed walking routes. The Regent's Park zoo, which began to order its collections by taxa in 1840, carried this

Figure 25.1 New Ostrich House, Berlin Zoo, built in the 1870s; postcard from the 1920s. Private collection.

strategy further than most, striving to realise what one guidebook reviewer described in 1855 as the goal of zoology itself: 'to furnish every possible link in the grand procession of organised life'.[3] The use of Linnean systematics to design such spaces was not universal; both the assignment of individual taxa and preferences among classification systems remained open questions. One popular guidebook to the Regent's Park zoo, published shortly after its opening, deliberately presented the animals kept there not 'in the order of their scientific arrangement', but in alphabetical order, so that 'the reader will be enabled to turn at once to the history of any animal'.[4]

Working at times in cooperation, but also in competition with the great natural history museums founded during the same period, zoos and aquaria aimed to collect as many species as possible.[5] Maximising species numbers was a matter of prestige for zoo directors, and also enabled them to depict an ordered Creation, in keeping with their generally conservative politics. Whether the scientific and pedagogical aims that drove such collecting were actually achieved is questionable. Guidebook sales rarely approached visitor numbers, and zoo visitors were surely no more likely to keep to prescribed walking routes than zoo or museum visitors are today.

Zoo directors were aware that exotic animals were not equally capable of surviving in Europe's climate; high death rates, when discovered, led to scandalous press reports. In response to this practical problem, and to explore potential economic uses for exotic creatures, 'acclimatisation' societies were established in Paris in 1854, later in London, Moscow, Berlin, Melbourne, Calcutta and other cities. Michael Osborne describes the *Jardin Zoologique d'Acclimatation*, founded in Paris in 1860, as the focal point of an 'epistemic community' grounded on 'a variant of mitigated Lamarckian transformism' emphasising the plasticity of individual animal forms and the potential heritability of adaptations.[6] These institutions sought to domesticate exotic breeds for their feathers, fur or meat, and also to expand the collections and improve the survival rate of existing zoos. Though these goals were at best partially achieved, the Paris institution still exists today.

Nineteenth-century zoos, like earlier menageries, were also convenient locations for animal observation by naturalists. A well-known example is Charles Darwin's encounter with the orangutan 'Jenny' in Regent's Park zoo in 1838, after which he noted the similarity of her emotional expressions to those of humans as evidence against the latter's claim to 'proud preeminence'.[7] Decades later he engaged the assistance of a keeper at the same zoo, named 'Mr Sutton', to study facial expressions in monkeys, and incorporated the results into *The Expression of the Emotions in Animals and Men* (1872), citing the keeper by name.[8] George John Romanes and Leonard Trelawney Hobhouse continued the tradition, documenting their observations in Regent's Park zoo in *Animal Intelligence* (1882) and *Mind in Evolution* (1901). Hobhouse went beyond behavioural observation to conduct experimental studies on chimpanzee problem solving using zoo animals.

Buffon and Cuvier had already debated whether animals behaved 'naturally' in cages.[9] With the rise of 'bourgeois' zoos claiming scientific status, the debate entered a new phase. (See Nyhart, Chapter 20, this volume.) German zoo directors were opposed to established zoologists' preference for measuring and safely dissecting dead creatures.[10] As the head of the zoo at Frankfurt am Main, David Friedrich Weinland, claimed in the journal he founded, *Der Zoologische Garten*, zoos offered the chance to study animals through 'long and repeated observation'; moreover, he added, 'Only the living animal is the entire animal.'[11] However, examination of the journal's authorship suggests that it served mainly as a forum for information exchange among zoo directors and enthusiastic visitors.[12] Similar sites for information exchange sprang up during the 'aquarium craze' in Germany during and following the 1880s. Most prominent in this movement was Alfred Brehm, director of the Hamburg zoo from 1863 to 1866, then founder and director of the Berlin Aquarium from 1869 to 1878. His multivolume bestseller,

Illustriertes Tierleben (1864–9), incorporated information from hunters, foresters and zookeepers as well as scientists. Though zoo directors could be academics, people with no formal training such as Alois Kraus at the Schönbrunn menagerie achieved respect through knowledge derived from long experience. Trained zoologists or veterinarians did not become the majority of zoo directors until long after 1900.

Socially diverse participation in live animal studies in zoos and aquaria did much to advance and distribute knowledge of exotic creatures. Academic zoologists like Ernst Haeckel participated in this trend, but public controversies and media attention were not to everyone's taste. American zoologist Charles Otis Whitman formulated this damning indictment in 1902: 'Heterogeneous collections of animals, exhibited for the amusement of people, are wholly unsuited to the purposes of investigation in time, place, and character'; instead, the investigator 'must have complete and permanent control of his quarters and the forms he is to study, and above all, *complete isolation from the public*'.[13]

Hagenbeck's 'revolution'

Carl Hagenbeck was an animal merchant with global reach who became a spectacularly successful showman of exotic animals and humans, from circus performances to allegedly ethnographic displays called *Völkerschauen*. (See Qureshi, Chapter 22, this volume.) These two business lines worked in tandem: the animal trade supplied the shows, which in turn advertised the trade, while the exotic humans Hagenbeck displayed were recruited by his agents and worked under contract. After experimenting with panoramic landscape displays incorporating humans and animals – for example, an 'Arctic panorama' with icescapes executed by scenery painters, Arctic animals displayed at various elevations against this backdrop, and their trainers disguised as 'Eskimos' – Hagenbeck obtained a patent for a 'Natural Scientific Panorama' in 1896.[14] The 'animal park' (*Tierpark*) based on that concept opened in Stellingen near Hamburg in 1907. The basic principles involved can be stated briefly: authentic exotic animals of varied species were to be displayed together without cages in 'suitable terrain', meaning rock-like cement constructions, landscaped with exotic plants and trees and artifical ponds. These carefully designed open spaces were to be surrounded by moats to prevent the animals escaping (Figure 25.2).

As Hagenbeck explained in his patent application, what made the design 'scientific' was not the use of zoological systematics as design guides for animal displays, but rather exact measurements of animals' jumping distances by experimental tests, to determine the breadth and depth of the moats and thus assure their practicability. It has often been claimed that Hagenbeck's design was a predecessor of today's

Figure 25.2 Hagenbeck's Animal Park, Stellingen, main display. Photo by Götz Berlik, 2005. Courtesy of Hagenbeck Archives.

immersion exhibits, discussed later.[15] However, Hagenbeck chose the name 'animal park' deliberately, dismissing the taxonomic organisation of zoos as too abstract. Mixing species that did not live together in nature broke with then-modern zoo design. Walking paths were organised in such a way that visitors could at times view the surrounding farm country while also looking at the park, thus juxtaposing immersive and distancing gazes.[16] The thrust of Hagenbeck's design concept was aesthetic; the goal was to excite in viewers a theatrical illusion of freedom and naturalness, not to reproduce actual scenes from or the organisation of 'nature'. Hagenbeck emphasised this imaginary quality by calling his park a recreated 'Eden', and even named it an 'animal paradise' (*Tierparadies*).[17] Seen historically, Hagenbeck's creation might thus be interpreted as a renewal of the *tableaux vivantes* in baroque menageries for mass entertainment purposes. To add to the excitement, *Völkerschauen* were placed adjacent to the grand animal displays. Visitors could thus experience the fascination of exotic animals and humans in the same space, combined with a *frisson* of fear – what if the lion actually leaped over the moat?

Due to its aesthetic appeal and unmistakable circus atmosphere, the park was an enormous popular and financial success. Even so, science was done there by zoologist Alexander Sokolowsky, a student of Ernst Haeckel and Karl Möbius. Sokolowsky participated in the development of the park's design, and claimed that it was conceived on 'biological' lines as defined by his teacher Haeckel. By this he appears to have meant that the open-air display of mixed species associated with a given landscape type, without restraints on their movement, was healthier for the animals, encouraging them to breed, and might even lead to interspecific mating with 'scientifically interesting results'.[18] In a pamphlet he described the displays as 'zoogeographical zone pictures', though he was surely aware that they did not depict any real place.[19] In 1908, he published a brief account of the 'psychology of primates', based upon travel accounts and his own observations at the 'animal park' (Figure 25.3).[20] This included what he termed 'physiognomical' descriptions of apes' emotional expressions and behaviour: orangs were 'phlegmatic', while chimpanzees were excitable and easily distracted. The anecdotal style and obvious anthropomorphism of this work earned Sokolowsky criticism from colleagues. His claim that primates, especially chimpanzees, were intelligent creatures capable of using tools was strongly disputed, because it was based solely on an analogy to human intelligence and had no experimental basis.[21] Sokolowsky himself acknowledged that observing animals in captivity was no substitute for field studies, and advocated the establishment of research stations to study primates in or near their 'homelands'.[22]

Hagenbeck was honoured in scientific circles for recognising that animals he brought back from his expeditions belonged to new species or varieties, sometimes by incorporating his name into their species names.[23] However, German zoo directors denounced his claim to 'scientific' status for the design of his 'animal park' as mere showmanship, called for a boycott, and strove to defend their scientific standards and design concept once it became clear that his invention posed an existential threat to the nearby Hamburg zoo.[24] But they could hardly fail to acknowledge the park's popularity, and by the 1920s zoo directors from Europe and America were going to Stellingen to learn how his displays were made.[25] The ideal of zoos as living natural history museums persisted, but zoos began adding more spacious habitat displays, for example at the Geo-Zoo in Munich (1928).

Ethology, zoo biology and mass culture

In the United States, showmanship had already won out before Hagenbeck, while the Bronx Zoo head William T. Hornaday and others continued to advocate a public education mission for zoos. In Europe and America zoo directors were quick to utilise new media like radio

Figure 25.3 Alexander
Sokolowsky, with research
subject, from
A. Sokolowsky,
*Beobachtungen über die
Psyche der Menschenaffen*
(Frankfurt am Main, 1908),
frontispiece.

for spreading the word about their animals, and zoos soon became
film venues. After 1945 zoo directors achieved immense popularity
thanks to television. Most prominent in the United States was Marlin
Perkins, director first of the Lincoln Park Zoo in Chicago from 1944 to

1962, and then of the St Louis Zoo. His television programme *Zoo Parade*, launched in 1950, had 11 million viewers by 1952; its successor, *Wild Kingdom*, ran from 1963 to 1985.[26] In post-war West Germany, the media leader was Bernhard Grzimek, zoo director in Frankfurt am Main from 1945 to 1974. Both men claimed to advance research and wildlife conservation with their efforts. Grzimek reissued *Brehm's Tierleben* in sixteen volumes under his general editorship (1968 onwards), and asserted that he was saving exotic species from extinction by bringing them back alive from the jungle or the savanna. One of his films, *Serengeti Must Not Die!* (1959), received an Academy Award for best documentary.

Parallel to this, two new disciplines emerged, both involving zoos: ethology and zoo biology. Though historical scholarship has focused on ethology, the two disciplines have a common history. Ethology began, inter alia, at the Berlin zoo, where ornithologists Oskar and Magdalene Heinroth combined results of their studies of hand-raised geese with information from numerous correspondents to produce *Die Vögel Mitteleuropas* ('Birds of Central Europe', 1924–34). This four-volume work aimed to use studies of behavioural repertoires in birds, including their social 'customs and rituals', in order to classify them systematically.[27] A close colleague of the Heinroths was Otto Antonius, director of the Vienna zoo since 1924. A zoologist and paleontologist interested in the history of domestication, with an emphasis on ancient horses, Antonius was in 1926 a founder of the international association of zoo directors. One justification the association offered for its work was the claim that research on animal behaviour in zoos could fill the gap between laboratory and field studies.[28] An example was the effort to save the European bison through captive breeding with the aid of a centrally kept stud book, which presaged current conservation programmes (see later). In a popularly written 1933 book entitled *Gefangene Tiere* ('Captive Animals'), Antonius introduced the idea that zoo animals co-create their own living conditions. Taking the part of an expert who gently but firmly confronts the naive sentimentality of a (female) zoo visitor, he opposed the conventional belief that animals longed for the 'freedom' of the wild. Citing a statement by Alfred Brehm about birds, he argued that, if designed to fulfil the biological needs of its residents, a cage could become an 'apartment' rather than a 'jail' for animals.[29]

Though the writings of Antonius, Konrad Lorenz, and others were important first steps toward the creation of zoo biology, the founder of the discipline as a distinct research field was Heini Hediger, zoo director at three Swiss locations: Dählhölzli (1938–43), Basel (1944–53) and Zurich (1954–73). In *Wild Animals in Captivity: An Outline of the Biology of Zoological Gardens* (1950, orig. 1942), Hediger united the elements of this field into an integrated system, which he then elaborated in *Studies of the Psychology and Behaviour of Captive Animals in Zoos and Circuses* (1955, orig. 1954) and several other books. Zoo biology might be seen as a sideline of ethology, since it too studies animal behaviour, but its significance goes far beyond the history of science. This was one of the first disciplines of any kind focused reflexively on the institution in which its research was carried out. It was a pioneering move in the direction taken somewhat earlier by library science, and decades later by museology.

No extended study of the history of zoo biology exists, but its key principles as articulated by Hediger are easily summarised.[30] First of all, he rejected the fanciful idea that wild animals are in any sense 'free'. Viewed biologically, he argued, animals are captives of their living conditions and their inherited behavioural schemata. Second, Hediger asserted that zoo animals establish habitats comparable with the territories they establish in the wild. He supported this claim with theoretical considerations drawn from Jakob von Uexküll's work on animal subjectivity and his own and others' detailed observations of territoriality, competition, social hierarchies and niche creation in zoo enclosures. Third, Hediger argued that the key to animals' survival and well-being in zoos is their accommodation to captivity, assured by measures taken to suppress the flight instinct. As he emphasised, zoo animals are tame but not domesticated. The purpose of zoo biology was thus to counteract domestication and preserve behaviour patterns of the wild state so far as possible by designing animals' quarters to enable these behaviours to occur. Fourth, achieving this required systematic study of the biology of animal–human relations in captivity, which for Hediger meant focusing on rather than excluding animals' psychological interactions with their keepers and visitors. As Hediger recognised, a central problem of animals in zoos is enforced inactivity; he therefore advocated training, which he called 'disciplined play', for exercise and occupation, risking criticism for producing circus-like performances. In later work he develped a classification of behavioural repertoires in zoo animals, and conducted species-specific experimental studies of approach and flight distances, which he elaborated into a set of distance concepts: flight distance, critical distance, individual and social distance, each of which he conceived to be measurable.[31]

Hediger wrote both for his colleagues and for a wider public, trying to overcome what he termed the stereotypes held by both groups about animal behaviour. The declared goal was to achieve better understanding of animal behaviour in captivity, in order to help meet animals' biological requirements and social needs. Zoo biology was therefore behavioural science devoted to the improvement of zoo management.

Into the twenty-first century

A long-term shift in the mission of zoos toward focusing on environmentalism and conservation began in the 1970s and has accelerated since the 1990s. Amongst numerous steps marking this shift are the Species Survival Plan adopted by the (American) Association of Zoos and Aquaria (AZA) in 1981, and the World Zoo Conservation Strategy, first published in 1991 by the World Zoo Organization (now World Association of Zoos and Aquaria, WAZA) and the Captive Breeding Specialty Group of the International Union for the Conservation of Nature (IUCN). The shift came in response to the fundamental challenge posed by the animal rights movement, and was also an effort to align zoos with the increasingly popular environmental movement without taking any specific political stance. Popular education and mass entertainment remain central missions for zoos, but the addition of environmentalist and conservation goals has made zoo design and science with animals in zoos more complicated than ever before.

The most obvious embodiments of this shift in mission are immersion exhibits such as 'Jungle World' at the Bronx Zoo or the 'rainforest' exhibits in the Cleveland Park and Vienna zoos, among others (Figure 25.4).[32] In such 'environments', visitors are not only invited to gaze at exotic creatures from a safe distance, but also encouraged to imagine that they share the same naturalised space with them. In 'rainforest' displays, the illusion of 'naturalness', evoked by tropical birds flying about unhindered and the use of climate control technology to achieve tropical temperatures and humidity, is achieved at the price of walling off the display from the open air. Seen historically, this seems like acclimatisation in reverse, only now visitors need to adapt (but not too much: building codes require that safety features like exit signs be visible). Earlier zoo displays were not inexpensive, but the costs now involved can run into many millions of pounds. The increased visitor revenues and corporate sponsorships needed to raise these levels of funding have enhanced the commercial feel of zoos, even when shopping spaces selling plush toy animals are separated from the animal displays. The similarities of holistic immersion exhibits to theme park dream-worlds has been widely noticed.[33]

Relevant here is the changing role of science in zoo architecture. Hagenbeck claimed that his displays were 'scientifically' designed to assure the animals' physical safety and visitors' unhindered gaze, but the species displayed in his panoramic spaces rarely belonged together in any scientific sense. In contrast, the authenticity of immersion exhibits is said to be based on detailed research on appropriate living conditions for the animals themselves. In this respect, immersion exhibits and 'zoogeographical' displays grouping animals by habitat or region are two research-based approaches to the simulation

of exotic animals' living conditions: by functional substitution of land-scape and other features in the case of 'zoogeographical' displays, and by physical replication in immersion exhibits.[34] As these displays have grown in number and size, the older drive to maximise species numbers has been replaced by a new ideal of 'species-specific' design; this means arranging exhibits to optimise satisfaction of animals' biological and social needs in the space at hand, even if this means reducing the number of species kept at a given zoo.

The conservation mission, like the others, has links with the past. The 'Noah's Ark' metaphor has always been central to the discourse of modern zoos, but the fact that the biblical story is a tale of salvation from a failed Creation was less prominent in the past. In recent zoo discourse, the 'end of days' aspect has taken first place. Zoo-based conservation projects combine two aims: to raise public awareness of threats to biodiversity, and also to save species from extinction, by keeping and breeding them in relative safety and in certain cases returning them to the wild.[35] Due in part to this new mission, topics for research in zoos have multiplied. Zoo management has become a science-based profession in its own right, with its own textbooks and journals; demand has arisen for advanced training not only of zoo

Figure 25.4 Visitors at the Rainforest House, Tiergarten Schönbrunn, Vienna. Photo by Daniel Zupanc. Published by permission of the Director, Tiergarten Schönbrunn.

directors, but also for zookeepers.[36] In this context, three types of scientific activity in zoos can be distinguished: fundamental research, applied research and science-based technologies.

Fundamental behavioural research with zoo animals has continued in the ethological line, for example in Desmond Morris's work at London Zoo on courtship rituals in pheasants (1957) and the behaviour of higher primates in captivity (1958, 1961), published before he carried over these observations to humans in *The Human Zoo* (1969).[37] Frans de Waal noted the logistical advantages of zoos for behaviour research in *Chimpanzee Politics* (1982, rev. edn 2008), a best-selling study based on work at Arnhem Zoo, which holds the largest population of these primates in the world. The book describes in detail the complex social interactions among chimpanzees, including conflicts and reconciliations, establishing that changes in dominance hierarchies in chimpanzee troupes depend on changing coalitions among troupe members, male and female. De Waal ascribed Machiavellian behaviour to the apes, thus challenging existing prohibitions against anthropomorphic terminology. Behaviour has also been observed in zoos in other species that might not have been detected otherwise; a case in point is the discovery of asexual reproduction in Komodo dragons.[38] Zoos widely report that they are sites for academic research training in zoology, veterinary medicine and other disciplines; however, comprehensive data on the extent of these activities is difficult to acquire.

Far more frequent than fundamental research is management-related applied research in the tradition of zoo biology, designed to improve zoo operations. This includes quantitative work on 'activity budgets' of the kind originally proposed by Heini Hediger, studies assessing the design and effectiveness of so-called environmental enrichment programmes for zoo animals, research on animal health and the incidence of disease aiming to improve survival and longevity rates, and social-scientific or communications-oriented visitor studies.[39] An example of health and disease research is a longitudinal study published in 2004 of mortality rates and causes of death in penguin species at Edinburgh Zoo, based on records of more than 1,000 post-mortem examinations kept for over 90 years.[40] Added to this are conservation and preservation projects, including in situ work to preserve existing zoo populations and ex situ field projects to save threatened species or protect habitats. The AZA currently lists nearly 500 projects under its Species Survival Program, whilst the WAZA names 238 approved projects.[41]

Though captive breeding is as old as the zoo, and research on the multiple factors influencing the success or failure of captive breeding programmes in zoos is now decades old, recently science-based technologies such as artificial insemination have become more important in such efforts. The 'Millennium Ark' concept, first advanced in 1986, goes further, proposing to use biotechnologies to preserve genetic material from captive species with the aim of repatriating them later.[42] A counterpart approach in botany is the Millennium Seed Bank at Kew Gardens in London. Whether such programmes are realistic – in particular whether the requisite technologies are being developed quickly enough to assure significant impacts in the desired time frame – is unclear.

Criticism of zoo-based science has taken many forms.[43] In addition to raising methodological issues such as sampling problems in small populations and the difficulty of carrying out controlled experiments in zoos, critics assert that there is no consensus on what counts as research in zoos, and that zoo-based science avoids scientific quality control, because many studies are published in informal self-published papers, in-house journals or pamphlets or conference proceedings rather than peer-reviewed journals. They also note a lack of coordination amongst varied projects, insufficient attention to the problems of maintaining genetic diversity in small captive populations and the likelihood that genetic deterioration in captivity may prevent successful repatriation. More fundamentally, critics of conservation projects note two mismatches. First, they argue, such work tends to concentrate on charismatic megafauna like elephants, big cats, rhinos, hippos, exotic horses or giant pandas, whether or not they are actually on international 'red lists' of threatened species, whereas working with less well-known species that may be in greater danger might be more effective. Second, they note a discrepancy between the small number of species selected for conservation work or repatriation to the wild and the far greater number of species disappearing from the Earth. The overall implication is that zoo-based conservation projects have public relations value, but little impact on biodiversity. Zoo advocates reply that even small-scale beginnings are better than doing nothing, and that such efforts have educational potential. However, visitor surveys have found that immersion exhibits do not necessarily add to visitors' existing awareness of environmental threats or give them an opportunity to participate seriously in workable solutions.[44]

Conclusion

Today, zoos are amongst the most popular cultural institutions anywhere. Their attendance seems little affected by criticism from cultural studies authors and animal rights advocates, much of which may be based on romantic assumptions about the 'freedom' animals allegedly need that zoo biologists refuted long ago. Public desire for actual contact with exotic wildlife remains strong; its fulfilment is apparently assisted rather than replaced by new media offerings, from nature shows on film or television to live-streamed videos of panda babies in zoos to blogs maintained on behalf of zoo animals. Spacious displays outside urban centres, for example at Whipsnade near London, and built 'environments' within zoos are responses to public demand for a more authentic zoo experience, which in turn forms part of a wider shift in leisure culture toward intensified experiences like extreme sports. Seen in still broader context, both immersion exhibits and shifts toward new ideals of species-specific animal care in zoos can be interpreted as part of a transformation in urban middle class values from quantity to quality, from 'having' to 'being'. Research with animals in such settings may produce knowledge that is valuable for itself, and may improve the management of zoos by improving the care of the animals held there, thus also raising visitor satisfaction; but surely this is not natural knowledge in the sense originally meant by the term. 'Nature' at this site, including both human and non-human animals, is now, and indeed has always been, a cultural creation.

Further reading

Ames, E., *Carl Hagenbeck's Empire of Entertainments* (Seattle, 2008).

Baratay, E. and Hardoin-Fugier, E., *Zoo: A History of Zoological Gardens in the West*, trans. O. Welsh (London, 2002).

Cowie, H., *Exhibiting Animals in Nineteenth-Century Britain: Empathy, Education, Entertainment* (Basingstoke, 2014).

Donahue, J. and Trump, E., *The Politics of Zoos: Exotic Animals and their Protectors* (DeKalb, IL, 2006).

Hanson, E., *Animal Attractions: Nature on Display in American Zoos* (Princeton, 2002).

Hochadel, O., 'Watching exotic animals next door: "scientific" observations at the zoo (ca. 1870–1910)', *Science in Context*, 24 (2011), pp. 183–214.

Rothfels, N., *Savages and Beasts: The Birth of the Modern Zoo* (Baltimore, 2002).

26   Provincialising global botany

In 1901, Jinzo Matsumura (1856–1928), professor at Tokyo Imperial University, claimed a scientific binomial *Prunus yedoensis* for the most beloved cherry in Japan. Matsumura received an imperial award for this feat in 1904, upon which he posed proudly for a photograph in a European-style imperial garment complete with a shiny sash and a plumed hat. This photo was later fittingly supplemented with a pink-hued cherry blossom background, adding a wisp of joy.[1] The Japanese botanist's ability to rename this familiar Japanese plant, which almost every Japanese person knew, in a language (Latin) that only a few understood, was celebrated as the indication of Japan's successful modernisation of its plant systematics.[2] It was a success resonating with the spirit of 'Escaping Asia towards Europe', the striking catchphrase of Meiji Japan (1868–1912). By imbuing the Japanese with a sense of national urgency for reform and action, Meiji leadership had swiftly transformed Japan into a strong empire capable of competing with Western imperial powers.[3]

The celebration of *Prunus yedoensis*, however, was short-lived. In 1906, a German botanist who had studied specimens of cherry sent by a French missionary based on the Korean island of Chechu claimed that *Prunus yedoensis* was native to that island instead of Japan.[4] Most Japanese botanists rejected this claim. However, neither their effort to deny the Korean cherry's identity with *Prunus yedoensis* nor their effort to find its Japanese home was successful. The situation became more complicated in 1916 when a Western authority invited by Matsumura suggested the possibility that this 'soul of Japan' was a hybrid cultivar created by a florist.[5] Japanese botanists could not make up their mind whether it was better for their soul to have originated in Korea, its colony since 1910, or to have been concocted in a Japanese flower shop.

Is this a quaintly provincial episode in Japan's modern botanical practice? Does it reveal the unique insecurity of its modernising path, which Komori Yoichi poignantly characterised as pre-emptive self-colonising dressed up in imperial attire?[6] The botanical establishment at Tokyo Imperial University, as will be shown here, followed a path of all-out Europeanisation. It chose to project itself into the globalising theatre of modern science, a space beginning to be peopled by more than just Europeans, in a fully European style. However, Japanese botanists' apparent need to claim intellectual ownership of things that they felt were properly Japanese shows that they were not ready to renounce the value of their own culture.

Natural history, a scientific practice popular in modernising Japan, was not unique in its moments of provincial attachment in the midst of its outwardly neat Europeanising path. The self-denying mantra of 'Escaping Asia towards Europe' could not thoroughly permeate the Japanese scientific community, as historians have noted. Many revolutionaries in science could not dissociate themselves from the cultural, political and intellectual traditions of Tokugawa Japan (1603–1867), the mostly peaceful and prosperous period ruled by the shogunate. These revolutionaries made use of these foundations when establishing themselves as modern elites. Also, Japanese scientists in every field seem to have contemplated unique Japanese experiences that could not be explained away by universal theories; disciplines such as seismology or meteorology – two essential disciplines for this earthquake and typhoon-ridden archipelago – made exemplary successes in bringing Japanese experiences to bear on Western theories.

The case of botanical taxonomy in Japan shows that its provincial moments, while aiming at domestic audiences, were also shaped by global interaction. By focusing on, first, the complicated relationship of Japanese botanists with dried plant specimens, the most powerful tool of modern taxonomy, and, second, their contemplations of the elusiveness of a universal standard in plant classification, this chapter reveals the tensions in globalising modern botany. It observes how this cross-cultural knowledge-making practice, originating from European expansion, provincialised most of its participants.

East Asian centres and immobile specimens

Plant specimens – collected in the field, dried to preserve their morphological qualities as much as possible, and then shipped to herbaria to be stacked up, compared and ordered – are indeed an ingenious tool. They embody the empirical, objective and possessive spirit characterising the modern way of studying nature. The botanical establishment of Meiji Japan, which secured an institutional home in the botanical gardens and herbarium associated with Tokyo Imperial University in 1877, was rightly enamoured of the power of dried plant specimens.

However, Japanese botanists first encountered the power of relocating the natural world by collecting nature as an insurmountable obstacle. That power, exerted by European centres through several very successful expeditions to Japan via the Dutch trading post in

Nagasaki since the late seventeenth century, had relocated much of the Japanese flora to various European centres. Numerous type specimens of Japanese plants, against which any new specimens had to be checked, were found not in Japan, but in European herbaria. As summarised in Philipp Franz von Siebold's thirty-volume *Flora Japonica* (1835-70), many Japanese plants had already been classified and named by European botanists. Japanese botanists were disheartened; Japan was a remote periphery even for the study of Japanese flora.[7]

Moreover, as mobile as these specimens were, circulating from field to herbaria, they had a certain immobility that frustrated Japanese botanists even more. Ryokichi Yatabe (1851-99), a Cornell graduate who had taken up the task of modernising Japanese botany as the sole professor in that subject at Tokyo Imperial University, claimed that not only those specimens already in European herbaria but also the new ones that he collected and sent to Europe for verification were not circulating well. In 1890, in an article addressed to 'European botanists', Yatabe wrote: '[from] some of [those to] whom I sent many valuable specimens, I have been so unfortunate as to have received no answer whatever even after the lapse of several years. Nothing, it will be admitted, is more trying and disappointing than this to an earnest worker.'[8] His way of overcoming this disappointing situation was to strengthen his Japanese centre with specimens of Japanese plants and references in order to justify his declaration of independence from European centres. He announced this independence in the same article: he 'decided to begin to give new names to those plants which [he] consider[ed] as new, without attempting in many cases to consult with European specialists'.[9]

Before he had had much success in naming Japanese plants on his own, Yatabe retired from his Japanese centre. Matsumura Jinzo, who had been collecting Japanese plants with Yatabe and succeeded him in the chair of Japanese botany in 1890 after studying in Germany, inherited his predecessor's dream. But his troubled experience with *Prunus yedoensis* quickly displayed the continued vulnerability of the Japanese centre in the circulation of plant specimens. Those specimens of cherry from Korea were sent to Europe, not Japan, a situation that Matsumura tried to redress. He embarked on serious botanical explorations to Japanese peripheries for the university herbarium, thereby re-enacting the quintessential strategy of European centres. He himself explored, or sent his associates to explore, all of the promising peripheries of Japan: Taiwan (1896), Okinawa (1897), Hokkaido (1899), Korea (1900, 1902), and Karafuto and the Kwantung Leased Territory (1906).[10]

At first glance, Matsumura's path appears to follow the successful imperial march of his nation, which colonised Taiwan in 1895,

Karafuto and the Kwantung Leased Territory in 1905, and Korea in 1910.[11] However, Matsumura's project was not merely a response to Japan's military and political advances: his pre-emptive moves towards Korea pre-dated Japan's colonisation there. Matsumura made it clearer that the imperial path was his conscious choice when he gave the specimens from Taiwan and Korea to two of his promising students, Hayata Bunzo (1874–1934) and Nakai Takenoshin (1882–1952), in 1903 and 1906 respectively, as thesis materials.[12] Although he was already receiving praise for his successful modernisation of Japanese botany, he saw that success was to be secured by a more serious engagement with colonial flora by Japanese botanists. Hayata and Nakai – who later inherited his professorship – met his expectations by becoming internationally renowned authorities in Taiwanese and Korean flora.

Dried specimens: too sterile or too captivating?

Hayata and Nakai, who worked side by side at the university from Nakai's entrance in 1903 until Hayata's death in 1934, thus shared a teacher, an institution (the Japanese botanical centre), and the centre's imperial path. Supported by their respective colonial governments, their core botanical achievements were also marked by multi-volume monographs on Taiwanese and Korean flora, *Icones of the Plants of Formosa* (1911–20) and *Flora Sylvatica Koreana* (1913–40) respectively, and the numerous new taxa they proposed in these works.

Despite these similarities, Hayata and Nakai took different stances regarding almost every aspect of the imperial path of Japanese botany. Their divergence became so great that in 1930, Nakai decided that he had to prohibit any further discussion of Hayata's taxonomic theory by discrediting it in public via an article written both in English and in Japanese for his *Flora Sylvatica Koreana*. Here Nakai declared Hayata's theory to be so worthless that he, Nakai, was now 'one of the Japanese naturalists who keenly regret having had the publication of such theory'.[13]

This dramatic eruption of dissonance between two colleagues at the Japanese centre of botany naturally involves many different desires, fears and hopes of two different personalities. Some of them are captured in their different takes on dried plant specimens, the basis for their Japanese centre's secure botanical successes. For Hayata, dried specimens seemed sterile; for Nakai, they were captivating. While one was sceptical about their power, the other was rather possessed by it.

Hayata, the senior colleague, began his botanical studies in the field on his own while working at a clothing shop. His first visit to Taiwan, at the invitation of a friend, was in 1897, the same year that he belatedly entered high school in Tokyo at the age of twenty-three. Although he left no records of any botanising on this first trip, Taiwan's verdant sub-tropical nature must have interested him. In 1900, the year that he received university admission, he made a three-month botanical trip to Taiwan.[14] When Hayata received his thesis materials from Matsumura in 1903, it was not their amazing potential to give order to Taiwanese flora that impressed him, so much as their striking difference from real plants.

His 1906 identification of a new genus, *Taiwania*, a huge conifer, provides an example. Identifying a tree from 'a few dry branches with cones' was challenging, he confessed. Since he found the dried specimens insufficient, he tried to secure 'some alcoholic materials for its histological study, and, if possible, to get a photograph of the whole plant'.[15] He wanted to know in what kind of habitat, and from which part of the tree, these branches were taken. Only after obtaining those materials with the help of his collectors in Taiwan could Hayata ascertain the tree's distinctiveness.[16]

This is not to say that Hayata dismissed the importance of specimens. He 'thought it very necessary for [his] work to see all the plants represented in the principal herbaria of the West' because the herbaria at Kew, Dahlem, Paris and St Petersburg had specimens of 'Chinese plants' that he had to compare with his collections from Taiwan. Noting that the circulation of those specimens at old centres was still limited, Hayata chose to visit them in 1910, sponsored by the colonial government. However, he was not content with merely seeing specimens. His detailed acknowledgement after the trip shows that he actively sought advice and information from other botanists.[17] Hayata tried to solve the limitations of specimens by placing himself in the network not just of specimen exchange but also of knowledge exchange.

Nonetheless, Hayata found this network not to be entirely satisfactory. In particular, the wealth of specimens at European centres seemed to hinder thorough exchange of knowledge about the plants: some botanists around those centres tended to provide shorter descriptions of plants, at times less than three lines long, simply noting the basics. Hayata thought that these botanists expected that others 'may go to see the types themselves' if their descriptions were insufficient. But for those who worked 'in the East, far from Western herbaria', that was of course impossible.[18]

This censure of 'Western' practice was Hayata's response to a criticism made of his own taxonomic practice. According to Hayata, he was accused of having proposed too many new species without adequate study of plants that had already been recorded; but such was not the case. He said that he had made a careful study of related species whenever he claimed a new one, but often could not relate his new plant to an existing one due to inadequate descriptions and his distance from the type specimens.[19]

Although he attributed the problem of his 'Eastern' practice to Western botanists' slack practices and his centre's remoteness from Western centres, he did not seek to divide. He instead tried to involve every botanical practitioner by reframing this as a common problem for the globalising botanical community: 'Eastern botanists are in no more urgent need of exhaustive descriptions of plants [for] which types are preserved in Western herbaria, than Western botanists are of those [for] which types are preserved in Eastern herbaria. The same can be said of all botanists either with reference to different countries or to different herbaria.'[20] In Hayata's view, all botanists joining in this global science were peripheral to somewhere. If botanists really meant to globalise their practice, all botanists had to solve their common provinciality together. His solution was to alter the standards of botanical practice, which had been established earlier under the assumption that everyone could access big European herbaria with ease, in order to reflect the globalised status of the field. Intended to make other botanists familiar with a plant 'without taking the trouble to look at types', his solution would make specimens superfluous through 'exhaustive descriptions' and illustrations.[21]

Hayata approached botanical illustrations, a variable but long-important practice in botany, with his concern about the limitation of dried specimens.[22] He decided to provide figures, 'reproduced from my field notes drawn from the living material'. He believed that these were 'more trustworthy in the exactness of the outline, than those drawn from dried specimens'. Throughout his series on Taiwanese flora, he tried many ways of making specimens superfluous, as shown in Figure 26.1. For plants such as ferns, his photographs were of specimens, but their flatness was compensated for by accompanying sketches.

Hayata's concern about the various limitations of dried specimens and his careful use of images to overcome their immobility and partiality may appear to be a typical concern of a botanist. But Nakai's case shows that such concerns were not shared by all botanists. Among various ways that Nakai revealed his secure confidence about dried specimens, the drawings in his series on Korean flora are most illustrative. As shown in Figure 26.2, Nakai's illustrators simply depicted dried specimens, even reproducing broken stems, with all leaves flatly layered one over the other as on the specimen sheet. Although later volumes in the series showed some changes by adding sectional views not visible in such depictions, his reliance on dried specimens, not limited to illustrations, proves consistent.

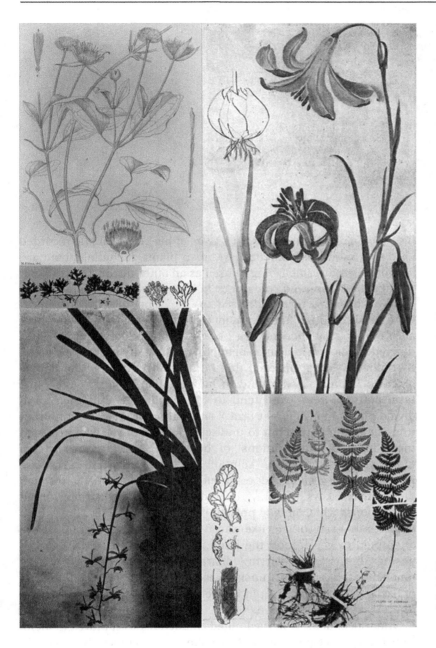

Figure 26.1 Images from Hayata's *Icones of the Plants of Formosa*. Plants of Taiwan, tai2.ntu.edu.tw.

Nakai was a typical 'armchair botanist', finishing his thesis on Korean flora in 1909 without having visited Korea. With the immense number of dried specimens at hand, seeing Korean plants in the field was not a priority for him. The first volume of his thesis, *Flora Koreana* (1909), written entirely in Latin, shows the efficiency of investing all of

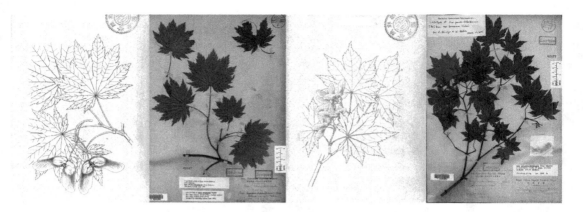

Figure 26.2 Images from Nakai's *Flora Sylvatica Koreana* and the associated type specimens. Type specimen images courtesy of the University Museum, University of Tokyo.

his time on dried specimens. In 3 years at his Tokyo lab, he recorded 1,970 species and 183 varieties of Korean plants, including about 200 new ones, easily establishing himself as an authority on Korean flora.

Nakai had displayed confidence in his approach relying on specimens in a 1908 article addressed to 'the venerable J. D. Hooker'. Here he showed his full acceptance of dried specimens, which had allowed his easy success as a botanist, as a perfect tool; he even defined the botanist's job as being that of 'classify[ing] a specimen'. By doing so, Nakai believed that he had solved an age-old debate between 'lumpers', who tend to lump similar plants as the same taxa, and 'splitters', who tend to focus more on differences and divide plants into different taxa.[23] Nakai thought that the confusion was due not to nature but to undisciplined botanists, who provided their 'indefinite descriptions' of botanical characteristics in too many varying styles. His solution was the provision of precise wordings that articulated each individual characteristic through dichotomous keys. Notably, he did not mention anything about illustrations, which might present 'so many characteristics which blend into each other' like those problematic descriptions. He attempted to get around the issue of mutability in nature by proposing to disregard varying characteristics and focus on only 'unchangeable or almost unchangeable' ones. That limited number of fixed characteristics would help end the debate about whether to lump or split, he imagined.[24]

Dried specimens, with the limited characteristics captured in them, were thus perfect guides for Nakai, not in spite but precisely because of their partiality and immutability. Although Nakai made eighteen mostly short field trips to Korea during his 30 years of service to the colonial government, his study of Korean plants relied on dried specimens sent by many capable collectors in Korea. He kept himself occupied by examining minute details on his specimens, such as the

shapes of veins and edges of leaves as well as the shapes of hairs on the
backs of the leaves, and splitting old taxa with his numerous new taxa
from Korea.

Nakai's longest exploration to the regions outside Japan was to
Europe and the United States from 1923 to 1925. It constituted
a belated study-abroad trip for Nakai, who had assumed the professorship in 1922. In this journey, Nakai 'saved every second' of his stay to
study more specimens in herbaria. By gaining permission to stay after
hours, he worked day and night at herbaria in Boston, Leiden, Paris
and London, examining as many East Asian specimens as possible. He
appears not to have had much conversation with anyone. In his work
immediately after the visit, he thanked several people mostly for
allowing him unlimited access to the herbaria.[25] For him, accumulated
dried specimens were all one needed to pursue botany.

The dynamic politics of taxonomic systems

The botanical practices of Hayata and Nakai were divergent in many
respects, partly reflecting their different takes on dried specimens.
As mentioned, their differences finally came to a head in 1930 when
Nakai harshly discredited Hayata's theory, which, since its publication
in 1921, had been translated into three Western languages.
Notwithstanding Nakai's critique, a Swedish botanist had just appreciated it as 'a most remarkable revolt against the traditional phylogenetic method of taxonomy'.[26] Phylogenetic systematics was the
standard in plant taxonomy at the time, shored up in the world plant
catalogue *Die Natürlichen Pflanzenfamilien* (1887–1915), edited by
Heinrich Gustav Adolf Engler and his German colleagues.
Importantly, despite his eagerness to suppress Hayata's revolt, Nakai
was no more content with Engler's order of the plant world. Hayata
and Nakai again possessed fundamentally diverging views, this time
about the current taxonomic order.

Hayata came to think that botanists' continuous misidentifications,
including his own, were not just about having limited access to specimens or information. He realised that there had never been any
consensus in the field about the 'natural system'. While the systems
of Linnaeus, Antoine-Laurent de Jussieu, De Candolle, Bentham-
Hooker and now Engler, had obtained many followers and functioned
as standards, no one system had ever succeeded in showing why its
criteria were indeed natural and should become the ultimate standard.
Furthermore, Engler's claim that his system expressed the natural
affinity of plants in the evolutionary or phylogenetic tree looked even
less well founded.[27]

In 1917, Hayata decided to look to nature to resolve this fundamental
issue. Adding to his experience in temperate and subtropical fields, he

chose to immerse himself in the tropical forests of Southeast Asia (then Indo-China). In 1921, he made a report on his observations, having spent 3 years thinking them through, for the last volume of his work on Taiwanese flora. 'Much against [his] will', his conclusion was revolutionary. He denied any fixed order to plant relationships and refuted not just Engler's system but more generally the possibility of finding phylogenetic relations among plants – the holy grail of plant taxonomy since the publication of Darwin's work. Instead, he proposed his 'dynamic system'.[28]

Hayata argued that Engler's system, 'at present regarded as the most natural', instead of capturing the 'natural affinity' of plants, revealed how botanists had inadequate means to demonstrate relationships between plants. While Hayata believed that 'natural affinity' should mean a kind of 'blood-relationship', what Engler addressed in considering natural affinity was just 'constitutional resemblance instead of blood-relationship'. For example, although Hayata agreed that it was neat to divide the dicotyledons into subclasses of *Archichlamydeae*, plants with flowers without petals or separate petals, and *Metachlamydeae*, plants with flowers with fused petals, as Engler did, he could not see any natural grounds for doing so. Engler's natural system seemed 'nothing more than a very convenient system'.[29]

Engler also claimed to arrange his species, genera, and families to display phylogeny, and Hayata was even less sure about this. He maintained that a 'blood-relationship' between plants was an extremely complicated affair, given plants' ability to cross-pollinate, at times even beyond their species boundaries. Hayata also believed that the mechanisms for species formation, such as crossing, mutation, and adaptation would be 'by no means independent, but closely interrelated', increasing the complexity in phylogeny. Since criss-crossing relations abounded in the plant world, a truly natural system had to show those multiple lineages dynamically.[30]

Hayata's system sought to accommodate a greater number of perspectives, targeted to different needs, in a flexible way. In criticising Hayata in 1930, Nakai revealed his strong distaste for such an open and flexible attitude to taxonomy. Nakai wrote: 'All creations of nature are in order. Each species, genera, families and so forth have their own characteristics. The characteristics do not appear casually in one plant or others but are inherited phyletically from their ancestors, and specifically, generically, or individually they diverge or converge and segregate.'[31] Nakai denied the possibility of any chance crossing or drastic divergences that would disturb a neat hierarchical lineage. He asserted that 'common characteristics of phyletically remote groups do not jump dynamically from one to other, but [are] inherited'. He also argued that 'newly acquired characteristics by segregation or mutation' could not show unexpected variations, because, again, they 'are not dynamical but phyletical'.[32] In Nakai's view, even speciation followed a predetermined path. Systematists should be able to reveal this well-fixed order.

Nakai, who had been in charge of the Japanese centre of botany since Hayata's illness in 1922, was not in fundamental disagreement with Engler, an armchair and, at times, imperial botanist whose ambition and approach were virtually identical with his own.[33] Nakai objected to Engler's standard only to protect his extreme attention to detail, often questioned by others, which he needed in order to assert the richness and distinctiveness of Korean flora in Engler's manner, and in turn to maintain his authority as a representative Japanese botanist.

As expected from his earliest work addressed to Hooker, Nakai scarcely found Engler's more widely accepted criteria any more natural than his own. Within his job of classifying specimens, he seemed to think the search for purely natural criteria was unnecessary, if not mistaken. He asserted that Engler's boundaries for plant geography, which divided and mixed regions of Korea, Japan, China and Russia based on natural grounds such as climate and geography, were unacceptable. Korea, Japan, China and Russia should remain as distinctive regions in plant geography, just as they were politically.

Nakai found the gesture of dissociating politics from botany misguided. To him, what made Engler's or Hooker's criteria more natural than his own criteria for classification was not nature but the power of Western centres. After failing to convert Hooker and others to his criteria, Nakai adopted a strategy revealing this belief. Nakai dubbed his challenged method 'Oriental' in 1914, 'Asiatic' in 1917 and 'East Asian' in 1927, in each case drawing it out as distinctive from the 'Great Western', 'European' and 'Euro-American' methods imposed by old centres. These changing claims as to the regional nature of his method, addressed only to Japanese audiences, displayed his wish to demarcate a scientific territory free from Western influence, and his ambition for botanical authority beyond the small colonial flora. His ambition naturally grew with Japan's imperial expansion. Around 1940, he intended to 'place the research originated from East Asia on the legitimate lineage of study'.[34] His classification, not the European one, should be on top of that lineage as the legitimate new standard. This time he dubbed his classification, which still emphasised small details, simply 'new', discarding the earlier regional modifiers that had never reflected true regional roots. The 'new' classification was East Asian but universal.

One may be tempted to say that Nakai immersed himself in imperial politics by asserting a new universal with his criteria, whereas Hayata stayed politically detached, letting nature be his guide while philosophising about confusion in plant systematics. Hayata's search in the tropics, however, was equally motivated by his political concerns about imperially charged exchanges. He chose to share his concerns in 1926 at the first international scientific congress in Japan; its aim of strengthening 'the bonds of peace among Pacific peoples' emboldened him. His agenda was to redress a 'belief in a ruthless struggle for existence based on the triumph of the strong and the crushing of the weak' with his dynamic system and its accompanying 'participation theory'. Hayata argued that nature did not support that kind of highly individualistic evolutionary theory; extremely variegated formations of plants that he observed in subtropical and tropical fields in 1917 suggested that 'there can be no severe struggle among different plants such as to cause the total extinction of a species, and that nature does not select plants as men do'.[35]

Since the 'belligerent theory' contradicting his own Buddhism-inspired pacifist view of nature was so popular, Hayata worked hard to put his participation theory on solid scientific ground. He utilised genetics, chemistry and physics, including quantum mechanics, as well as Johann Wolfgang von Goethe's work on plant metamorphoses. For Hayata, his participation theory was a formula for making science into 'a common knowledge both for the selected and also for the rejected', the likes of those whom he saw living in the Japanese-colonised atoll off Taiwan in an admirable 'spirit of mutual friendship', in spite of their poverty.[36]

Hayata was pre-eminently a global practitioner, who did not seek to divide. He tried to work with botanists all over the world, each of whom he believed to be as provincial as he was. He never characterised Goethe's idea of ideal forms, quantum mechanics or chemistry as Western. Even the 'belligerent theory' was not Western, when he criticised it in 1926, but some idea 'first advocated by the West', 'approved even by the people of the Far East' and taught in every Japanese school as 'a veritable truth', thus becoming universal.[37] He also presented his Buddhist idea in a cosmopolitan way; he mentioned it as 'Tendai's doctrine' without using the possibly oriental and religious-sounding term 'Buddhism'.[38] It was 'a theory based on the eternal truth initiated into mankind', written simply by a man from the Orient, not an oriental truth for oriental people.[39]

In the last years of his life, Hayata abandoned this tactic of asking everybody's interest and participation; he sought a regional division while specifically appealing to the Japanese audiences that had been so unreceptive to his idea. He addressed people 'who were born in the Orient' as 'we', and told them that current evolutionary theory was 'a conclusion obtained by seeing living organisms through Western perspectives utilizing Western ideas'. If 'we' saw the evolution 'from outside', through Eastern perspectives utilising Eastern ideas, the result would be his participation theory. He urged that Japan, 'the absolute

East', had to become the seedbed for this Eastern theory.[40] Japan, which congratulated itself on its successfully 'Escaping Asia towards Europe' would be outside the West once again, becoming 'absolutely Eastern' to reflect an Eastern light back onto the West. These statements, despite Hayata's anti-imperialism, resonated eerily with the Asianist ideology that had flexibly accompanied Japanese imperialism. Copying Nakai, Hayata had politically provincialised his globally shaped theory.

Provincialising globalisation

These Japanese botanists had not aimed to provincialise their botanical practice so that it would seem relevant only or mostly to Asian or even just Japanese botanists, as can be seen from the subdued and passing nature of their provincial claims. Fascinated by the ingenuity of dried specimens and the universal guidelines promised by the concrete and global approach of powerful centres, the Japanese botanical establishment voluntarily embraced all the trappings of modern taxonomy as its own and largely strengthened them. The unexpected provincial turns in Japanese botanists' serious undertaking of modern botany thus reveal various provincialising forces innate in the globalisation of modern botany rather than some unique provinciality of Japanese botanists or Japanese culture.

Above all, this Japanese story suggests that it was the claim to universality of European modern botany and the concomitant unidirectional terms of globalisation that drove these Japanese provincialities. The premature assertion of the universality of European standards, as built from confined European centres on a limited number of dried specimens uprooted from their habitats, was a quintessentially provincial act that apparently inspired similar responses. Hayata's and Nakai's claims about the East Asian, Asian, Oriental, absolutely Eastern, or dynamic or universal nature of their practices were largely responses to that imposing claim of universality emanating from European centres. Although Hayata and Nakai differed in many respects, they shared confusions and frustrations regarding this imposition of Western standards, which proved to be neither clear nor empirically, logically or politically feasible for their regional yet decently globalised re-enactment of European practices.

Notably, Nakai and Hayata took divergent turns to similar frustrations. Nakai's provincial turns, charged with his desire to maintain imperial interactions that allowed his successful botanising as a representative Japanese botanist unchecked by other imperial centres, followed the European path; he declared all others provincial by claiming his institution a new centre and asserting various Japanese universals. His path shows that there was nothing intrinsically

European about such activities – including their imperial origins. Hayata's absolutely Eastern turn reveals his wariness about such provincialities: he seems to have worried that botanising relying on hard-to-access dried specimens and built on subjugation of other lands might hinder thorough examination of plants in the field and exchanges of information among botanists, confining them to the unpromising search for one hierarchical order of nature. Showing that there was nothing intrinsically Eastern about his concerns, his ideas found more resonance outside Japan than within it.

The multiple reproductions of modern botany and indeed other sciences beyond Europe have had the effect of muffling and trivialising the serious tensions that occurred in the globalisation of modern science. As described here, even the alleged model student Japan was not immune to the frustrations and confusions that went with 'accepting' European standards and rules of exchange. Japanese botanists' variously provincial responses highlight those tensions and show how globalisation concealed them largely through mutual and politically sustained provincialisations.

Further reading

Bartholomew, J. R., *The Formation of Science in Japan* (New Haven, 1993).

Clancey, G., *Earthquake Nation: The Cultural Politics of Japanese Seismicity, 1868–1930* (Berkeley, 2006).

Fukuoka, M., *The Premise of Fidelity: Science, Visuality and Representing the Real in Nineteenth-Century Japan* (Stanford, 2012).

Low, M., *Building a Modern Japan: Science, Technology and Medicine in the Meiji Era and Beyond* (New York, 2005).

Miller, I. J., *The Nature of the Beasts: Empire and Exhibition at the Tokyo Imperial Zoo* (Berkeley, 2013).

Mizuno, H., *Science for the Empire: Scientific Nationalism in Modern Japan* (Stanford, 2009).

Moore, A. S., *Constructing East Asia: Technology, Ideology and Empire in Japan's Wartime Era, 1931–1945* (Stanford, 2013).

Morris-Suzuki, T., *The Technological Transformation of Japan: From the Seventeenth to the Twenty-First Century* (Cambridge, 1994).

Nakayama, S., Goto, K. and Yoshioka, H., *A Social History of Science and Technology in Contemporary Japan* (Melbourne, 2001–6).

Walker, B. L., *The Conquest of Ainu Lands: Ecology and Culture in Japanese Expansion, 1590–1800* (Berkeley, 2001).

27 Descriptive and prescriptive taxonomies

In 1992, Lord Dainton rose to address Britain's House of Lords, to introduce the results of an investigation into the state of systematic biology – which he defined as the description, naming and classification of organisms. His findings were alarming. Major scientific institutions where such research was done were being starved of funds and there were fewer jobs for classifiers. Yet British universities were nevertheless unable to train sufficient people to fill even these. The great national collections were neither properly maintained nor used. Similar problems were common in most countries.[1]

Dainton was a distinguished scientist who knew precisely why systematics mattered, but in addressing the Lords he acknowledged that for some of his listeners 'the words "systematic biology" had little significance'. So he explained that the science was vital to conservation and was in the process of being revolutionised by 'DNA studies, molecular biological tools, computers and information technology'. Yet despite this innovation, Dainton acknowledged, systematics was losing out in competition with 'the more glamourous branches of biology such as molecular biology', because 'it is rather patronisingly perceived as "comparatively unoriginal"'. Dainton's speech summarised the twin problems of low status and limited funding that beset systematic biology throughout the twentieth century; at the same time, he inadvertently provided a vivid insight into why those problems had persisted. In an effort to get his audience's attention, he reminded them of 'those few school and college contemporaries who liked nothing better than to collect and mount wild flowers or insects (especially it seemed to me beetles, butterflies and moths) and then proceed to name them according to a binomial system indicating their genus and species'.[2] No doubt, many a noble lord nodded in happy recollection of some eccentric contemporary, but what Dainton described hardly sounded like cutting-edge science.

The problems highlighted by Dainton and his committee were not new ones, and they are a reminder that classification is pre-eminently a set of practices: collecting, preserving, storing and organising.[3] In the mid 1950s, the US Biology Council published a report on the state of American museum collections that identified issues similar to those identified by Dainton.[4] In the same decade, Britain's Royal Society set up a committee to consider 'the need for taxonomists and provision of taxonomic training'.[5] They reported that training and funding were inadequate, and noted, as Dainton would 30 years later, that Britain was losing what had once been a global pre-eminence in the field. The report complained that taxonomy struggled to compete with newer branches of biology and needed more money.[6]

Twentieth-century taxonomy was largely shaped by such complaints about low status and inadequate funding, and the perception of classification as an old-fashioned, intellectually undemanding practice. If physics is traditionally the most prestigious of sciences, classification heads the list of those dismissed as mere 'stamp collecting'.[7] Its practitioners often responded to these challenges by seeking to import advanced techniques from other branches of biology, but these innovations sometimes exacerbated the very problem they were intended to resolve.[8] Dainton's argument in the 1990s that systematics was changing rapidly because of recent advances in related sciences was very similar to those made by the Royal Society, 30 years earlier, about a 'revolution within taxonomy itself' built on a novel and 'wide range of experimental data' together with a 'thorough rethinking of its philosophy'.[9] However, these reports were talking about two different revolutions (numerical taxonomy, or phenetics, in the 1960s, and phylogenetic systematics, or cladistics, in the 1990s).[10] Both phenetics and cladistics were revolutionary new approaches, intended to free classification from its long-standing association with dusty museums in which idiosyncratic cataloguers fought to impose order on nature. Yet paradoxically, these (and other) revolutions often prompted bitter disputes that exacerbated the perception that systematic biology was too subjective and unstable to be taken seriously. Unambiguously named species were, as many systematists asserted, the fundamental building blocks upon which biology was built, yet the supposed experts seemed unable to agree on the basic methods for describing them.

The problems of low funding and status were thus linked to a long-running dispute about the fundamental purpose of systematics. Some considered it a simple matter of providing a stable set of names for other scientists to use; neither glamorous nor exciting work, but if a classification did not produce a stable nomenclature, it surely served no purpose. By contrast, others within the systematics community were equally confident that this lack of intellectual ambition doomed systematics to low status: unless they set themselves more ambitious – and more explicitly scientific – goals, taxonomists deserved to be underfunded and undervalued. The central paradox that emerged from this contrast between stability and science was that every effort to make taxonomy more stable made it appear even less scientific, while every effort to make it more scientific seemed to make it less stable.

Describing or prescribing

Taxonomy's 'science vs stability' contest began (like many aspects of modern biology), with Charles Darwin, who knew the world of systematics well (having devoted eight years to detailed taxonomic work on barnacles). Since at least the late eighteenth century, much day-to-day classification had been done using convenient but 'artificial' systems. Meanwhile naturalists sought a truly 'natural system' of classification, one that described the order within nature itself. However, the source of the natural order was much disputed; for some it represented God's plan for the Creation, while others sought strictly naturalistic descriptions, based on less-than-clear concepts such as the 'affinity' that apparently linked groups of organisms. In the *Origin*, Darwin announced a solution to this long-standing debate: 'the natural system is founded on descent with modification', therefore 'all true classification is genealogical'. Common descent was 'the hidden bond which naturalists have been unconsciously seeking, and not some unknown plan of creation'. Darwin claimed at one and the same time to have finally freed systematics from its shaky metaphysical foundations, and to have liberated his fellow classifiers from the banal business of 'putting together and separating objects more or less alike'.[11] Classification would finally have clear, positive goals. In short, it would become a real science.

However, Darwin's potentially revolutionary announcement was accompanied by a deeply conservative one. Despite promising 'a considerable revolution in natural history', he assured systematists that they would 'be able to pursue their labours as at present'.[12] Darwin's conservative revolution appealed to many systematists, not least because it did not necessitate renaming every specimen in their museums or herbaria. Instead, they could carry on as before, confident that they now had a more impressive, philosophical, rationale for their work.[13]

The elegant compromise embodied in Darwin's view was, however, unstable. The claim that 'all true classification is genealogical' was profoundly ambiguous. For some, it was merely descriptive; classification would gradually converge on a stable consensus because the 'hidden bond' of evolutionary descent underlay those similarities. A descriptive reading of Darwin meant there was really no need to change taxonomic practice; the natural (evolutionary) order would emerge in time. However, others saw Darwin's phrase as prescriptive, and took it to mean that classification was a tool with which to investigate phylogeny, i.e., the historical relationships between species. In effect, classification could now become a powerful tool with which to understand evolution, a mission that not only allowed but required taxonomists to abandon their traditional tasks of merely describing and naming.

Darwin's divided legacy was apparent in 1927, when a distinguished taxonomist, Francis Arthur Bather, argued that when it came to classification, practicality was paramount, a goal that was 'served best by a classification based on superficial and obvious characters, selected and arranged in an arbitrary manner'.[14] Such an artificial classification might have practical uses, but Bather contrasted it with a classification that grouped those things 'that possess in common the greatest number of attributes', which he represented as more philosophically profound. 'Such a classification', he asserted, 'may be called SCIENTIFIC.'

Bather admitted that stable nomenclature must be classification's main purpose, while also acknowledging that only a scientific classification, 'by enabling us to predict', provided 'a means of advancing science'. Yet such advances seemed to come at the expense of stability, which led him to ask, 'Can any classification serve both purposes equally well?' He took his listeners on an erudite tour of the history of taxonomy, from Aristotle onward, elucidating the philosophical twists and pragmatic turns along the way. He credited the eighteenth-century French naturalist Michel Adanson with being the first to try and cut taxonomy's Gordian knot by sheer force, by classifying unknown plants using each of their characters in turn. This produced a set of different classifications, and Adanson assumed that those groups that reappeared in several of the classifications were genuinely natural ones. However, Adanson's method was not widely adopted; most nineteenth-century taxonomists tried to discover principles for deciding which of an organism's characteristics were the most important, and to base their classifications on these. This idea, which became known as the subordination of characters, argued that – for example – the colour of a flower was much less important to the plant than the anatomy of its reproductive system, hence the latter provided a more useful guide to classification. This concept could, Bather noted, be readily adapted to an evolutionary interpretation, and taxonomists had been trying to form one ever since Darwin. Yet while 'many paid lip-service to evolution', no actual method for detecting evolutionary patterns had been proposed, much less implemented. Somewhat despairingly, he concluded that 'important though phylogeny is as a subject of study, it is not necessarily the most suitable basis for classification'.

Nevertheless, Bather suggested that in the twentieth century 'the last word is with the experimental geneticist', on the assumption that 'the ultimate element . . . of biological classification is the gene'. He therefore proposed that 'some modern analytical palaeontologist . . . should join forces with a mathematical geneticist, and see how far the two sets of facts can be harmonized'. These comments reflect the excitement that gripped early-twentieth-century biology, when many hoped that the modern laboratory sciences would come to the rescue of taxonomy. Various factors, such as the rediscovery of Mendel's work, Hugo de Vries's mutation theory, and early breakthroughs in biochemistry, had produced new funding for what would become known as the Science of Life.[15] These innovative forms of biology made taxonomy look increasingly outdated.[16] However, while some were moving nature into the lab, others were trying to take the lab out into nature. Ecologists sought to bring some of the mathematical and experimental rigour of the laboratory sciences to analysing wild populations.[17] Many taxonomists saw such work as close to their own concerns, and tried to apply their traditional field- and herbarium-based approaches to taxonomic questions in tandem with the methods of modern ecology.[18] Practitioners of what would soon become known as experimental taxonomy traced the origins of their approach to Frederic E. Clements's *Research Methods in Ecology* (1905).[19] Clements, a US Professor of Plant Physiology, promised them that 'the general use of experiment will leave much less opportunity for the personal equation than is at present the case'.[20] The key problem for botanists was that plants grew in different ways in different environments (so stunted alpines collected on a mountain might be the same species as those collected at sea level). Without direct access to the plant's genes, how were genuine species to be distinguished from what were becoming known as 'ecotypes', non-heritable adaptations to local habitats?

Transplantation experiments promised a solution to this dilemma. Clements's enthusiasm for them was shared by Harvey Monroe Hall, unofficial curator of the herbarium at the University of California. Hall and Clements obtained funding from the Carnegie Institution for a series of experiments using transplantation methods devised by the Swedish naturalist Göte Wilhelm Turesson, which involved growing plants from diverse habitats in a uniform environment, to identify genuine inherited variations.[21] Clements and Hall pioneered reciprocal transplantation: swapping two plants between stations, placing each into the hole left by the other, to ensure 'an exact reversal of habitats'. Their research developed into a more formal, long-term programme in the late 1920s, led by Hall, who brought the Danish geneticist and ecologist Jens Clausen to California, as part of a Carnegie-funded interdisciplinary group at Stanford that included taxonomist David Keck and physiologist William Hiesey.[22]

Clausen, Keck and Hiesey used three field stations – Stanford (30 ft/ 9 m above sea level), Mather (4,600 ft/1,400 m) and Timberline (10,000 ft/3,000 m) – to transplant various plants, particularly of yarrow (*Achillea*) species (Figure 27.1). The team used a combination of notes, photographs, numbered stakes and maps to keep track of their plants and relied on 'careful yearly weeding' to prevent 'contamination' from surrounding plants. Despite the relatively straightforward methods involved, these early experiments proved too difficult to

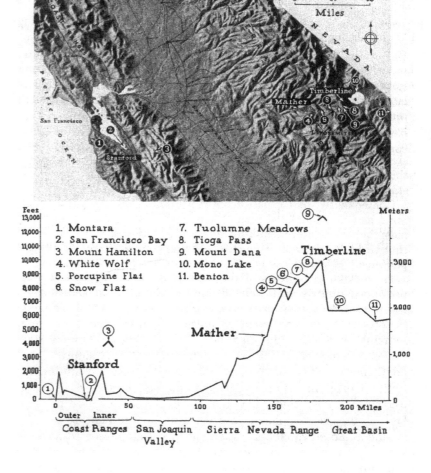

Figure 27.1 Map of Clausen, Keck and Hiesey's field stations, showing elevations. From J. Clausen, D. D. Keck and W. M. Hiesey, *Experimental Studies on the Nature of Species*, vol. I (Washington, DC, 1940).

sustain; many specimens were lost either to shock from transplantation or from the depredations of 'campers and grazing animals'. Alternatives were tried, such as moving whole plants and their soil ('sod transplants'), but these proved too difficult to transport over mountain roads and it remained hard to identify plants. Finally, they turned to cloning plants in the lab, to create a genetically identical stock, which could then be grown in a variety of habitats.[23]

For a period of about 30 years, experimental taxonomy was widely practised. But transplantation experiments were complex, time-consuming and expensive; even with the combined resources of the Carnegie Institution and Stanford University, Clausen and his colleagues had to reduce the number of experimental sites by 1940. Moreover, they had to admit that their work rested on 'close familiarity with the plants themselves in the different environments'. As a result, 'experiments like these can never replace laboratory tests under complete control'.[24] Despite the attempt to replace idiosyncrasy with experimental rigour, individual expertise and the tacit knowledge needed to conduct successful fieldwork remained essential; taxonomy still appeared more like connoisseurship than science.

Although experimental taxonomy would eventually lead to important work on such topics as plant evolution, it did not seem to resolve taxonomic difficulties.[25] Taxonomists had long been divided into 'lumpers' (who define taxonomic groups broadly in an effort to reduce their number) and 'splitters' (who multiply groups and names in the pursuit of greater precision); experimental taxonomy largely failed to resolve such disputes, since there were still no objective criteria for defining species (let alone higher taxonomic groups).[26] Among the practical objections was that new methods involved a new terminology which, opponents claimed, created more confusion than clarity, for example in Turesson's coinage of terms like 'genecological', 'coenospecies', 'ecospecies' and 'ecotype'.[27] Finally, experimental work was slow by comparison with the analysis of herbarium specimens. As one taxonomist noted, such studies were 'in the case of trees, perhaps a work of centuries. One can hardly expect taxonomists and phytogeographers to wait so long!'[28]

Experimental taxonomy, which took place in both labs and at field stations, was seized on as an example of the rapprochement between different biological traditions.[29] It caught the attention of Julian Huxley, who coined the name 'modern evolutionary synthesis' for this joint endeavour. Although not a systematist himself, he agreed to edit the collection *The New Systematics*, noting its rather 'presumptuous' title: 'For the new systematics is not yet in being: before it is born, the mass of new facts and ideas which the last two or three decades have hurled at us must be digested, correlated, and synthesized.'[30]

Huxley's introduction to this revolution in progress conveyed the excitement gripping systematics which, 'even a quarter of a century ago' appeared wholly 'empirical and lacking in unifying principles'. By contrast, it was now 'one of the focal points of biology', which detected 'evolution at work' and hence 'the world is our laboratory, evolution itself our guinea pig'.[31] Population genetics provided tools for understanding the power of natural selection, and made it possible to reject the claim that all taxonomic units were merely artificial.[32] However, despite the rhetoric (embodied in the term 'new' in the book's title, but 'systematics' was also intended to sound more dynamic), the old problem of how to read the relationship between evolution and classification remained.[33] Among the volume's authors some took the prescriptive view – that phylogenetics was the only valid approach.[34] Others, by contrast, argued that all classification was primarily practical, shaped by the purpose for which it had been devised, so there could be no single, correct or best classification.[35]

As Keith Vernon has argued, the major problem with the new approaches was that they made no tangible difference to the field's prospects or funding. Indeed, some saw the emphasis on evolution as undermining the significance of traditional, museum-based classificatory work. Huxley's definition of systematics as 'detecting evolution at work' reduced its scope 'to the narrowest possible dimensions' and ignored 'the grand tasks of collecting, describing, naming, comparing and grouping the results of one thousand million years of evolution'.[36]

Phenetics and cladistics

Even among those who accepted phylogeny as the basis for classification, substantial disagreements over methods persisted.[37] One approach to reducing the role of individual judgement was to devise mathematical methods for quantifying and measuring resemblance. The goal of mathematical taxonomy (which soon became known as phenetics) was to replace individual expert judgement with robust, objective methods. By aggregating the results of numerous different classificatory schemes, and eschewing a priori (and potentially circular) judgements about how to subordinate characters, the sheer weight of data would iron out any idiosyncrasies in specific taxonomists' work, creating a consensus classification that could be gradually improved with the addition of further data.[38]

Among the founders of this new approach was the British microbiologist Peter Henry Andrews Sneath, a doctor who hoped to improve the then-chaotic state of bacterial classification. Inspired by the numerical approach, but 'being averse to hack work', Sneath tried punch cards and sorting machines, then by chance made contact with a pioneering British computer company looking for new applications for their machines. Sneath was soon using computers for classificatory work.[39] Meanwhile, in the US, an Austrian-born biologist, Robert Reuven Sokal, had done a PhD on classification and also studied statistical methods in biology. He became frustrated by what he saw as a lack of rigour in existing methods, leading to a now-celebrated bet with a fellow student, over a six-pack of beer – that he could do a better job classifying organisms by statistical means than by the traditional approach.[40] Within a few weeks, Sokal had a purely algorithmic method of classification. He and Sneath, learning of each other's work, collaborated on a book, *Principles of Numerical Taxonomy* (1963) that explained the new approach.[41]

Some pursued the goal of objectivity with almost fanatical intensity. Paul Ehrlich promoted Sokal's method vigorously, predicting in 1961 that electronic data-processing equipment would soon be the systematist's most important tool. When one taxonomist asked indignantly: 'You mean to tell me that taxonomists can be replaced by computers?', Ehrlich is said to have responded, 'No, some of you can be replaced by an abacus.'[42] Paradoxically, this attempt to raise the status of taxonomy by making it more objective threatened to lower the status of taxonomists themselves; like many white-collar workers, they faced the prospect of being replaced or deskilled by computers. For some, this was a fresh threat to the future of systematics; mathematical methods might provide the stability systematics had long sought, but at the expense of individual expertise, thus arguably making classification look more like office work than science.

Ernst Mayr, a leading figure among those who saw taxonomy as a key tool for investigating evolution, spoke for many critics when he wrote a long, negative review of *Principles of Numerical Taxonomy*.[43] His ire focused on what he saw as Sokal and Sneath's extremism. Given that they 'criticize just about every axiom and method of classical taxonomy', one would presume that 'all previously proposed classifications ought to be complete chaos': yet numerical methods had 'almost invariably . . . confirmed the orthodox classifications'.[44] Mayr also objected to the pheneticists' refusal to weight the characters they were using, which seemed to go against the logic of taxonomy. Classification could not be a single, logical procedure that could be applied to any arbitrary set of objects; according to Mayr, 'a classification of organisms that deliberately ignores their *historical* information content is prone to be misleading or at best inefficient and uneconomical'.[45] Mayr reasserted the prescriptive reading of Darwin's claim that 'all true classification is genealogical'. That reading underlay another taxonomic revolution that emerged in parallel (and sometimes in reaction) to phenetics: phylogenetic systematics, or cladistics.[46]

The conventional history of cladistics traces its origins to Willi Hennig, an East German entomologist who took the claim that classifications ought to represent phylogeny more seriously than perhaps any previous taxonomist. However, Hennig realised that there was no simple method for converting the complexity of phylogenetic development into traditional classifications, exemplified by the Linnaean hierarchy. He instead attempted to identify groups which shared a common ancestor by analysing the distribution of characters. Characters of ancient origin (shared primitive characters, or symplesiomorphies in cladistic terminology) were too widely distributed to serve in identifying evolutionary relationships. Cladists therefore focused on more recently evolved characters (shared derived characters, or synapomorphies) in order to identify sister groups which shared a common ancestor more recently than with any other group. Such analyses produced hypothetical diagrams of relationships,

known as cladograms (Figure 27.2).[47] Although Hennig's first works
appeared in the early 1950s, his German was heavy going, even for
native speakers, hence his ideas made little impact before they
appeared in English in 1966.[48] Yet, despite Hennig's importance as
a founding father of cladistics, very similar methods were developed
independently by systematists in Italy, Britain, Australia and New
Zealand, the USA and Argentina.[49] Taxonomic communities worldwide shared similar long-standing concerns with the nature and status
of their discipline (and, of course, many practical problems), and
cladistics was a common response, which aimed to make taxonomy
a tool for investigating evolution.

Historians love describing a good fight, and controversies within
classification have been somewhat exaggerated as a result. (And the
taxonomic community seems to boast more than its fair share of
combative and uncompromising characters.) Nevertheless, it is possible to discern a preference for stable taxonomic categories among
the pheneticists, whereas many of the cladists seemed to place
a higher value on scientific innovation.[50] Although twentieth-century
taxonomists largely agreed that evolution underpins classification,
pheneticists tended to exclude evolutionary theorising from their
work, confident that evolutionary patterns would necessarily emerge,
while cladists took a prescriptive line, building theory into their work
from the outset. Yet despite these differences, there was considerable
common ground between phenetics and cladistics. For example, taxonomic methodologies began to converge as computers became cheap
and ubiquitous. By the end of the twentieth century, virtually all
taxonomists were using purely algorithmic methods. The result was
that, despite the apparent triumph of cladistics in the twenty-first
century, its daily practice owes a great deal to phenetics, not least in
being highly automated. One key output generated by cladistic software is known as a 'consensus cladogram', in that it is the best
hypothesis based on the available data – but perhaps it represents
consensus at a more profound level?[51]

Conclusion

For systematists, the twentieth century ended more or less as it had
begun: in 2002, the British government's Select Committee on Science
and Technology complained that systematics was still inadequately
funded.[52] Apparently, little had changed since the Dainton Report.

As this chapter has shown, one common feature of the history of
twentieth-century taxonomy was attempts to raise its status.[53]
Improving taxonomic ideas and methods, on which this chapter
focuses, was crucial but issues such as institutional affiliations and
links to other sciences – especially medicine – were also important.[54]

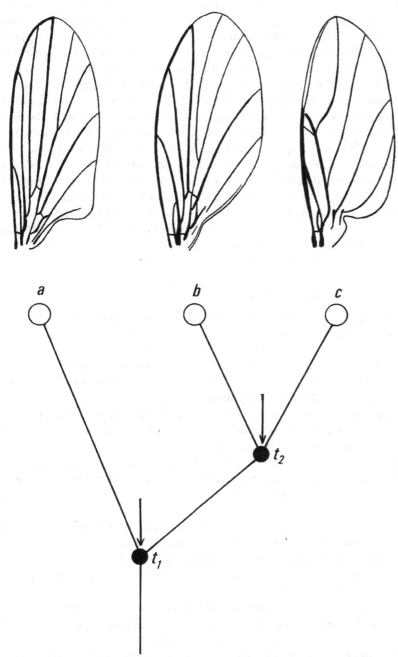

Figure 27.2 A cladogram, illustrating the relationships between three species of flies. From W. Hennig, *Phylogenetic Systematics* (Champaign, IL, 1966), p. 164. © Board of Trustees of the University of Illinois, 1966, 1979. Used with permission of the University of Illinois Press.

Different groups argued that taxonomy could raise its status by tackling causal questions, becoming less idiosyncratic, or utilising less ambiguous data, such as molecular or chemical characters in place of traditional morphological ones. Even in this brief overview, it is clear that several schools of classification embodied more than one of these impulses. Although considerable work remains to be done, the abiding tension between stability and science seems clear. Stable classifications might not be glamorous, but they make classification essential, whereas innovative methods make potentially more exciting but also more fragile claims. This dispute remains unresolved, but a degree of consensus has emerged because of a common practical goal – to improve the status and funding of systematics. Given this goal, fractious public disputes about underlying philosophical and methodological principles perhaps appeared counterproductive. As Bather argued: 'All of us, like Aristotle, classify objects in various ways according to the end in view.'[55] While those with different ends in view can never reach agreement, once ends are agreed, controversies over means tend to fade.

Further reading

Craw, R., 'Margins of cladistics: identity, difference and place in the emergence of phylogenetic systematics, 1864–1975', in P. E. Griffiths (ed.), *Trees of Life: Essays in Philosophy of Biology* (Dordrecht, 1992), pp. 65–106.

Dean, J. P., 'Controversy over classification: a case study from the history of botany', in B. Barnes and S. Shapin (eds.), *Natural Order: Historical Studies of Scientific Culture* (Beverly Hills, 1979), pp. 1–30.

Endersby, J., 'Classifying sciences: systematics and status in mid-Victorian natural history', in M. Daunton (ed.), *The Organisation of Knowledge in Victorian Britain* (Oxford, 2005), pp. 61–86.

Gee, H., *Deep Time: Cladistics, the Revolution in Evolution* (London, 2001).

Hagen, J. B., 'Experimentalists and naturalists in twentieth-century botany: experimental taxonomy, 1920–1950', *Journal of the History of Biology*, 17:2 (1984), pp. 249–70.

Hull, D. L., *Science as a Process: An Evolutionary Account of the Social and Conceptual Development of Science* (Chicago, 1988).

O'Hara, R. J., *The History of Systematics: A Working Bibliography, 1965–1996*, (1998). Available at SSRN: https://ssrn.com/abstract=2541429 or http://dx.doi.org/10.2139/ssrn.2541429.

Smocovitis, V. B., 'Botany and the evolutionary synthesis, 1920–1950', in M. Ruse (ed.), *The Cambridge Encyclopedia of Darwin and Evolutionary Thought* (Cambridge, 2013), pp. 313–21.

Stevens, P. F., 'Metaphors and typology in the development of botanical systematics 1690–1960, or the art of putting new wine in old bottles', *Taxon*, 33:2 (1984), pp. 635–59.

Vernon, K. 'Desperately seeking status: evolutionary systematics and the taxonomists' search for respectability, 1940-60', *British Journal for the History
of Science*, 26:2 (1993), pp. 207-27.

Winsor, M. P., 'The English debate on taxonomy and phylogeny, 1937-1940',
History and Philosophy of the Life Sciences, 17 (1995), pp. 227-52.

28   Imperilled crops and endangered flowers

Keeping a wary eye out for saltwater crocodiles, Carlos Magdelena waded into a long, narrow lake, deep in the wilderness of the Kimberley, Australia. Within a few strides, he had secured a sample of a waterlily that he believed was both a new species and likely to be endangered. Magdelena and his fellow explorers discovered more populations of the waterlily in subsequent days, enough to furnish scientists in Australia and at Magdelena's home institution, the Royal Botanic Gardens at Kew, with live specimens to cultivate in their collections. It was 2015, and in contrast to plant exploration missions of prior centuries, the collection of this rarity was thoroughly permitted. Prior to the expedition, Kew had acquired legal permission from various Australian authorities, including the negotiation of agreements required under the 1992 Convention on Biological Diversity. Once the plants had arrived safely at their new homes, scientists would undertake DNA analysis to confirm the waterlilies' taxonomic classification and begin efforts to propagate the plant ex situ and harvest seeds for long-term storage. Success in the latter would be considered a massive victory: though endangered in the wild, it would be safely stored in Kew's seed bank where, in theory, it could be regenerated at any time.[1]

According to many accounts, plant exploration reached its zenith in the age of European colonial empires. Daring men, sponsored by power-hungry European governments and supported by aspiring scientific institutions, traversed the globe in search of plants that would provide potent new therapeutics, launch profitable agricultural and industrial operations, and satisfy the ever-changing tastes and crazes embraced by growers back home. The success of plant exploration and exchange, made possible through the knowledge and participation of local people, underpinned imperial expansion, not least by enabling the development of export-based agricultural economies in far-flung colonies. It also transformed global biota. Through plant exchanges, Africans became increasingly dependent on crops from the Americas, British gardeners grappled with a staggering number of new plants available for cultivation, and weedy introductions edged out local species in varied landscapes around the world.[2]

Yet as Magdelena's story reminds us, plant hunting is alive and well in the twenty-first century, even if accounts of its recent history are more scattered and fragmented. In recent decades, plant collectors scoured the planet for a variety of reasons, many of which echoed the motivations of much earlier expeditions. They hoped to discover new species, to find plants that would be useful to agricultural production, or to locate known rare plants to complete collections at their sponsoring institutions. There are differences, of course: for example, in the tools used to locate and store seeds and specimens, as well as in the greater participation of local peoples and institutions as recognised, if not equal, partners.

Two differences are of overriding importance. First, plant exploration and collection over the course of the twentieth century was increasingly characterised by awareness of the loss of biological diversity globally and especially the possibility that particular plant species and cultivated varieties could be lost forever. Second, new international legal regimes were put in place to govern the activities of collecting and exchange – regimes that arose from a growing awareness of the inequities arising from centuries of plant transfers in which countries of the Global North had profited from the resources of the Global South.

Plant exploration and agricultural improvement

Botanical gardens, long understood as key sites for the creation and utilisation of knowledge about the flora of the world and well known to have been involved in extensive plant exploration and collection, took on new identities in the early decades of the twentieth century. Once the unquestioned centres of botanical investigations, they experienced ever-greater competition from universities and other research institutions. Once central instruments of imperial expansion and economic development, they increasingly took on education and entertainment as supplementary roles alongside continued research.[3]

Although botanical gardens were receding from the limelight, other institutions concerned with the cultivation of plants gained ground. In many countries, agricultural research attracted greater attention as an activity crucial to economic growth in the closing decades of the nineteenth century. The investigation of potential crops had been a core function of many botanical gardens, especially those situated in European colonies, but by the early decades of the twentieth century this work was increasingly the purview of institutions dedicated to agricultural concerns. Agricultural experiment stations, for example, became permanent features of the scientific and agricultural landscape in a number of countries. The first such station was founded in 1851 in Saxony. The subsequent development of stations in other

German states – and their perceived achievements – inspired the founding of analogous institutions elsewhere in the world.[4] Researchers at such government-funded agricultural institutions were tasked with investigations that would increase farm productivity such as developing or testing fertilisers, investigating livestock diseases, and distributing crop varieties (Figure 28.1).

The latter activity – the distribution of crop varieties in the form of improved breeds or new introductions – entailed continued state attention to the activities of plant collection and exploration. In the United States, a country notably devoid of native plant species of significant economic value, government sponsorship of plant introduction began as early as 1839 and subsequently formed part of the mandate of the US Department of Agriculture (USDA) on its establishment in 1862. The task of finding useful plants was of such importance that the USDA created a Section of Seed and Plant Introduction in 1898 to coordinate exploration and the introduction of new species and varieties. Typical missions sponsored in its early years included travels across Russia to locate varieties of wheat and barley, to northern Africa in search of cotton types, date palms and fodder crops, and around China and Japan to acquire citrus fruits and ornamental plants. These missions, combined with independent donations, kept the division awash in varieties, which it regularly published in its *Plant Inventory* and sent out on request.[5]

Meanwhile, in the Russian Empire, the creation of a government Bureau of Applied Botany in 1894 signalled the growing authority of agricultural scientists as well as state interest in research on Russia's key cultivated crops. The bureau's plant exploration activities began from behind a desk, in the form of a mass mailing to Russian provinces requesting barley varieties, but soon it was sending researchers across the empire to gather cultivated plants for evaluation. In 1920, its ambitious head, Nikolai Ivanovich Vavilov, envisioned an even greater scope for collecting in the new Soviet Union: 'I'd like to gather the varietal diversity from all over the world, bring it to order, turn the Department into the treasury of all crops and other floras, and launch the publishing of "Flora Culta," the botanical and geographical study of all cultivated plants.'[6] Vavilov went on to establish the broadest and deepest collection of cultivated varieties anywhere in the world at his institute, reorganised in 1924 as the All-Union Institute of Applied Botany and New Crops. The chief purpose of this massive collection, which included some 250,000 samples of crop varieties by the end of the 1930s, was to support state-led agricultural modernisation efforts by providing Soviet breeders with the resources they needed to create and disseminate improved strains (Figure 28.2).[7]

In short, as the perceived centrality of botanical gardens to economic growth and imperial expansion diminished, the tasks that had

Figure 28.1 Education and outreach to farmers was another important part of agricultural experiment station activities; here, a farmer views an exhibit of sorghum types at a station in Akron, Colorado, United States, *c.*1939. Courtesy of US Library of Congress, LC-USF33-003375-M2.

secured them this centrality – the exploration and collection of plant species of potential economic importance – became increasingly routine endeavours of national agricultural institutions.

Notwithstanding this continuity in the importance attributed to collecting, there were significant differences in how it was conducted. A shift in the objectives of plant exploration tracked changes in ideas about the mechanisms of heredity and the best methods for plant

Figure 28.2 Collections await further organisation at the Institute of Applied Botany, *c*.1923. Courtesy of the N. I. Vavilov Institute of Plant Genetic Resources (VIR) archives.

improvement. Early plant explorers had often sought wholly new species or novel varieties that could be cultivated alongside or in place of established breeds, with botanical gardens serving as spaces for acclimatising anything not immediately cultivable. After 1900, greater adherence to a hardened view of heredity, in which traits remained unchanged from one generation to the next, and the embrace of breeding techniques that promised to quickly capture and fix the most desirable traits in a given variety generated new targets for collection. Agricultural explorers increasingly sought out plants with specific traits, such as resistance to particular diseases or tolerance of drought. These were to serve as source material for further improvement via methods such as selection, pure-line breeding or hybridisation.[8]

Explorers seeking diverse varieties to mine for valuable traits were provided a further tool in the 1920s with the publication of Nikolai Vavilov's theory of the centres of origin of cultivated plants. Drawing on his first-hand experiences gathering and studying crops from around the world, Vavilov made a case that the diversity of a given crop species would be greatest in the region of its origination. By producing maps of these 'centres of origin', he effectively provided a guide for fellow and future plant explorers to recover the greatest possible diversity on their ever-expanding excursions.[9]

Inspired by the vision of creating more productive crops through new breeding methods and guided by Vavilov's theory in their pursuit of novel genetic material, national and imperial plant exploration and collection missions abounded in the interwar years. In Britain, wheat breeders at the School of Agriculture in Cambridge, where the new discipline of genetics had taken strong hold, began after World War I to gather varieties from markets around the world with help of the London Board of Trade – a project that eventually generated more than 7,000 samples.[10] Following a similar trajectory, the British Imperial Agricultural Bureaux in 1939 sponsored a collecting mission in South America to secure materials with which to improve potato cultivation across the British Empire; its collectors returned with about 1,200 samples of both wild and cultivated types.[11] Comparable German collecting activities began in the 1920s; these gained urgency with the rise of National Socialism and the emphasis of Nazi leaders on achieving autarky in food production. Significant material for central German collections was gathered in Ethiopia, India, Afghanistan and Iran in the 1930s and subsequently augmented by more violent means: after the invasion of the Soviet Union, German officials moved swiftly to take control of agricultural institutes, prizing in particular Soviet seed collections.[12]

Improved breeds, imperilled plants

The output of plant improvement programmes, along with the exchange of varieties across countries, regions and continents, reshaped agricultural landscapes in many parts of the world. This transition was subtler than that which had accompanied the plant introductions associated with export-based agricultural economies in earlier centuries. It involved not the introduction and rapid domination of a novel crop – as after the arrival of coffee and sugar cane in South America, cocoa in Africa or rubber in Southeast Asia – but instead the replacement of manifold locally adapted varieties (today called landraces) with a smaller number of improved lines produced by professional plant breeders. In late nineteenth-century France, for example, most wheat growers cultivated either local varieties or imported landraces adapted for local conditions. From the 1910s to the 1960s, these diverse landraces were gradually supplanted by pedigreed wheat lines adapted to a wider range of growing conditions, which had been developed via methods such as pure-line breeding.

Whereas farmers in different *départements* in the nineteenth century would almost certainly have been growing distinct landraces, the single cultivar *Etoile de Choisy* constituted half of the acres planted to wheat in southern France by 1950.[13]

Some of the first individuals to notice this transition and worry about its potential consequences were plant explorers. One often-cited example is that of the American agronomist Harry Harlan. As head of barley investigations for the USDA from the 1910s through the 1940s, Harlan travelled to Eastern Europe, Russia, northern Africa, Ethiopia, and elsewhere to collect barley varieties that might be of use to American breeders. His travels led him to believe that many potentially valuable landraces were disappearing as the farmers who traditionally cultivated these transitioned to varieties created through the efforts of professional breeders.

Like any crop variety, a landrace persists through the efforts of those who plant and harvest it each year, stewarding seed from one season to the next. If neither farmers, breeders nor seed producers bother to maintain a line, it will die out. Harlan and his USDA colleague Mary Martini described this unfolding process in 1936, noting that 'when new barleys replace those grown by the farmers of Ethiopia or Tibet, the world will have lost something irreplaceable'. Because these diverse, long-cultivated landraces were considered to be sources of useful traits, their potential disappearance from cultivation – and therefore their likely extinction – was believed to pose a significant threat to future plant breeding programmes.[14]

The worry articulated by Harlan and Martini was not unique. It was expressed, for example, by German agronomists and breeders from the turn of the twentieth century. But only occasionally did worries about the absolute loss of varieties (as opposed to the agricultural demands of nation or empire) provide the catalyst for significant collecting activities.[15] Although the immediate post-war years saw the further growth of national crop collections and the creation of collections for key crops within international agricultural research programmes, the primary motivation for these collections was that which had inspired most earlier varietal collections: they were to serve as resources for plant breeders engaged in the creation of new varieties and, as such, support economic development in the short- and medium-term (Figure 28.3).[16]

By the 1960s, a growing concern of breeders, geneticists, and other agriculturalists over the imminent extinction of many landraces (and the potentially disastrous consequences of these losses for future agricultural production) fostered urgent calls for plant exploration missions to locate and gather endangered crop varieties and for secure storage facilities to ensure their survival. Behind these calls lay a perception that farmers in many parts of the world were transitioning

Figure 28.3 A collection of Mexican maize varieties, Mexico, c.1954. Rockefeller Foundation photo collection, series 323, box 88, folder 1767. Courtesy of Rockefeller Archive Center.

more rapidly than ever before to new modes of agricultural production. 'Traditional' farming practices were giving way – often at the behest of international aid programmes – to the use of new high-yielding crop varieties alongside synthetic fertilisers, pesticides and increased mechanisation.[17]

One of the clearest articulations of this consensus came via the joint efforts of the UN Food and Agriculture Organization (FAO) and an international research initiative called the International Biological Programme (IBP). As the initial report of a 1967 FAO/IBP conference described, 'The genetic resources of the plants by which we live are dwindling rapidly and disastrously . . . The efforts to explore and collect these invaluable resources are, on a world scale, inadequate . . . Nor are there concerted, or even internationally agreed, efforts to preserve the material which has been assembled.'[18] Faced with this dearth of activity, participants charted conservation concerns related to 'plant genetic resources', surveyed current practices and planned future international activities.[19] The essential follow-up actions they envisioned included conducting a worldwide survey of existing collections, gathering threatened varieties that remained 'in the field', documenting all of these 'in a standard and international form', and arranging for their long-term conservation. The key recommendation of the 1967 conference was the creation of 'international seed storage facilities', in particular an 'international gene bank . . . available to all nations'.[20]

While scientists associated with the FAO/IBP programme attempted to foster international action to explore, collect and conserve plant genetic resources before their predicted irreversible loss, national conservation initiatives expanded. These were often motivated by similar concerns, though directed more to national needs. For example, a National Seed Storage Laboratory opened in Japan in 1966, the purpose of which was 'to collect and to maintain huge numbers of materials' for use by Japanese breeders and geneticists and 'to prevent the loss of valuable genetic resources which are disregarded at present'.[21] In the United States in 1970, an epidemic of corn leaf blight linked to genetic uniformity within hybrid corn varieties provided impetus for the expansion of a national programme for the conservation of crop germplasm; this expansion included among other things the creation of a National Plant Genetic Resources Board, greater resources for an already established National Seed Storage Laboratory and increased funding for plant exploration missions.[22]

Meanwhile, regional initiatives attempted to pool national resources to achieve shared plant exploration, collection and conservation goals. In Western Europe, breeders and biologists began agitating within the European Association for Research on Plant Breeding (EUCARPIA) for the creation of regional 'germplasm stations' in the early 1960s. The first proposal for a European station centred on potatoes, 'since the potato was of such great economic importance in Europe', and included plans for a South American Gene Centre Station from which collecting missions would be launched and a European Potato Introduction Station to facilitate the transfer of these varieties to European breeders.[23] Little came of this or other proposals within EUCARPIA until the 1970s, when several regional European banks were finally established.[24]

It proved similarly difficult to coordinate international activities along the lines imagined at the 1967 FAO/IBP conference. In 1973, early steps were taken with the creation of an International Board for Plant Genetic Resources, an institution whose mandate was 'to promote an international network of genetic resources activities to further the collection, conservation, documentation, evaluation and utilization of plant germplasm'.[25] This board was not the tightly coordinated programme of collecting missions and closely integrated network of conservation centres envisioned in the earlier FAO/IBP proposals, and its operations were contested from the outset. Yet it did serve as a conduit for directing agricultural aid money to collection and conservation activities. By 1991, when it underwent a transition in governance to become the International Plant Genetic Resources Institute, it had sponsored more than 400 collecting missions in over 100 countries and facilitated the development of genetic resources repositories in a number of countries.[26]

The conservation of biological diversity

The spread of genetically uniform crop varieties was one source of anxiety about the survival of plant species and varieties – and thus one spur to greater plant exploration, collection and conservation – but it was not the only source. The late 1960s and the 1970s were a time of growing environmental awareness in many countries. Once-disparate concerns about issues such as nuclear fallout, air pollution, resource depletion and species loss coalesced into a mainstream movement. Many individuals questioned the very survival of human societies in light of environmental degradation. In the midst of these changing perceptions of the robustness of the planet and its inhabitants, a new mission arose, or in some cases was reinvigorated, for botanical gardens and other living plant collections: the conservation of endangered wild species. If national and international agricultural research institutions were the logical stewards of the world's threatened crop varieties, then botanical gardens were first in line to claim a role in protecting the rest of the world's flora.

Campaigns for the protection of rare or endangered species gathered steam in the 1960s and 1970s, with long-standing concerns about the diminishment of certain animal and bird populations due to overhunting and habitat destruction now heightened by broader environmental concerns. New international organisations, conservation initiatives and international agreements emerged as a result. The early 1960s saw the International Union for the Conservation of Nature (IUCN) produce its first Red Lists (beginning in 1964) and publish its first Red Data Books (beginning in 1966), compilations of conservation data that charted the changing status of rare and endangered species worldwide. Between 1963 and 1973 the IUCN also fostered the development of an international convention regulating and restricting trade in wild species, the Convention on International Trade in Endangered Species of Wild Fauna and Flora.[27]

Amidst these developments, greater attention began to be given to wild plants – long neglected in wildlife conservation – as objects of concern. Transforming alarm about the disappearance of plant species into action proved difficult, largely because scientists' knowledge of rare plants was far from adequate for this task. In 1970, the IUCN published its first Red Data Book on flowering plants. Whereas the lists compiled earlier for mammals and birds had aimed at comprehensiveness, covering all known rare and endangered species, botanists knew well the impossibility of achieving the same. Ronald Melville, compiler of the initial entries, estimated in 1970 that perhaps 10 per cent of 'global flora' was endangered; at the then-current estimate of more than 200,000 flowering plants, this meant some 20,000 plant species ought to be covered in a comprehensive Red Data Book on flowering plants. The first release, however, featured just sixty-eight species. This lack of information made the further exploration of global plant populations imperative, to determine the nature and extent of endangerment and to devise conservation strategies.[28]

There was also the question of how best to safeguard endangered plant species once these were identified. Organisations including the IUCN had long advocated in situ conservation through the creation of protected areas – nature reserves, national parks and the like – as the principal measure needed to protect endangered species. But parks and preserves were not the only contenders for sites to ensure the long-term survival of endangered plants. Botanical gardens in many parts of the world had for centuries collected and cultivated plants and served in particular as places where 'rare' and 'exotic' types might be encountered. Not surprisingly, many envisioned themselves as key participants in the conservation of global plant diversity through the ex situ maintenance of threatened species. As a report of a 1975 conference at the Royal Botanic Gardens at Kew commented, many curators and scientific staff of botanical gardens who participated in the event 'now look upon conservation as one of their most significant functions'.[29]

Conservation activities felt to be within the remit of botanical gardens included the provision of botanical knowledge to scientists and governments, public education on conservation issues and the exploration, collection, and ex situ conservation of plants (Figure 28.4).

The latter set of tasks was perhaps the most contested, and not simply because preservation in the wild was the assumed ideal. There was the question of whether and when it was appropriate to gather seeds, cuttings or other materials from wild populations known to be threatened. There were also concerns about foreign expeditions encroaching on the work of local institutions or indeed ignoring permitting and other requirements for collecting. Finally, there was a question of where collected material was most appropriately maintained: by the collecting institution or in the country in which it was collected? These were sensitive issues, not least because plant exploration and collection of prior decades and centuries had at times led to

the depletion or destruction of wild populations, had rarely been conducted with regard to local interests, and had always been primarily for the benefit of institutions in imperial centres or, later, industrialised nations. Through continued conferences and in collaboration with the IUCN, administrators and curators worked determinedly from the late 1970s onwards to address these and other concerns about the participation of botanical gardens and other similar institutions in conservation.[30]

By the end of the 1970s, then, the most urgent aims of plant exploration and collection included increasingly rare landraces of agricultural crops and their wild relatives destined for seed banks as well as threatened wild plants thought to be good candidates for conservation at botanical gardens or other ex situ conservation sites.

In subsequent decades, the emergence of novel technologies for genetic manipulation and the development of pharmaceutical research created further incentives, especially for commercial actors,

Figure 28.4 A team of researchers on a Royal Botanic Gardens (RGB) at Kew expedition to Cameroon in 1995; between 1995 and 2003, several RBG Kew–National Herbarium of Cameroon teams combed the fields and forests in search of critically endangered plants, supported by local conservation NGOs and Earthwatch volunteers. © RBG Kew and used by permission.

to seek crop landraces and wild plants of the world. In agriculture, the increasing sophistication of tools for manipulating genes in conjunction with an expansion of the intellectual property rights that could be claimed in plants spurred interest in surveying plants (and indeed other organisms) for useful genetic traits that could be transferred into established high-yielding lines using recombinant DNA technologies. Meanwhile, pharmaceutical firms had intensively sought out plants with active compounds in order to develop commercial products since the nineteenth century. These activities had grown over time with the increasing sophistication of techniques for surveying and screening plants and other organisms of interest. (See Duarte, Chapter 29, this volume.) One indication of the importance attributed to such activities by the early 1990s was the announcement of an agreement in September 1991 between the pharmaceutical giant Merck & Co and the National Biodiversity Institute of Costa Rica. The initial agreement stipulated that Merck would provide US$1.14 million along with training and technical resources to support a two-year screening effort targeting potentially valuable chemical extracts from the country's immense diversity of wild plants, insects and microorganisms.[31]

Amidst these technical developments, with their attendant commercial and economic consequences, a new sensibility about the use and ownership of plants emerged. It was increasingly clear that the world's biodiversity was not only under threat as a result of environmental change but also immensely valuable to both individual companies and national economies. It was equally obvious that the benefits derived from biological diversity – whether this circulated as novel crop plants, new pharmaceuticals or still other products – did not accrue in those countries where such diversity was typically found.

Within the realm of agriculture, growing discontent and anger over the ways in which European and American companies used plant genetic resources obtained from the Global South to produce new crop lines – and then limited access to these lines through intellectual property protection – came to a head at discussions at the FAO in 1981. Delegates from Mexico, supported by representatives of other developing countries as well as international NGOs, played a key role in the events that led to the FAO's 1983 International Undertaking on Plant Genetic Resources. The Undertaking set out rules 'to ensure that plant genetic resources of economic and/or social interest, particularly for agriculture, will be explored, preserved, evaluated and made available for plant breeding and scientific purposes'. Rejecting the established international intellectual property system for plant cultivars, it specified that genetic resources were 'a heritage of mankind and consequently should be available without restriction' – an assumption resoundingly rejected by the United States and other industrial nations.[32]

The Undertaking itself did not immediately generate significant change in the then-standard norms for plant exploration and exchange, nor did it undercut international intellectual property regimes, but its development marked the starting point of a vigorous and ongoing international debate over how the circulation of agricultural plants ought to be governed. That debate resulted in a 2001 International Treaty on Plant Genetic Resources for Food and Agriculture and to the implementation of Standard Material Transfer Agreements (contracts between providers and recipients of certain plant materials) as a central means of ensuring both the 'conservation and sustainable use' of plant genetic resources as well as 'the fair and equitable sharing of the benefits arising out of their use'.[33]

Heightened awareness of inequity in the distribution of benefits resulting from the use of plants and other biological resources similarly informed the 1992 Convention on Biological Diversity. This convention, which sought coordinated global action to achieve reductions in biodiversity loss, took as a central principle the idea that the 'benefits arising out of the utilization of genetic resources' must be 'fair and equitable'. The Convention aimed to stem the tide of biodiversity loss not through international regulation (as in the case of the Convention on International Trade in Endangered Species) but instead through the development of national action plans and affirmation of the sovereignty of states over their biological resources. The latter provided the basis for what many hoped would be an effective system for managing biological diversity that also supported the economic interests of the countries within which this diversity was found: the use of bi-lateral agreements to secure authorised access for those who wished to collect, explore, or use biological resources in a given country and to generate benefits for the nations or peoples from whom the biodiversity was to be collected.[34]

Conclusion

By the end of the twentieth century, new international regimes had been established to govern the exploration, collection, circulation and conservation of both wild and domesticated plants. Government permits, prior informed consent, access-and-benefit-sharing agreements – these and other legal and contractual documents were standard components of field collecting of landraces and wild plants and sometimes even the transfer of materials already in collections from one country to another. One significant spur to the development of these regimes had been awareness of the histories of past exploitation, especially of the global inequities arising from long centuries of botanical extraction and exchange.

Scholars took part in this dialogue. Lucile Brockway's *Science and Colonial Expansion*, a study of the British Royal Botanic Gardens and its central role in imperial growth, appeared in 1979 and proved a particularly influential characterisation of the deep entanglement of botany and empire. An equivalent touchstone for the history of agricultural breeding was Jack Kloppenburg's 1988 *First the Seed*, which investigated the political economies of plant genetic resources. These and other works brought attention to how crucial access to plants had been to the development of national and imperial economies – and how important this access continues to be in the contemporary world.[35] Such narratives circulated widely beyond the academy, too. Twenty-first-century genetic 'bio-piracy', for example, was firmly linked to preceding centuries of exploitation in many accounts. 'Today bioprospectors working for biotech firms, pharmaceutical and agribusiness giants are off to the far corners of the earth hoping to come back with the (living) goods', described a 2002 article in *New Internationalist* before assessing, 'It seems suspiciously like the old colonial story of glass beads for gold.'[36]

The awareness generated through these narratives in turn informed efforts to shift the global balance of power so that all countries and peoples might benefit equally. The various standards and protocols that emerged from international agreements such as the Convention on Biological Diversity and the International Treaty on Plant Genetic Resources for Food and Agriculture have neither resolved the problem of global inequities associated with the use of biological resources nor stemmed the tide of biodiversity loss; in some cases, individuals and institutions have charged that the standards and protocols these put in place in fact undermine these goals. Yet they are a powerful reminder of the status accorded to these issues and of the central place of plant exploration, collection and conservation in the contemporary world.

Further reading

Adams, W., *Against Extinction: The Story of Conservation* (London, 2004).

Akoi, K., *Seed Wars: Cases and Controversies on Plant Genetic Resources and Intellectual Property* (Durham, NC, 2008).

Barrow, M. V., Jr, *Nature's Ghosts: Confronting Extinction from the Age of Jefferson to the Age of Ecology* (Chicago, 2009).

Farnham, T., *Saving Nature's Legacy: Origins of the Idea of Biological Diversity* (New Haven, 2007).

Fullilove, C., *The Profit of the Earth: The Global Seeds of American Agriculture* (Chicago, 2017).

Hayden, C., *When Nature Goes Public: The Making and Unmaking of Bioprospecting in Mexico* (Princeton, 2003).

Kloppenburg, J. R., Jr, *First the Seed: The Political Economy of Plant Biotechnology, 1492–2000*, 2nd edn (Madison, WI, 2004).

Pistorius, R. and Van Wijk, J., *The Exploitation of Plant Genetic Information: Political Strategies in Crop Development* (Wallingford, UK, 1999).

29 Networks of natural history in Latin America

At the close of 2001, the bioprospecting project International Cooperative Biodiversity Groups (ICBG) halted its activities in the wake of a heated controversy. Brent Berlin, an ethnobiologist with the University of Georgia, had been heading ICBG efforts, since 1998, to identify and collect botanical specimens in Chiapas, Mexico, using Mayan traditional knowledge. Working in the field alongside scientists, these indigenes contributed the fruits of their observations and experience, amassed over several generations. The intention of the Maya ICBG effort was to describe and systematically classify plant species found in this biodiversity hotspot and ultimately to produce surveys and multilingual handbooks on ethnobiology, ethnomedicine and ethnopharmacology, findings that could inform agricultural, pharmacological and medical innovation. Other participants included El Colegio de la Frontera Sur, a centre for research and higher education located in Chiapas; the Welsh pharmaceutical company Molecular Nature Ltd; and the Promaya foundation, an NGO set up to represent the indigenous communities that were voluntarily participating in the project.

The negotiations were rife with legal uncertainties. Yet the daily practices at Maya ICBG evinced a scenario of turmoil in which actors defended diverse and complex interests. Not long after the activities began, the Consejo Estatal de Parteras y Médicos Indígenas Tradicionales de Chiapas (National Council of Traditional Indigenous Midwives and Doctors of Chiapas) and the Canadian NGO Rural Advancement Foundation International (RAFI) accused the project of imperialist biopiracy. The RAFI launched an all-out Internet campaign and the controversy was heard around the world. Under this pressure, the research project was forced to shut down.[1]

This incident, which took place on the threshold of a new millennium, encapsulates key facets of the history of natural history in Latin America. Even though natural history has in recent memory often been disparaged as a lesser field – a remnant from a past that pre-dates the development of academic biology and appropriate for amateurs – it remains a vibrant practice. The field engages not only in identification and classification but in observation and experimentation; as the Maya ICBG controversy reminds us, it has opened up to fruitful dialogue with more recent specialities including molecular biology, genetics, biochemistry, biomedicine, biogeography, ecology, botany and experimental zoology. Natural history in Latin America has involved and continues to attract a wide array of participants with many motives. The Maya ICBG controversy involved myriad actors and institutions within a multifaceted network in which actors possessed overlapping and divergent interests, values and attitudes toward nature: indigenous peoples and traditional communities, Latin American and foreign scientists, laboratories and universities, global institutions, local agencies and NGOs. Still other themes emerge from the ICBG narrative: Latin America's active role in the production of scientific knowledge, built at the busy intersection of the regional and the global, and ongoing clashes and struggle over power relations. These factors, all visible in the Maya ICBG controversy, hold sway when analysing the history of natural history in Latin America.

This essay suggests that natural history in Latin America during the twentieth century affords a prime vantage point from which to understand the circulation of knowledge within the region within the broader scope of global history, by shining a light on Latin America's active role in practices of knowledge construction. It is a departure from a merely national approach because it leaves it clear that the history of these countries is part and parcel of complex transnational connections.[2] It affirms the political and social nature of natural history, which is shaped as scientists and researchers wage disputes in arenas of power. It aims neither to craft an overall narrative, like a puzzle where the pieces fit neatly together, nor to propose a single synthetic model. With this in mind, this chapter focuses on three main topics: the complexity of national, regional and global aspects in the creation of natural historical and biological knowledge; the relations between natural history, biogeography and conservation; and the present-day accumulation of biological collections.

Between the national and the global

Under the impact of the 'global turn', historians of science have increasingly emphasised the itineraries, and the relationships and conflicts that shape 'global knowledge on the move'.[3] This knowledge is now understood as the product of interactions among and the circulation of objects, texts, people, concepts and so on. This global circulation is necessarily irregular, since not everything travels in the same way or at the same speed or scope, or in the same direction. It reflects a world revolutionised by new means of communication and transportation, where routes are safer, time frames more predictable, and information and knowledge enjoy broad dissemination. This form of analysis adopts a global perspective while still revealing specific local features and interactions, as well as historical connections and disconnections. Understanding the development of natural history in Latin American profits from taking just such a perspective, for it places at the centre the ever-present, ever-increasing movement of experts, knowledge and objects of study to and from Latin America in exploring and explaining the vibrancy of natural history in the region.

The search for ornithological knowledge and zoo animals in 1920s Mexico offers a case in point and a useful starting place (Figure 29.1). From April to August 1923, José Maria Gallegos, representative of Mexico's federal Department of Agriculture, and also of its Office of Biological Studies; Laurence Huey and Carroll Scott, both with the San Diego Natural History Museum; and Ralph Hoffman, of the Santa Barbara Museum of Natural History, travelled through the Coronados Islands, regions of the Colorado River Delta inside Mexico, Laguna Hanson and Guadalupe Island. It was breeding season for birds and thus a favourable time to observe their habitats and characteristics and to collect and photograph them. The foreign members contributed US navy boats, automobiles for inland transportation, animal traps of varying sizes and sophisticated photographic equipment. In many areas, the group's observations showed that birds and sea lions were in danger of extinction, suggesting that the Mexican government needed to protect them better from tourists and hunters.

The expedition also brought four large wood and iron cages, furnished by Harry M. Wegeforth, director of the Zoological Society of San Diego. The cages were intended for use in capturing four elephant seals, two of which would be sent to the San Diego Zoo while the other two would be given to the Chapultepec Zoo. The expeditioners caught four Californian sea lions (*Zalophus californianus*) instead, two of which succumbed to the rigours of transportation. The survivors were dispatched to the San Diego Zoo, which made up for the losses by shipping two African lion cubs to its Mexican counterpart. Gallegos, Huey and Hoffman returned to these regions during bird breeding season the following year and enjoyed yet another prolific round of scientific exploration.[4]

The lion cubs were part of the first group of animals housed at the Chapultepec Zoo, inaugurated in 1924 with biologist Alfonso Herrera at the helm. The zoo had a threefold mission: to conduct zoological research; to evoke the 'roots' of Mexican nationhood, as an allusion to the collection of animals that had once belonged to the Aztec emperor Moctezuma; and to entertain and educate Mexicans in zoology by bringing them into contact with native and exotic specimens. Herrera was an eminent scientist who was actively engaged with the international scientific community. In Mexico, he worked to disseminate knowledge produced at centres abroad; he lobbied the government

to enact protection laws, acclimatise species and fight a variety of pests. While at conferences overseas and through scientific correspondence, Herrera made an effort to inform the world about the science being produced in Mexico. He moved between the national and the global, although not without some tension. In his book for young students, *Zoologia* (1924), he argued that Mexico should produce a body of work focused on the plants and animals within its own territory – the plants and animals that Mexicans lived with and that needed protecting. He advocated a renewal of the teaching of the natural sciences in his homeland and avowed that the study of biology would encourage a universalist posture, pushing scientific reason toward a worldwide state beyond nations, languages and religions, and also toward an appreciation of life and evolution across the planet.[5]

Other Latin American zoologists likewise played a role in conjoining natural history, museums, and the international exchange of and commerce in living specimens, demonstrating the initiative and creativity of naturalists at Latin American institutions. A case in point was the zoological garden of Buenos Aires, founded in 1875 by President Domingo Faustino Sarmiento and charged with the task of disseminating science, aesthetics and civic education. Naturalists of renown like Eduardo Ladislao Holmberg and Clemente Onelli numbered among its first directors. Both men fought to implant the conception of a zoo not just as a place of leisure but also as a scientific and educational institution. Holmberg founded the *Revista del Jardín Zoológico de Buenos Aires* in 1893, a journal 'dedicated to the natural sciences and especially to the interests of the Botanical Garden'. Onelli lived at the zoo for 20 years, keeping systematic notes on the unique habits and characteristics of its animals. He monitored their faeces, diets, behavioural changes, diseases, mortality rates and autopsies. He wrote a section on ethology for the zoo's magazine, entitled 'Individual idiosyncrasies of the inhabitants of the Zoological Garden'. When an animal died – be it an orangutan, elephant or bear – its corpse was eagerly studied by scientists at laboratories of pathology and embryology and by faculty members at schools of medicine and veterinary science. Lastly, the zoo would forward the specimen to the Museo Argentino de Ciencias Naturales Bernardino Rivadávia for possible inclusion in its holdings. In other words, the zoo interacted with various Argentinian institutions in the life sciences.[6]

The network of global scientific and commercial relations that linked those working in natural history was extensive. The National Zoo in Washington was one of the institutions that traded animals with the Buenos Aires zoo, as indicated in correspondence between Onelli and Frank Baker, director of the Washington facility. They discussed the exchange of specimens, transportation logistics and the best ways to protect animal health during travel. In addition to being the object of exchange, live animals were also precious commodities, captured in different corners of the world and shipped across continents, seas, and oceans. Carl Hagenbeck, of Hamburg, Germany, was one of the biggest merchants of live animals; he sold polar bears, gnus, zebras, antelopes, leopards, tigers, lemurs, elephants and cassowaries to the Buenos Aires zoo, to cite just a few species. (See Ash, Chapter 25, this volume.) Explorers of the seas south of Patagonia brought back animals as well. In 1910, naval captain Guillermo J. Nunes returned from a voyage to South Georgia Island with fourteen king penguins (*Aptenodytes patagonica*), which he donated to the zoo. Despite celebrations over the fact that they arrived in Buenos Aires alive – and imbued with nationalist symbolism because of the dispute between Argentina and England over that and other islands – they died shortly thereafter.[7]

The scholars at institutions who took part in exchange became mediators between various worlds; they became people-in-between. Through these histories, we can glimpse what historian Neil Safier has called 'itineraries of peoples, objects, and ideas relating to the natural world'.[8]

Natural history, biogeography and conservation

A global perception of nature and the need to protect it became one of the most striking features of the contemporary natural sciences. If nineteenth-century travelling naturalists in Latin America had enriched collections at European museums by sending them stuffed and dried specimens from continents around the world, twentieth-century researchers began focusing on voluntary or passive flows of life, scrutinising the extensive and intricate spatial dynamics of the planet's living creatures. These scholars contemplated the relentless movement of the ground traversed by humans, animals and plants and experienced a dizzying sense of time and space, nourishing the notion that the history of human societies and nations represented but a split second in the vast history of Earth. The same natural historical methods and tools that yielded knowledge of species laid bare the perils of their extinction.

For example, in the mid twentieth century, Brazilian arachnologist Cândido de Mello Leitão declared that advances in the means of communication and faster, safer travel were revolutionising the study of life, because they facilitated the endeavour to survey global biological diversity as never before. Technological resources improved knowledge of new species such as the microscopic creatures and animals that dwelled in previously inaccessible locations, like the fauna inhabiting the abyssal zone of the ocean. Increasing contact with the diversity of life also expanded perception of space. As Mello Leitão described, scholars of biogeography, then a nascent science, were enhancing knowledge of migration, nomadism and the dissemination and dispersion of plants and animals across the globe and through the ages (Figure 29.2).[9] Mello Leitão was a Latin American naturalist devoted to studying the nature of his homeland. He lived in a nationalist context of authoritarian, populist governments that sang the vainglorious praises of their country's land and natural wealth, and his work eventually came to be marked by an intriguing paradox. Although he was first motivated by nationalism, his object of study refused to obey political borders, and his findings relativised and even negated nationalist outlooks. Concomitantly, historical biogeography showed how the time frame of nations was absolutely meaningless against the greater backdrop of the history of the Earth. Like other staff members at the natural history museums maintained by populist governments in Latin America, Mello Leitão developed a view of nature that would not conform to nationalist assumptions.[10]

In October 1948, Mello Leitão was the Brazilian delegate to a conference that UNESCO organised in Fontainebleau, France. The conference led to the foundation of the International Union for the Protection of Nature (IUPN, later the International Union for

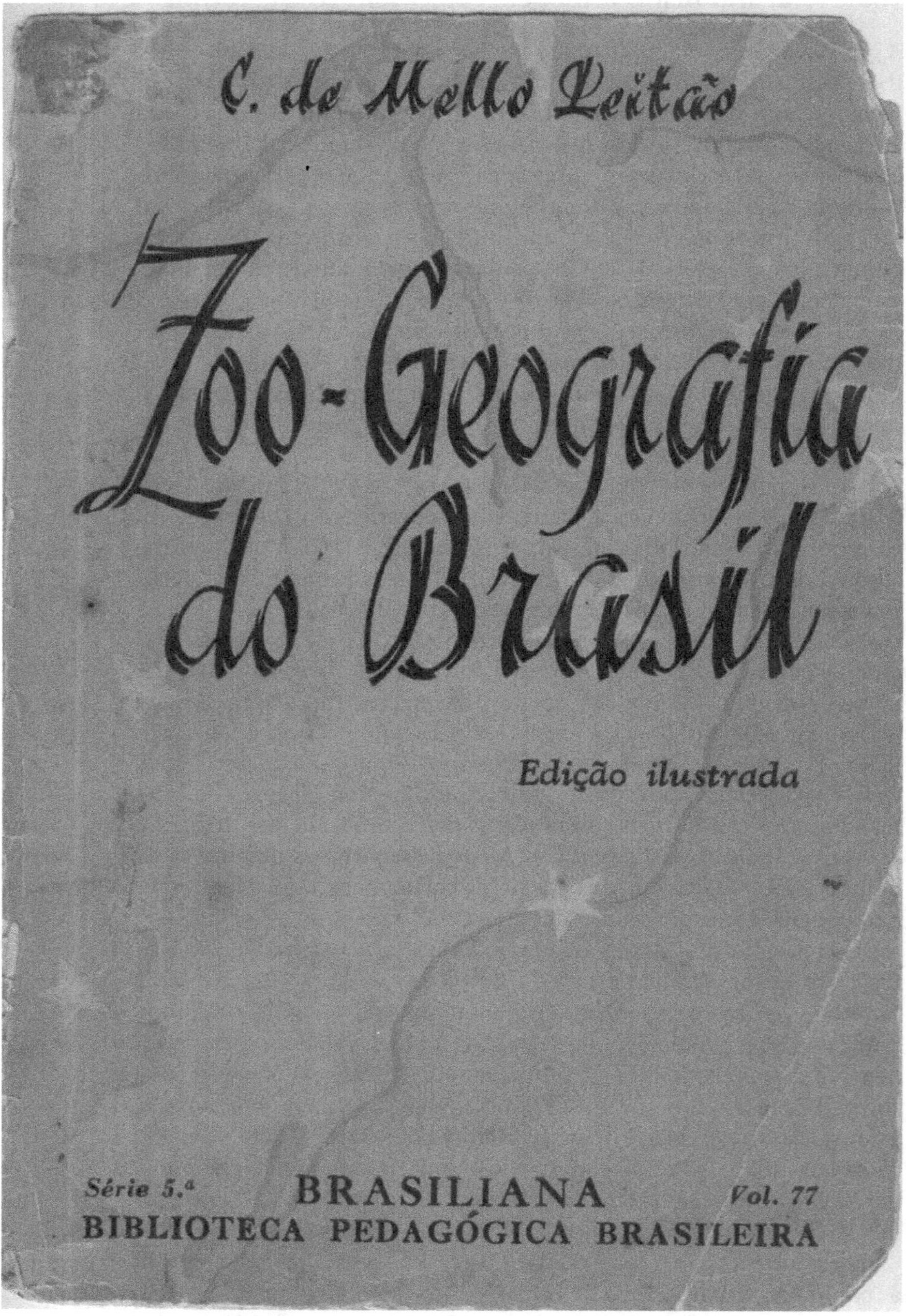

Figure 29.2 Book by Candido de Mello Leitão on Brazilian zoogeography, published in 1947. Photo by author.

Conservation of Nature and Natural Resources, or IUCN), which eventually evolved into, as it claims, a 'global authority on the status of the natural world and the measures needed to safeguard it'.[11]

Mello Leitão was also there representing the National Commission for the Protection of South American Fauna. His dual role – as a Brazilian and a South American – reflected his complex position with regard both to the continent's fauna, geography and history, and to the challenges of conserving this natural world. A significant number of Latin Americans were in attendance at the event, including delegates from Argentina, Brazil, Panama, Venezuela, Peru, the Dominican Republic, Mexico, Nicaragua and Bolivia.[12] Among them were other scholars of natural history like William H. Phelps Jr, José Yepes and Jean-Albert Vellard. Like Mello Leitão, they forged scientific careers exemplary of a 'creole science', hybrid, multicentric and diverse from a native, local *criollo* science. Dissolving the sharp distinction between 'imperial' and 'nationalist', they became members of international networks of scientists and institutions even while they established scientific practices specific to Latin America.[13]

William H. Phelps Jr went to Fontainebleau in the name of both the Venezuelan government and the Sociedad Venezoelana de Ciencias Naturales. He was born in Caracas to a New York ornithological enthusiast who had moved to Venezuela, where he devoted himself to birds and a gamut of businesses, including the coffee trade. The younger Phelps followed in his father's footsteps and in 1938 threw himself into amassing the ornithological collection currently preserved by the Phelps Foundation, a repository of nearly 100,000 specimens from the Caribbean, Amazonia and various regions of Venezuela, collected over the course of more than ninety expeditions. A businessman by profession, Phelps's status as an amateur naturalist did not keep him from playing a decisive role in Latin American natural history. He was a graduate of Princeton University and a highly productive research associate for the American Museum of Natural History; in 1941, he joined the American Ornithologist's Union. He published many important papers on the birds of Venezuela and lived in Caracas until his death in 1998.[14]

José Yepes, zoologist at the Universidad de Buenos Aires and head of the vertebrate zoology section at the Museo Argentino de Ciencias Naturales Bernardino Rivadávia, was elected member of the first executive board of the IUPN, at Fontainebleau, in 1948. He had written a respected book on zoogeography, *Mamíferos Sud-americanos* ('South American Mammals'), published in 1940 in co-authorship with Spanish-born Ángel Cabrera, who naturalised in 1925 so that he could work at the Museo de La Plata. The prize-winning book, which saw several editions, contained a map of the zoogeographic regions of South America that traced out eleven regions, none of which corresponded to national borders. Much like Mello Leitão, Yepes's studies in biogeography transformed his view of nature in his country. In their scholarship, both went beyond the study of life inside the territorial boundaries of their native lands to pursue a broader perspective: the biogeography of the American continent and its relations to the rest of the planet through the ages.[15]

The second Latin American representative elected to the first executive board of the IUPN was Jean-Albert Vellard, then director of the Museo de Historia Natural in Lima, Peru. Born in Tunisia in 1901, Vellard earned his medical degree in 1924 and took an interest in reptiles and arachnids. He travelled to Brazil, where he studied reptiles at the Butantã Institute, working alongside scientist Vital Brasil. In 1935, he accompanied anthropologist Claude Lévi-Strauss on his expeditions through the interior of Brazil. In the 1940s, Vellard was employed by a number of Bolivian institutions as a naturalist. The museum in Lima hired him in 1946 and he served as its director from 1947 to 1956. Years later he relocated to Buenos Aires, where he passed away in 1996. Over the course of his life, he moved between such diverse fields of knowledge as herpetology, arachnology, anthropology and ethnobiology, among other things publishing on the preparation and use of curare and other venomous substances by South American indigenes. Vellard approached natural history from a cultural perspective that found an intriguing interface between biology and anthropology. In terms of today's bioprospecting initiatives, we can view him as a forerunner. He was fascinated by venom from both animal and plant sources. His studies of curare went beyond his ethnographic interests, because he was convinced that knowing the substance's precise composition would be valuable to researchers in chemistry and physiology.[16]

Gathering in Fontainebleau for the founding of the IUPN, Mello Leitão, Phelps, Yepes and Vellard embodied the dynamic role that Latin American natural history played in the world context and represented a time in which the conservation of nature on the planet was coming to be designated as a global challenge.

Biological collections

Although bioprospecting is associated with the most modern technological innovations available at laboratories specialising in genetics, molecular biology and pharmaceuticals, there has been a long history of international experimentation with plant and animal species and varieties from Latin America and the Caribbean, dating back to 1492, including potatoes, corn, tomatoes, cacao, tobacco and rubber. Over the centuries, this global circulation of living things has left significant marks on the Latin American landscape, with the arrival of sugar, coffee, bananas, cattle, horses and many other plants and animals starting to threaten native species.[17] In recent decades, scientific interest in tropical biomes has triggered another wave of biological exchange. The recent history of bioprospecting shows how such initiatives have refashioned and updated natural history practices and integrated them into broader realms of the biological sciences. In recent decades, bioprospecting has boosted the circulation of objects and knowledge to a new level and has intensified contacts between natural history and the laboratory, between collections and experimentation. Against this backdrop, some scientific institutions in Latin America have launched innovative initiatives.

One example is the Instituto Nacional de Biodiversidad, or INBio, in Costa Rica, founded in 1989 with the support of the MacArthur Foundation and the Swedish International Development Cooperation Agency. INBio is a not-for-profit public interest organisation created to inventory the country's biodiversity, find ways to conserve and use it sustainably and conduct systematic bioprospecting research. From the outset, the initiative has brought scientists from a number of nations together, and has engaged in international cooperation with businesses, universities, scientific institutions and conservation agencies. INBio has earned prizes of note at home and abroad. Its collections currently hold over 3.5 million specimens and serve to document cycles of animal and plant life. After each specimen is labelled and bar-coded, it is digitised for databank entry. The institute's website offers information on Costa Rica's ecosystems, species, demography and conservation, distribution, taxonomy, references, photos, maps and drawings. INBio trains people from rural areas to serve as 'para-taxonomists', who gather and process specimens under the guidance of academic taxonomists and curators, thereby creating jobs and ideally sparking interest in the pursuit of science as a career. Of course, all this activity does not take place without some feathers being ruffled. There is a rivalry over specimen collection and guardianship between INBio and the National Museum of Costa Rica. There have also been protests about the private status of the institute, along with criticisms stemming from the tenuous legal boundaries between bioprospecting and biopiracy.[18]

In 1992, Mexico created both the Comisión Nacional para el Conocimiento y Uso de la Biodiversidad (CONABIO) – charged with promoting, coordinating, supporting and conducting activities related to biodiversity inventories, research, conservation and sustainability – as well as the national and global information network Sistema Nacional de Informaciones sobre Biodiversidad (SNIB). CONABIO systematises the efforts of a variety of institutions, with an emphasis on collections. In 1999, it coordinated 193 collections housed at 69 institutions; by 2016, the figures had reached 696 collections and 237 institutions (see Plate 18). The commission also works to 'repatriate' information on Mexican specimens stored at museums and herbaria abroad. In 1995, a survey of collections put the estimated number of specimens in Mexico at 10 million, with another 100 million specimens that had been gathered in Mexico held abroad. Institutions located in England, Spain and the United States (including Texas, New York, California and Minnesota) have cooperated by offering information and images to the CONABIO databank.[19]

Founded in Colombia in 1993, the Instituto de Investigación de Recursos Biológicos Alexander von Humboldt inventories biodiversity and works in conservation and sustainable use in collaboration with government agencies, universities and NGOs. Its twelve collections contain more than 500,000 objects that have been catalogued and organised taxonomically. They encompass an herbarium; collections of insects, non-insect invertebrates, vertebrates, tissues and eggs; and environmental sound recordings. All data are linked to a single, national register managed by the institute, which coordinates records on 203 Colombian collections, which hold over 6 million specimens of fauna, flora and microorganisms; it also posts data and reports on its website where they are accessed by a global audience. In addition to older collections that were incorporated through donation, these holdings are steadily augmented by expeditions and impasses sometimes arise between the actors involved. In May 2016, clashes with indigenous communities were fuelled by activities taking place in the Colombian department of Putumayo, where the Inga people, who belong to the Quechua group, denounced bioprospecting activities and claimed that samples or tissues were being sent abroad. The institute responded by asserting that it was merely trying to build up its scientific collections and produce genetic information on Colombia's biodiversity; it subsequently delayed a planned expedition to Putumayo until these social and political issues could be cleared up.[20] As in the Maya ICBG controversy cited previously, this conflict evinces discrepant interests and perspectives: on the one hand, the status of the Colombian institution can only be maintained globally through its participation in such practices; on the other, indigenous communities denounce the 'sending abroad' of biological material as a type of misappropriation.

The collections amassed and maintained at Latin American and Caribbean institutions have inventoried biodiversity and set in motion research, scientific exchange, and technological, industrial and agricultural innovations. These collections constitute a 'zone in between': between field work and the laboratory, between landscapes and 'labscapes'. They facilitate points of contact between distinctly different practices and knowledge, driving forward this 'border crossing' between field and lab while likewise interweaving the social and the political. Examples of the myriad fruit of these initiatives include recent efforts by the Humboldt Institute's conservation genetics laboratory; INBio's diagnosis of plant diseases; and research on the use of algae for food, industrial, or medicinal purposes, conducted in Mexico by CONABIO. In the view of some scientists excited by the emergence of an integrative biology, the big challenge of the twenty-first century 'should be in the training of biologists to again be naturalists'.[21] In this regard, some Latin American institutions are already ahead of the game.

Nature as a contested terrain

From 1993 to 1996, the anthropologists Manuela C. da Cunha and Mauro B. de Almeida and the zoologist Keith S. Brown Jr coordinated a research project funded by the MacArthur Foundation. This initiative brought together anthropologists, biologists, agronomists, sociologists, rubber tappers and members of indigenous communities. Results included the 2002 publication of *Enciclopédia da Floresta*, a compendium of studies on the Alto Juruá, a 32,090 km² region on the border between Brazil and Peru. Alto Juruá is home to great biological and cultural diversity and comprises Serra do Divisor National Park, nineteen indigenous areas, and three extractive reserves. The book addresses various aspects of the region's history, its fauna and flora and the lives of its traditional peoples. It also analyses the day-to-day lives of rubber tappers and the indigenous Kaxinawá, Ashaninka and Katukina communities – how they live, farm, hunt, cook, eat and use lianas for a variety of purposes. The project sought to engender communication between different ways of acquiring knowledge about nature, from scientific knowledge through types of traditional knowledge.[22]

The major assumption underpinning the encyclopaedia is that the region's remarkable biodiversity does not flourish in spite of human settlement but on the contrary is directly connected to cultural diversity and traditional forms of relations with the forest and its resources. From this perspective, human action has a diversifying effect, with certain kinds of anthropic actions, on a limited scale, tending to augment diversity further at different levels of the biological system. According to this logic, the presence of traditional peoples would therefore not only be compatible with the preservation of biodiversity but also decisive to its growth. Since the 1980s, a number of Latin American countries have undergone processes of re-democratisation and witnessed the advance of both social movements and neoliberalism. In this context, the development of the concept of traditional peoples, which emphasises these peoples' historical nature and their ability to manage fauna, flora and soils, has had a marked impact on anthropological research and concerns, fuelled by heated interdisciplinary debate.

The *Enciclopédia da Floresta* provides detailed information on how the region's various traditional inhabitants classify animals and offers a thorough regional dictionary of plants and animals. However, the authors decided not to transcribe a portion of what they discovered about traditional knowledge of nature because they were leery that this knowledge might be misappropriated by economic interests foreign to the communities that produced it.[23] The book is thus a natural history work that assembles diverse knowledge about the Alto Juruá region and, most importantly, constructs a political argument by trenchantly defending the continued presence of traditional peoples within areas of tropical forest preservation. Its narrative is also the story of the struggles of associations of rubber tappers, small farmers and the region's indigenous peoples.

The view propounded by the encyclopaedia was and remains highly controversial. Many zoologists, botanists, ecologists and some social scientists have made the case that tropical forests should be preserved free of any human settlements. They allege that anthropologists over-value and mythicise the knowledge possessed by indigenous and traditional populations, constructing a false notion of a 'good savage' in total harmony with nature. They argue that the supposed increase in biodiversity is deceptive, merely a by-product of the opportunistic growth of invasive species that take advantage of the extinction of endemic species even before these have been recognised and described by scientists. Such commentators assert that human populations are increasingly caught up in the market economy, seduced by the profits to be earned from marketing natural resources and feeling pressure from timber merchants, cattle raisers, soybean farmers and even drug traffickers, all of which have a growing impact on their native biomes.[24]

The diversity of endemic species represents both the wealth and the fragility of Latin America's tropical forests. Once these woods have disappeared, whether before or after science has described them, they are gone forever. Researchers devoted to discovering, describing and classifying them taxonomically represent themselves as in a race against time. In 2016, scientists working in the Amazon region of Brazil, in the Madeira River Basin, described the fish *Hyphessobrycon procyon* and simultaneously announced its imminent extinction. In fragments of Brazil's Atlantic Forest, other researchers discovered the amphibian *Brachycephalus sulfuratus*, likewise endangered.[25] Such examples give us an inkling of what is seen by scientists studying tropical biomes where indices of environmental destruction are skyrocketing, lending special meaning to the debate over whether human populations should remain in areas where environmental protection is urgently needed.[26] This conflict reflects the divergent ways in which scientists view the relations between society and nature and is fed by what are often opposing political aspirations. And this debate is far from over. It also involves countless private and governmental actors who have ties to Brazil's expanding agricultural and livestock-raising, timber exploitation, mining, hydroelectric power plant construction and tourism industries.

Natural history in Latin America has survived the twentieth-century transformations of biology, staying active and capable of innovation. It currently faces thorny challenges over how it will deal with bioprospecting, the formation and maintenance of biological collections, and the challenges of international conservation. All of these elements of natural history in Latin America have been, and continue to be, rendered complex as a result of the involvement of manifold actors from around the world with conflicting interests and aims. The field remains caught between the national and the global. International power relations are filled with tension in light of the somewhat fuzzy boundaries between bioprospecting as a source of returns and biopiracy as virtual plunder. The complex nature of the environmental question demands much more than the simple identification of supposedly 'good' or 'bad' practices or the definition of which actors are better fit to make decisions and enforce policies. In their vast complexity, little is straightforward about the relations between Latin American societies and the environment.

By focusing these pathways of natural history in Latin America, this chapter affirms that the processes of constructing knowledge about Latin America's natural history were and are interwoven with political, cultural and social history. Latin American natural history was and remains structured as a network science, forged in the process of linking and communicating across continents, oceans and centres of knowledge production, amidst international clashes and collaborations. All of these issues highlight how natural history offers a prime field of knowledge in the writing of the history of Latin America from a global perspective.

Further reading

Cañizares-Esguerra, J., *Nature, Empire, and Nation: Explorations of the History of Science in the Iberian World* (Stanford, 2006).

Duarte, R. H., *Activist Biology: The National Museum, Politics, and Nation Building in Brazil* (Tucson, 2016).

Glick, T., 'Science and society in twentieth-century Latin America', in L. Bethell (ed.), *The Cambridge History of Latin America* (Cambridge, 1994), vol. VI, pp. 463–535.

Kohler, R. E., *Landscapes and Labscapes: Exploring the Lab-Field Border in Biology* (Chicago, 2002).

McCook, S., *States of Nature: Science, Agriculture and Environment in the Spanish Caribbean, 1760–1940* (Austin, 2002).

Quintero Toro, C., *Birds of Empire, Birds of Nation: A History of Science, Economy, and Conservation in United States-Colombia Relations* (Bogotá, 2012).

Safier, N., 'Global knowledge on the move: itineraries, Amerindian narratives, and deep histories of science', *Isis*, 101:1 (2010), pp. 133–45.

Sivasundaram, S., 'Sciences and the global: on methods, questions, and theory', *Isis*, 101:1 (2010), pp. 146–58.

30 The unnatural history of human biology

In 1959, the biologist Garrett Hardin was worried about the future of his species. In *Nature and Man's Fate*, he attempted to clarify biological theories of evolution as they pertained to human welfare. Echoing concerns of fellow public intellectuals like Rachel Carson and Barry Commoner, Hardin explained that the stresses created by humans – the threat of nuclear war and ionising radiation more generally, the body burden of pesticides, the resource and psychic strains of population explosion – could be neither ignored nor reduced to matters of national concern, for they threatened the survival of the human species. Hardin contended that this problem was at once technical, political and moral: 'How is man to control his own evolution? . . . we see no answer to this problem. The worst of it is, we have forced ourselves into a position in which we *have* to give an answer . . . Having disposed of all his predators, man preys on himself.'[1]

Several years later, in his book *Man Adapting*, the microbiologist Rene Dubos proposed a solution: 'What is needed is nothing less than a new methodology to acquire objective knowledge concerning the highest manifestation of life – the humanness of man.'[2] By the end of the 1960s, such unselfconsciously gendered calls to action had contributed to intense interest in the natural history of the human in a world that increasingly seemed unnatural, overtaken by the unintended by-products of modernity. Human biology, as this broad domain of inquiry came to be called, braided together practices of epidemiology, evolutionary biology and anthropology. It gained momentum as an effort to diagnose the contemporary human condition by salvaging and repurposing older natural historical traditions to serve new aims in a molecular age. It also involved the salvage of data from human bodies seen as closer to nature, often referred to as 'primitive', to serve as a precious resource to benefit members of the so-called 'modern' world.

This chapter explains how anxieties about contamination and disequilibrium gave shape to human biology during the Cold War. I focus, in particular, on an initiative called the International Biological Program (IBP). The IBP, which ran from 1964 to 1974, comprised seven sections, one of which – human adaptability – focused exclusively on humans as functional agents in the environments in which they lived.

This large-scale international effort to assess biological baselines in the mid twentieth century provides an exemplary case in which to examine human biology as a domain of natural history shaped by Cold War cultures of fear surrounding environmental change.[3] The decision to cast members of indigenous groups as baselines for the study of human biology repurposed the Enlightenment ideal of the 'noble savage' for the nuclear age.

The end of World War II, marked by the dropping of atomic bombs on Hiroshima and Nagasaki, serves as a fulcrum for examining the uptake of research projects that were an admixture of natural history and experiment, of biology and anthropology, of the deep past and the uncertain future. As the Cold War set in, new hope was invested in biology, specifically in a field-based ecological biology, as a science of salvation.[4] The urgency posed by imminent environmental degradation stimulated efforts to 'take stock' of the planet's human resources. In the case of IBP, the experts in sciences of life would be instrumental in redeeming a mode of existence made perilous by their counterparts in physical and chemical sciences. They would focus their attention on indigenous peoples who were understood to be insulated from radiation and urbanisation and therefore could be regarded as closer to nature. Mid twentieth-century human biology was shaped by both older ideas about natural history and unique concerns that to be modern was to be unnatural. In the balance of this chapter, I describe this set of beliefs and how they informed the intellectual agenda of the IBP. I then examine the scientific practices of a specific IBP-affiliated endeavour known as the Harvard Solomon Islands Project before offering some conclusions about the importance of this history for present-day practices of anthropology and genomics.

Biology's human

The human biology that flourished after World War II grew, in part, from seeds planted by the biologist Raymond Pearl earlier in the 1920s. Pearl was among the first to argue that biology – previously the study of life properties and processes in non-human organisms – should also look at the normal human. It was unnatural, Pearl argued, that the study of humans should be solely the domain of medicine, which was focused on the pathological body.[5] This opened the possibility for asking biological-based questions about humans' 'normal' life trajectories, aging processes and nutrition, as well as their evolution and adaptation to a diverse array of environments. Pearl guided *Human Biology*, a journal he had founded, away from anatomy towards population biology, and explored this interest himself in a late work, *Natural History of Population*.[6]

During the same period, the so-called 'evolutionary synthesis' – which blended Darwinian ideas about natural selection with Mendelian genetics – yielded the new field of theoretical population genetics.[7] The architects of this synthesis dubbed it as such because they felt their efforts had achieved the holy grail of unity in biology, even as they disregarded fields like embryology and development. Experimental non-human organisms such as fruit flies (some collected in the field, others bred in the lab) were the kinds of bodies upon which basic premises of population genetics were established.

Scientists like Russian émigré and Rockefeller University Professor Theodosius Dobzhansky gained confidence and recognition for this new way of accounting for biological variation. He and others began to contemplate extending the population genetic approach to include humans, who could not be manipulated and bred in controlled conditions.[8] This was a hybrid research culture in that it both extended practices developed to study non-humans to humans and mixed values of natural history with those of experimentalism.[9]

In their eagerness to investigate processes of natural selection and adaptation in humans, certain human geneticists and biomedical researchers became fieldworkers in the anthropological mould. This meant seeking out human groups they understood to be on a different evolutionary trajectory from 'modern' humans. At the same time, eager to slough off a tarnished legacy of typological thinking, physical anthropologists embraced biology. They shifted their focus from races to populations, becoming proficient in biomedical language and basic laboratory practices. In the process, both human genetics and physical anthropology were transformed. The courtship between these two intellectual realms intensified during the Cold War. Anxieties about how to manage the exploding populations of the developing world, as well as how to assess the effects of mutagens on bodies in developed nations helped legitimate the biological study of humans and their variation.

In the early 1950s, physical anthropologists had begun to remodel understandings of human biology through studies of fossils from ancient humans, and the behaviour of hunting and gathering societies and primates. This work, part of what the anthropologist Sherwood Washburn called 'The New Physical Anthropology', was meant to destablise older essentialising ideas about race and replace them with purportedly anti-racist concepts of population and plasticity.[10] Many of those invested in this new approach had participated in writing the 1950 and 1951 UNESCO statements on race, which were hailed as the end of scientific racism. These statements sought to draw boundaries around the appropriate and inappropriate uses of race, and promoted the idea that biological race is not legitimate grounds for discrimination. The uptake of multivariate statistics and new techniques of molecular analysis and the shift from race to population also contributed to scientising and re-legitimating the biological study of human bodies as an anti-racist enterprise.

Human biologists sought to replace essentialist notions of race with evolutionary perspectives on time, mediated by technology and place. Increasingly, human biologists focused their attention on groups they understood to be homogenous 'primitive' ancestors of their more cosmopolitan, admixed selves. In fact, making physical anthropology biological and human population genetics anthropological merely contributed to the redistribution and evolution of ideas about race. The ability to split living members of the human species into temporally and technologically distinctive populations, which scientists began referring to as 'stone age' and 'atomic age', was an ironic by-product of efforts of post-war biological humanists to construct a universal and post-racial human. This was the agenda that the IBP helped to formalise and put into practice. It was how race found its way back into biology's human.

Human adaptability in the International Biological Program

The IBP was inspired by a similarly broad project called the International Geophysical Year, which happened in 1957 and united scientists from both sides of the Iron Curtain in an effort to understand questions of shared relevance, such as climate. When the IBP was first proposed, in 1958, it quickly became clear that much more than a year would be necessary to adequately take stock of the biosphere. One strategy that was developed was to carve the planet into a series of six biomes and to assign different international teams of scientists to characterise them. Because none of these biomes incorporated humans, it was soon agreed that a seventh section, to be called 'human adaptability', would be created.

Planning for the 'human adaptability' component of the IBP began in the early 1960s. Joseph Sydney Weiner, a British anthropologist and physiologist based at Oxford University, reached out to fellow scientists to form working groups representing disciplines including anthropology, demography, genetics, human biology and physiology.[11]

In the United States, the 'human adaptability' effort centred around three integrated research programmes, or IRPs. Each offered a different interpretation of and practical approach to studying human adaptability. They included an 'International Study of Eskimos', headed by William Laughlin, an anthropologist at the University of Wisconsin; 'Population Genetics of the Amerindians', headed by the geneticist James Neel at the University of Michigan; and the 'Biology of Human Populations at

High Altitude', headed by Paul Baker, an anthropologist at Penn State University. A brief summary of these three projects conveys the diversity of methods perceived as valuable in understanding the natural history of humans.

Laughlin's research, which examined how native populations dealt with the stresses of extreme cold, was the most intensely collaborative and relied on a wide variety of methods. Working in circumpolar regions, it engaged with scientists from a number of northern European countries. Their multidisciplinary study employed the full gamut of techniques in human biology, from studies of growth, physiology and nutrition to blood analysis. There was, of course, a strategic dimension to the choice of research site. Not only did the northern circumpolar regions link the United States to the USSR, via Alaska, they also represented terrain upon which military engagements might take place should the Cold War tensions between the two nations become hot.[12]

Baker's research on Peruvian highlanders' adaptation to the stress of extreme altitude relied on anthropometric and exercise physiologic methods. In order to understand the extent to which highlanders had adapted to the stress of altitude, research subjects were asked to undergo tests of lung capacity that included riding on exercise bikes in oxygen tents. Researchers involved with this programme also sought to map energy flows in the environment by attempting to chart how calories were cultivated through agriculture and metabolised by communities of hunter–gatherers. This attempt to integrate systems ecology into human population biology tried to speak to ecologists undertaking similar work that did not address humans.

By comparison, Neel's research with Amerindian groups such as the Xavante was dominated by a focus on population genetics and a search for evidence of natural selection in humans. His project, already in progress when IBP began, aimed to establish a baseline for human evolutionary genetics by sampling so-called primitive populations in the Amazon.[13]

If any single characteristic linked the methodological approach of these three IRPs, it was the focus on populations perceived as being isolated, primitive and in danger of disappearing. Scientists characterised these groups simultaneously in terms of past stressors which were environmental (cold, altitude and heat, respectively) and future stressors which were cultural (the disintegration of traditional patterns). Although the projects of the human adaptability component were not focused exclusively on so-called primitive groups, they were a dominant feature of the enterprise. As one of the leaders of the human adaptability section later recalled, 'emphasis was placed on the need to intensify the study of simple societies still living under "natural" conditions. Such groups would provide object lessons of the actual adaptability achievable by man when relying largely on his biological endowment.'[14]

The Harvard Solomon Islands Project

What did it mean in practice for human biologists to seek out what Weiner called 'object lessons' of human adaptability – evidence obtained from people thought to be living differently in evolutionary time? Importantly, it involved first creating an enduring compendium of methods which drew on contributions from approximately one hundred human biologists and described fifty separate procedures, old and new. Through this publication, Weiner ensured that even if the adaptability agenda lacked conceptual coherence, the practices and approaches innovated by its researchers would be documented and circulated.[15]

The legitimacy of human biology as a field depended upon a network of experts using techniques that were 'most likely to give inter-comparable results when applied to different human groups living in environments which range from the Poles to the Equator'. This emphasis on the standardisation of practices meant that when human biologists attempted to salvage biological traces of the 'primitive', such as blood, urine and stool samples, hair, fingerprints, anthropometric measurements and photographs, they would do so in a way that allowed samples collected in one place to be compared to those collected in another. The samples accumulated by human biologists would be made meaningful through a combination of tried and true practices of physical anthropology and new techniques in molecular science.

Both of these kinds of approaches were present in the Harvard Solomon Islands Project, a longitudinal study undertaken as part of the human adaptability component of the IBP. Led by the physician and physical anthropologist Albert Damon, it involved surveys of eight different indigenous communities in the Solomon Islands between 1966 and 1972 (Figure 30.1).

The Solomon Islands, a British protectorate from 1893 to 1978, encompass more than 900 Melanesian islands and atolls located to the northeast of Australia on the 'ring of fire' in the Pacific Ocean. It was in this region that Alfred Russel Wallace forged his theories about natural selection through his study of birds in the 1860s. These ideas, along with those of Darwin, shaped a British anthropological tradition of human biogeography in the 1920s.[16] When evolutionary theory was spliced with genetics in the mid twentieth century, biologists returned to the region with renewed interest. These island communities held great scientific appeal in that they appeared to be natural laboratories for understanding human biogeography.

Damon had previously worked with datasets culled from studies of Harvard undergraduates. As he explained to his dean, comparisons of these with data drawn from members of 'primitive' groups could contribute to knowledge about human biological potential and its relationship to the environment.[17] It is important to emphasise that Damon was not absolutely certain of what, exactly, he and his colleagues would find through an intensive study of Solomon Islanders. However, he was convinced that this comparative work should be undertaken as soon as possible. It was urgent that data be collected from these communities before they were overtaken by the forces of modernity.

Yet the bounds of the modern world already extended far from the mainland United States. The infrastructure created by the American military presence in the Pacific during World War II offered crucial support for human biological research in the region in subsequent decades. Airstrips, fuel stations and electric power grids made it possible for researchers – many of whom had relatively little experience of working outside the clinic or the lab – to make expeditions to otherwise remote communities in Melanesia and Polynesia. Groups living in the Pacific islands whom Damon sought to study were already

Figure 30.1 The Harvard Solomon Islands Project team during its first field season, 1966. The man with the pipe and anorak is Lot Page, a physician and friend of Albert Damon who stands beside him in white button down shirt. The Solomon Islanders kneeling in front of the Harvard team signal the enduring colonial logic of the enterprise, as they served both as hosts and as research subjects. Unprocessed Papers of Jonathan Friedlaender. Reprinted with permission of the American Philosophical Society.

enthralled by the apparent magic and bounty that accompanied Americans. They formed what were known as 'cargo cults', in which ritual practices were undertaken with the express purpose of accruing the perceived benefits of modernity.[18]

In a grant proposal submitted several years later to the US National Institutes of Health, titled 'Medical-Anthropological Studies in the Solomon Islands', Damon invoked the ways in which collecting anthropological information from such communities was dependent on and could be useful to colonial authorities.[19] He explained that 'our malaria smears are read by Administration malariologists as a guide to eradication and control programs'.[20] Medicine itself was to be a strategic resource in working with these colonial subjects. As a doctor, Damon was able to provide such care. He included on his expeditions several of his physician friends, who were excited to get out of the clinic for exotic adventures in the field.

The biomedical team for the Solomon Islands Project typically comprised between six or seven physicians, who applied their expertise in cardiovascular disease, haematology, ophthalmology, radiology, paediatrics and aspects of internal medicine. Each expedition also included a dentist or dental anthropologist, and three or four physical anthropologists, who took body measurements, somatotype photos, finger prints and blood samples. These examinations were extensive and incorporated many of the techniques described in Weiner's methodological handbook, including those that relied on access to blood for molecular studies.

What makes the Solomon Islands project such a revealing case study for understanding the mid-twentieth century history of human biology is just how varied and comprehensive its collecting activities were. Damon used a detailed two-page 'Harvard Questionnaire' that had to be completed for each individual subject specifically for the Harvard Solomon Islands Project (Figure 30.2). The first page of the data sheet included fields for documenting the subject's biological parents (and their respective birthplaces) and grandparents, siblings, children and medical history. It also had spaces to record detailed anthropometric measurements, including such general information as height and weight, as well as measurements of head length, breadth, 'total face h[eigh]t', 'nose h[eigh]t' and 'nose breadth'.

The second page, entitled, 'Harvard-Solomons: Observations, Genetics' asked for extensive detail about hair (texture, form – straight, loose helix, medium helix, tight helix); the amount and location of hair distributed across the body, baldness, greyness, the direction of any whorls; as well as physiognomy, including the size of brow ridges, eye folds, chin size, eyebrow thickness, forehead slope and nostril size.

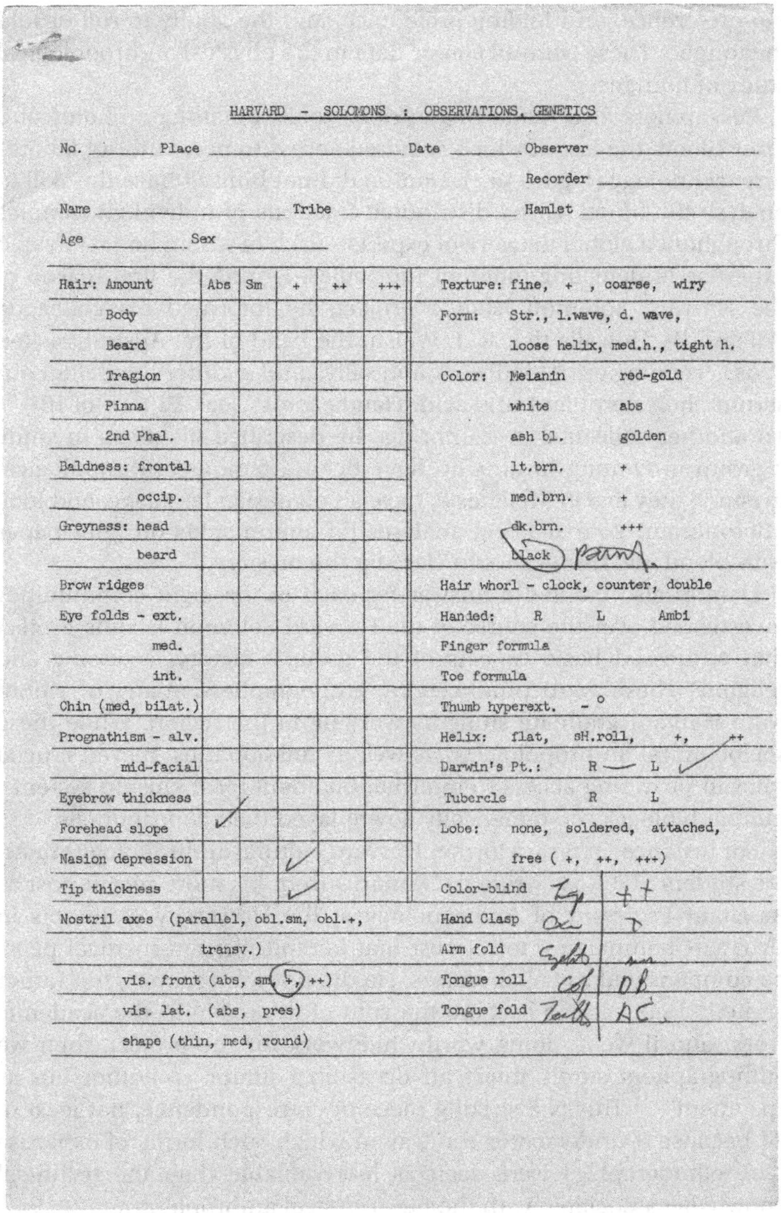

HARVARD - SOLOMONS : OBSERVATIONS, GENETICS

No. ______ Place ____________________ Date ________ Observer ______________

Recorder ______________

Name ______________ Tribe ______________ Hamlet ______________

Age __________ Sex __________

Hair: Amount	Abs	Sm	+	++	+++	Texture: fine, + , coarse, wiry
Body						Form: Str., l.wave, d. wave,
Beard						loose helix, med.h., tight h.
Tragion						Color: Melanin red-gold
Pinna						white abs
2nd Phal.						ash bl. golden
Baldness: frontal						lt.brn. +
occip.						med.brn. +
Greyness: head						dk.brn. +++
beard						black
Brow ridges						Hair whorl - clock, counter, double
Eye folds - ext.						Handed: R L Ambi
med.						Finger formula
int.						Toe formula
Chin (med, bilat.)						Thumb hyperext. - °
Prognathism - alv.						Helix: flat, sH.roll, +, ++
mid-facial						Darwin's Pt.: R — L ✓
Eyebrow thickness	✓					Tubercle R L
Forehead slope						Lobe: none, soldered, attached,
Nasion depression			✓			free (+, ++, +++)
Tip thickness			✓			Color-blind
Nostril axes (parallel, obl.sm, obl.+,						Hand Clasp
transv.)						Arm fold
vis. front (abs, sm, +, ++)						Tongue roll
vis. lat. (abs, pres)						Tongue fold
shape (thin, med, round)						

Figure 30.2 The second page of Damon's questionnaire for the Harvard Solomon Islands Project. Albert Damon Papers, Solomon Islands, Folder 1.24, Box 1 (tables) © President and Fellows of Harvard College, Peabody Museum of Archaeology and Ethnology PM 995-17-00/1.6.1.2 (digital file no. 99350265).

Other categories included handedness, colour-blindness, hand clasping preference, arm folding preference, and the ability to roll or fold the tongue. These were all classic data in the physical anthropological study of humans.

Researchers also made fingerprints, took somatotype photos and drew blood, the last of which required access to new kinds of laboratory technology (Figure 30.3). Damon did not himself have the skill to analyse the blood, so he distributed fractions of individual samples throughout a global network of experts – each of whom had cultivated expertise in detecting different molecules. During the first season of the Harvard Solomon Islands Project, he informed his colleague William W. Howells that R. J. Walsh, the head of the Australian Red Cross, 'is doing our blood work and saliva and another Australian our serum cholesterol and uric acid. Hands across sea. All part of IBP.'[21] To another colleague in Cambridge he described his plans to study a group in Ontong Java, 'who have been a famous anthropological mystery: they live in Melanesia, have a Polynesian language, and look Micronesian! Your magical analyses [of amino acids on filter paper with blood and urine] should clear up the mystery.'[22]

Damon also created a dossier for each of the eight communities investigated over the course of the Harvard Solomon Islands Project that addressed basic aspects of the group's history, economy and religion. These were produced by ethnographers, many of whom were Harvard graduate students working in the region. While these sociocultural anthropologists, as well as missionaries, played crucial roles in providing access to information about local kinship systems, human biologists systematically downplayed their contributions.

For instance, in 1974, a former Harvard cultural anthropology graduate student Hal Ross wrote to Damon – from his more recent post as Assistant Professor of Anthropology at the University of Illinois in Urbana-Champaign – to request that he and his non-medical peers be compensated as collaborators. He did not want money, but rather argued: 'Publications lists are the coin of the realm in the academic work; and if we're doing worthwhile work for the project, then we ethnographers surely merit an occasional junior co-authorship as "payment".'[23] This is a striking piece of correspondence, not least of all because it underscores the way in which such forms of expertise (i.e., ethnographic) were seen as less valuable than the technical knowledge associated with the new physical anthropology.

Damon was not able to reply to Ross because he died of cancer in 1973. It took more than a decade, until 1987, for Harvard Solomon Islands Project researchers to publish conclusions based on the data collection when they conceded that 'most of the biological variation cannot be systematized'.[24] This should not have been surprising given the vague goals that animated the data collection activity.

Nevertheless, the Harvard Solomon Islands Project should not be regarded as a failure if for no other reason than that the practices of human biology it employed as part of the IBP helped create an infrastructure that supported later studies in medical anthropology and genomics. In particular, blood that was collected as part of the Harvard Solomon Islands Project, before the ability to analyse DNA, has persisted in freezers where it has since been thawed for innovative studies of human variation, evolution and epidemiology.[25]

Figure 30.3 Jonathan Friedlaender, shown in white being observed by young villagers with an unidentified Melanesian research subject in the mid 1960s, was a biological anthropology graduate student when the Harvard Solomon Islands Project began. He accompanied the team on its first field season where he oversaw several aspects of data collection, including dermatoglyphics (finger printing). Unprocessed papers of Jonathan Friedlaender. Reprinted with permission of the American Philosophical Society.

Conclusion

Under the mandate of a holistic ecology, seen as imperative for a civilisation in crisis, IBP researchers took and preserved stock, in the form of standardised quantities of biological substances such as blood and standardised measurements of human bodies. The techniques applied to the analysis of all of these materials, including data collection sheets, IBM punch cards, and early computer programs – all of which were honed during the years of the IBP – dramatically contributed to the process of turning nature to number, transforming fleshy, variable human organisms into standardised (or at least inter-comparable) samples.

However, in the process, 'human adaptability' researchers built 'unnatural' and now often invisible assumptions into the data that persist today as a resource for human biology.[26] Their decision to focus on members of communities they cast as both closer to nature and destined to disappear perpetuated a divide within a science of human biology that claimed to seek unity. Furthermore, the way that cultural knowledge – including indigenous peoples' own accounts as well as ethnographic analysis of social and economic forces – was situated as less significant than biological knowledge cast those who were enrolled as subjects in this form of science as research objects.

While mid twentieth century human biologists believed themselves to be salvaging important sources of information about their own species for the use of future scientists, it is less clear what was gained by those who participated in human biological research. At some times, human biology, and the medical care that travelled with it, were interpreted as a form of sorcery or witchcraft.[27] At others, it was seen as an avenue towards improving living conditions which had been undermined by the arrival of mining and similar projects of industrialisation.[28]

Today, blood and other forms of data collected at mid century persist as 'entangled objects', material manifestations of the ways in which Europeans and Americans are all 'caught up in international relations of production and appropriation which stretch across the spaces separating us' from the indigenous peoples of the Pacific or even the Amazon.[29] These bodily traces have, in some instances, become the source of intense disputes over the appropriate use of human remains.[30] In September of 2015, blood samples collected by researchers affiliated with the IBP from members of an indigenous community in South America, known as the Yanomami, were repatriated. Davi Kopenawa, a Yanomami shaman and philosopher who has since become an internationally recognised advocate for indigenous causes, was quoted as saying, 'The white man didn't tell us . . . "We're going to store your blood in the cold, and even if a long times goes by, even if you die, this blood is going to remain here" – he didn't tell us that! Nothing was said.'[31] This was, above all, an accusation that members of the community had not been adequately informed about what the researchers intended to do with their collections over time and therefore had no right to retain them. It was also a statement about how practices of human biological data extraction had created a seemingly unnatural form of life itself – blood that persisted outside the body.

As human biologists have reckoned with the history of their field, they have sought to redefine their approach to research by emphasising principles of inclusion and, increasingly, collaboration with members of the communities they study. At the same time, human biologists continue to innovate new uses for old materials, including those accumulated during the IBP. These vital legacies of human nature are the product of mid century fears about a toxic future in which the death of indigenous peoples could be cast as a resource for the survival of others. Recognising this unnatural history is a crucial step in ensuring that the racist assumptions of the past stay there, even as its collections endure.

Further reading

Anderson, W., *The Collectors of Lost Souls: Turning Kuru Scientists into Whitemen* (Baltimore, 2008).

Bamford, S., *Biology Unmoored: Melanesian Reflections on Life and Biotechnology* (Berkeley, 2007).

Haraway, D., 'Remodeling the human way of life: Sherwood Washburn and the new physical anthropology', in G. W. Stocking (ed.) *Bones, Bodies, Behavior: Essays on Biological Anthropology* (Madison, 1988).

Little, M. and Kennedy, K. R. (eds.), *Histories of American Physical Anthropology in the Twentieth Century* (New York, 2010).

Radin, J., *Life on Ice: A History of New Uses for Cold Blood* (Chicago, 2017).

Reardon, J., *Race to the Finish: Identity and Governance in an Age of Genomics* (Princeton, 2005).

Smocovitis, V. B., 'Humanizing evolution', *Current Anthropology*, 53, supplement 5 (2012), pp. S108–25.

Spencer, F. (ed.), *A History of American Physical Anthropology, 1930–1980* (New York, 1982).

TallBear, K., *Native American DNA: Tribal Belonging and the False Promise of Genomic Science* (Minneapolis, 2013).

Turner, T., *Biological Anthropology and Ethics: From Repatriation to Genetic Identity* (Albany, 2005).

31   Fieldwork out of place

In the naturalist tradition, knowledge of nature has been closely tied to the embodied experience of place. Indeed, it is the direct and unmediated encounter with particular places and the natural objects found within them that has often been held to distinguish naturalists from other students of the natural world, particularly laboratory scientists. In his 1994 memoir *Naturalist*, for example, E. O. Wilson writes of a formative childhood summer spent at Florida's Paradise Beach encountering jellyfish, stingrays, and other forms of sea life: 'Hands-on experience at the critical time, not systematic knowledge, is what counts in the making of a naturalist.' Similar claims for the importance of emplaced, embodied experience in the establishment of naturalists' identities and expertise have been embraced by historians of natural history and of the range of specialised field sciences that have emerged from it over the past two centuries, from geology to ecology. From this perspective, the scientist's placeless and universalist form of expertise may well flourish in the sterile atmosphere of the laboratory, but it can never rival the holistic, experiential and embodied knowledge acquired by the naturalist in direct contact with the natural world in all its messy complexity.[1]

It does not matter much for my purposes here whether this characterisation of natural history and naturalists is accurate. Clearly the borders are blurred. There have been many scientists who have not described themselves as naturalists but who have nonetheless worked very hard, through close engagement with particular places, to make their findings relevant to the wider world beyond the laboratory. At the same time, there have been many self-described naturalists – indeed, some of the most influential – who have conducted most of their research in offices, museums and even laboratories. I am more interested in the question of how the practices and values associated with being emplaced have changed over time, along with the conceptualisation of emplacement itself. Instead of seeing these practices and values as transhistorical givens, equally valid in all times and places, we can learn more by asking how place and emplacement have come to matter for naturalists in different ways over time and how such understandings have become intertwined with disputes over who may speak authoritatively for nature.

A compelling example of changing modes of emplacement and their role in authorising claims about nature can be found in the history of the study of bird migration in the twentieth century. At the beginning of the century, studies of bird migration were based primarily on observations of the seasonal arrival or departure of birds of particular species at particular places. By the end of the century, technologies of remote and automated observation had made it possible for research-ers to acquire vast amounts of detailed and frequently updated data on the movements of individual birds and on the environments through which they moved. These data were, moreover, often collected in the absence of any human observer at all, except the one located in an office, museum or laboratory possibly half a world away. Although fieldwork remained essential for many purposes, researchers were now capable of learning more intimate details about the lives of individual birds and of gaining a broader view of the environments through which they moved through the use of satellite-tracking and remote-sensing techniques than through personal observation in the field. This inversion challenges us to rethink the importance of emplaced, embodied experience in the history of natural history.[2]

Bird banding: the naturalist in the network

It is not easy to study bird migration. Birds have been a popular subject of study for experts and non-experts alike in part because they are more readily identified and less elusive than many other kinds of animals. Most are also highly mobile, however, and human observers can find it difficult to keep up. Attempts to mark individ-ual birds with tags, dyes and other artificial markers so that they can be re-identified over the course of repeated encounters in the field have a long history, but until the twentieth century they were never pursued on a large enough scale to generate significant insights into migration. Before the twentieth century, the most common approach to studying bird migration involved not the tracking of individual birds but rather the observation of the timing of the appearance of birds of a given species in different places. If birds of a particular species were observed in Canada in the summer and in Mexico in the winter, and if, moreover, birds of that species were observed flying northward in the spring and southward in the fall, it was assumed that the birds observed at each of these places were in fact the very same birds, and that they were migrating seasonally from place to place. This approach involved synthesising a number of locally emplaced observations to make claims about the long-distance movements of species or populations.[3]

Figure 31.1 A brown thrasher (*Toxostoma rufum*) being held in position for banding in the 1910s. S. Prentiss Baldwin, 'Bird banding by means of systematic trapping', *Abstract of the Proceedings of the Linnaean Society of New York*, 31 (1919), pp. 23–56, plate II, figure 4.

The use of the technique of bird banding on a widespread, coordinated basis at the beginning of the twentieth century in Europe and North America introduced, for the first time, a highly effective means of studying the long-distance movements of individually identifiable birds rather than of species or populations. First popularised in the late 1890s by the Danish ornithologist Hans Christian Cornelius Mortensen, bird banding consisted of capturing birds, affixing light-weight metal bands stamped with serial numbers or other unique identifying marks to their legs, and then releasing them (Figure 31.1). Initially these banded birds tended to be easy-to-catch nestlings, but by the 1910s steady improvements in techniques of live capture had made it feasible to band large numbers of adults. Once a bird had been banded, a report of the time and place of banding, the serial number and the bird's species, gender, age class and other characteristics was sent to a central registry or clearing house. When the banded bird was recaptured either dead or alive – many bands were recovered from ducks and geese by hunters – the report filed with the registry could be used to reconstruct its movements. Taken together, the records of thousands and eventually millions of banded birds made it possible by the middle of the twentieth century to map migration patterns in extraordinary detail.[4]

The growth of bird banding over this period also helped to transform the relationship between place and the production of ornithological knowledge. Because migration is a phenomenon that by definition extends beyond any single locality, its study has almost always involved the coordination of the knowledge of multiple observers. In that sense, the reliance of bird banding on numerous geographically distributed volunteers – both banders and those who reported the bands but did not attach bands themselves – was not new. Where it did break new ground, however, was in the absolute necessity, rather than merely the desirability, of the coordination of these observers within a single, well-defined, carefully managed network. Whereas the observation of the date of the annual arrival of a migratory species in a particular locale could be made and published whether or not any-one else was carrying out similar observations elsewhere, banding a bird in the absence of observers in other places who were ready and willing to report the discovery of the band either to the original bander or to a centralised registry was about as effective as casting a message in a bottle into a world where no one could read.

A coordinated, continent-wide network of observers animated by the faith that the latent value of their emplaced labour would be realised through the contributions of geographically and socially distant strangers did not emerge spontaneously. Persistent effort and significant resources were required. In the United States, the earliest efforts to coordinate bird banding on a local or regional scale took place in the first decade of the twentieth century. In 1908, for example, Leon J. Cole launched a bird banding effort centred at the New Haven Bird Club. Participating club members agreed to use standardised bands and sought to raise public awareness of the importance of reporting bands once they had been discovered. The results of this and other early efforts were largely unsatisfactory, however, in part because the networks that they constructed were of too limited a geographical scope – in the case of the New Haven Bird Club, a small stretch of the eastern coast of North America – to produce data on continent-wide bird migrations. In 1909, convinced of the necessity of building a network of observers as far-reaching as the phenomenon it was intended to study, Cole persuaded members of the American Ornithological Union to found the American Bird Banding Association. The association began operating on a national scale in 1910, distributing standardised and uniquely num-bered bands at no cost to volunteers and serving as a central clearing house for reports of banded or recaptured birds from its base in the offices of the Linnaean Society of New York.[5]

Although the American Bird Banding Association distributed tens of thousands of bands in the decade after its founding and collected thousands of corresponding reports, it was hampered by its lack of financial resources and by growing frustration on the part of volunteer banders with the paucity of results from their efforts. Since the value of one's emplaced labour was dependent on a band being discovered and reported, the extraordinarily low rate of recovery in this early period – well under 3 per cent of banded birds were recovered – made it difficult for banders to feel that their labours were valuable. The entrance of the US federal government into bird banding proved to be a crucial turning point. In 1920, in the wake of the signing of the Convention for the Protection of Migratory Birds by Canada and the United States in 1916, the US Biological Survey established a new Bird Banding Office under the direction of ornithologist Frederick C. Lincoln, which took over the task of coordinating a national bird banding effort from the American Bird Banding Association. With a small but permanent professional staff at his disposal, and with the cooperation of amateur ornithologists and volunteer banders across the country, Lincoln was able to do what the association could not: construct a national network that was sufficiently dense, well trained and uniform to produce abundant and reliable data on bird migration. By the 1930s, Lincoln was able to use these data to map the 'flyways' of distinct populations of migratory birds across North America, which transformed the way waterfowl, in particular, were managed (Figure 31.2).[6]

One might dispute whether or not Lincoln should be described as a naturalist, and it is a term he only occasionally used himself. But it would be difficult to deny the appellation to the thousands of volunteers who banded birds and dutifully filed reports with the Bird Banding Office. Deeply engaged with the natural histories of the places where they lived and often enamoured of the birds they handled in the process of banding, they were people who would probably have agreed with E. O. Wilson that hands-on experience and direct, embodied encounters with particular places and living beings were essential. In many ways, their participation in the national bird banding networks established first by the American Bird Banding Office and then by the US Biological Survey deepened their connection to place by giving them new reasons to go into the field, to trap and handle individual birds, and to establish relationships with like-minded people in their communities. But it also changed the significance of place and of emplaced observations. For the study of migration, the practice of banding a bird only accrued meaning when someone else – perhaps a stranger living hundreds or thousands of miles away – found and reported the band to Lincoln's office in Washington. The value of one's local observations and emplaced labour thus became contingent on one's position within a large and geographically distributed network of observers working in accordance with shared standards.

Radio tracking: the naturalist on the move

Bird banding was and remains an effective way of studying migration, but it has its limits. Most birds that are banded are never seen again. Except under extraordinary circumstances, such as in populations that

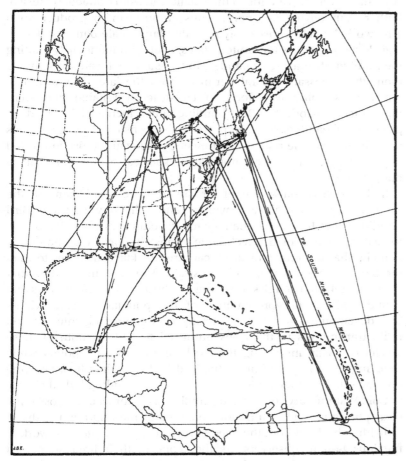

FLIGHTS MADE BY BANDED COMMON TERNS

The straight lines connect points of banding and recovery while the broken lines indicate theoretical routes that probably were followed. (Photograph from Biological Survey)

Figure 31.2 Migratory
paths mapped by the Bird
Banding Office on the basis
of volunteers' banding
reports in the 1920s.
F. C. Lincoln, 'Bird banding
in America', in *Annual
Report of the Board of
Regents for the Smithsonian
Institution Showing the
Operations, Expenditures,
and Condition of the
Institution for the Year
Ending June 30 1927*
(Washington, DC, 1928),
pp. 331–54, plate 4.

are heavily hunted, the proportion of recovered and reported bands
rarely exceeds 5 per cent. Even when a recovered band is reported to
the appropriate authority, the amount of data generated about migration is usually limited to two pairs of time and place, one corresponding to the initial capture and one to the recapture or killing. The further
the places or times between these two data points are removed from
each other, the more the uncertainties multiply. Moreover, the quality
of the data depends on the density and distribution of the networks of
banders and of people willing and able to report recovered bands.
Lincoln's discovery of continental flyways in the 1920s and 1930s was
contingent on the existence of a network of human observers

stretching from the Atlantic to the Pacific and from Canada to Mexico. Rather than relying on spontaneous enthusiasm to produce such a network, Lincoln consciously recruited large-scale banders at strategic locations across the United States, while simultaneously limiting the number of small-scale banders in less suitable locales by denying them the necessary federal permits. From the late 1930s onwards, partly to simplify its efforts to construct a reliable and well-distributed network, the Bird Banding Office increasingly relied on professional banders, often employees of state fish and game agencies or of federal wildlife refuges. Volunteers, particularly those interested in songbirds and other non-game species, were increasingly marginalised.[7]

If the professionalisation of bird banding resolved some of the problems of geographical coverage and reliability that the Bird Banding Office had faced when reliant on volunteers alone, it did little to overcome one of the technique's most important limitations – namely, that the bird had to be captured or killed in order to read the band number, which both constrained the amount of data that could be collected and risked altering the bird's behaviour in ways that would make it unrepresentative of the population as a whole. The development of wildlife tracking tags containing miniaturised radio transmitters in the 1960s promised to address both of these concerns, vastly increasing the total number of data points while reducing the number of times the bird had to be handled. Drawing on new sources of technological expertise and funding linked closely to Cold War military concerns, ornithologists embraced the possibility of acquiring virtually continuous data about the migration paths of individual birds without the need to recruit and manage a network of fractious human observers. As the scope of their interest in bird migration extended across the Pacific and around the world, including into regions where Americans were unlikely to succeed at building large networks of human observers, the appeal of a technique that did not depend on such networks increased even further.[8]

As with bird banding, the use of radio tracking to study migration helped to transform naturalists' relationship to place. Whereas the widespread adoption of bird banding had had the effect of increasing the amount of time that professional ornithologists spent in the office sorting and analysing data – even as it gave amateur and volunteer naturalists new reasons to get out into the field – radio tracking brought professional ornithologists back into the field and provided them an opportunity to embrace their identities as naturalists. To detect a radio-tagged bird, one had to be within range of its radio signal with a receiver tuned to the proper frequency. That range varied with the landscape – woods and hills, for instance, could attenuate or block radio signals – but it was rarely

more than a few miles. Without the proper equipment and techniques, which included an understanding of the way radio signals
propagated across different landscapes, the signal could neither be
detected nor localised. Amateurs without the proper equipment
and training were thus excluded from contributing meaningfully
to the generation of knowledge about radio-tagged birds, while
naturalists who did have such equipment and training could only
acquire data on bird movements by being present in the places
where their objects of study were found (Figure 31.3). Indeed, this
was one of the most appealing aspects of radio tracking for some
post-war ornithologists: it gave them an excuse to get away from
desks piled high with reports and statistics, including those generated by bird banding, and spend some time outdoors.[9]

By making it possible to repeatedly locate an individual bird on the
wing, radio tracking not only gave ornithologists a reason to go into the
field but also changed the nature of their engagement with place. For
those interested in studying migration, an inherently non-local phenomenon, tracking a bird in and around a single place was clearly
insufficient. In addition to developing an intimate familiarity with
particular places and the living beings encountered within them,
ornithologists thus needed to develop an ability to move quickly and
effectively across the landscape, just as the migratory birds they studied did. In the early 1960s, William Cochran, an engineer affiliated with
the Illinois Natural History Survey who designed some of the first
effective wildlife radio tags, pioneered the practice of conducting high-
speed chases after radio-tagged migrating birds in a station wagon that
he had customised with a rooftop antenna and radio receiver. For bird
banders, the ultimate value of banding a bird in the place they were
most familiar with depended on the labours of other observers, each
immersed in his or her own local place. For Cochran, in contrast, the
significance of his observations of a radio-tagged bird at one place
depended on his ability to move rapidly, accurately and safely from
place to place in pursuit of that individual. Success depended both on
knowledge of birds – how fast and far a bird of a given species, age and
sex was likely to fly in a day, for instance – and on knowledge of the
landscapes across which both radio waves and souped-up station
wagons could travel.[10]

Because of its cost and complexity, radio tracking helped to sharpen
the lines dividing professional ornithologists from amateur naturalists
and the interested public, even as the former embraced the technique
in part as a way of recovering aspects of naturalist practice that earlier
innovations had taken away from them. Whereas bird banders had
depended on the fact that some of the birds they banded would be
recaptured or killed by other people to render their emplaced labours
meaningful, researchers who used radio tagging hoped that no one

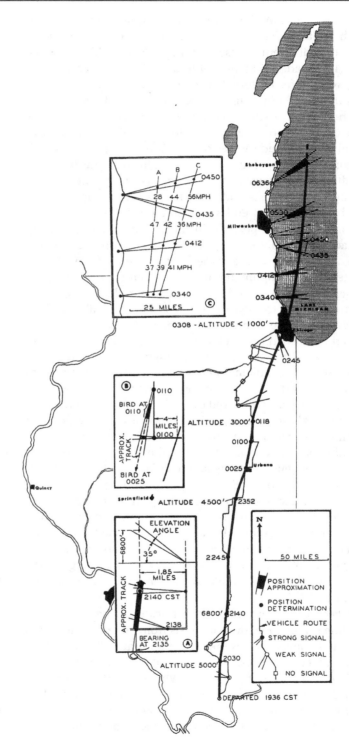

Figure 31.3 The flight path of a veery (*Catharus fuscescens*) radio tracked by automobile on 15 May 1969. W. W. Cochran, 'Long-distance tracking of birds', in S. R. Galler, K. Schmidt-Koenig, G. J. Jacobs and R. E. Belleville (eds.), *Animal Orientation and Navigation* (Washington, DC, 1972), pp. 39–59, figure 6, p. 49.

would interfere with the birds they were studying. They could hardly prevent a hunter from shooting a radio-tagged duck or goose, but they had little to gain from such a hunter's cooperation, since the single point of data that they would acquire from the report of a dead radio-tagged bird was trivial compared to the hundreds of additional data points they might have acquired if the bird had been allowed to live. Moreover, while there was still value in the information that a local naturalist might provide about the conditions under which a bird had been recaptured or killed, the relative value of that information shrank as radio tracking increased the total number of data points along the migration path. The increasing use of the technique thus gradually broadened the divide between those with intimate knowledge of particular places and the birds that lived in and passed through them and the radio-tracking researchers whose engagement with place was no less profound but was, by virtue of the technologies they used and the aims they pursued, becoming increasingly mobile and transitory.

Remote sensing: the naturalist in the virtual field

Implementing the kinds of mobile or transitory forms of emplacement associated with radio-tracking studies of bird migration was neither easy nor cheap. Birds continued to be hard to keep up with, even with the help of radio-tracking equipment, an automobile, good roads and a casual attitude toward traffic laws. More generally, while radio tracking made it possible to repeatedly locate an individual bird over the course of weeks, months or even years without having to recapture it each time, the method nonetheless remained extremely labour-intensive. Indeed, some desk-weary ornithologists embraced it precisely because it gave them a reason to spend time in the field. For studies of migration, however, the necessity of remaining within receiving range of the tag over the course of days or weeks demanded a level of commitment from the researcher beyond that of even the most assiduous bird bander. Moreover, some migratory paths, such as those that crossed mountain ranges or large bodies of water, were virtually impossible for researchers to follow in person. For these reasons, the kind of migration studies pioneered by Cochran in the 1960s remained rare, and radio tracking did not transform the study of migration as radically as some of its early proponents had hoped. The transitory mode of emplacement that it required of researchers was simply too demanding.[11]

The introduction of radio tags that could be automatically tracked by orbiting satellites promised to overcome these hurdles of labour and geography. First proposed by Cochran and others in the early 1960s, satellite-based animal tracking systems faced numerous technical challenges. With support from the Smithsonian Institution and the National Aeronautics and Space Administration, experiments with the satellite tracking of elk and other large animals began in earnest in the 1970s, but the increased power required to transmit a signal to an orbiting satellite and the complexity of the associated electronics meant that early tags were too heavy to be used to track birds of any species, let alone those capable of flight. However, the launching of the Argos satellite system for retrieving data from oceanographic and meteorological buoys and balloons in the late 1970s, along with the development of increasingly power-efficient, reliable and lightweight satellite tags over the course of the 1980s, eventually made possible the widespread use of the technique to study bird migration. Even then, the tags continued to be too heavy for many smaller bird species. Scientists succeeded, however, in using Argos tags and data loggers – electronic tags that recorded location and environmental data for later recovery – to produce detailed maps of the movements of individual albatrosses, hawks, geese, storks, swans, cranes and other large birds on continental and global scales and in areas such as the high seas where human observers could operate only with great difficulty. They did so, moreover, without requiring researchers to spend any time whatsoever in the field, except that required to capture the birds and attach the tags.[12]

The development of satellite tags that were small and light enough to be used on flying birds thus enabled a new mode of emplacement that differed both from that of bird banding and from that of conventional radio tracking. There are clear similarities between satellite-based animal tracking and radio tracking, including the fact that both involve attaching radio transmitters to animals and that both widen the divide between amateur birdwatchers and professional ornithologists by demanding high levels of technical expertise and financial resources. In both cases, researchers are required to go out into the field to capture birds and attach the tags; whenever possible, they also return to the field to retrieve the tags, which in the case of satellite-based tracking can be very expensive. To do so successfully, they must develop a certain amount of emplaced knowledge of how to capture the birds and attach the tags in such a way as not to significantly injure or impair them. There are, however, significant differences in the forms of emplaced research practice that these two methods enable. Whereas conventional radio tracking requires researchers to spend extended time in the field after the tagged bird has been released, as well as – if they are interested in migration – to develop demanding new practices of transitory or mobile emplacement, satellite-based tracking requires the researcher to return to the office as soon as the tagging is done. From there, the movements of multiple tracked animals can be monitored from afar in a way that is impossible in the field. With the introduction of satellite-based tracking, the transitory emplacement of the researcher chasing after the radio-tagged bird was thus replaced by a form of research in which the researcher's embodied, emplaced labour in the field consisted not primarily in observing and collecting data but rather in installing and maintaining devices of remote surveillance.

At the same time, the office became the site of a kind of virtual field – that is, an information environment in which the researcher could not only monitor the movements of tagged birds in near real time but also bring those movements into relation with environmental data collected through satellite imaging and other forms of remote sensing. After the National Aeronautics and Space Administration launched its Landsat satellite-imaging service in 1972, biologists and ecologists developed new ways of monitoring ecological processes remotely on the basis of multispectral satellite images. Over the course of the 1980s and 1990s, the amount and variety of remote-sensing data available for ecological research expanded dramatically, as did the sophistication of the models used to analyse them. Combined with detailed animal-movement data provided through the Argos system or through GPS-based data loggers, they made it possible to situate an animal's movements in relation to the surrounding landscape without ever having to set foot on or fly over that landscape oneself (see Plate 19). Together, these technologies allowed the office-bound ornithologist to learn more than even the most observant and dedicated field naturalist about the movements of an individual bird and the environmental context through which it moved, including weather, day–night cycles and vegetation cover. What could be encountered and studied in the virtual field was constrained by the kinds of sensors and models that were available, and there were some kinds of data that could still be collected only through fieldwork. For studying global patterns of bird migration, however, the advantages of the virtual field were difficult to match.[13]

In the late twentieth century, ornithology was not alone in its embrace of the virtual field and the new relationships to place that it enabled. Across environmental sciences such as geology, oceanography, climatology and ecology, researchers increasingly relied on global infrastructures to collect and transmit data from particular places to the laboratory or the office. These included satellite-based remote sensing as well as networks of automated thermometers, stream gauges, hydrophones, location trackers and other instruments. While a few researchers relied solely on the data produced by automatic instruments that were designed, constructed and installed by others, most continued to go out into the field to ensure that the devices were collecting precisely the data they needed and to collect data themselves when such devices could not do so automatically. Like many naturalists before them, they continued to see engagement with place as essential to their work, warranting the claims they made by referencing data collected in particular places and situating their explanations in relation to those places rather than seeking to establish universal principles applicable everywhere and at all times. But the way they engaged with those places had changed. They increasingly found that they could acquire more data about varied places and the living beings that lived in and moved through them by staying in the office than by spending time in the field. Whether this increase in data led inevitably to an increase in knowledge is another question, as is whether the researchers who embraced the possibilities of the virtual field should still be seen as naturalists. What is clear is that the significance of place and of embodied, emplaced labour to the production of natural-historical knowledge has changed and will continue to change in relation to the available technologies and social relations of data collection.

The power of emplacement

Claims to emplacement and situatedness, no less than claims to placelessness and the 'view from nowhere', have become means of legitimising forms of knowledge that inform consequential decisions about how we relate to our environments and to the other living beings found within them. That is one of the reasons why it matters how we understand the nature and limits of emplacement in the practice of natural history and the image of the naturalist as they have changed over time. When naturalists or historians of natural history claim that embodied experience of particular places is essential to the formation of a naturalist and the production of natural-historical knowledge, they are identifying and celebrating an important value that has often been overshadowed in accounts of science that emphasise wide-ranging theories and systems of classification. At the same time, they are also, intentionally or not, making a negative claim about the incapacity of people without this particular kind of emplaced, embodied experience to produce natural-historical knowledge and to participate fully in deliberations over how that knowledge should be used. Given the ascendance over the past two centuries of universalist sciences with ambitions that were imperial in every sense of the term, the value of defending embodied knowledge of particular places against its detractors is clear. Indeed, one of the virtues of the history of natural history is that it reminds us how essential this form of knowledge was and remains. Mounting such a defence in a way that depends on decontextualised and transhistorical understandings of place and emplacement, however, risks simply replacing imperialist universalism with nostalgic particularism. More durable defences require more subtle distinctions. As the history of the study of bird migration over the twentieth century suggests, technologies and modes of emplacement do not stay still any more than birds do, and there is much to be learned from tracing their paths through time.[14]

Further reading

Barrow, M. V., Jr, *A Passion for Birds: American Ornithology after Audubon* (Princeton, 1999).

Benson, E. S., *Wired Wilderness: Technologies of Tracking and the Making of Modern Wildlife* (Baltimore, 2010).

De Bont, R., *Stations in the Field: A History of Place-Based Animal Research, 1870–1930* (Chicago, 2015).

Gabrys, J., *Program Earth: Environmental Sensing Technology and the Making of a Computational Planet* (Minneapolis, 2016).

Kohler, R. E., *Landscapes and Labscapes: Exploring the Lab–Field Border in Biology* (Chicago, 2002).

Vetter, J., *Field Life: Science in the American West during the Railroad Era* (Pittsburgh, 2016).

Wilson, R. M., *Seeking Refuge: Birds and Landscapes of the Pacific Flyway* (Seattle, 2010).

32   Wild visions

Imagine a journey across the surface of the Earth. Fly over its rain-forests, deserts, mountains and oceans. Consider its poison dart frogs and blind mole rats. See the fossils of extinct ammonites encased in limestone, the flash of turquoise on the underside of a hummingbird's wing and the tiny embryo of a red kangaroo developing inside its mother's pouch. Compare the prehensile tails of marmosets and howler monkeys as they swing between the treetops, and the dexterity of chimpanzees and gorillas as they fashion tools on the forest floor. Now consider how it is you have come to know these diverse animals and environments, and the fragile interconnections that bind them together.

This is the world of *Life on Earth* (1979), a thirteen-part television series on the evolution of life, produced by the Natural History Unit of the British Broadcasting Corporation (BBC) and written and presented by Sir David Attenborough. This series did not so much visualise wild animals in a new way as present them in an entirely new light. For the first time, viewers were shown a global vision of wildlife and nature. To audiences in 1979 it must have seemed as if the image of the blue planet, a fragile Earth floating in deep space, first captured during the Apollo space missions of the 1960s, was suddenly endowed with new depth and clarity. *Life on Earth* enabled audiences to see a broad survey of animal life and to feel connected to a global ecology. It was the first BBC landmark wildlife series, a genre that continues today with *David Attenborough's Great Barrier Reef* (2016) and *Planet Earth II* (2016) and has changed the way millions of people in the industrialised world look at wild animals.[1] Rather than focusing on a particular species or exploring the ecology of a particular environment, as many wildlife documentaries had done before, landmarks had the space to develop and dramatise complex scientific ideas, weaving together footage of different species and environments to create a more elaborate narrative about disciplines such as evolutionary biology, ecology or ethology.

In this chapter, I explore the discursive and ideological networks behind the extraordinary global vision that landmark wildlife series made possible. I trace the prehistory of these series in the storytelling

518

practices pioneered by Attenborough in the unique context of public service broadcasting in Britain, examine why Attenborough's landmarks have remained fixated on the narrow scientific paradigm of natural history, and analyse how more complex explorations of environmental issues have played out on screen in recent years. The evolution of these series over nearly 40 years offers a productive lens through which to examine the commodification of wildlife television and the enduring popular appeal of the landmark series. It also highlights how, as these series were transformed into international commodities, they largely glossed over the complexities of environmental politics to offer increasingly spectacular and technologised visions of nature.

Storytelling

When the television critic Clive James reviewed the first two episodes of *Life on Earth* in January 1979, he emphasised that its success depended as much on the pictures as on the words. 'Attenborough has all the resources of technology at his disposal,' he wrote, 'but the chief attribute he brings to this titanic subject is his own gift for the simple statement that makes complexity intelligible.'[2] What set Attenborough apart from previous science presenters was his ability to distil complex scientific concepts into meaningful and easily comprehensible narratives, without over-simplifying or dampening his innate enthusiasm for the natural world. 'Here is proof', argued James, with his characteristic humour, 'that someone can be passionate about science and still look and sound like an ordinary human being.'[3] Attenborough's skill as a presenter and narrator was honed in the early 1950s when the BBC's television programmes were, of necessity, transmitted live. This played a key role in his ability to remember long, carefully crafted narratives and deliver them in a seemingly effortless way. By the time *Life on Earth* was first broadcast in 1979, he had already totted up over a quarter of a century of experience making television programmes on subjects as diverse as folk music, tribal art, Japanese culture and, of course, natural history. Nonetheless, it was his experience as a commissioner that was instrumental to the birth of landmark television series.

Attenborough joined the BBC's television Talks department in December 1952 as an assistant producer. His first presenting role was in the long running *Zoo Quest* (1954–63) series, which accompanied collecting expeditions for the London Zoo to remote and exotic locations like Sierra Leone, Guiana and Madagascar. Each series took the form of a quest to find and potentially capture a species of particular significance for the Zoo's collection, or to record a rarely seen animal. In the absence of a more visually appealing subject, the first series to Sierra Leone had gone in search of *Picatharthes gymnocephalus*, commonly known as the bald-headed rock crow. These were the days when even a glimpse of a little-known animal, referred to by its Latin name and broadcast in black and white, was enough to hold the attention of viewers.

The programmes in early *Zoo Quest* series were rudimentary. The format was driven more by the limitations of available filming and broadcast technologies than anything else. There was no synchronous sound to accompany the film clips shot on location, as this was not possible with the lightweight and manoeuvrable 16 mm cameras favoured by wildlife cinematographers. Powerful long-focus lenses were also yet to be developed, so any close-ups of the animals, including the much sought after *Picatharthes*, whose appearance was saved for the final episode, had to be provided in the studio. There, they were flooded with lights and filmed from only a few feet away. The format relied on Attenborough's ability to weave together these disparate clips with live studio sequences in which he pointed out interesting details about the anatomy or behaviour of the captive animals. Unscripted elements and the unpredictable actions of the animals subjected to this treatment provided moments of light relief. As Attenborough points out, the possibility of 'a bite or two maybe, perhaps an escape' was part of the appeal of the programmes.[4] Nevertheless, the series was the most popular wildlife programme of its time and established Attenborough's career as a presenter.

Attenborough had not been the first choice to present the series. Jack Lester, the leader of the expedition and the curator of the Zoo's reptile house, was originally intended to be the presenter, but he became ill after the first programme went to air and Attenborough was asked to take over. Lester recovered in time to accompany the second *Zoo Quest* expedition to British Guiana (now Guyana), but he took ill again on his return to London, and died soon after of an unidentified tropical illness in 1955. Lester's illness and premature death, in effect, forced Attenborough out of the control gallery, where live programmes were mixed, in front of the cameras, where he was to remain.

At the time, the BBC had a monopoly on wildlife programming thanks to tight regulations around public service broadcasting. In 1955, the BBC was joined by the first commercial television network in Britain, Independent Television (ITV), but ITV's first wildlife documentary programme, the long-running *Survival* series, only began in 1961. So when *Zoo Quest* began in 1954, natural history programming in Britain was exclusively produced and broadcast by the BBC.

Wildlife programming evolved from several offshoots within the BBC. By the mid 1950s it was split between the BBC Talks department in London and BBC West Region in Bristol, where Desmond Hawkins had pioneered a long tradition of natural history programming. In 1957, in line with the BBC's policy of encouraging specialisation in its regional centres, it was decided that the BBC Natural History Unit (NHU) should be established in Bristol. Attenborough continued to produce and present programmes from London, where he established the Travel and Exploration Unit within the Talks department. In 1965 he accepted an offer to become the controller of BBC2, the broadcaster's second television network. He was eventually promoted to the director of programmes, a position he held from 1969 to 1972, where he oversaw programming for both the BBC's television networks.

During his tenure as the controller of BBC2, Attenborough commissioned the first BBC landmark series, *Civilisation: A Personal View by Lord Clark* (1969), in which Sir Kenneth Clark explained the development of Western Europe through its art, and its successor series, *The Ascent of Man* (1973), in which Dr Jacob Bronowski explored imaginative leaps in the history of science. *Civilisation*, as its monumental title indicates, was unashamedly highbrow and patrician in its outlook. Armed with a budget unprecedented in the BBC's history, Attenborough was tasked with devising a spectacular series that would celebrate the arrival of colour television in Britain. His first choice as presenter was Sir Kenneth Clark, who, as an eminent art historian and the director of the National Gallery in London, possessed the necessary gravitas to give the series intellectual credibility. Rather than simply narrating the scripts, Clark took viewers on an intellectual journey, guiding them across time and space. It was this innovation, in which Clark based his arguments on 'things seen – towns, bridges, cloisters, cathedrals, palaces', that gave the series its distinctive edge.[5] The approach of taking viewers on a physical as well as an intellectual journey would later be a key feature of *Life on Earth*. Attenborough also ensured that *Civilisation* was designed with a number of in-built features to make it more attractive to US broadcasters. It was shot on 35 mm film, the gauge used in the cinema, in an effort to instil it with a lush, feature-film quality, while each of its thirteen episodes was fifty minutes long, a length that coincided with an American TV hour with space for commercials.

Landmarks are, by their very nature, collaborative productions, but the fact that Attenborough commissioned the first landmark series, and then went on to become the chief protagonist of this style of programming, is significant. However, unlike the presenters of previous landmark series, who were selected on the basis of their academic authority in a given field, Attenborough was no academic expert. He had earlier studied zoology and geology at the University of Cambridge, obtaining an undergraduate degree in natural sciences, but decided against studying for a doctorate in zoology as he found the

field to be too 'laboratory-bound'.[6] As Attenborough himself acknowledged, he was a 'storyteller' rather than a 'scientist': 'I see myself as
a storyteller – telling a traveller's tale. I am clearly not a scientist
making original observations, however much I wish I were.'[7] His
expertise lay elsewhere: in his experience of producing and presenting
television programmes and in the commissioning culture of the BBC
he had helped to craft.

During his time in BBC management, Attenborough made it
a stipulation of his contract that he 'was able to make a programme
of some kind every eighteen months or so just to keep in touch with the
latest technologies and techniques'.[8] The last programme he worked
on before resigning as director of programmes in December 1972 was
Eastwards with Attenborough (1973). This series of six 30-minute episodes, made by the NHU and filmed in Sumatra and Borneo, was
a direct precursor to *Life on Earth*. Encouraged by the series producer,
Richard Brock, to include his subjective reactions to some of the
dramatic locations featured in the programmes, Attenborough honed
the touches of breathless but understated excitement that would
become characteristic of his style of presentation in *Life on Earth*.
In particular, he recounts a 2-minute sequence in which he spoke
about the volcanic eruptions at Anak Krakatoa, listing many facts
and figures without interruption while volcanic ash rained down
upon him, as the point at which he came of age as a presenter.[9]
Viewers of this series may well recall a more memorable sequence,
in which Attenborough, primed to react with eloquence to his surroundings, delivered a speech about the sonar system of bats while
standing on top of a giant pile of guano within a cave. Choking on the
fumes, as the bats screeched and wheeled around the roof of the cave
and a carpet of cockroaches clambered across his feet, his narration
was measured and precise.

Throughout his career, Attenborough was able not just to overcome
the limitations of early filming technology, but to harness the skills that
live broadcasting had demanded from its presenters and transform
these constraints into one of his chief assets (Figure 32.1). During the
production of *Civilisation*, Kenneth Clark had used an autocue.
Attenborough required no such props. He was able to move effortlessly
from his subjective reaction to a particular environment, using this
moment of connection to propel his argument forward, linking his
various observations together into a broader thesis about evolutionary
biology. By the late 1970s, when *Life on Earth* went into production,
filming technologies had improved dramatically and were better suited to filming wild animals in their natural habitats. But it was
Attenborough's skill as a storyteller, forged in the era of live broadcasting, which provided the thread that held the wider scientific narrative of the series together.

Technology and science

Life on Earth wove together footage of over 650 different species from around the globe. Filmed in 39 different countries, it took 3 years to produce and the film crews involved in the project covered more than a million miles. Each sequence was carefully planned in consultation with biologists and other scientific experts from over 500 universities and scientific institutions.[10] The sheer scale of the series was jaw dropping. It was the most expensive natural history series to date, with a budget of £1 million, excluding the cost of BBC staff.[11] Its style of presentation was also revolutionary. Attenborough would begin explaining the adaptations or behaviour of a particular species in one location, before cutting to another location on the other side of the world to complete his illustration. Perhaps more than any other wildlife series, *Life on Earth* was shaped by the expanding visual possibilities of technology.

The contrast between *Life on Earth* and Attenborough's first television programmes was stark. While trying to film a sequence for *Zoo Quest* in a rainforest in Sierra Leone in 1954, cameraman Charles Lagus had lamented the impossibility of filming beneath the shaded canopy. 'The only way we can get enough light to film by here', he had

Figure 32.1 David Attenborough takes centre stage during filming for the BBC series *Life in the Freezer* (1993) on location in South Georgia. On the left is Alastair Fothergill, a producer and long-time collaborator on Attenborough's series. © Ben Osborne/naturepl. com.

remarked, 'is to chop down the trees.'[12] Produced just 25 years later, *Life on Earth* benefitted from improved 35 mm film stock that enabled it to show some of the sharpest and most colourful wildlife footage to date. By the 1970s, cameras were much smaller and more manoeuvrable, zoom lenses made it possible to film wild animals from greater distances, and macro-photography meant it was possible to film organisms too small to see with the naked eye. The series also employed highly inventive filming techniques, honed in a wildlife television industry that had rapidly professionalised around the NHU in Britain in the 1960s and 1970s, incorporating independent and commercial production companies and specialist wildlife cinematographers. But despite its technological innovations, the series remained fixated on a very particular brand of science.

Science had always been at the centre of the BBC's wildlife programming. From its origins, the BBC Natural History Unit had fostered a close association between the scientific community and the overlapping community of amateur naturalists who contributed much of the source material for its early programmes. In a report on the first 5 years of the unit's operation in 1962, Desmond Hawkins argued that science should be the driving force behind the BBC's natural history programming:

The spirit of scientific enquiry must have pride of place. In handling this subject we expose ourselves to the critical scrutiny of scientists, and their approval is an important endorsement. Moreover, it is their work that throws up the ideas and instances and controversies from which programmes are made. We look to them as contributors, as source material and as elite opinion on our efforts. In short, we need their good will.[13]

Early NHU programmes, such as *Look* (1955–65), regularly featured interviews with biologists and other scientific experts, and scientists were often directly involved in the making of programmes produced and commissioned by the NHU. *Signals for Survival* (1968), for example, which explained the complex communication system used by a colony of lesser black-backed gulls, was produced by renowned ethologist Niko Tinbergen.

However, as the wildlife television genre professionalised and more polished programmes gradually replaced earlier presenter-led formats, the direct involvement of scientists in the production of programmes and as on-screen experts began to decline. In the late 1960s, the blue-chip format, which depicted wild animals in a pristine wilderness devoid of people, became the industry standard. This was primarily because it was a format that allowed footage filmed at different times and places to be more easily edited together. It also had the added advantage of allowing programmes to be more readily adapted for international sale, as with no on-screen presenters they could be redubbed into different languages without the need for re-editing. Blue-chip programmes produced by the NHU or commissioned by the BBC for strands such as *The World About Us* (1967–83), *Wildlife on One* (1977–2005) and *The Natural World* (1983–present), many of them narrated by David Attenborough, have been sold internationally since the late 1960s. The BBC's blue-chip programmes retained the spectacular appeal of commercial formats, but their key distinction was that they placed science at the centre of more dramatic sequences. The BBC's preference for scientific narratives, as opposed to anthropomorphic narratives like those favoured by Disney's *True-Life Adventure* films (1948–60) for instance, allowed it to craft a unique niche in the international television market.

The dominance of the blue-chip format, however, had a profound impact on the way that science was communicated in the majority of the BBC's natural history programmes. More complex depictions of science as a dynamic and contested field, including the 'controversies' that Hawkins valorised as a key component of 'the spirit of scientific enquiry', were suppressed in favour of spectacular footage and largely totalising narratives that left little room for debate. In keeping with the narrow paradigm of zoology and natural history, the BBC's wildlife programmes have, since the late 1960s, remained fixated on 'a very particular brand of science: that which is already proven and beyond doubt. Safe science.'[14] This view accords with Michael Jeffries' assessment that 'the science of natural history not only occupies its own broadcasting niche; it works to a different paradigm'. Jeffries contrasts natural history programmes with science documentaries, particularly those in the BBC's flagship science strand *Horizon* (1964–present), which 'represent the world (and the rest of the universe) as changeable, challenging, contingent'.[15] Attenborough's landmarks differed from the BBC's blue-chip programmes in their construction of a broader scientific narrative, linking the episodes in each series together, and in their use of Attenborough as an on-screen presenter. But in keeping with the BBC's blue-chip programmes, they also retained a relatively narrow focus on evolutionary biology, ecology and later ethology. The science in Attenborough's series remained 'safe'. His narration offered viewers concise explanations, illustrating points of scientific consensus, rather than highlighting differing opinions or offering competing explanations from scientists whose work, in any case, was entirely hidden from view.

Attenborough was determined that the first landmark wildlife series should focus on evolution. Before he resigned from BBC management, he was approached by Christopher Parsons, a senior producer in the NHU, about the idea of a natural history series similar to *Civilisation* and *The Ascent of Man*. 'I was quite sure how it should go', Attenborough recalled, 'I would start at the beginning of life, trace the long story of its evolution as both plants and animals spread from the sea and onto the land in increasingly complex forms.'[16] In keeping with the systematising intent of the great naturalists like Buffon, Cuvier, Linnaeus and Darwin, *Life on Earth* used living species to chart the evolution of life chronologically, focusing on classification and anatomical variation as a means of drawing associations between different species. In this respect, the global vision at the centre of the series strove for breadth, featuring more than 650 different species, rather than depth. This approach initially drew the opprobrium of other producers within the NHU, who criticised the series as 'nineteenth century natural history at its most boringly conventional'.[17] They were critical of the fact that the first landmark wildlife series seemed to disregard the latest scientific discoveries in fields like genetics and ethology.

If *Life on Earth* were remade today, there is no question that it would have required a different structure. Contemporary viewers would not be expected to wait until the ninth episode, 'The Rise of Mammals', before they witnessed detailed portraits of charismatic mammals. But in 1979, Attenborough's chronological focus on evolution won out. The series began in South American forests, the same forests that had so enthralled a young Charles Darwin in 1832 when he accompanied the voyage of the *Beagle*. As Attenborough explained in his opening narration: 'There are some four million different kinds of plants and animals in the world, four million different solutions to the problem of staying alive. This is the story of how a few of them came to be as they are.' It was Attenborough's skill as a presenter that guided the audiences through this sometimes disparate visual array, as he explained how 'all these varied creatures and plants form one complex mosaic with immense variety within it'.[18]

There was another reason why *Life on Earth* focused on evolutionary biology rather than the emerging science of ethology. Although film technologies had advanced immeasurably since Attenborough first began making wildlife television programmes, the cameras and techniques available in the mid 1970s were simply not up to the task of capturing detailed sequences of animal behaviour. It would be another decade before advances in film stock and video, capable of filming in conditions of very low light, made it possible to film complex behavioural sequences at night, when most mammals are active.

The key technical innovation of *Life on Earth*, as Attenborough argues, was the coordination of communication and travel on a worldwide basis:

I believe that any of the programmes I have done, at any rate, have been to some considerable extent driven by technology. You could actually argue that *Life on Earth* was driven by technology because a decade before *Life on Earth* –

before we did it – you could not have done it. Simply because airlines were no good … By the time we made *Life on Earth* airlines had enmeshed the world and you could say, I wish to be in the upper Amazon on January 7th and you could be pretty sure, I mean we never failed, that you would be.[19]

It was this ability to plan and time filming expeditions, in consultation with scientists around the world, which enabled the production team to capture some of the most remarkable scenes in the series. In the closing sequence of the second episode on 'Building bodies', for example, Attenborough is shown strolling along a beach in Delaware Bay on the east coast of the United States, at the precise moment when an army of primitive looking horseshoe crabs emerges from the surf to mate and lay eggs in the sand. Capturing this one scene involved numerous communications between BBC producers and scientists, and many research and location scouting trips. It was based on a prediction by biologist Carl Shuster that the horseshoe crabs would arrive en masse on 1 June 1977. Fortunately, Shuster's prediction turned out to be correct. But a six-week filming schedule had to be planned around this one event.[20] Scenes such as this demonstrated that the BBC was prepared to take risks, gambling its budgets on expensive sequences that might not pay off, with deference to scientists and their expert opinions. Other sequences in *Life on Earth*, such as a male Darwin's frog disgorging froglets from its mouth, were less spectacular but had never before been filmed. This single shot, painstakingly filmed by Roger Jackman of Oxford Scientific Films, lasted only a second and the twenty-four frames had to be optically stretched so that it could be observed in the series.[21] It was on this combination of cutting-edge technology and filming firsts, together with the drama and spectacular appeal of sequences depicting large-scale natural events, that *Life on Earth* was to stake its claim as a popular yet inherently scientific series (even if the science was decidedly 'nineteenth century').

Attenborough's skill as a presenter ensured that his broader narrative about evolutionary biology was not subsumed by spectacular footage. Throughout the series, he rarely interacts with wild animals, but the one sequence in which he directly engages with animals perfectly illustrates his ability to telescope from a subjective moment to a broader scientific argument (see Plate 20). Sitting in the Rwandan rainforest, with a silverback gorilla a few feet away, Attenborough remarks:

There is more meaning and mutual understanding in exchanging a glance with a gorilla than any other animal I know. We are so similar. Their sight, their hearing, their sense of smell are so similar to ours that we see the world in the same way as they do. They live in the same sort of social groups with largely permanent family relationships. They walk around on the ground as we do, though they are immensely more powerful than we are. So if ever there was a possibility of escaping the human condition and living imaginatively in another creature's world it must be with the gorilla.[22]

It is easy to dismiss his remarks as merely anthropomorphic projection, but Attenborough's shared glance with a gorilla demonstrates his expertise as a scientific storyteller. He had originally intended to deliver a speech about the evolution of the opposable thumb, but, reacting to the presence of the gorilla, he spoke instead of the connection between humans and other primates in an entirely ad-libbed speech.[23] Jonathan Burt suggests that this sequence reminds us that, as viewers, 'we are looking from within nature and not at nature'.[24] Rather than seeing landmark series as replacements for reality, icons of our alienation from the lives of other animals, he observes, 'they seem more like the point of entry for our engagement with the natural world: an active moral gaze made possible, even structured, by the technology of modernity'.[25]

In the United Kingdom, the success of *Life on Earth* was unprecedented. The transmission of the first episode on BBC2 in January 1979 attracted an audience of around 9 million. By the end of the series, combined viewing figures for the twice-weekly broadcasts were reaching 15 million.[26] The global audience for the series is impossible to calculate, but estimates suggest that over 500 million viewers have watched it. The popularity of *Life on Earth* paved the way for the production of new landmark wildlife series. The first three full-length series seemed to cover the history of the Earth in its entirety. *Life on Earth* (1979) told the story of evolution, *The Living Planet* (1984) examined the world's ecosystems and *The Trials of Life* (1990) looked at animal behaviour. As Attenborough's landmarks continued to evolve, it was advances in filming technologies rather than science that ultimately provided the impetus for each new series (Figure 32.2). New technologies allowed old topics to be explored in new ways. These series often incorporated new scientific discoveries and filming firsts, or the ability to capture footage of animals or behaviour that had never before been filmed, like the beguiling sequence of a rare snow leopard hunting in the remote Karakoram ranges in *Planet Earth* (2006), filmed with the help of newly developed high-definition cameras. But the series rarely strayed from the paradigm of natural history.

Environmentalism

There is a particular moment during filming for *Life on Earth* that Attenborough recalls with vivid detail. As he stepped out of the Pacific Ocean on the edge of the Great Barrier Reef and the helicopter from which he was being filmed zoomed away, he was overtaken by

a sudden emotion. In that moment he saw for the first time that 'it was possible for a single unit, and a single man, and a single viewer, to get a comprehensive view of the planet – as a planet'.[27] *Life on Earth* made it possible for audiences to conceive of the world as a global and inter-connected ecology. As Attenborough argues, the series, as a result of its global viewership, undoubtedly contributed to the growth of environmental consciousness: 'I think the environmental movement does owe quite a lot to television – that people are aware of what's happening in Africa. I mean, actually you've only got to look back to my father or my grandfather, who didn't even know – the world was limited to a few miles.'[28] But the global vision at the heart of *Life on Earth* and Attenborough's subsequent series, underpinned by a combination of technological prowess and a new style of scientific storytelling, also makes them compelling examples of the ambivalence that structures the depiction of wild animals in visual culture. With each successive technological innovation, landmark wildlife series offered viewers ever more intimate and spectacular visions of wild animals and natural environments. These were things that most people had never seen before, provided at a time when the last relics of wilderness – those environments untouched by the encroaching urban sprawl of large cities, the deforestation of arable land for crops or the patchwork of

Figure 32.2 Filmed using infrared cameras on location in northern Botswana for the BBC series *Planet Earth* (2006), a pride of African lions feed on the carcass of an African elephant. Advances in infrared filming technologies enabled depictions of nocturnal mammalian behaviour to be filmed in intricate detail for this series. © Ben Osborne/naturepl.com

roads that increasingly carved up remote locations – were dwindling. Yet in common with the BBC's earlier blue-chip wildlife programmes, they generally excluded cars, roads and people (with the exception of Attenborough himself), while his narration avoided drawing attention to the broader impacts of human culture and industry. When environmental issues were mentioned at all, they were almost invariably corralled into the final scene of the last episode, in which Attenborough turned to the camera, locking the viewer with an imploring gaze as he spoke of the need to conserve the environment.

Largely stripped of environmental concern, save for Attenborough's final statements, landmarks continued to depict the natural world as abundant and diverse even as successive waves of environmental problems from mass extinctions to human-induced climate change gathered force. With the exception of *The Living Planet* (1984), which devoted a whole episode to human impacts on the environment, it was only after 2000 that the broader environmental consequences of a rapidly expanding human population, overdevelopment, unrestrained resource consumption and climate change were finally given the landmark treatment. The turning point came with *The State of the Planet* (2000), a three-part series entirely financed by the BBC, which broke with generic convention to feature on-screen interviews with scientists and environmentalists together with depictions of environmental destruction. This series touched briefly on global warming and climate change, but stopped short of definitively stating the extent of the threat. In his narration, for example, Attenborough was careful to include caveats: 'There is one kind of pollution, however, that *could* have worldwide consequences – that is the global warming that results from human activities that pump carbon dioxide into the atmosphere.' In *Frozen Planet* (2011) and *David Attenborough's Great Barrier Reef* (2016), Attenborough finally talked about the realities of climate change without including caveats about its devastating consequences.

The reasons behind the absence of environmental issues in earlier landmark series are complex, but they can be summed up by their continued focus on the narrow paradigm of natural history, which, in allowing spectacular wildlife footage to be given maximum airtime, helped to attract the international co-production funding necessary to support these series. But, while a focus on natural history proved to be economically expedient, it ultimately ensured that controversial issues were suppressed.[29] It was only as environmental issues gained widespread popular acceptance, and were subject to uncontroversial media treatment, that they were finally explored in Attenborough's landmarks. Yet as the multimillion pound budgets for these series have increased, so too has the need for production investment and the pressures from the BBC's mostly American co-production partners. This has led to tussles over editorial control. In 2011, for example, it was rumoured that the

Discovery Channel, the principal co-producers of *Frozen Planet*, were planning not to broadcast the final episode of the series in the United States. Although Discovery later backtracked on this decision following media pressure, their refusal to broadcast this episode would have meant that American viewers were not exposed to Attenborough's nuanced take on climate change.[30]

As the longest-serving television presenter in the world, Attenborough's legacy is to have communicated the variety and wonder of life to countless millions of viewers (Figure 32.3). The global vision inherent in his series, befitting its roots in Victorian natural history, gave rise to an immensely diverse, but still orderly and ultimately classifiable vision of the natural world. Attenborough's landmarks embody both the potential and the tensions of a global vision that is at once both totalising, enabling viewers to conceive of our world as a whole, and partial, leaving out key aspects that might disrupt the clarity and simplicity of this spectacular global vision. Over the last decade as Attenborough has cut back on his role as an on-screen presenter, landmark wildlife series have nevertheless continued to evolve. They will no doubt survive, albeit in new forms, in a post-Attenborough world.

Figure 32.3 Sir David Attenborough observing a male Oustalet's chameleon on location in Madagascar during filming for the BBC series *Life in Cold Blood* (2006). © Miles Barton/naturepl.com

Further reading

Bousé, D., *Wildlife Films* (Philadelphia, 2000).

Brockington, D., *Celebrity and the Environment: Fame, Wealth and Power in Conservation* (London, 1999).

Burt, J., *Animals in Film* (London, 2002).

Chris, C., *Watching Wildlife* (Minneapolis, 2006).

Darley, A., 'Simulating natural history: walking with dinosaurs as hyper-real edutainment', *Science as Culture*, 12:2 (2004), pp. 227–56.

Mitman, G., *Reel Nature: America's Romance with Wildlife on Film* (Cambridge, MA, 1999).

Richards, M., 'Greening wildlife documentary', in L. Lester and B. Hutchins (eds.), *Environmental Conflict and the Media* (New York, 2013), pp. 171–85.

Epilogue

HELEN ANNE CURRY AND JAMES A.
SECORD

Natural history and its histories in the twenty-first century

In September 2000, the ecologists David Wilcove and Thomas Eisner predicted a looming extinction. Such events could hardly be considered surprising at the turn of the twenty-first century. Reports from scientists and conservation organisations now regularly suggest that the current rate of extinction is one hundred or more times higher than the expected 'background rate' – that is, the rate we would see if humans weren't also around.[1] But Wilcove and Eisner's threatened form of life was indeed an unusual object of conservation concern: they mourned the disappearance of natural history. 'In schools and universities, in government agencies and research foundations, natural history has fallen out of favor', they claimed, bemoaning the loss of everything from science field trips for school children to prerequisite knowledge in natural history for postgraduate study in ecology.[2]

Wilcove and Eisner were not the only scientists to worry about what they called 'the impending extinction of natural history'. The past three decades have seen many expressions of concern about the survival of the knowledge and practices long associated with natural historical work. Researchers have called attention to the declining status of and funding for systematics and taxonomy, the substitution of computer modelling and remote sensing for field observation in ecology and conservation or evolutionary biology, the slow ebb of university course offerings in organismal biology, the neglect of museum collections, and more.[3]

For these researchers, the consequences of such shifts have appeared stark. Many claim that the overall quality of biological knowledge diminishes as certain ways of knowing nature become less common. 'Without years of bug-bitten trudging through hollows and bogs, how can a biologist be expected to separate biological truth from computer fabrication?' wondered the conservation biologist Reed Noss.[4] Poorer knowledge in turn means poorer conservation decision-making and more challenges to stewarding resources for the future and protecting wildlife and wild places. For those working in natural history museums, the loss of collections means missed opportunities to mine these for knowledge about the past that may be useful in the future, including the changing distribution of species, the circulation of diseases, and even historical contamination events. Other observers see a more generalised decline in appreciation of the natural world as the real crisis. So long as 'the demise of natural history goes unnoticed', predicted Wilcove and Eisner, 'future generations of schoolchildren will spend even more time indoors, clicking away on their plastic mice, happily viewing images of the very plants and animals they could be finding in the woods, streams, and meadows they no longer visit'.[5]

Natural history is, of course, not dead, nor are its practices likely to disappear any time soon. While the changing circumstances identified by these authors are likely very real – it is hard to deny cuts in funding, modifications to course offerings, technological transitions or shifting hierarchies of scientific disciplines – the consequences do not imply so much the extinction event proclaimed by Wilcove and Eisner as yet another set of transformations in how humans have observed and charted the natural world around them and how they have envisioned this practice as a rigorous, profitable, influential and even indispensable activity.

Where is natural history in the twenty-first century? The practices of observing, collecting, comparing and systematising natural objects are found in all the places one might expect, given the histories charted in this volume. Perhaps most obviously, they are relied upon by those working in zoos, arboretums, botanical gardens and natural history museums, and in university departments of plant and animal sciences, evolutionary biology, ecology, Earth sciences – as well as archaeology, anthropology, geography and still others. They are applied to produce material change in the world, whether the protection of wildlife or restoration of ecosystems by conservationists or the enhanced productivity of sites as diverse as farm fields, timber forests and tar sands. They are also found in living rooms, whether described in books or viewed on television, and experienced in backyards, national parks and the urban jungle, by birdwatchers, outdoor enthusiasts and guerrilla gardeners.

Natural history is also found in less expected places. Laboratories of experimental biology, long described as the central force driving natural history to its demise, can also be seen as sites where the natural historical practices of collecting and comparing have not only thrived but even, in a genomic era, come to dominate. As the historian Bruno Strasser argues, for example, the massive genetic sequence database GenBank, created via the laboratory activities of a generation of experimental biologists and now 'the largest and most frequently accessed collection of experimental knowledge in the world' can be understood as, fundamentally, a natural historical tool. It allows scientists to systematically compare genetic sequences from across a range of organisms or species in order to determine their structures, functions and histories. As Strasser concludes, 'Today, natural historical practices centered on the collection and comparison of data are increasingly recognized as legitimate ways to produce knowledge in the experimental life sciences.'[6]

They are also increasingly present on the Internet. The sociologist Christine Hine has charted the incorporation of computer and communications technologies into the practices of systematists in the later twentieth century. The centuries-old tasks of cataloguing, classifying and storing natural specimens now take place with the aid of DNA barcoding, phylogenetic software and digital databases. As Hine reminds us, many aspects of taxonomic work remain largely as they were before the introduction of computer technologies, with 'virtual collections ... deployed in forms continuous with practices based on use of material collections' while descriptions of phylogenetic analyses insist on 'computer packages ... as manifestations of theoretical propositions'.[7]

One thing information technologies do seem to have changed is the visibility of systematicists' work, and especially of the collections upon which they rely. Increasing public interest in biodiversity in the 1990s created new opportunities to argue for the importance of biological collections – and especially the importance of making these as accessible as possible – at roughly the same as the Internet boom. It was newly possible, for example, to imagine digitising the many millions of specimens tucked away in herbarium cabinets and hidden in museum stores and making these available via massive online databases. The New York Botanical Garden (NYBG) began digitising its vast herbarium collection in 1995, starting with transcription of data from specimen labels into a database and subsequently incorporating images of specimens to accompany these electronic records. It launched the C. V. Starr Virtual Herbarium in 1998, which made digitised herbarium records and images available via the Web. By 2016, some 2.5 million specimens had been made available via the database.[8] In addition to being searchable via the NYBG's own website, the virtual herbarium is integrated into still larger databases, for example the US Integrated Digitized Biocollections database, iDigBio. Initiated in 2011 as 'a permanent national resource to integrate data from biological research collections', iDigBio fosters digitisation of natural history collections across the country and renders these accessible via a single search portal.[9]

Computer and communications technologies promised greater opportunities to bring together taxonomic information, species descriptions and photographs not just nationally but internationally, and in so doing create a single, continuously updated, global access point for this information. As the biologist E. O. Wilson described in 2007, speaking about an idea then already in implementation, computer technologies had created the opportunity to create the 'Encyclopedia of Life'. This would be 'an encyclopedia that lives on the Internet and is contributed to by thousands of scientists around the world ... It has an indefinitely expandable page for each species ... It makes all key information about life on Earth accessible to anyone, on demand, anywhere in the world.'[10]

The Encyclopedia of Life (EoL) launched in May 2007 as a US$50 million collaboration among the Smithsonian Institution, Field Museum (Chicago), Harvard University, Biodiversity Heritage Library and Missouri Botanical Garden with additional support from the MacArthur and Sloan Foundations.[11] In its first decade, it grew steadily, from an initial collection of 30,000 entries to more than 1.3 million webpages in March 2017, each representing a different taxon.[12] Its institutional collaborators in 2017 stretched around the world – including Australia, China, Denmark, Mexico, Costa Rica, India, Egypt, France, the Netherlands, Norway, South Africa, Taiwan and the United States – and its content contributors included representatives from many more countries still.[13] Of course, with the vast majority of species on Earth still to be described, let alone have significant data generated about them, EoL surely will not ever, as Wilson hoped, 'complete the great Linnaean enterprise' – but it does carry the long-standing ambition of charting the entirety of nature forward into a new global endeavour.[14]

The promise and appeal of these online databases (of which there are many examples worldwide) lay especially in their rendering previously hard-to-reach materials widely accessible – and not just to researchers. From the outset, iDigBio was imagined as a collection for professionals and laypeople alike to use. It would 'help researchers identify gaps in scientific knowledge' and 'assist government agencies and others making decisions related to climate change, conservation, invasive species, biodiversity and other biological issues'. It would also let any person sort through hundreds of specimens previously in closed store, for whatever purpose they imagined. As one of the co-investigators on the original National Science Foundation grant for iDigBio explained, 'Ten years from now, if a kid plowing up a field in Iowa finds a fossil horse tooth, and he wants to compare it to fossil specimens in real museum collections, he will be able to download images and other information from his home computer.'[15]

The digitisation and online publication of museum collections is only one among a number of ways in which the computer technologies have enabled a re-imagining of the public face of natural history. Through varied initiatives, museum professionals have invited 'citizen scientists' to assist in making natural historical knowledge. In some cases, this means carrying out tasks linked to digitisation and online curation. From 2013 to 2016, the Notes from Nature project of the Natural History Museum, London, asked online volunteers to transcribe handwritten records documenting its bird collection – an effort that produced 370,000 new transcriptions.[16] Amateur observers are also recruited to help researchers generate data about organisms in the field. The *Svenska Artprojektet* (Swedish Taxonomy Initiative), which aims to inventory all multicellular species in Sweden, launched in 2001, with a 20-year horizon for completion. This work regularly incorporates amateur observation and volunteer labour alongside that of professional biologists and taxonomists.[17] Such collaboration is intended both to provide better data for research and nature management as well as stimulate public interest. For example, the *Artportalen* (Swedish Species Observation System), which catalogues observation of Swedish flora and fauna, attracts contributions from and can be searched by any user. As the website declares, 'By making it a pleasure to share one's observations, we hope that more people make a trip into the countryside to search for, to find, and to report their sightings to the Swedish Species Gateway – and thus contribute to better understanding of the biodiversity of Sweden and improved environmental management.'[18] Such programmes have been remarkably successful in recruiting participants. The Chinese Field Herbarium, a database supported by the Chinese Academy of Sciences, enables a wide range of users, lay and professional, to contribute to a catalogue of Chinese flora; as of January 2017, its more than 13,000 registered users had uploaded over 7 million photos documenting their field observations.[19]

It is clear, then, that natural history is not dead, and that its past practices and the knowledge and materials these practices produced are appreciated – though perhaps not to the extent that many might hope, especially in certain professional contexts. The growing political and social imperatives to chart biological diversity amidst widespread acknowledgement of environmental change have played no small part in ensuring this continued, even renewed, appreciation.

Environmental change has also sparked interest in rethinking the natures catalogued in natural historical collections and their presentation to the public. In the age of the Anthropocene, many people have argued that there is no longer nature 'out there', independent of human influence. Intentionally and unintentionally, *Homo sapiens* has indelibly altered the planet and all it contains – and museums are working to incorporate this into the accounts of the natural world they present. The Fossil Hall of the National Museum of Natural History in Washington, DC closed for renovation in 2015 with a promise that the renewal will not just update the exhibition display but change the geological narrative presented there. According to one account, 'Alongside the typical displays of *Tyrannosaurus rex* and *Triceratops*, there will be a new section that forces visitors to consider the species that is currently dominating the planet.'[20]

It is not only geology wings that are being reconsidered. The Carnegie Museum of Natural History in Pittsburgh, Pennsylvania, is both creating a permanent gallery on the Anthropocene and searching through its extensive existing collections to find evidence of the Anthropocene to exhibit.[21] Finds detailed on the project's blog include a bald eagle 'inadvertently shot out of the sky during the battle of Gettysburg', a preserved salamander specimen recorded as having choked to death on a marshmallow and a penguin that spent its entire life in the otherwise unexpected environs of San Diego, California thanks to the Sea World theme park.[22] The perspective that our categories of human-made and natural might no longer apply has also prompted museums of technology to explore alternative presentations of their classic materials. It was the Deutsches Museum, Munich, the world's largest science and technology museum, that opened the first major exhibition devoted to the idea of the Anthropocene in 2014. This brought together artefacts associated with the causes of rapid global change – industry, transport, fuel extraction – with those representing the 'new natures' of today, whether disappearing ecosystems, urban wildlife, high-yielding crop varieties, crocheted coral reefs, sneakers made of cane toad skin, or evolving generations of razor blades.[23]

New biotechnological capabilities have similarly inspired new visions of the remit of natural historical collections. The Center for Postnatural History, also in Pittsburgh (and collaborator in the Carnegie Museum's Anthropocene exhibition) concerns itself with all those deliberate changes, defining postnatural history as 'the study of the origins, habitats, and evolution of organisms that have been *intentionally* and *heritably* altered by humans'.[24] Their specimen vault includes things like agricultural crops, inbred laboratory rats, sterile male screwworm generated through radiation treatment and used in pest control, and transgenic goats that produce a spider silk protein in their milk. The centre's founder, Richard Pell, characterises it as extending the remit of earlier natural history museums: 'Museums of natural history typically end with the domestication of plants and animals ... The entire collection of the Smithsonian has a single genetically modified specimen in it ... And this is really where the meaning of postnatural history as we use it, comes from. It's not about an end of "nature", but rather what happens after "natural history".'[25]

If these various re-imaginings and reworkings of natural history – digital, crowd-sourced, Anthropocenic – indicate where natural history is in the twenty-first century, what about the history of natural history today? Here, no one is talking about disciplinary extinction, as the field is thriving in contexts ranging from museum displays and academic monographs to websites and television documentaries. Some of the fascination of the history of natural history doubtless involves nostalgia for lost days of childhood bug-hunting and vanished scenes of pristine nature. But the subject also has the potential to do more than provide a respite from the pressures of modern life. It can open up fresh perspectives on some of the most intractable problems we face, encouraging a new understanding of the place of humans in nature.

As the chapters in this volume suggest, the history of natural history is rapidly changing and widening its scope. Key questions that were beginning to be asked two decades ago are now central. We know far more about audiences for natural history, the reasons for pursuing it and the relations between the field, laboratory and museum. We are beginning to understand the extraordinary range of ways that people have attempted, over many centuries, to comprehend the diversity of nature. Above all, the history of natural history is now at the centre of general historical debates about globalisation, circulation, empire and exchange. In part, this has involved a widening of the community of historians dealing with natural history in different parts of the world; in part, it has involved a rethinking of the origins of natural history itself as a form of contact between different cultural traditions. Historical perspectives are here helping to understand, in a critical way, the contemporary politics involved in defining the relations between the 'global' and the 'local'.

As the contributions to this volume show, natural history as a set of practices and doctrines developed during centuries of exploration and imperial expansion, as a way of cataloguing novelties, charting unfamiliar territories and inventorying potentially useful resources. It was a hybrid subject forged in trading cities and through the great mercantile routes across the Atlantic and into the Pacific. Practices, specimens and conceptual frameworks were exchanged, borrowed, stolen, gifted and appropriated in an (often unequal) dialogue with trading partners and colonised peoples. They were also lost, forgotten and destroyed, as new ways – typically those forged in the West – replaced or modified long-standing modes of interacting with the non-human world.

These interactions and exchanges transformed natural history itself. Natural history had emerged largely in late medieval and Renaissance Europe as a learned activity pursued primarily by physicians or in the context of court display. From the eighteenth century onwards, the subject became increasingly tied to larger priorities of the state, through publicly funded museums, expeditions and surveys. By the early twentieth century practitioners tended to be paid employees of the state, with others dismissed as 'mere' amateurs. Not only did natural history become increasingly the province of dedicated professionals, it also became more directly implicated in economic policy, an instrument in developing new crops and locating likely sources of energy.

But at least as important, natural history played a key role in reconfiguring the relations between different places. From the earliest cabinets of curiosities, collections made it possible to define the 'exotic' by seeing it 'at home'. By the early decades of the twentieth century those studying nature were as likely to spend time in Brussels as in the Congo, in New York as in Mongolia. Researchers came to examine the millions of specimens – insects, dried plants, fossil bones, bird skins, mineral specimens – that had poured into the treasuries of the world's great storehouses of nature. Natural history collections displayed imperial ideals to wider publics, with the diversity of items on show demonstrating the range, efficiency and authority of control afforded by science.

In the ongoing transformations of natural history in the present day, these collections of specimens, books and periodicals from earlier centuries are the most visible reminders of the continuing significance of history for current practice. Zoologists and botanists have always depended on this accumulated material heritage, without which our understanding of everything from identifying new species to determining patterns of extinction would be impossible. Current online gatherings build directly on this legacy, so that the accumulations created by past collectors and collections are the foundations of our contemporary understanding. It is thus not surprising that the Biodiversity Heritage Library (www.biodiversitylibrary.org) – a collaboration involving natural history libraries and museums from across the world – has been a pioneer in compiling and making freely available the print legacy of a scientific field.

Historic collections also shape the views of wider publics. Thus, visitors to the Berlin Natural History Museum can marvel at what for many years was the largest dinosaur on display anywhere in the world, a 25 metre long *Brachiosaurus* (now *Giraffatitan*); this was part of a 230 ton haul of fossil remains transported by local labourers at Tendaguru Hill in German East Africa between 1909 and 1913. Like collections of human art and artefacts, such specimens were originally imperial trophies, demonstrating to urban viewers that Germany, a latecomer in the nineteenth-century 'scramble for Africa', was on its way to joining Britain and France as a colonial power. Natural history in such cases provided a source of wonder, illustrating imperial prowess, divine power and the comparative insignificance of humans in the face of millions of years of geological history. Today the same bones are a continuing source of wonder, not least for the millions who see them aided by state-of-the-art computer-enhanced binoculars, which reveal the latest scientific understandings of their internal physiology, habitat and herding behaviour.[26] And more assiduous visitors to the Natural History Museum learn at least something about where, how and why they were dug up in Africa. Such issues were at the heart of another recent exhibition in Berlin, as the German Historical Museum included pressed plants and geological specimens along with guns, helmets and chains, as part of the wider story of German colonialism in the late nineteenth and early twentieth centuries.[27] In such displays, the spectacle of nature is shown to be part of the spectacle of a very human, and often tragic, past.

Like historical accounts of varying kinds, such displays help us to see humans as part of nature, rather than as something separate and acting upon it. By showing science in this way, history can demonstrate the continuous labour – by curators, professors, collectors, local informants, administrators and popularisers – that has gone into the making of natural historical knowledge ever since the cabinets of curiosity, pharmaceutical exchanges and voyages of exploration of the Renaissance. It can reveal the wider, but often hidden structures of expertise, authority and convention that have shaped our understandings and attitudes. At the same time, the history of the sciences suggests that although scientific knowledge is made by people, it is not individual opinion, easily obtained or easily overturned.

Nowhere does the labour that goes into making of contemporary facts about nature need to be more evident than in current discussions of biodiversity and climate change, where the validity of a consensus among scientists has been undermined and attacked. It is not often enough stressed that this consensus relies upon knowledge about the living world that has been patiently built up over many centuries and through encounters between many different cultures. The environmental crisis is commonly understood to have at its root the tendency to take short-term perspectives, so that 'nature' is widely viewed as an 'externality' not only in economic calculations but also in ordinary speech. There is a danger that we will realise that human life is woven into the fabric of nature only when it is too late to act upon that knowledge. Changing this future will require many things, but among them is surely a better understanding of the past, and in particular the remarkable variety of ways that people have understood their place in nature.

Further reading

Blagoderov, V. and Smith, V. S. (eds.), 'No specimen left behind: mass digitisation of natural history collections', special issue, *ZooKeys*, 209 (2012), pp. 1–267.

Davis, J., *The Birth of the Anthropocene* (Berkeley and Los Angeles, 2016).

Haraway, D. J., *Staying with the Trouble: Making Kin in the Chthulucene* (Durham, NC, 2016).

Heise, U. K., *Imagining Extinction: The Cultural Meanings of Endangered Species* (Chicago, 2016).

Hunt. L., *Writing History in the Global Era* (New York, 2014).

Kolbert, E., *The Sixth Extinction: An Unnatural History* (London, 2014).

Latour, B., *Facing Gaia: Eight Lectures on the New Climatic Regime* (New York, 2017).

Radkau, J., *Nature and Power: A Global History of the Environment* (Cambridge, 2008).

Tsing, A. L., *The Mushroom at the End of the World: On the Possibility of Life in Capitalist Ruins* (Princeton, 2017).

Van Dooren, T., *Flight Ways: Life and Loss at the Edge of Extinction* (New York, 2016).

Vidal, F. and Dias, N. (eds.), *Endangerment, Biodiversity and Culture* (Abingdon, UK, 2016).

Waterton, C., Ellis, R., and Wynne, B., *Barcoding Nature: Shifting Cultures of Taxonomy in an Age of Biodiversity Loss* (Abingdon, UK, 2013).

Notes

Worlds of history

1. J. Needham et al., *Science and Civilisation in China*, vol. VI, *Biology and Biological Technology*, Parts 1–5 (Cambridge, 1984–2015).
2. For an overview of current scientific and social roles of natural history, see J. J. Tewksbury, 'Natural history's place in science and society', *BioScience*, 64:4 (2014), pp. 300–10.
3. G. Bachelard, *The Formation of the Scientific Mind*, trans. M. McAllister Jones (1930; Manchester, 2002); T. Kuhn, *The Structure of Scientific Revolutions* (Chicago, 1962); M. Foucault, *The Order of Things: An Archaeology of the Human Sciences* (1966; London, 1970).
4. E. Bloch, 'Nonsynchronism and the obligation to its dialectics', trans. M. Ritter, *New German Critique*, 11 (1932; 1977), pp. 22–38.
5. P. Fontes da Costa, *The Singular and the Making of Knowledge at the Royal Society of London in the Eighteenth Century* (Newcastle upon Tyne, 2009); B. Seaton, *The Language of Flowers: A History* (Charlottesville, 1995); H. Waters, 'Botanists finally ditch Latin' (2011), available at https://blogs.scientificamerican.com/cultur ing-science/botanists-finally-ditch-latin-and-paper-enter-21st-century/.
6. N. Jardine, J. A. Secord and E. C. Spary (eds.), *Cultures of Natural History* (Cambridge, 1996).
7. For excellent overviews of the development and themes of this cultural history, see L. Hunt, ed., *The New Cultural History* (Berkeley, 1989); P. Burke, *What Is Cultural History?* (Cambridge, 2004).
8. R. Williams, *Keywords: A Vocabulary of Culture and Society* (London, 1976). See also A. Kuper, *Culture: The Anthropologists' Account* (Cambridge, MA, 1999).
9. For applications of this kind of book history to the sciences, see A. Johns, *The Nature of the Book: Print and Knowledge in the Making* (Chicago, 1998); M. Frasca Spada and N. Jardine (eds.), *Books and the Sciences in History* (Cambridge, 2000).
10. On textual genres, see K. Chemla and J. Virbel (eds.), *Texts, Textual Acts and the History of Science* (Dordrecht, 2016); L. Taub, *Science Writing in Greco-Roman Antiquity* (Cambridge, 2017). On types and uses of natural historical imagery, see S. Kusukawa, *Picturing the Book of Nature: Image, Text and*

Argument in Sixteenth-Century Human Anatomy and Medical Botany (*Chicago*, 2012); K. Nickelsen, *Draughtsmen, Botanists, and Nature: The Construction of Eighteenth-Century Botanical Illustrations* (Dordrecht, 2006); D. Bleichmar, *Visible Empire: Botanical Expeditions and Visual Culture in the Eighteenth-Century Hispanic World* (Chicago, 2012); J. Pimentel, *The Rhinoceros and the Megatherium: An Essay in Natural History*, trans. P. Mason (Cambridge, MA, 2017).

11. See G. Pomata (ed.), *Historia: Empiricism and Erudition in Early Modern Europe* (Cambridge, MA, 2005); B. Ogilvie, *The Science of Describing: Natural History in Renaissance Europe* (Chicago, 2006); K. Acheson, *Visual Rhetoric in Early Modern English Literature* (Farnham, 2013), chapter 4.

12. D. Hicks, 'The material-cultural turn: event and effect', in D. Hicks and M. C. Beaudry (eds.), *Oxford Handbook of Material Culture Studies* (Oxford, 2010), pp. 25–98.

13. See L. Taub, 'Introduction' to special section on 'The history of scientific instruments', *Isis*, 102:4 (2011), pp. 689–96; A. Guerrini, 'The material turn in the history of the life sciences', *Literature Compass*, 13:7 (2016), pp. 469–80; U. Klein and E. Spary (eds.), *Materials and Expertise in Early Modern Europe: Between Market and Laboratory* (Chicago, 2010), introduction; T. Bennett, *Material Powers: Cultural Studies, History, and the Material Turn* (Florence, 2013); M. Wintroub, 'Taking a bow in the theater of things', *Isis*, 101:4 (2010), pp. 779–93; B. Jardine, 'State of the field: paper tools', *Studies in History and Philosophy of Science*, 64 (2017), pp. 1–11.

14. On the spatial turn, see D. A. Finnegan, 'The spatial turn: geographical approaches in the history of science', *Journal of the History of Biology*, 41:2 (2008), pp. 369–88; K. Raj, 'Introduction: circulation and locality in early modern science', *British Journal for the History of Science*, 43:4 (2010), pp. 513–17; S. McCook, 'Focus: global currents in national histories of science: the "global" turn and the history of science in Latin America', *Isis*, 104:4 (2013), pp. 773–6; L. Roberts, 'The circulation of knowledge in early modern Europe: embodiment, mobility, learning and knowing', *History of Technology*, 31 (2012), pp. 47–68; J. A. Secord, 'Knowledge in transit', *Isis*, 95:4 (2004), pp. 654–72. Fine instances of global approaches to the history of science are to be found in S. Schaffer, L. Roberts, K. Raj and J. Delbourgo (eds.), *The Brokered World: Go-Betweens and Global Intelligence, 1770–1820* (Sagamore Beach, MA, 2009), and P. Manning and D. Rood (eds.), *Global Scientific Practice in an Age of Revolutions, 1750–1850* (Pittsburgh, 2016).

15. On such hybridity, see B. Latour, *Science in Action: How to Follow Scientists and Engineers through Society* (Milton Keynes, 1987),

and P. Galison, *Image and Logic: A Material Culture of Microphysics* (Chicago, 1997). An engaging early essay in the exploration of hybridity is M. Callon, 'Some elements of a sociology of translation: domestication of the scallops and the fishermen of Saint Brieuc Bay', in J. Law (ed.), *Power, Action and Belief* (London, 1986), pp. 196–233.

16. See J. Cañizares-Esguerra, *Nature, Empire and Nation: Explorations of the History of Science in the Iberian World* (Stanford, 2006); J. Delbourgo and N. Dew (eds.), *Science and Empire in the Atlantic World* (New York, 2008).

17. Representative examples are A. Cooper, *Inventing the Indigenous: Local Knowledge and Natural History in Early Modern Europe* (Cambridge, 2007); L. Schiebinger, *Plants and Empire: Colonial Bioprospecting in the Atlantic World* (Cambridge, MA, 2007); A. Winterbottom, *Hybrid Knowledge in the Early East India Company World* (Basingstoke, 2016).

18. For a balanced overview of such turns, see L. K. Nyhart, 'Historiography of the history of science', in B. Lightman (ed.), *A Companion to the History of Science* (New York, 2016), pp. 7–22.

19. On the ontological turn in anthropology, see M. Holbraad and M. A. Pedersen, *The Ontological Turn: An Anthropological Exposition* (Cambridge, 2017). For reflections on applications of this approach to the sciences, see the special issue *Ontology in Science and Technology Studies, Social Studies of Science*, 43:3 (2013). Works relating to natural history that well exemplify the ontological turn include U. Klein, 'Shifting ontologies, changing classifications: plant materials from 1700 to 1830', *Studies in History and Philosophy of Science*, 36:2 (2005), pp. 261–329; A. L. Tsing, *The Mushroom at the End of the World: On the Possibility of Life in Capitalist Ruins* (Princeton, 2015).

20. F. Kermode, 'Talking about doing: the deconstructionists and the new historicists', in his *The Uses of Error* (London, 1991), pp. 119–29.

21. A. te Heesen, *The World in a Box: The Story of an Eighteenth-Century Picture Encyclopedia*, trans. A. M. Hentschel (Chicago, 2002 [1997]); J. Secord, *Victorian Sensation: The Extraordinary Publication, Reception and Secret Authorship of* Vestiges of Creation (Chicago, 2000).

22. D. Bleichmar, *Visible Empire: Botanical Expeditions and Visual Culture in the Hispanic Enlightenment* (Chicago, 2012), chapter 4; K. Raj, *Relocating Modern Science: Circulation and the Construction of Knowledge in South Asia and Europe, Seventeenth to Nineteenth Centuries* (New Delhi, 2006), conclusion.

23. On such prejudicial effects, see G. Wilder, 'From optic to topic: the foreclosure effect of historiographic turns', *American Historical Review*, 117:3 (2012), pp. 723–45.

24. See R. Rosenzweig, *Clio Wired: The Future of the Past in the Digital Age* (New York, 2007); T. Weller, ed., *History in the Digital Age* (London, 2013); and P. Manning, *Big Data in History* (Basingstoke, 2013).

25. For a markedly scientistic version, see J. Guldi and D. Armitage, *The History Manifesto* (Cambridge, 2014); for a robust response, see D. Cohen and P. Mandler, '*The History Manifesto*: a critique', *The American Historical Review*, 120:2 (2015), pp. 530–42.

26. D. J. Haraway, *Crystals, Fabrics and Fields: Metaphors of Organicism in Twentieth-Century Developmental Biology* (New Haven, 1976). R. E. Kohler, 'A generalist's vision', *Isis*, 92:2 (2005), pp. 224–9, suggests themes common to knowledge production on which such larger narratives may be built.

27. On the links between practices and agendas, see I. Hacking, 'Language, truth and reason', in S. Lukes, ed., *Rationality and Relativism* (Oxford, 1982), pp. 48–66, and, elaborating on Hacking's insights, N. Jardine, *The Scenes of Inquiry: On the Reality of Questions in the Sciences* (Oxford, 1991).

28. This substantial literature includes, for the early modern period, R. Drayton, *Nature's Government: Science, Imperial Britain, and the Improvement of the World* (New Haven, 2000), D. Margocsy, *Commercial Visions: Science, Trade, and Visual Culture in the Dutch Golden Age* (Chicago, 2014) and J. Delbourgo, *Collecting the World: The Life and Curiosity of Hans Sloane* (London, 2017); for the modern age, E. Benson, *Wired Wilderness: Technologies of Tracking and the Making of Modern Wildlife* (Baltimore, 2010), R. E. Kohler, *All Creatures: Naturalists, Collectors and Biodiversity in the Twentieth Century* (New York, 2006) and R. de Bont, *Stations in the Field: A History of Place-Based Animal Research, 1870–1930* (Chicago, 2015).

29. Such an approach is explored in J. Secord, 'Thoughts on "Knowledge in transit": seven questions in history of science', forthcoming.

30. On natural history and heritage, see R. J. Wilson, *Natural History: Heritage, Place and Politics* (London, 2017); E. Dorfman (ed.), *Intangible Natural Heritage: New Perspectives on Natural Objects* (New York, 2012); S. Macdonald (ed.), *The Politics of Display: Museums, Science, Culture* (London, 1998). On the digitisation of museum objects, see R. Parry, *Recoding the Museum: Digital Heritage and the Technologies of Change* (Abingdon, 2007), chapter 2. For some sites that have explored the natural history collection as symbolic artefactualisation of 'nature', see, for example,

https://curiositas.org/cabinet or www.msn.unifi.it/en/collezioni/ceroplastica-2/.

31. On artisan naturalists, see A. Secord, 'Artisan botany', in Jardine et al. (eds.), *Cultures of Natural History*, pp. 378–93; on antiquarians and natural history, N. Heringman, *Sciences of Antiquity: Romantic Antiquarianism, Natural History, and Knowledge Work* (Oxford, 2013); on 'citizen scientists', T. Charvolin, A. Micoud and L. K. Nyhart, *Des Sciences citoyennes? La question de l'amateur dans les sciences naturalistes* (La Tour d'Aigues, 2007).

32. On past standards of testimony, see P. G. Adams, *Travelers and Travel Liars, 1660–1800* (Berkeley, 1962); R. Serjeantson, 'Testimony and proof in early-modern England', *Studies in History and Philosophy of Science*, 30:2 (1999), pp. 195–236; R. Kennedy, *A History of Reasonableness and Authority in the Art of Thinking* (Rochester, 2004).

33. See L. Daston, 'Type specimens and scientific memory', *Critical Inquiry*, 31:1 (2004), 153–82.

34. See C. Jarvis, *Order Out of Chaos: Linnaean Plant Names and their Types* (London, 2007).

35. See R. Laudan, 'Redefinitions of a discipline: histories of geology and geological history', in L. Graham, W. Lepenies and P. Weingart (eds.), *Functions and Uses of Disciplinary Histories* (Dordrecht, 1983), pp. 79–104; N. Jardine, 'Legitimation and history', in *The Scenes of Inquiry*, chapter 6; J. Skopek, 'Principles, exemplars and uses of history in early twentieth-century genetics', *Studies in History and Philosophy of Biological and Biomedical Sciences*, 42:2 (2011), pp. 210–25.

1 Visions of ancient natural history

1. C. Gessner, *Historiae animalium lib. I. de quadrupedibus viviparis* (Zurich, 1551), sig. A[5]r.

2. J. Gerard, *The Herball or Generall Historie of Plantes* (London, 1597), sig. [B4]r.

3. A.-M. Leroi, *The Lagoon: How Aristotle Invented Science* (New York, 2014); M. Manquat, *Aristote naturaliste* (Paris, 1932).

4. M. Belozerskaya, *The Medici Giraffe and Other Tales of Exotic Animals and Power* (New York, 2006), chapter 2.

5. 1 Kings 4:33.

6. See C. G. Nauert, Jr, *Humanism and the Culture of Renaissance Europe* (Cambridge, 2006), for an overview, and A. Nagel and C. S. Wood, *Anachronic Renaissance* (New York, 2010), for an

analysis of the more complex approach to the past taken by some German humanists.

7. See S. A. McKnight, *The Modern Age and the Recovery of Ancient Wisdom: A Reconsideration of Historical Consciousness, 1450–1650* (Columbia, MO, 1991).

8. Gerard, *Herball*, 'To the Reader'.

9. J. Parkinson, *Paradisi in Sole Paradisus Terrestris, or, A Garden of all sorts of pleasant flowers which our English ayre will permitt to be noursed vp: with A Kitchen garden of all manner of herbes, rootes, & fruites, for meate or sause vsed with vs, and An Orchard of all sorte of fruitbearing trees and shrubbes fit for our land together With the right orderinge planting & preseruing of them and their vses & vertues* (London, 1629).

10. R. French, *Ancient Natural History: Histories of Nature* (London and New York, 1994), argues otherwise, but French gathered under the rubric 'natural history' books that ancient authors would not have recognised as being closely related. For a fuller version of my argument here, see B. W. Ogilvie, *The Science of Describing: Natural History in Renaissance Europe* (Chicago and London, 2006), chapter 3.

11. For details on these texts and their authors, see S. Hornblower, A. Spawforth and E. Eidinow (eds.), *The Oxford Classical Dictionary* (Oxford, 2012).

12. This text is often translated into English under the title *History of Animals*, but in context, Aristotle's Greek *historiai* is better translated as 'inquiries'.

13. E. Wotton, *De differentiis animalium libri decem* (Paris, 1552).

14. J. Monfasani, 'The pseudo-Aristotelian *Problemata* and Aristotle's *De animalibus* in the Renaissance', in A. Grafton and N. Siraisi (eds.), *Natural Particulars: Nature and the Disciplines in Renaissance Europe* (Cambridge, MA and London, 1999), pp. 205–47.

15. See L. D. Reynolds (ed.), *Texts and Transmission: A Survey of the Latin Classics* (Oxford, 1983) and L. D. Reynolds and N. G. Wilson, *Scribes and Scholars: A Guide to the Transmission of Greek and Latin Literature*, 4th edn (Oxford, 2014).

16. U. Aldrovandi, *De animalibus insectis libri septem* (Bologna, 1602), pp. 6, 10–12.

17. W. Harvey, *The Works of William Harvey* (London, 1847; reprinted Philadelphia, 1989), pp. 152, 153, 162.

18. R.-A. Ferchault de Réaumur, *Mémoires pour servir à l'histoire des insectes* (Paris, 1734–42), vol. I, pp. 25–8.

19. Réaumur, *Mémoires*, vol. I, pp. 28–9.

20. On Buffon's attention to history, see J. Roger, 'Buffon et l'introduction de l'histoire dans l'*Histoire naturelle*', in J. Gayon et al.

(eds.), *Buffon 88. Actes du Colloque international pour le bicenten-aire de la mort du Buffon* (Paris, Montbard, Dijon, 14–22 juin 1988) (Paris, 1992), pp. 193–205.

21. G.-L. Leclerc, comte de Buffon, *Histoire naturelle, générale et particulière, avec la description du cabinet du Roy*, 36 vols. (Paris, 1749–1804), vol. I, pp. 43–4.

22. Buffon, *Histoire naturelle*, vol. II, p. 85.

23. J. Monfasani, 'Aristotle as scribe of nature: the title-page of MS Vat. Lat. 2094', *Journal of the Warburg and Courtauld Institutes*, 69 (2006), pp. 193–205.

24. J. Feros Ruys, J. O. Ward and M. Heyworth (eds.), *The Classics in the Medieval and Renaissance Classroom: The Role of Ancient Texts in the Arts Curriculum as Revealed by Surviving Manuscripts and Early Printed Books* (Turnhout, 2013).

25. W. McCuaig, *Carlo Sigonio: The Changing World of the Late Renaissance* (Princeton, 1989), pp. 152–3; see also P. Godman, *From Poliziano to Machiavelli: Florentine Humanism in the High Renaissance* (Princeton, 1998), pp. 31–79.

2 Gessner's history of nature

1. On the classical sources of natural history, see Ogilvie, this volume, Chapter 2 and his *The Science of Describing: Natural History in Renaissance Europe* (Chicago, 2006), pp. 87–138. For the life and works of Gessner, I draw on U. B. Leu, *Conrad Gessner (1516–1565): Universalgelehrter und Naturforscher der Renaissance* (Zurich, 2016), pp. 127–65.

2. For recent rediscoveries of his drawings, see F. Egmond, 'A collection within a collection: rediscovered animal drawings from the collections of Conrad Gessner and Felix Platter', *Journal of the History of Collections*, 24 (2012), pp. 1–22, and U. B. Leu, 'The rediscovered third volume of Conrad Gessner's *Historia plantarum*', in A. Blair and A. Goeing (eds.), *For the Sake of Learning: Essays in Honor of Anthony Grafton* (Leiden, 2016), pp. 415–22. Few objects from his museum have survived: W. Etter, 'Conrad Gessner als Paläontologe', in U. B. Leu and M. Ruoss (eds.), *Facetten eines Universums: Conrad Gessner 1516–2016* (Zurich, 2016), pp. 175–83, esp. pp. 176, 182.

3. C. Gessner, *Historia animalium*, 4 vols. (Zurich, 1551–8), vol. I, β1r. For the dangers of anachronistic 'zoological' evaluation, see K. A. E. Enenkel, 'Zur Konstituierung der Zoologie als Wissenschaft in der frühen Neuzeit: Diskursanalyse zweier Grossprojekte (Wotton, Gessner)', in K. A. E. Enenkel and P. J. Smith (eds.), *Early Modern Zoology: The Construction of*

Animals in Science, Literature and the Visual Arts (Leiden, 2007), pp. 15–74.

4. P. Findlen, 'Courting nature', in N. Jardine, J. A. Secord and E. C. Spary (eds.), *Cultures of Natural History* (Cambridge, 1996), pp. 57–74.

5. For the reception of Aristotle's works on animals, see C. Steel, G. Guldentops and P. Beullens (eds.), *Aristotle's Animals in the Middle Ages and Renaissance* (Leuven, 1999) and S. Perfetti, *Aristotle's Zoology and its Renaissance Commentators (1521–1601)* (Leuven, 2000).

6. A. R. Cunningham and S. Kusukawa, *Natural Philosophy Epitomised: Books 8–11 of Gregor Reisch's Philosophical Pearl (1503)* (Aldershot, 2010), pp. 135–47.

7. I. Maclean, 'White crows, graying hair and eyelashes: problems for natural historians in the reception of Aristotelian logic and biology from Pomponazzi to Bacon', in G. Pomata and N. Siraisi (eds.), *Historia: Empiricism and Erudition in Early Modern Europe* (Cambridge, MA, 2005), pp. 147–79.

8. Here I draw on N. G. Siraisi, *Late Medieval and Early Renaissance Medicine* (Chicago, 1990), pp. 143–52; K. M. Reeds, 'Renaissance humanism and botany', *Annals of Science*, 33 (1976), pp. 519–42; R. Palmer, 'Medical botany in northern Italy in the Renaissance', *Journal of the Royal Society of Medicine*, 78 (1985), pp. 149–57.

9. S. Kusukawa, *Picturing the Book of Nature: Image, Text and Argument in Sixteenth-Century Human Anatomy and Medical Botany* (Chicago, 2012), pp. 101–23.

10. H. J. Cook, 'Physicians and natural history', in N. Jardine, J. A. Secord and E. Spary (eds.), *Cultures of Natural History* (Cambridge, 1996), pp. 91–105; for publications on animals, see further L. Pinon, *Livres de Zoologie de la Renaissance: Une anthologie (1450–1700)* (Paris, 1995); P. Delaunay, *La Zoologie au XVIe siècle* (Paris, 1962).

11. C. Gessner, *Historia animalium*, vol. I, α2r.

12. Gessner's use of '*historia*' is fully discussed in L. Pinon, 'Conrad Gessner and the historical depth of Renaissance natural history', in Pomata and Siraisi (eds.), *Historia*, pp. 241–67, at pp. 243–50. For the status of '*historia*' as knowledge, see G. Pomata and N. Siraisi, 'Introduction', in Pomata and Siraisi (eds.), *Historia*, pp. 1–38.

13. Gessner, *Historia animalium*, vol. I, α3r; C. Gessner, *Physicarum meditationum, annotationum et scholiorum libri V* (Zurich, 1586), p. 1.

14. See Aristotle, *Parts of Animals*, trans. A. L. Peck (London, 1937), p. 53. For the different views of Renaissance commentators on how *paideia* and *episteme* were related, see Perfetti, *Aristotle's Zoology*, pp. 48–51, 75–80, 132–5, 198–201.

15. Gessner, *Historia animalium*, vol. I, α1r.

16. Aristotle, *Parts of Animals* 1.5, p. 97; Maclean, 'White crows', p. 157.

17. Also noted by Pinon, 'Conrad Gessner and the historical depth of Renaissance natural history', p. 247; what follows is a summary of Gessner, *Historia animalium*, vol. I, α2v–α4r.

18. Gessner, *Historia animalium,* vol. I, γ1v–3r. W. B. Ashworth Jr, 'Emblematic natural history of the Renaissance', in Jardine, Secord and Spary (eds.), *Cultures of Natural History*, pp. 15–37; Enenkel, 'Zoologie als Wissenschaft', pp. 51–7.

19. Kusukawa, *Picturing the Book of Nature*, pp. 59–60; Leu, *Gessner*, p. 97.

20. Gessner, *Historia animalium,* vol. I, β4v–γ1r. The number of books he had access to was far greater than that in his own library, see U. B. Leu, R. Keller and S. Weidmann, *Conrad Gessner's Private Library* (Leiden, 2008).

21. Gessner, *Historia animalium,* vol. IV, title page. For the extent of Gessner's copying in this volume, see F. Egmond and S. Kusukawa, 'Circulation of images and graphic practices in Renaissance natural history: the example of Conrad Gessner', *Gesnerus*, 73 (2016), pp. 29–72, p. 33.

22. A. Blair, 'Conrad Gessner's paratexts', *Gesnerus*, 73 (2016), pp. 73–122. For Gessner's extensive contacts, see U. B. Leu, 'Conrad Gessners Netzwerk', in Leu and Ruoss (eds.), *Facetten eines Universums*, pp. 61–74. For his management of correspondents, see C. Delisle, 'Accessing nature, circulating knowledge: Conrad Gessner's correspondence networks and his medical and naturalist practices', *History of Universities*, 23 (2008), pp. 35–58.

23. Gessner, *Historia animalium*, vol. I, pp. 785, 944; vol. III, p. 611. On Mattioli's sense of community, see P. Findlen, 'The formation of a scientific community: natural history in sixteenth-century Italy', in A. Grafton and N. Siraisi (eds.), *Natural Particulars: Natural Philosophy and the Disciplines in Early Modern Europe* (Cambridge, MA, 2000), pp. 369–400.

24. Gessner, *Historia animalium*, vol. III, p. 774.

25. For Aristotelian reliance on testimony of others, see Maclean, 'White crows', p. 164; Ogilvie, *The Science of Describing*, pp. 252–6.

26. This has been summarised from Gessner, *Historia animalium*, vol. I, β1v, pp. 1, 834, 952; vol. II, appendix p. 19; vol. III, pp. 488, 742.

27. N. G. Siraisi, *History, Medicine and the Traditions of Renaissance Learning* (Ann Arbor, 2007). For the historical nature of Gessner's enquiry, I am indebted to Pinon, 'Gessner'.

28. Gessner, *Historia animalium*, vol. I, pp. 935–8. For the status of monsters, see further Lawrence, this volume, Chapter 6.

29. Pinon, 'Gessner', p. 248; Gessner, *Historia animalium*, vol. I, β2r.

30. I borrow this phrase from Pomata and Siriasi, 'Introduction', p. 29.

31. Gessner, *Historia animalium,* vol. I, γ1v.

32. S. Kusukawa, 'The sources of Gessner's pictures for the *Historia animalium*', *Annals of Science,* 67 (2010), pp. 303–28; Egmond and Kusukawa, 'Circulation of images and graphic practices'. For the similarity of Gessner's approach to his text and images, see Ogilvie, *The Science of Describing,* pp. 236–40; for Gessner's note-taking practice, see A. Blair, *Too Much to Know: Managing Scholarly Information before the Modern Age* (New Haven, 2010), pp. 212–25.

33. C. Gessner, *Icones* (Zurich, 1560), p. 130. P. J. Smith, 'On toucans and hornbills: readings in early modern ornithology from Belon to Buffon', in Enenkel and Smith (eds.), *Early Modern Zoology,* pp. 75–117, at pp. 87–8; P. Mason, 'André Thevet, Pierre Belon and *Americana* in the embroideries of Mary Queen of Scots', *Journal of the Warburg and Courtauld Institutes,* 78 (2015), pp. 207–21.

34. Gessner, *Historia animalium,* vol. II, appendix pp. 19–20.

35. Kusukawa, *Picturing the Book of Nature,* p. 76. For colour printing, see A. Stijnman and E. Savage (eds.), *Printing Colour 1400–1700: History, Techniques, Functions and Receptions* (Leiden, 2015).

36. Lines and colour are important for Gessner's appreciation of the external form of Wentzel Jamnitzer's nature cast, which he conflates with a Platonic Form. See S. Kusukawa, 'Conrad Gessner on an "*ad vivum*" image', in P. H. Smith, H. J. Cook and A. R. W. Meyers (eds.), *Ways of Making and Knowing: The Material Culture of Empirical Knowledge* (Ann Arbor, 2014), pp. 330–56. See also V. Pugliano, 'Ulisse Aldrovandi's color sensibility: natural history, language and the lay color practices of Renaissance virtuosi', *Early Science and Medicine,* 20 (2015), pp. 358–96.

37. Kusukawa, *Picturing the Book of Nature,* pp. 139–61.

38. C. Gessner, *De rerum fossilium, lapidum et gemmarum maxime, figuris et similitudinibus liber* (Zurich: J. Gesnerus, 1565), Aa7v, AA3r, 4v, 166v. Kusukawa, *Picturing the Book of Nature,* pp. 175–6. For Gessner's views on fossils, see M. J. S. Rudwick, *The Meaning of Fossils: Episodes in the History of Palaeontology* (Chicago, 1972), pp. 1–35.

39. P. Glardon, *L'Histoire naturelle aux XVIᵉ siècle: Introduction, étude et édition critique de La Nature et diversité de Pierre Belon (1555)* (Geneva, 2011); P. Findlen, *Possessing Nature: Museums, Collecting, and Scientific Culture in Early Modern Italy* (Berkeley, 1994).

40. See for example, U. Aldrovandi, *Musaeum Metallicum* (Bologna, 1648), pp. 758–63.

3 Natural history in the apothecary's shop

This research was supported by Wellcome Trust grant WT100278MA.

1. F. Egmond, 'Apothecaries as experts and brokers in the sixteenth-century networks of the naturalist Carolus Clusius', in M. Feingold (ed.), *History of the Universities* (Oxford, 2008), vol. XXIII/2, pp. 59–91.

2. E. Zilsel, 'The sociological roots of science', *The American Journal of Sociology*, 47 (1942), pp. 544–62; P. Rossi, *Philosophy, Technology and the Arts in the Early Modern Era* (London, 1970), esp. pp. 1–62.

3. P. H. Smith and P. Findlen (eds.), *Merchants and Marvels: Commerce, Science and Art in Early Modern Europe* (London, 2002).

4. S. Shapin, 'The image of the man of science', in L. Daston and K. Park (eds.), *The Cambridge History of Science* (Cambridge, 2006), vol. III, pp. 175–93.

5. V. Pugliano, 'Botanical artisans: apothecaries and the study of nature in Venice and London, 1550–1610', PhD thesis (Oxford, 2012); F. Egmond, *The World of Carolus Clusius: Natural History in the Making, 1550–1610* (London, 2010).

6. B. W. Ogilvie, *The Science of Describing: Natural History in Renaissance Europe* (Chicago, 2006).

7. A. M. Brasavola, *Examen omnium simplicium medicamentorum, quorum in officinis usus est* (Lyon, 1537). On this early period, see K. M. Reeds, *Botany in Medieval and Renaissance Universities* (New York, 1971).

8. Biblioteca Universitaria Bologna (BUB), Aldrovandi MS 38/2–3, fos. 101r, 110r, 181r, 233r.

9. T. Garzoni, *La piazza universale di tutte le professioni del mondo* (Venice, 1587), p. 664.

10. Brasavola, *Examen*, p. 8. BUB, Aldrovandi MS 38/2–3, fo. 14r (Fenari to Aldrovandi, 19 December 1571).

11. BUB, Aldrovandi MS 38/2–3, fo. 241r (Orselini to Aldrovandi, 23 October 1554).

12. G. Melichio, *Avvertimenti nelle compositioni de' medicamenti per uso della spetiaria* (Venice, 1605), fo. 69v.

13. BUB, Aldrovandi MS 38/2–3, fo. 66r (Cantarini to Aldrovandi, 20 September 1562).

14. S. Kusukawa, 'The role of images in the development of natural history', *Archives of Natural History*, 38:2 (2011), pp. 189–213.

15. These instructions were often included in the trade manuals used by apothecaries: see *Ricettario Fiorentino* (Florence, 1574), pp. 2–13.

16. G. Fantuzzi, *Memorie della vita di Ulisse Aldrovandi* (Bologna, 1774), p. 13; M. Cermenati (ed.), *Francesco Calzolari da Verona e le sue lettere ad Ulisse Aldrovandi* (Rome, 1910).

17. A similar situation is recognisable in western and southern Europe, with botanical apothecaries operating from entrepots like Antwerp, Amsterdam, London, Seville and Marseilles (see Egmond, *Carolus Clusius*).

18. P. A. Mattioli, *I Discorsi nelli sei libri di Pedacio Dioscoride Anazarbeo della Materia Medicinale* (Venice, 1581), b4r; V. Pugliano, 'Pharmacy, testing and the language of truth in Renaissance Italy', *Bulletin of the History of Medicine*, 91:2 (2017), pp. 233–73; C. N. Fabbri, 'Treating medieval plague: the wonderful virtues of theriac', *Early Science and Medicine*, 12:3 (2007), pp. 247–83.

19. G. Pona, *Monte Baldo descritto* (Venice, 1617), p. 2. On this reconfiguration of the apothecary's identity, see Pugliano, 'Botanical artisans', pp. 104–50.

20. BUB, Aldrovandi MS 38/2-3, fo. 5r (12 January 1571). For the community's code of conduct, see A. Goldgar, *Impolite Learning: Conduct and Community in the Republic of Letters, 1680–1750* (New Haven, 1995).

21. BUB, Aldrovandi MS 38/2-3, fo. 176r (Fulcheri to Aldrovandi, 15 November 1571).

22. On apothecary collecting: Pugliano, 'Botanical artisans', pp. 227–85; E. Stendardo, *Ferrante Imperato: collezionismo e studio della natura a Napoli tra Cinque e Seicento* (Naples, 2001). On early modern collecting generally: P. Findlen, *Possessing Nature: Museums, Collecting, and Scientific Culture in Early Modern Italy* (Berkeley, CA, 1994), and the chapters by Lawrence and Felfe, this volume.

23. F. Calzolari, *Il viaggio di Monte Baldo* (Venice, 1566), C1v. Their claim resonated with the wider rhetoric documented across early modern artisan writing; see J. S. Amelang, *The Flight of Icarus: Artisan Autobiography in Early Modern Europe* (Stanford, CA, 1998).

24. Calzolari, *Il viaggio*; Pona, *Monte Baldo*. On the apothecaries' publications, see Pugliano, 'Pharmacy, testing and truth'.

25. Erlangen University Library, Briefsammlung Trew, Pona to Ludwig Jungermann (13 August, n/a).

26. Pugliano, 'Botanical artisans', pp. 286–334.

27. Melichio, *Avvertimenti*, b2v; J. Shaw and E. Welch, *Making and Marketing Medicine in Renaissance Florence* (Amsterdam, 2011).

28. Quoted in F. De Vivo, 'Pharmacies as centres of communication in early modern Venice', *Renaissance Studies*, 21:4 (2007), pp. 505–21.

29. BUB, Aldrovandi MS 38/2-3, fo. 110r (Fulcheri to Aldrovandi, 23 February 1569) and MS 38/2-1, 211r (Nobili to Aldrovandi, 18 February 1567); C. Fahy, *Printing a Book at Verona in 1622: The Account Book of Francesco Calzolari Jr* (Paris, 1993), p. 18.

30. P. Findlen, 'The formation of a scientific community: natural history in sixteenth-century Italy', in A. Grafton and N. Siraisi

(eds.), *Natural Particulars: Nature and the Disciplines in Renaissance Europe* (Cambridge, MA, 1999), pp. 369–400.

31. BUB, Aldrovandi MS 38/2, I, fo. 23r (Leoni to Aldrovandi, 30 October 1560).

32. G. B. Olivi, *De reconditis et praecipuis collectaneis ab Francisco Calceolario* (Verona, 1584), p. 3.

33. Quoted in D. A. Franchini et al., *La scienza a corte. Collezionismo eclettico, natura e immagine a Mantova fra Rinascimento e Manierismo* (Rome, 1979), p. 50.

34. Findlen, *Possessing Nature*, p. 226.

35. Pugliano, 'Botanical artisans', p. 219.

36. BUB, Aldrovandi MS 382/1, fo. 164r (Gentile della Torre to Aldrovandi, 6 Jul. 1554).

37. R. Palmer, 'Pharmacy in the Republic of Venice in the sixteenth century', in A. Wear, R. K. French and I. M. Lonie (eds.), *The Medical Renaissance in the Sixteenth Century* (Cambridge, 1985), pp. 100–17, at p. 100.

38. Quoted in Findlen, *Possessing Nature*, p. 174.

4 Horticultural networking and sociable citation

1. T. Sprat, *The History of the Royal-Society of London for the Improving of Natural Knowledge* (London, 1667), p. 72.

2. On the formation of early modern publics, see the web archive of *Making Publics* (http://makingpublics.mcgill.ca), an international research project, now concluded, which investigated new forms of association in Europe between 1500 and 1700. Research for this chapter arose initially from the author's affiliation with *Making Publics*.

3. J. Parkinson, *Paradisi in Sole Paradisus Terrestris* (London, 1629), cited parenthetically by page number throughout. Copies consulted include that in Early English Books Online and the facsimile produced as number 758 of The English Experience series (Theatrum Orbis Terrarum; Walter J. Johnson, 1975). On early gardening manuals in English, see R. Bushnell's *Green Desire: Imagining Early Modern English Gardens* (Ithaca, 2003).

4. On these ways of reading similar materials, see S. Greenblatt, *Renaissance Self-Fashioning: From More to Shakespeare* (Chicago, 1980) and S. Shapin, *A Social History of Truth: Civility and Science in Seventeenth-Century England* (Chicago, 1994). More recently, M. Swann's *Curiosities and Texts: The Culture of Collecting in Early Modern England* (Philadelphia, 2001) framed seventeenth-century collecting practices (including the collection of *naturalia*) in relation to the formation of genteel identities by collectors who curated not

only objects but relationships with one another. As Swann writes, 'collectors creatively combined literary and material forms as they fabricated new identities – and enhanced social status – for themselves' ('*The Compleat Angler* and the early modern culture of collecting', *English Literary Renaissance*, 37:1 (2007), pp. 100–17, p. 106). For a related mapping of related communities of scientific practitioners (local to London yet drawing on immigrant and international networks), see D. E. Harkness, '"Strange" ideas and "English" knowledge: natural science exchange in Elizabethan London', in P. H. Smith and P. Findlen (eds.), *Merchants and Marvels: Commerce, Science and Art in Early Modern Europe* (New York and London, 2002), pp. 137–62. E. Yale, *Sociable Knowledge: Natural History and the Nation in Early Modern Britain* (Philadelphia, 2016) offers a large-scale examination of the textual, material and social exchanges of a wide range of seventeenth-century knowledge-makers. See also Egmond's chapter, this volume.

5. See R. E. Duthie, 'English florists' societies and feasts in the seventeenth and first half of the eighteenth centuries', *Garden History*, 10:1 (1982), pp. 17–35, p. 18.

6. J. Bainbridge, 'To the Excellent Herbarist Mr. John Parkinson', in J. Parkinson, *Theatrum Botanicum* (London, 1640), sig. a1r.

7. J. Parkinson, 'To the Qveenes most Excellent Maiestie', *Paradisi in Sole Paradisus Terrestris* (London, 1629), sig. **2r and 'Epistle to the Reader', sig. **4r; page numbers cited parenthetically henceforth.

8. On cultural tensions embodied in cross-border relations between peoples and plants in Parkinson's time, see B. S. Robinson, 'Green seraglios: tulips, turbans and the global market', *Journal for Early Modern Cultural Studies*, 9:1 (2009), pp. 92–122. For another recent study of the racialised conceptual interpenetration of early modern plants and peoples, see J. Feerick, 'Botanical Shakespeares: the racial logic of plant life in *Titus Andronicus*', *South Central Review*, 26:1–2 (2009), pp. 82–102.

9. On the Catholic influence on Parkinson's childhood, see A. Parkinson, *Nature's Alchemist: John Parkinson, Herbalist to Charles I* (London, 2007), pp. 32–4.

10. C. M. Hibbard, 'Henrietta Maria (1609–1669)', *Oxford Dictionary of National Biography* (Oxford, 2004), available online at dx.doi.org/10.1093/ref:odnb/12947.

11. See R. Laroche, *Medical Authority and Englishwomen's Herbal Texts, 1550–1650* (Farnham, 2009).

12. On Tradescant, see A. L. Tigner, 'The Tradescants' culinary treasures', *Gastonomica*, 12:4 (2012), pp. 74–83.

13. P. Findlen. 'The formation of a scientific community: natural history in sixteenth-century Italy', in A. Grafton and N. Siraisi (eds.), *Natural Particulars: Renaissance Natural Philosophy and the Disciplines* (Cambridge, MA, 1999), pp. 369–400, pp. 373, 379.

5 European exchanges and communities

The writing of this chapter would not have been possible without the financial support of The Netherlands Organisation for Scientific Research (NWO) during successive research projects based at the University of Leiden. I would like to thank Peter Mason and the editors of this volume for their valuable comments and suggestions.

1. All examples concerning Clusius and friends are taken from F. Egmond, *The World of Carolus Clusius: Natural History in the Making* (London, 2010). Seminal work on natural history exchanges by G. Olmi remains untranslated. See his '"Molti amici in varii luoghi": studio della natura e rapporti epistolari nel secolo XVI', *Nuncius. Annali di Storia della Scienza*, 6:1 (1991), pp. 3–31 and *L'Inventario del mondo. Catalogazione della natura e luoghi del sapere nella prima età moderna* (Bologna, 1992).

2. Older seminal works are O. Impey and A. MacGregor, *The Origins of Museums: The Cabinets of Curiosities in Sixteenth and Seventeenth-Century Europe* (Oxford, 1985); and P. Findlen, *Possessing Nature: Museums, Collecting and Scientific Culture in Early Modern Italy* (Berkeley, 1994). More recently, see especially P. H. Smith and P. Findlen (eds.), *Merchants and Marvels: Commerce, Science and Art in Early Modern Europe* (New York, 2002) and A. MacGregor, *Curiosity and Enlightenment: Collectors and Collections from the Sixteenth to the Nineteenth Century* (New Haven, 2007).

3. For recent studies on Habsburg naturalia collecting, see A. Pérez de Tudela and A. Jordan Gschwend, 'Renaissance menageries: exotic animals and pets at the Habsburg courts in Iberia and central Europe', in K. Enenkel and P. Smith (eds.), *Early Modern Zoology: The Construction of Animals in Science, Literature and the Visual Arts* (Leiden, 2007), pp. 419–47; A. Jordan Gschwend, *Hans Khevenhüller at the Court of Philip II of Spain* (London, 2018); and S. Haag (ed.), *Echt tierisch! Die Menagerie des Fürsten* (Vienna, 2015). On the Medici, see esp. L. Tongiorgi Tomasi, 'The study of the natural sciences and botanical and zoological illustration in Tuscany under the Medicis from the sixteenth to the eighteenth centuries', *Archives of Natural History*, 28 (2001), pp. 179–94.

4. Quoted from B. Bukovinská, 'The known and unknown Kunstkammer of Rudolf II', in H. Schramm, L. Schwarte and J. Lazardzig (eds.), *Collection, Laboratory, Theater: Scenes of Knowledge in the 17th Century* (New York, 2005), pp. 199–227, p. 205.

5. See esp. S. Kusukawa, 'The role of images in the development of Renaissance natural history', *Archives of Natural History*, 38:2 (2011), pp. 189–213; S. Kusukawa, *Picturing the Book of Nature: Image, Text and Argument in Sixteenth-Century Human Anatomy and Medical Botany* (Chicago, 2012).

6. See M. Hochmann (ed.), *Villa Medici. Il sogno di un cardinale. Collezioni e artisti di Ferdinando de' Medici* (Rome, 1999); M. Bath, *Emblems for a Queen: The Needlework of Mary Queen of Scots* (London, 2008); Bath, *Renaissance Decorative Painting in Scotland* (Edinburgh, 2003), on Earshall, pp. 253–6.

7. See P. Mason, 'Eighty Brazilian birds for Johann Georg', *Folk Journal of the Dansk Ethnographic Society*, 43 (2001), pp. 103–21.

8. S. Kusukawa, 'The sources of Gessner's pictures for the *Historia animalium*', *Annals of Science*, 67:3 (2010), pp. 303–28; F. Egmond and S. Kusukawa, 'Circulation of images and graphic practices in Renaissance natural history: the example of Conrad Gessner', *Gesnerus*, 73:1 (2016), pp. 29–72.

9. R. Palmer, 'Medical botany in northern Italy in the Renaissance', *Journal of the Royal Society of Medicine*, 78 (1985), pp. 149–57, at p. 149. Still very relevant is F. D. Hoeniger and J. W. Shirley (eds.), *Science and the Arts in the Renaissance* (Cranbury and London, 1985). The recent standard work is B. Ogilvie, *The Science of Describing: Natural History in Renaissance Europe* (Chicago, 2006).

10. See esp. Ogilvie, *The Science of Describing*. On standards and styles of visual representation, see Kusukawa, *Picturing the Book of Nature*; F. Egmond, *Eye for Detail: Images of Plants and Animals in Art and Science, 1500–1630* (London, 2017).

11. On scientific community, see P. Findlen, 'The formation of a scientific community: natural history in sixteenth-century Italy', in A. Grafton and N. Siraisi (eds.), *Natural Particulars: Nature and the Disciplines in Renaissance Europe* (Cambridge, MA, 1999), pp. 369–400; M. Biagioli, 'Scientific revolution, social bricolage and etiquette', in R. Porter and M. Teich (eds.), *The Scientific Revolution in National Context* (Cambridge, 1992), pp. 11–54; Biagioli, 'Etiquette, interdependence and sociability in seventeenth-century science', *Critical Inquiry*, 22 (1996), pp. 193–238.

12. See A. M. Raugei (ed.), *Pinelli, Gian Vincenzo - Dupuy, Claude: Une correspondance entre deux humanistes*, 2 vols. (Florence,

2001). No survey of Pinelli's network or of his further correspondence has been published to my knowledge, nor have I come across literature surveying the sixteenth-century history of 'private' or informal societies operating in the field of natural history.

13. See S. Dupré, B. De Munck, W. Thomas and G. Vanpaemel (eds.), *Embattled Territory: The Circulation of Knowledge in the Spanish Netherlands* (Ghent, 2016); K. Davids, 'Dutch and Spanish global networks of knowledge in the early modern period: structures, connections, changes', in L. Roberts (ed.), *Centres and Cycles of Accumulation in and around the Netherlands during the Early Modern Period* (Zurich and Berlin, 2011), pp. 29–52.

6 Making monsters

1. D. Wilson, *Signs and Portents: Monstrous Births from the Middle Ages to the Enlightenment* (London, 1993), p. 6; A. W. Bates, *Emblematic Monsters: Unnatural Conceptions and Deformed Births in Early Modern Europe* (Amsterdam, 2005), pp. 12–13.

2. Wilson, *Signs and Portents*, pp. 75, 152; S. Davies, 'The unlucky, the bad and the ugly: categories of monstrosity from the Renaissance to the Enlightenment', in A. S. Mittman and P. Dendle (eds.), *Ashgate Research Companion to Monsters and the Monstrous* (Farnham, 2013), pp. 49–77.

3. C. van Duzer, '*Hic sunt dracones*: the geography and cartography of monsters', in Mittman and Dendle (eds.), *Ashgate Research Companion*, pp. 387–91; H. Strickland, 'Monstrosity and race in the late Middle Ages', in Mittman and Dendle (eds.), *Ashgate Research Companion*, p. 386; Davies, 'The unlucky, the bad and the ugly', p. 49.

4. J. Cañizares-Esguerra, 'Demons, stars, and the imagination: the early modern body in the tropics', in B. Isaac, M. Eliav-Feldon and Y. Ziegler (eds.), *Racism in Western Civilisations before 1700* (Cambridge, 2009), pp. 314, 320; P. Adams, *Travellers and Travel Liars, 1660–1800* (Berkeley, 1962).

5. K. A. Enenkel, 'The species and beyond: classification and the place of hybrids in early modern zoology', in K. A. Enenkel and P. J. Smith (eds.), *Zoology in Early Modern Culture: Intersections of Science, Theology, Philology, and Political and Religious Education* (Leiden, 2014), pp. 108–9; J. Cañizares-Esguerra, *Nature, Empire and Nation: Explorations of the History of Science in the Iberian World* (Stanford, CA, 2006), pp. 73–89.

6. J. de Jong, 'Drawings, ships and spices: accumulation in the Dutch East India Company', in L. Roberts (eds.), *Centres and*

Cycles of Accumulation in and around the Netherlands during the Early Modern Period (Berlin, 2011), pp. 177–204; D. Margócsy, *Commercial Visions: Science, Trade and Visual Culture in the Dutch Golden Age* (London, 2014), p. 25.

7. S. Schaffer, L. Roberts, K. Raj and J. Delbourgo (eds.), *The Brokered World: Go-Betweens and Global Intelligence, 1770–1820* (Sagamore Beach, MA, 2009).

8. P. J. Smith, 'On toucans and hornbills: readings in early modern ornithology from Belon to Buffon', in Enenkel and Smith (eds.), *Early Modern Zoology*, p. 77; N. Lawrence, 'Assembling the dodo in early modern natural history', *British Journal for the History of Science*, 48:3 (2015), pp. 387–408.

9. P. Mason, *Before Disenchantment: Images of Exotic Animals and Plants in the Early Modern World* (London, 2009), pp. 124–35.

10. P. Swadling, *Plumes from Paradise: Trade Cycles in Outer Southeast Asia and their Impact on New Guinea and Nearby Islands until 1920* (Papua New Guinea, 1996), introduction and chapter 2.

11. A. Pigafetta, quoted in H. E. J. Stanley (ed.), *The First Voyage Round the World by Magellan* (New York, 2010), p. 143.

12. M. Transylvanus, 'A letter from Maximilianus Transylvanus to the Most Reverend Cardinal of Salzburg', in Stanley, *First Voyage Round the World*, pp. 205–6.

13. Aristotle, *History of Animals*, 9 vols., trans. D. Wentworth Thompson (1910; republished by University of Adelaide Ebooks, 2015), vol. I, pt. 1, available at https:ebooks.adelaide.edu.au/a/aristotle/history/.

14. C. Gessner, *Historiae animalium*, vol. III, *De avium natura* (Zurich, 1555), pp. 636–9; J.-M. Massing, '*Paradisaea apoda*: the symbolism of the bird of paradise in the sixteenth century', in J. A. Levenson (ed.), *Encompassing the Globe: Portugal and the World in the 16th and 17th Centuries* (Washington, DC, 2007), pp. 29–37.

15. U. Aldrovandi, quoted in F. Willughby and J. Ray, *The Ornithology of Francis Willughby* (London, 1678), p. 91.

16. H. J. Cook, *Matters of Exchange: Commerce, Medicine and Science in the Dutch Golden Age* (New Haven, 2007), pp. 120–1: recent Dutch independence from Habsburg dominion in 1581 made competition especially fierce.

17. E. Fucikova, 'The collection of Rudolf II at Prague: cabinet of curiosities or scientific museum?', in O. R. Impey and A. MacGregor (eds.), *The Origins of Museums: The Cabinet of Curiosities in Sixteenth- and Seventeenth-Century Europe* (Oxford, 1985), pp. 47–53; T. daCosta Kaufmann, *The School of Prague: Painting at the Court of Rudolf II* (Chicago, 1988).

18. C. Clusius, *Exoticorum libri decem* (Leiden, 1605), p. 359; C. Swan, 'Exotica on the move: birds of paradise in early modern Holland', *Art History*, 38 (2015), pp. 620–5.

19. A. Pagden, *European Encounters with the New World* (New Haven and London, 1993), pp. 51–9.

20. Clusius, *Exoticorum*, p. 359.

21. Clusius, *Exoticorum*, pp. 360–1.

22. M. Wintroub, 'The looking glass of facts: collecting, rhetoric and citing the self in the experimental natural philosophy of Robert Boyle', *History of Science*, 35 (1997), pp. 189–217; S. Shapin, 'Pump and circumstance: Robert Boyle's literary technology', *Social Studies of Science*, 14:4 (1984), pp. 481–520; Pagden, *European Encounters*, pp. 51–9; T. Cave, *The Cornucopian Text: Problems of Writing in the French Renaissance* (Oxford, 1985); M. Campbell, *The Witness and the Other World: Exotic European Travel Writing, 400–1600* (Ithaca, 1988).

23. Clusius, *Exoticorum*, pp. 360–1; S. Greenblatt, 'Resonance and wonder', in I. Karp and S. Lavine (eds.), *Exhibiting Culture: The Poetics and Politics of Museum Display* (Washington, DC, 1991); S. Greenblatt, *Marvelous Possessions: The Wonder of the New World* (Oxford, 2003), pp. 31–4; L. Daston and K. Park, *Wonders and the Order of Nature: 1150–1750* (New York, 1998); P. Findlen, 'Inventing nature, commerce, art and science in the early modern cabinet of curiosities', in P. Smith and P. Findlen (eds.), *Merchants and Marvels: Commerce, Science and Art in Early Modern Europe* (London, 2001), pp. 297–323; K. Whitaker, 'The culture of curiosity', in N. Jardine, J. A. Secord and E. C. Spary (eds.), *Cultures of Natural History* (Cambridge, 1996), pp. 75–90.

24. Clusius, *Exoticorum*, pp. 360–1; P. Findlen, 'Jokes of nature, jokes of knowledge: the playfulness of scientific discourse in early modern Europe', *Renaissance Quarterly*, 43:2 (1990), pp. 292–331.

25. F. Egmond, *The World of Carolus Clusius: Natural History in the Making, 1550–1610* (London, 2010), pp. 184–5.

26. Clusius, in Willughby and Ray, *Ornithology*, p. 94; Clusius, *Exoticorum*, p. 361; Adams, *Travellers and Travel Liars*.

27. Clusius, in Willughby and Ray, *Ornithology*, p. 94; Clusius, *Exoticorum*, p. 361; Adams, *Travellers and Travel Liars*.

28. Clusius, *Exoticorum*, pp. 360–1; Heemskerck in E. Stresemann, 'Die Entdeckungsgeschichte der Paradiesvogel', *Journal of Ornithology*, 95 (1954), pp. 263–91, p. 268.

29. J. Bontius, *Historiae naturalis et medicae Indiae Orientalis libri sex*, pp. 64–5. This work was published posthumously as part of *De Indiae utriusque re naturali et medica* (Amsterdam, 1658) by the Dutch physician Willem Piso, from a manuscript Bontius produced while stationed in the Indies.

30. Willughby and Ray, *Ornithology*, pp. 55, 90–7.
31. Two studies that examine similar cases of emblematic exotic nature are M. Dove and C. Carpenter, 'The "poison tree" and the changing vision of the Indo-Malay realm', in R. L. Wadley (ed.), *Histories of the Borneo Environment: Economic, Political and Social Dimensions of Change and Continuity* (Leiden, 2005), pp. 183–212, and J. A. Boon, *Affinities and Extremes: Crisscrossing the Bittersweet Ethnology of East Indies History, Hindu-Balinese Culture, and Indo-European Allure* (Chicago, 1990).
32. L. de Camões and E. R. Hodges (eds.), *The Lusiad: Or, The Discovery of India*, 5th edn, trans. W. J. Mickle (London, 1776), p. 311.
33. W. Eisler, *The Furthest Shore: Images of Terra Australis from the Middle Ages to Captain Cook* (Cambridge, 1995), p. 74.
34. Swadling, *Plumes from Paradise*, pp. 41–3; Bontius, quoted in Willughby and Ray, *Ornithology*, p. 91.
35. J. Jonston, *Historiae naturalis de quadrupedibus libri* (Amsterdam, 1657), pp. 117–18.
36. The Mauritshius, The Hague, the Netherlands; J. R. Marcaida, 'Rubens and the bird of paradise: painting natural knowledge in the early seventeenth century', *Renaissance Studies*, 28 (2014), pp. 112–27.
37. Kunsthistorisches Museum, Vienna.
38. Museo del Prado, Madrid.
39. Mittman and Dendle, *Ashgate Research Companion*, p. 9; van Duzer, '*Hic sunt dracones*', p. 393.
40. H. Smith and L. Wilson (eds.), *Renaissance Paratexts* (Cambridge, 2011), introduction.

7 Indigenous naturalists

I am grateful to Allison Caplan, Marcy Norton and Guilhem Olivier for fruitful exchange and advice in the preparation of this chapter.

1. On the Tlatelolco workshop, see D. Magaloni Kerpel, *The Colors of the New World: Artists, Materials and the Creation of the Florentine Codex* (Los Angeles, 2014), pp. 1–8. The Mexica were the inhabitants of Mexico-Tenochtitlan before the Spanish conquest; the name 'Aztec' was popularised by Alexander von Humboldt in the nineteenth century to stress their origins in the northwestern territory of Aztlan.
2. G. Olivier, '¿Modelos europeos o concepciones indígenas? El ejemplo de los animales en el Libro XI del *Códice florentino* de fray Bernardino de Sahagún', in J. R. Romero Galván and

P. Máynez (eds.), *El universo de Sahagún: pasado y presente. Coloquio 2005* (Mexico City, 2007), pp. 127–8.

3. Magaloni, *Colors of the New World*, p. 27.

4. D. Robertson, 'The sixteenth century Mexican encyclopedia of Fray Bernardino de Sahagún', *Cahiers d'Histoire Mondiale*, 9 (1966), pp. 617–27; A. Hernández, 'La *Historia general* de Sahagún a la luz de las enciclopedias de tradición grecorromana', in *Bernardino de Sahagún. Quinientos años de presencia* (Mexico City, 2002), pp. 41–59.

5. R. French, *Ancient Natural History: Histories of Nature* (London, 1994), p. 230.

6. D. Robertson, *Mexican Manuscript Painting of the Early Colonial Period* (New Haven, 1959); E. H. Boone, 'Pictorial documents and visual thinking in postconquest Mexico', in E. H. Boone and T. Cummins (eds.), *Native Traditions in the Postconquest World: A Symposium at Dumbarton Oaks, 2nd through 4th October 1992* (Washington, DC, 1998), pp. 149–200.

7. For another example of the use of the micro-historical approach in analysing knowledge transfers, see M. Norton, 'The quetzal and *The Ornithology of Francis Willughby*: microhistory, Mesoamerican knowledge and early modern natural history', in R. Bauer and J. Marroquín Arredondo (eds.), *Translating Nature: A Transcultural History of Science in the Early Modern Atlantic* (Philadelphia, forthcoming).

8. A. López Austin, 'The research method of Fray Bernardino de Sahagún: the questionnaires', in M. S. Edmonson (ed.), *Sixteenth-Century Mexico: The Work of Sahagún* (Albuquerque, 1974), pp. 111–49.

9. M. León Portilla, *Bernardino de Sahagún: First Anthropologist* (Norman, OK, 2002), p. 161.

10. I. Palmeri Capesciotti, 'La fauna del Libro XI del *Códice Florentino* de Fray Bernardino de Sahagún. Dos sistemas taxonómicos frente a frente', *Estudios de Cultura Náhuatl*, 32 (2001), pp. 189–221; P. Escalante Gonzalbo, 'Los animales del Códice Florentino en el espejo de la tradición occidental', *Arqueología Mexicana*, 6:36 (1999), pp. 52–9.

11. Biblioteca Medicea Laurenziana. MS Mediceo Palatino 218–220: Sahagún, *Historia general*, vol. III, book 11, fos. 211r, 190r, 62v.

12. Sahagún, *Historia general*, vol. III, book 11, fo. 151v.

13. A. Blanco, G. Pérez, B. Rodríguez, N. Sugiyama, F. Torres and R. Valadez, 'El zoológico de Moctezuma ¿Mito o realidad?', *AMMVEPE*, 20:2 (2009), pp. 28–39; H. B. Nicholson, 'Montezuma's Zoo', *Pacific Discovery*, 8:4 (1955), pp. 3–11; R. Martín del Campo, 'El más antiguo parque zoológico de América', *Anales del Instituto de Biología* 14 (1943), pp. 635–43.

14. Sahagún, *Historia general*, vol. II, book 8, chapter 4, fo. 30rv.

15. A. Russo, 'Plumes of sacrifice: transformations in sixteenth-century Mexican feather art', *RES: Anthropology and Aesthetics*, 42 (2002), pp. 226–50.

16. E. H. Boone, 'Incarnations of the Aztec supernatural: the image of Huitzilopochtli in Mexico and Europe', *Transactions of the American Philosophical Society*, 79:2 (1989), pp. i–iv and 1–107, p. 4; A. Hvidfeldt, *Teotl and Ixiptlatli. Some Central Concepts in Ancient Mexican Religion* (Copenhagen, 1958).

17. H. Cortés, *Carta tercera de relación* (Seville, 1523), n. p.

18. A. O. Lovejoy, *The Great Chain of Being: A Study of the History of an Idea* (Cambridge, MA, 1936).

19. Sahagún, *Historia general*, vol. III, book 11, fo. 151v.

20. Here I follow the version of Y. González Torres, 'Seler y Huitzilopochtli', in R. von Hanffstengel and C. Tercero Vasconcelos (eds.), *Eduard y Caecilie Seler: Sistematización de los estudios americanistas y sus repercusiones* (Mexico City, 2003), pp. 127–36, p. 127.

21. E. Seler, 'Huitzilopocthli, the talking hummingbird', in C. P. Bowditch, E. S. Thomson and F. B. Richardson (eds.), *Collected Works in Mesoamerican Linguistics and Archaeology* (Culver City, CA, 1996 [1923]), pp. 93–9, at p. 93. One recent exception is M. López Hernández, 'El colibrí como símbolo de la sexualidad masculina entre los mexicas', *Itinerarios*, 21 (2015), pp. 79–100.

22. On torpor see, for instance, S. Hiebert, 'Energy costs and temporal organisation of torpor in the rufous hummingbird (*Selasphorus rufus*)', *Physiological Zoology*, 63 (1990), pp. 1082–97.

23. Sahagún, *Historia general*, vol. III, book 11, fo. 24r.

24. A. de Molina, *Vocabulario en lengua castellana y mexicana* (Mexico City, 1571), fo. 149r and fo. 155r. I am grateful to Allison Caplan for alerting me to this inconsistency.

25. John Carter Brown Library, MS Codex Ind 2: '*Codex Tovar*', fo. 148v.

26. The main versions of the migration visual narrative are found in 'Códice Boturini', in J. Corona Núñez (ed.), *Antigüedades de México, basadas en la recopilación de Lord Kingsborough* (Mexico, 1964), vol. I; *Codex Azcatitlan*, trans. L. López Luján and D. Michelet, annot. R. H. Barlow and M. Graulich (Paris, 1995); E. Mengin, 'Commentaire du Codex mexicanus no. 23–24 de la Bibliothèque Nationale de Paris', *Journal de la Société des Américanistes*, 41–2 (1952), pp. 387–498.

27. *Codex Azcatitlan*, pp. 72–5.

28. D. Durán, *The History of the Indies of New Spain*, trans. D. Heyden (Norman, OK, 1994 [*c.*1581]), p. 119.

29. On Hernández, see S. Varey, R. Chabrán and D. B. Weiner (eds.), *Searching for the Secrets of Nature: The Life and Works of Dr Francisco Hernández* (Stanford, 2000); G. Somolinos

D'Ardois, *Vida y obra de Francisco Hernández* (Mexico City, 1960), vol. I; S. Varey (ed.), *The Mexican Treasury: The Writings of Dr. Francisco Hernández* (Stanford, 2000).

30. F. Hernández, *Rerum medicarum Novae Hispaniae thesaurus seu Nova plantarum, animalium et mineralium Mexicanorum historia*, ed. Nardo Antonio Recchi (Rome, 1651), pp. 321–2.

31. Hernández, *Rerum medicarum*, p. 322.

32. I. Montero Sobrevilla, 'Transatlantic hum: natural history and the itineraries of the torpid hummingbird, *c.* 1521–1790', PhD thesis, University of Cambridge (2015), chapter 2.

33. F. Ximénez, *Quatro libros de la naturaleza y virtudes medicinales de las plantas y animales de la Nueva España* (Mexico City, 1615), p. 188v.

34. N. Grew, 'A query put by Dr N. Grew concerning the food of the humming bird', *Philosophical Transactions*, 17 (1693), p. 815.

35. Sahagún, *Historia general*, vol. III, book 11, fo. 238v, quoted in Magaloni, *Colors of the New World*, p. 14.

8 Insects, philosophy and the microscope

1. Most pages of the album survive in the National Gallery of Art, Washington, DC. See L. Hendrix and T. Vignau-Wilberg, *Mira Calligraphiae Monumenta: A Sixteenth-Century Calligraphic Manuscript Inscribed by Georg Bocskay and Illuminated by Joris Hoefnagel* (Malibu, 1992), pp. 15–28; T. Vignau-Wilberg (ed.), *Archetypa Studiaque Patris Georgii Hoefnagelii, 1592: Natur, Dichtung und Wissenschaft in der Kunst um 1600* (München, 1994), pp. 17–29.

2. B. W. Ogilvie, 'Order of insects: insect species and metamorphosis between Renaissance and Enlightenment', in O. Nachtomy and J. E. H. Smith (eds.), *The Life Sciences in Early Modern Philosophy* (Oxford, 2014), pp. 222–45.

3. S. Kusukawa, *The Transformation of Natural Philosophy: The Case of Philip Melanchthon* (Cambridge, 1995); E. Jorink, *Reading the Book of Nature in the Dutch Golden Age, 1575–1715* (Leiden, 2010).

4. Pliny, *Naturalis historia*, ed. and trans. H. Rackham (London and Cambridge, MA, 1938–52), book XI, p. i.

5. Jorink, *Book of Nature*, p. 20; emphasis added.

6. G. Hoefnagel, *Archetypa studiaque patris Georgii Hoefnagelii* (Frankfurt, 1592); see also Vignau-Wilberg (ed.), *Archetypa*.

7. J. Ray, *The Wisdom of God Manifested in the Works of the Creation* (London, 1691), p. 130.

8. W. B. Ashworth, Jr, 'Emblematic natural history of the Renaissance', in N. Jardine, J. A. Secord and E. C. Spary (eds.), *Cultures of Natural History* (Cambridge, 1996), pp. 17–37.

9. D. Harkness, *The Jewel House: Elizabethan London and the Scientific Revolution* (New Haven, 2007).

10. T. Moffet, *Insectorum sive minimorum animalium theatrum* (London, 1634); British Library (BL) MS Sloane 4014.

11. E. Ruestow, *The Microscope in the Dutch Republic: The Shaping of Discovery* (Cambridge, 1996); C. Huygens, *Mijn jeugd* (Amsterdam, 1993), p. 132.

12. On Descartes in the Dutch Republic, see D. Clarke, *Descartes: A Biography* (Cambridge, 2006); H. J. Cook, *Matters of Exchange: Commerce, Medicine and Science in the Dutch Golden Age* (New Haven, 2007), pp. 226–66.

13. M. Cobb, *The Egg and Sperm Race: The Seventeenth-Century Scientists Who Unravelled the Secrets of Sex, Life and Growth* (London, 2006).

14. Jorink, *Book of Nature*, pp. 201–8.

15. B. W. Ogilvie, 'Willughby on insects', in T. Birkhead (ed.), *Virtuoso by Nature: The Scientific Worlds of Francis Willughby FRS (1635–1672)* (Leiden, 2016).

16. T. Birch, *The History of the Royal Society of London*, 4 vols. (London, 1756–7), vol. I, p. 270.

17. J. T. Harwood, 'Rhetoric and graphic in *Micrographia*', in M. Hunter and S. Schaffer (eds.), *Robert Hooke: New Studies* (Woodbridge, 1989), pp. 119–47; J. Bennett, M. Cooper, M. Hunter and L. Jardine, *London's Leonardo: The Life and Work of Robert Hooke* (Oxford, 2003).

18. *Oeuvres complètes de Christiaan Huygens*, 22 vols. (The Hague, 1888–1950), vol. V, p. 305.

19. Robert Hooke, *Micrographia: or, some Physiological Descriptions of Minute Bodies made by Magnifying Glasses with Observations and Inquiries Thereupon* (London, 1665) pp. 171–2.

20. Cobb, *Egg and Sperm Race*; B. Ogilvie. 'Insects in John Ray's natural history and natural theology', in K. Enenkel and P. J. Smith (eds.), *Natural History and the Arts from the Perspective of Religion and Politics (16th–17th centuries)* (Leiden, 2014), pp. 234–57; Ogilvie, 'Willughby on insects'; Jorink, *Book of Nature*, pp. 239–68.

21. J. Swammerdam, *Bybel der natuure of historie der insecten. Bibla naturae, sive historia insectorum* (Leiden, 1737–8), p. 669.

22. Cobb, *Egg and Sperm Race*, pp. 63–93.

23. M. Cobb, 'Malpighi, Swammerdam and the colourful silkworm: replication and visual representation in early modern science', *Annals of Science*, 59 (2002), pp. 111–47.

24. Ray to James Petiver, BL MS Sloane 4063, fo. 210. According to a letter of Ray to Petiver, BL MS Sloane 4064, fo. 9, Ray owned the French translation of Swammerdam's *Historia*.

25. Much has been written on van Leeuwenhoek; see Ruestow, *Microscope in the Dutch Republic*, pp. 146–200; F. Henderson, 'Making 'the good old man' speak English: the reception of Antonie van Leeuwenhoek's letters at the Royal Society, 1673–1723', in H. J. Cook and S. Dupré (eds.), *Translating Knowledge in the Early Modern Low Countries* (Berlin, 2014), pp. 243–68.

26. G. A. Lindeboom (ed.), *The Letters of Jan Swammerdam to Melchisédec Thévenot* (Amsterdam, 1975), p. 106.

27. *Alle de brieven van Antoni van Leeuwenhoek/The Collected Letters of Antoni van Leeuwenhoek* (Amsterdam, 1939-), vol. I, pp. 29–39, 42–61.

9 The materials of natural history

1. A. Seba, *Locupletissimi rerum naturalium thesauri* (Amsterdam, 1734–65), vol. I, preface, quoted (with modifications) in E. Jorink, *Reading the Book of Nature in the Dutch Golden Age*, trans. P. Mason (Leiden, 2010), p. 335.

2. P. Mason, *Before Disenchantment: Images of Exotic Plants and Animals in the Early Modern World* (London, 2009), pp. 129–30; H. J. Cook, *Matters of Exchange: Commerce, Medicine, and Science in the Dutch Golden Age* (New Haven, 2007), p. 129; Jorink, *Reading the Book of Nature*, pp. 265–6.

3. J. Camus, 'Historique des premiers herbiers', *Malpighia*, 9 (1895), pp. 283–314; A. Arber, *Herbals: Their Origins and Evolution* (1938; rev. edn Cambridge, 1986), pp. 138–43; L. Tongiorgi Tomasi, 'Dall'essenza vegetale agglutinata all'immagine a stampa: il percorso dell'illustrazione botanica nei secoli XVI–XVII', *Museologia scientifica*, 8 (1992), pp. 271–95. Aldrovandi's discussion of his improvements to the herbarium can be found in L. Frati, 'La vita d'Ulisse Aldrovandi', in A. Baldacci et al., *Intorno alla vita e alle opera di Ulisse Aldrovandi* (Bologna, 1907), p. 24; B. Antonino (ed.), *L'erbario di Ulisse Aldrovandi* (Milan, 2003).

4. M. de Lobel and P. Pena, *Stirpium adversaria nova* (London, 1571), p. 271.

5. C. Clusius, *Exoticorum libri decem* (Leiden, 1605), pp. 87, 90. See S. Kusukawa, 'The uses of pictures in printed books: the case of Clusius's *Exoticorum libri decem*', in F. Egmond, P. Hoftijzer and R. Visser (eds.), *Carolus Clusius: Towards a Cultural*

History of a Renaissance Naturalist (Amsterdam, 2007), pp. 221–46, p. 240.

6. S. Quiccheberg, *The First Treatise on Museums: Samuel Quiccheberg's Inscriptiones 1565*, trans. M. A. Meadow and B. Robertson (Los Angeles, 2013), p. 65. For Daniel Frösch's description of the bird of paradise, see K. Schulze-Hagen, F. Steinheimer, R. Kinzelbach and C. Gasser, 'Avian taxidermy in Europe from the Middle Ages to the Renaissance', *Journal für Ornithologie*, 144 (2003), pp. 459–78, p. 471.

7. P. Belon, *L'histoire de la nature des oiseaux* (Paris, 1555), vol. I, p. 8; Schulze-Hagen et al., 'Avian taxidermy', pp. 459, 464–5, 471; K. H. Dannenfeldt, 'Egyptian mumia: the sixteenth-century experience and debate', *The Sixteenth Century Journal*, 16 (1985), pp. 163–80.

8. Schulze-Hagen et al., 'Avian taxidermy', p. 66.

9. Schulze-Hagen et al., 'Avian taxidermy', p. 82.

10. Mason, *Before Disenchantment*, p. 136. Salviani, quoted in P. Findlen, 'The death of a naturalist: knowledge and community in Renaissance Italy', in C. Klestinec and G. Manning (eds.), *Professors, Physicians and Practices in the History of Medicine: Essays in Honor of Nancy Siraisi* (New York, 2017), pp. 127–67.

11. R. Alexandratos, '"With the true eye of a lynx": the paper museum of Cassiano dal Pozzo', in David Attenborough, et al., *Amazing Rare Things: The Art of Natural History in the Age of Discovery* (New Haven, 2015), pp. 72–105.

12. N. Grew, *Musaeum Regalis Societatis, or a Catalogue & Description of the Natural and Artificial Rarities Belonging to the Royal Society and Preserved at Gresham Colledge* (London, 1681), p. 58; J. E. Symonds, *Fluid Preservation: A Comprehensive Reference* (Lanham, MD, 2014), pp. 8–11.

13. R. Boyle, *Some Considerations Touching the Usefulness of Experimental Natural Philosophy* (Oxford, 1663), essay I, p. 26.

14. R. M. Peck, 'Preserving nature for study and display', in S. A. Prince (ed.), *Stuffing Birds, Pressing Plants, Shaping Knowledge in North America, 1730–1860* (Philadelphia, 2003), p. 14.

15. O. Thomas, 'On the probable identity of certain specimens, formerly in the Lidth de Jeude collection, and now in the British Museum, with those figured by Albert Seba in his "Thesaurus" of 1734', *Proceedings of the Zoological Society of London*, 22 (1892), pp. 309–18.

16. Cook, *Matters of Exchange*, pp. 268–81; D. Margócsy, *Commercial Visions: Science, Trade, and Visual Culture in the Dutch Golden Age* (Chicago, 2014), pp. 114–31.

17. H. Sloane, *A Voyage to the Islands Madera, Barbados, Nieves, S. Christophers and Jamaica*, 2 vols. (London, 1707–25), vol. II,

p. 346. See A. MacGregor (ed.), *Sir Hans Sloane: Collector, Scientist, Antiquary, Founding Father of the British Museum* (London, 1994); M. Hunter, A. Walker and A. MacGregor (eds.), *From Books to Bezoars: Sir Hans Sloane and his Collections* (London, 2012).

18. J. Woodward, *Brief Instructions for Making Observations in All Parts of the World* (London, 1696), pp. 10, 14; D. Price, 'John Woodward and a surviving British geological collection from the early eighteenth century', *Journal of the History of Collections*, 1 (1989), pp. 79–95.

19. J. Petiver, *Brief Directions for the Easie Making and Preserving Collections of all Natural Curiosities* (c.1700), n. p.; R. P. Stearns, 'James Petiver: promoter of natural science, c.1663–1718', *Proceedings of the American Antiquarian Society*, 62 (1953), pp. 243–365, p. 285 n99.

20. Sloane, *Voyage*, vol. II, preface.

21. MacGregor (ed.), *Sir Hans Sloane*, p. 120.

22. A.-J. Dezallier d'Argenville, *La Conchyliologie* (Paris, 1780), vol. I, p. 199.

23. Walker et al., *From Books to Bezoars*, pp. 10, 69.

24. C. Linnaeus, *Philosophia botanica* (Stockholm, 1751), pp. 291, 309; Linnaeus, *Instructio musei rerum naturalium* (Uppsala, 1753), p. 15.

25. C. Jarvis, 'A concise history of the Linnean Society's Linnaean Herbarium, with some notes on the dating of the specimens it contains', *The Linnean*, 7 (2007), pp. 5–18; R. M. Peck, 'Alcohol and arsenic, pepper and pitch', in Prince (ed.), *Stuffing Birds, Pressing Plants*, pp. 37–8; W. T. Stearn and G. Bridson, *Carl Linnaeus (1707–1778): A Bicentenary Guide to the Career and Achievements of Linnaeus and the Collections of the Linnean Society* (London, 1978), 16; S. Müller-Wille, 'Linnaeus' herbarium cabinet: a piece of furniture and its function', *Endeavour*, 30:2 (2008), pp. 60–4.

26. P. Mauriès, *Cabinets of Curiosities* (New York, 2002), p. 194.

27. J. M. MacKenzie, *Museums and Empire: Natural History, Human Cultures and Colonial Identities* (Manchester, 2009), p. 1; P. Findlen, *Possessing Nature: Museums, Collecting and Scientific Culture in Early Modern Italy* (Berkeley, 1994), p. 403.

28. E. C. Spary, *Utopia's Garden: French Natural History from Old Regime to Revolution* (Chicago, 2000), p. 11; S. Conn, *Museums and American Intellectual Life, 1876–1926* (Chicago, 1998), p. 39.

29. C. Argot, 'Changing views in paleontology: the story of a giant (Megatherium, Xenarthra)', in M. Dagosto and E. J. Sargis (eds.), *Mammalian Evolutionary Morphology* (Amsterdam, 2008), p. 38.

30. M. J. S. Rudwick, *Bursting the Limits of Time: The Reconstruction of Geohistory in the Age of Revolution* (Chicago, 2005), p. 357.

31. R. Klein and L. Rexer, *American Museum of Natural History: 125 Years of Expedition and Discovery* (New York, 1995), p. 25.

32. W. T. Alderson, *Mermaids, Mummies and Mastodons: The Emergence of the American Museum* (Washington, DC, 1992), p. 26.

33. G.-L. Leclerc de Buffon, *Historie naturelle des oiseaux*, 9 vols. (Paris, 1771), vol. II, p. 3; L. E. Robbins, *Elephant Slaves and Pampered Parrots: Exotic Animals in Eighteenth-Century Paris* (Baltimore, 2002), p. 66.

34. Peck, 'Preserving nature', p. 23 n21. For the French techniques that inspired Peale, see P. L. Farber, 'The development of taxidermy and the history of ornithology', *Isis*, 68 (1977), pp. 550–66; Farber, *Discovering Birds: The Emergence of Ornithology as a Scientific Discipline 1760–1850* (Baltimore, 1997).

35. K. Sloan, 'Sir Hans Sloane's pictures: the science of connoisseurship or the art of collecting?', *Huntington Library Quarterly*, 78:2 (2015), pp. 381–415, p. 383.

10 Experimental natural history

1. R.-A. F. de Réaumur, *Mémoires pour servir à l'histoire des insectes*, 6 vols. (Paris, 1734–42), vol. I, p. 351. In *Mémoires*, vol. VI, p. lxxix, Réaumur used *physicien* and *naturaliste* interchangeably.

2. J.-A. Nollet, *Discours sur les dispositions et sur les qualités qu'il faut avoir pour faire du progrès dans l'etude de la physique expérimentale* (Paris, 1753), p. 9.

3. Nollet, *Discours*, p. 24.

4. Nollet, *Discours*, p. 24.

5. Nollet, *Discours*, p. 30.

6. 'Expériences à faire faire par M. l'Abbé Nollet', Archives de l'Académie des Sciences (Paris), Fonds Réaumur, carton 5, dossier 47.

7. R.-A. F. de Réaumur, 'Expériences sur les différents degrés de froid qu'on peut produire, en mêlant de la Glace avec différents Sels', *Mémoires de l'Académie royale des sciences*, 1734, pp. 186–8.

8. On temperature experiments, see M. Terrall, *Catching Nature in the Act: Réaumur and the Practice of Natural History in the Eighteenth Century* (Chicago, 2014), pp. 52–3.

9. Réaumur, *Mémoires*, vol. VI, pp. 10–11.

10. Réaumur, *Mémoires*, vol. I, p. 152.

11. Réaumur, *Mémoires*, vol. III, p. 45.

12. Réaumur, *Mémoires*, vol. III, p. 53.

13. Réaumur, *Mémoires*, vol. III, pp. 74–80.

14. Réaumur, *Mémoires*, vol. III, explanation of the vignette on p. xl.

15. On this episode, see Terrall, *Catching Nature in the Act*, pp. 98–101.

16. Réaumur, *Mémoires*, vol. II, pp. xxxix–xl.

17. J. T. Needham, 'A summary of some late observations upon the generation, composition, and decomposition of animal and vegetable substances', *Philosophical Transactions*, 45 (1748), p. 647.

18. J.-A. Nollet, *Leçons de physique expérimentale*, 6 vols. (Paris, 1745–8), vol. I, pp. 61–3.

19. Réaumur, *Mémoires*, vol. VI, p. xlvii.

20. Terrall, *Catching Nature in the Act*, chapter 5. Eventually, Lyonet saw aphids mating before going dormant for the winter.

21. V. Dawson, *Nature's Enigma: The Problem of the Polyp in the Letters of Bonnet, Trembley and Réaumur* (Philadelphia, 1987); M. Ratcliff, 'Trembley's strategy of generosity and the scope of celebrity in the mid-eighteenth century', *Isis*, 95 (2004), pp. 555–75.

22. Trembley to Réaumur, 8 June 1741, in M. Trembley (ed.), *Correspondance inédite entre Réaumur et Trembley* (Geneva, 1943), pp. 78–9.

23. C. Bonnet, *Traité d'Insectologie; ou observations sur quelques especes de vers d'eau douce* (Paris, 1745), p. 4.

24. Bonnet, *Traité d'Insectologie*, pp. 42–3.

25. Bonnet, *Traité d'Insectologie*, p. 81.

26. Lyonet to Réaumur, 11 May 1742, Bibliothèque de Genève, MS Trembley 5.

11 Spatial arrangement and systematic order

1. This text was first published in the nineteenth century as *Joannis Bodini Colloquium heptaplomeres de rerum sublimum arcanis abditis*, ed. L. Noack (1857; Stuttgart, 1966). On the *pantotheca*, see pp. 2–3. Its authorship was debated: see N. Malcolm, 'Jean Bodin and the authorship of the *Colloquium heptaplomeres*', *Journal of the Warburg and Courtauld Institutes*, 69 (2006), pp. 95–150; K. F. Faltenbacher, 'Stand der Forschung. Eine kurze Entgegnung auf Noel Malcolm', in K. F. Faltenbacher (ed.), *Der kritische Dialog des* Colloquium Heptaplomeres: *Wissenschaft, Philosophie und Religion zu Beginn des 17. Jahrhunderts* (Darmstadt, 2009), pp. 44–50.

2. Bodin, *Colloquium heptaplomeres*, p. 3; on the *pantotheca* in context of the *ars memoriae* tradition, see J. P. Brach, 'Sur quelques notations arithmologiques dans le *Colloquium Heptaplomeres*', in K. F. Faltenbacher (ed.), *Magie, Religion und Wissenschaft im* Colloquium Heptaplomeres (Darmstadt, 2002), pp. 157–61.

3. On early modern *ars memoriae*, see F. A. Yates, *The Art of Memory* (London, 1966); J. Godwin, *Athanasius Kircher's Theatre of the World* (London, 2009). More generally, on relations between *ars memoriae* and collecting, see L. Bolzoni, 'Das Sammeln und die *ars memoriae*', in A. Grote (ed.), *Macrocosmos in Microcosmo. Die Welt in der Stube. Zur Geschichte des Sammelns 1450 bis 1800* (Opladen, 1994), pp. 129–68.

4. P. Findlen, 'Courting nature', in N. Jardine, J. A. Secord and E. C. Spary (eds.), *Cultures of Natural History* (Cambridge, 1996), pp. 57–74.

5. J. Woodward, *Essay toward a Natural History of the Earth* (London, 1692).

6. L. Daston and K. Park, *Wonders and the Order of Nature, 1150–1750* (New York, 1998), esp. pp. 267–301.

7. M. Kemp, 'Wrought by no artist's hand: the natural, the artificial, the exotic and the scientific in some artifacts from the Renaissance', in C. Farago (ed.), *Reframing the Renaissance: Visual Culture in Europe and Latin America 1450–1650* (New Haven, 1995), pp. 177–95, esp. p. 179.

8. On the *Kunstkammer* as a specific type of early modern collection, see A. MacGregor, *Curiosity and Enlightenment: Collections and Collectors from the Sixteenth to the Nineteenth Century* (New Haven, 2007), pp. 9–69.

9. See A. van Suchtelen and B. van Beneden (eds.), *Room for Art in Seventeenth-Century Antwerp* (Zwolle, 2009); E. Vegelin van Claerbergen (ed.), *David Teniers and the Theatre of Painting* (London, 2006).

10. P. Findlen, 'The market and the world: science, culture and collecting in the Venetian Republic', in B. Aikema and M. Seidel (eds.), *Zur Geschichte des Sammelns in Venedig und in Venetien zur Zeit der Serenissima* (Venice, 2005), pp. 55–68.

11. E. M. Beekman (ed.), *The Amboinese Curiosity Cabinet: Georgius Everhardus Rumphius* (New Haven, 1999).

12. See R. Felfe, 'Collections and the surface of the image: pictorial strategies in early modern *Wunderkammern*', in H. Schramm, L. Schwarte and J. Lazardzig (eds.), *Collection - Laboratory - Theatre: Scenes of Knowledge in the 17th Century* (Berlin, 2005), pp. 228–66.

13. See M. J. S. Rudwick, *The Meaning of Fossils: Episodes in the History of Paleontology* (Chicago, 1985), pp. 3–13.

14. See A. Cooper, 'The museum and the book: the *Metallotheca* and the history of an encyclopaedic natural history in early modern Italy', *Journal of the History of Collections*, 7:1 (1995), pp. 1–23.

15. This discussion is based on the manuscript catalogue drawn up in 1741 by Gottfried August Gründler, where the stones occupy over

one hundred pages. See T. J. Müller-Bahlke and K. E. Göltz, *Die Wunderkammer. Die Kunst- und Naturalienkammer der Franckeschen Stiftungen zu Halle* (Halle [Saale], 1998), p. 44.

16. D. J. Meijers and B. van de Roemer, 'Ein "gezeichnetes Museum" und seine Funktion – damals und heute', in B. Buberl and M. Dückershoff (eds.), *Palast des Wissens. Die Kunst- und Wunderkammer Zar Peters des Große* (Munich, 2003), vol. II, pp. 168–82.

17. S. Quiccheberg, *Inscriptiones vel Tituli Theatri Amplissimi* (Munich, 1565), 'Digressiones et Declarationes', ll. 257–68.

18. L. C. Sturm[?], *Des geöffneten Ritter-Platzes dritter Theil … Besonders was bey Raritäten- und Naturalienkammern … Hauptsächliches und Remarquables vorfället* (Leipzig, 1705), p. 21.

19. D. Freedberg, *The Eye of the Lynx: Galileo, his Friends, and the Beginnings of Modern Natural History* (Chicago, 2002).

20. See C. Meier, 'Virtuelle Wunderkammern. Zur Genese eines frühneuzeitlichen Sammelkonzepts', in R. Felfe and A. Lozar (eds.), *Frühneuzeitliche Sammlungspraxis und Literatur* (Berlin, 2006), pp. 29–74.

21. See A. M. Roos, 'The art of science: a 'rediscovery' of the Lister copperplates', *Notes and Records of the Royal Society of London*, 66 (2012), pp. 19–40.

22. See L. Tarp, 'Marble and marvel: Ole Worm's globe and the reception of 'nature's art' in seventeenth-century Denmark', *Kritische Berichte*, 3 (2013), pp. 9–23.

23. On Gessner, see Kusukawa, this volume, Chapter 2; on Scheuchzer, see R. Felfe, *Naturgeschichte als kunstvolle Synthese. Physikotheologie und Bildpraxis bei Johann Jakob Scheuchzer* (Berlin, 2003), esp. pp. 173–99.

12 Linnaean paper tools

This chapter partly builds on a much longer article that I have contributed to E. Aronova, C. von Oertzen and D. Sepkoski (eds.), *Historicizing Big Data* (Chicago, 2017). I would like to thank the members of the working group 'The Sciences of the Archive' at the Max Planck Institute for the History of Science, Berlin, and of the Data Studies Group at the University of Exeter for frequent opportunities to discuss my ideas.

1. My periodisation of 'classical natural history' is inspired by J.-M. Drouin, 'De Linné à Darwin: les voyageurs naturalistes', in M. Serres (ed.), *Éléments d'histoire des sciences* (Paris, 1989), pp. 321–35. It partly overlaps, but is not identical, with Michel Foucault's *âge classique*, which extends from the late seventeenth

century to around 1800; cf. M. Foucault, *Les Mots et les choses* (Paris, 1966), pp. 140–4.

2. S. J. M. M. Alberti, 'Placing nature: natural history collections and their owners in nineteenth-century provincial England', *British Journal for the History of Science*, 35 (2002), pp. 291–311; B. Dietz, 'Making natural history: doing the Enlightenment', *Central European History*, 43 (2010), pp. 25–46; D. Phillips, *Acolytes of Nature: Defining Natural Science in Germany, 1770–1850* (Chicago, 2012). For an overview of the use of media, especially drawings, in sixteenth- and seventeenth-century natural history, see Felfe, this volume, Chapter 11.

3. L. Schiebinger and C. Swan (eds.), *Colonial Botany: Science, Commerce and Politics in the Early Modern World* (Philadelphia, 2005); U. Klein and E. C. Spary (eds.), *Materials and Expertise in Early Modern Europe: Between Market and Laboratory* (Chicago, 2010).

4. H. J. Cook, *Matters of Exchange: Commerce, Medicine and Science in the Dutch Golden Age* (New Haven, 2007); S. Schaffer, L. Roberts, K. Raj and J. Delbourgo (eds.), *The Brokered World: Go-Betweens and Global Intelligence, 1770–1820* (Sagamore Beach, MA, 2009).

5. See, for example, D. P. Miller, ' Joseph Banks, empire, and "centres of calculation" in late Hanoverian London', in D. P. Miller and P. H. Reill (eds.), *Visions of Empire: Voyages, Botany and Representations of Nature* (Cambridge, 1996), pp. 21–37; M. Stuber, S. Hächler and L. Lienhard, *Hallers Netz. Ein europäischer Gelehrtenbriefwechsel zur Zeit der Aufklärung* (Basel, 2005).

6. On Kew Gardens, see R. Drayton, *Nature's Government: Science, Imperial Britain and the 'Improvement' of the World* (New Haven, 2000); on the Muséum national d'Histoire naturelle, see E. C. Spary, *Utopia's Garden: French Natural History from Old Regime to Revolution* (Chicago, 2000). For an example outside Europe, see J. R. Marcaida and J. Pimentel, 'Green treasures and paper floras: the business of Mutis in New Granada (1783–1808)', *History of Science*, 52 (2014), pp. 277–96.

7. See, for example, J. Endersby, *Imperial Nature: Joseph Hooker and the Practices of Victorian Science* (Chicago, 2008).

8. D. Outram, 'New spaces in natural history', in N. Jardine, J. A. Secord and E. C. Spary (eds.), *Cultures of Natural History* (Cambridge, 1996), pp. 249–65.

9. P.-Y. Lacour, *La République naturaliste: collections d'histoire naturelle et Révolution française, 1789–1804* (Paris, 2014).

10. S. Sörlin, 'Ordering the world for Europe: science as intelligence and information as seen from the northern periphery', *Osiris*, 15 (2000), pp. 51–69.

11. J. Cañizares-Esguerra, *How to Write the History of the New World: Histories, Epistemologies and Identities in the Eighteenth-Century Atlantic World* (Stanford, 2001).

12. See, for example, G. R. McOuat, 'Species, rules and meaning: the politics of language and the ends of definitions in 19th century natural history', *Studies in History and Philosophy of Science*, 27 (1996), pp. 473–519; H. Cowie, 'A Creole in Paris and a Spaniard in Paraguay: geographies of natural history in the Hispanic world (1750–1808)', *Journal of Latin American Geography*, 10 (2011), pp. 175–97.

13. See, for example, J. Drouin, 'Principles and uses of taxonomy in the works of Augustin-Pyramus de Candolle', *Studies in History and Philosophy of Biological and Biomedical Sciences*, 32 (2001), pp. 255–75.

14. For a promising start, see T. Kinukawa, 'Learned vs commercial? The commodification of nature in early modern natural history specimen exchanges in England, Germany, and the Netherlands', *Historical Studies in the Natural Sciences*, 43 (2013), pp. 589–618; N. Güttler and I. Heumann (eds.), *Sammlungsökonomien* (Berlin, 2016).

15. M. Terrall, 'Following insects around: tools and techniques of eighteenth-century natural history', *British Journal for the History of Science*, 43 (2010), pp. 573–88, and B. Dietz, 'Mobile objects: the space of shells in eighteenth-century France', *British Journal for the History of Science*, 39 (2006), pp. 363–82.

16. F. A. Stafleu, *Linnaeus and the Linnaeans: The Spreading of their Ideas in Systematic Botany, 1735–1789* (Utrecht, 1971), esp. pp. 337–9; D. R. Headrick, *When Information Came of Age: Technologies of Knowledge in the Age of Reason and Revolution, 1700–1850* (Oxford, 2000), chapter 2.

17. B. H. Soulsby, *A Catalogue of the Works of Linnaeus (and Publications More Immediately Relating Thereto) Preserved in the Libraries of the British Museum (Bloomsbury) and the British Museum (Natural History) (South Kensington)*, 2nd edn (London, 1933).

18. L. Daston, 'Type specimens and scientific memory', *Critical Inquiry*, 31 (2004), pp. 153–82; J. Witteveen, 'Suppressing synonymy with a homonym: the emergence of the nomenclatural type concept in nineteenth century natural history', *Journal of the History of Biology*, 49 (2015), pp. 135–89.

19. J. L. Heller, *Studies in Linnean Method and Nomenclature* (Marburg, 1983).

20. S. Müller-Wille, 'Systems and how Linnaeus looked at them in retrospect', *Annals of Science*, 70 (2014), pp. 305–17.

21. My distinction between labels and containers is inspired by A. te Heesen, 'Boxes in nature', *Studies in History and Philosophy of*

Science, 33 (2000), pp. 381–403; S. Leonelli, 'Packaging small facts for re-use: databases in model organism biology', in P. Howlett and M. S. Morgan (eds.), *How Well Do Facts Travel? The Dissemination of Reliable Knowledge* (Cambridge, 2010), pp. 325–48.

22. On the concept of paper tools, see U. Klein, 'Paper tools in experimental cultures', *Studies in History and Philosophy of Science*, 322 (2001), pp. 265–302; A. te Heesen, 'The notebook: a paper-technology', in B. Latour and P. Weibel (eds.), *Making Things Public: Atmospheres of Democracy* (Cambridge, MA, 2005), pp. 582–9; M.-N. Bourguet, 'A portable world: the notebooks of European travellers (eighteenth to nineteenth centuries)', *Intellectual History Review*, 20 (2010), pp. 377–400; V. Hess and J. A. Mendelsohn, 'Case and series: medical knowledge and paper technology, 1600–1900', *History of Science*, 48 (2010), pp. 287–314.

23. A. M. Blair, *Too Much to Know: Managing Scholarly Information before the Modern Age* (New Haven, 2010); R. Yeo, *Notebooks, English Virtuosi and Early Modern Science* (Chicago, 2014).

24. A. te Heesen, 'Accounting for the natural world: double-entry bookkeeping in the field', in L. Schiebinger and C. Swan (eds.), *Colonial Botany: Science, Commerce and Politics in the Early Modern World* (Philadelphia, 2005), pp. 237–51; I. Charmantier and S. Müller-Wille, 'Worlds on paper: an introduction', *Early Science and Medicine*, 19 (2014), pp. 379–97.

25. S. Müller-Wille and I. Charmantier, 'Natural history and information overload: the case of Linnaeus', *Studies in History and Philosophy of Biological and Biomedical Sciences*, 43 (2012), pp. 4–15. Interestingly, Linnaeus's botanical textbook *Philosophia botanica* functioned in analogous ways; see M. D. Eddy, 'Tools for reordering: commonplacing and the space of words in Linnaeus's *Philosophia botanica*', *Intellectual History Review*, 20 (2010), pp. 227–52.

26. C. Linnaeus, *Critica botanica* (Leiden, 1737), p. 204.

27. S. Müller-Wille and S. Scharf, 'Indexing nature: Carl Linnaeus and his fact gathering strategies', *Svenska Linnésällskapets Årsskrift* (2012), pp. 31–60; K. Böhme and S. Müller-Wille, '"In der Jungfernheide hinterm Pulvermagazin frequens". Das Handexemplar des *Florae Berolinensis Prodromus* (1787) von Carl Ludwig Willdenow', *NTM Zeitschrift für Geschichte der Wissenschaften, Technik und Medizin*, 21 (2013), pp. 93–106.

28. See Spary, *Utopia's Garden*, chapter 3, esp. p. 82; M.-N. Bourguet, 'Voyage, collecte, collections. Le catalogue de la nature (fin 17e – début 19e siècles)', in D. Lecoq and A. Chambard (eds.), *Terre à decouvrir, terres à parcourir: exploration et connaissance du monde, XIIe–XIXe siècles* (Paris, 1999), pp. 185–207.

29. B. Dietz, 'Contribution and co-production: the collaborative culture of Linnaean botany', *Annals of Science*, 69 (2012), pp. 551–69, p.

560. See also A. Secord, 'Coming to attention: a commonwealth of observers during the Napoleonic wars', in L. Daston and E. Lunbeck (eds.), *Histories of Scientific Observation* (Chicago, 2011), pp. 421–44.

30. I. Charmantier and S. Müller-Wille, 'Carl Linnaeus's botanical paper slips (1767–1773)', *Intellectual History Review*, 24 (2014), pp. 1–24.

31. See Stafleu, *Linnaeus and the Linnaeans*, p. 338; B. Dietz, 'Linnaeus' restless system: translation as textual engineering in eighteenth-century botany', *Annals of Science*, 73 (2016), pp. 143–56.

32. M. Houttuyn, 'Voorreden', in *Natuurlyke historie of uitvoerige beschryving der dieren, planten en mineraalen, volgens het samenstel van den Heer Linnæus*, Del 1, Stuk 1, (Amsterdam, 1761), unpaginated [pp. 9–13].

33. P. L. S. Müller, 'Vorbericht', in P. L. S. Müller, *Des Ritters Carl von Linné vollständiges Natursystem … Erster Theil. Von den säugenden Thieren* (Nürnberg, 1773), unpaginated [pp. 1–6].

34. Müller, 'Vorbericht', in *Des Ritters Carl von Linné vollständiges Natursystem … Supplements- und Registerband* (Nürnberg, 1776), unpaginated [pp. 2, 6–8].

35. See G. N. Raspe, 'Nachricht des Verlegers', in P. L. S. Müller, *Vollständiges Natursystem … Supplements- und Registerband*, unpaginated.

36. See Soulsby, *Catalogue*, nos. 96–100, p. 577.

37. J. F. Gmelin, 'Ratio hujus novae editionis', in *Caroli a Linné. Systema Naturae per Regna Tria Naturae … Editio decima tertia, aucta, reformata* (Leipzig, 1788), unpaginated [pp. 3–4].

38. See Müller, 'Vorbericht', in *vollständiges Natursystem … Erster Theil*, unpaginated [p. 2].

39. J. C. Adelung, *Grammatisch-kritisches Wörterbuch der hochdeutschen Mundart*, 4 vols. (Leipzig, 1793–1802), vol. I, p. 1735.

40. See Charmantier and Müller-Wille, 'Carl Linnaeus's botanical paper slips', pp. 13–16.

41. J. R. Forster, *A Catalogue of British Insects* (Warrington, 1770), vol. II; emphases in the original.

42. The notion that specimens of the same species were expendable duplicates seems not to have existed in pre-Linnaean natural history; see G. Olmi, 'From the marvellous to the commonplace: notes on natural history museums (16th–18th centuries)', in R. G. Mazzolini (ed.), *Non-Verbal Communication in Science prior to 1900* (Florence, 1993), pp. 235–78, on pp. 252–61; C. Swan, 'From blowfish to flower still life paintings', in P. Smith and P. Findlen (eds.), *Merchants and Marvels: Commerce, Science, and Art in Early Modern Europe*, ed. (London, 2001), pp. 109–36, on p. 118; J. Delbourgo, 'Collecting Hans Sloane', in A. Walker,

A. MacGregor and M. Hunter (eds.), *From Books to Bezoars: Sir Hans Sloane and his Collections* (London, 2012), pp. 9–23, on p. 14.

43. J. R. Forster, *A Catalogue of British Insects* (Warrington, 1770), Staatsbibliothek Berlin, Abteilung Historische Drucke, call no. Lt 12373R.

44. *Caroli a Linne . . . Prælectiones in Ordines Naturales Plantarum*, ed. P. D. Giseke (Hamburg, 1792), p. 625.

45. S. Müller-Wille and I. Charmantier, 'Lists as research technologies', *Isis*, 103 (2012), pp. 743–52; see also A. J. Cain, 'Linnaeus's *Ordines naturales*', *Archives of Natural History*, 20 (1993), pp. 405–15.

46. M. P. Winsor, 'Darwin and taxonomy', in M. Ruse (ed.), *The Cambridge Encyclopedia of Darwin and Evolutionary Thought* (Cambridge, 2013), pp. 72–9; D. Sepkoski, 'Towards "a natural history of data": evolving practices and epistemologies of data in paleontology, 1800–2000', *Journal of the History of Biology*, 46 (2013): pp. 401–44; N. Güttler, *Das Kosmoskop. Karten und ihre Benutzer in der Pflanzengeographie des 19. Jahrhunderts* (Göttingen, 2014).

47. R. H. Finnegan and R. Horton (eds.), *Modes of Thought: Essays on Thinking in Western and Non-Western Societies* (London, 1973).

13   Image and nature

I am grateful to Staffan Müller-Wille (Exeter), Caterina Schürch (Munich) and Robert-Jan Wille (Munich) for helpful criticism on a first draft of this chapter – and to the editors of this volume, in particular Emma Spary, for their comments and support.

1. C. J. Trew to J. A. Beurer (22 December 1731). See T. Schnalke, 'Das genaue Bild. Das schöne Bild. Trew und die botanische Illustration', in T. Schnalke (ed.), *Natur im Bild. Anatomie und Botanik in der Sammlung des Nürnberger Arztes Christoph Jacob Trew. Eine Ausstellung aus Anlass seines 300. Geburtstages 8 November–10 Dezember 1995* (Erlangen, 1995), p. 104, for the German original.

2. L. Daston and P. Galison introduced the term 'atlas' for this literary (and epistemic) genre. See L. Daston and P. Galison, *Objectivity* (New York, 2007), chapter 1.

3. Quoted in C. E. Raven, *John Ray, Naturalist: His Life and Works* (Cambridge, 1950), p. 213.

4. J. S. Kerner, *Abbildungen ökonomischer Pflanzen*, 8 vols. (Stuttgart, 1786–96), vol. I, preface, p. 5.

5. J. Miller, *An Illustration of the Sexual System of Linnaeus* (London, 1777), preface.

6. S. Müller-Wille and K. Reeds, 'A translation of Carl Linnaeus's intro-duction to *Genera plantarum*', *Studies in History and Philosophy of Biological and Biomedical Sciences*, 38 (2007), pp. 563–72, p. 568. On Linnaeus and visual representations, see: I. Charmantier, 'Carl Linnaeus and the visual representation of nature', *Historical Studies in the Natural Sciences*, 41 (2011), pp. 365–404.

7. See also S. Müller-Wille, 'Text, Bild und Diagramm in der klas-sischen Naturgeschichte', in *kunsttexte*.de, 4 (2002), available at https://edoc.hu-berlin.de/handle/18452/7563.

8. A. Cooper, *Inventing the Indigenous: Local Knowledge and Natural History in Early Modern Europe* (Cambridge, 2007).

9. W. Curtis, *Flora Londinensis*, 6 vols. (London, 1777–98), unpagin-ated preface.

10. The original Latin is *Anthoxanthum spica ovato-oblonga, flosculis subpedunculatis arista longioribus*. C. Linnaeus, *Species plant-arum* (Stockholm, 1753), p. 28.

11. Daston and Galison, *Objectivity*, chapter 2.

12. P. J. Redouté, *Les Liliacées*, 8 vols. (Paris, 1802–16), vol. I, preface, p. 1.

13. J. S. Kerner, *Beschreibung und Abbildung der Bäume und Gesträuche, welche in dem Herzogthum Wirtemberg wild wachsen*, 9 vols. (Stuttgart, 1783–92), preface, p. 5.

14. C. J. Trew to A. Haller (9 April 1746). For the German original, H. Steinke, *Der nützliche Brief. Die Korrespondenz zwischen Albrecht von Haller und Christoph Jacob Trew, 1733* (Basel, 1999), pp. 94–5.

15. Correspondence edited by C. Andree, 'Quaenam est differentia inter vegetabilia et animalia? Über das Verhältnis zwischen Tier und Pflanzenreich ... an Hand des ... Briefwechsels von Linnaeus mit dem schottischen Arzt David Skene', in H. Goerke (ed.), *Carl von Linné. Beiträge über Zeitgeist, Werk und Wirkungsgeschichte* (Göttingen, 1980), pp. 51–80.

16. W. Curtis, *Flora Londinensis*, unpaginated preface.

17. See C. Nissen, *Die botanische Buchillustration. Ihre Geschichte und Bibliographie*, 2 vols. (Stuttgart, 1966), p. 140. The lavish new edition of Duhamel Du Monceau's *Traité des Arbres et Arbustes*, published between 1804 and 1819, provides another example of this attitude of the French state at the time.

18. Translated from C. Linnaeus, *Philosophia botanica* (Stockholm, 1751), paragraph 332, p. 263.

19. J. J. Dillenius, *Hortus Elthamensis* (London, 1732), preface, p. vii. Linnaeus was very fond of Dillenius and his work, and dedicated the *Critica botanica* (Leiden, 1737) to him.

20. Around 20,000 letters from Trew's correspondence, together with 2,500 botanical drawings and a large collection of books, survive at the University of Erlangen (E. Schmidt-Herrling, *Die*

Briefsammlung des Nürnberger Arztes Christoph J. Trew (Erlangen, 1940).

21. See, for example, W. Kemp, *'Einen wahrhaft bildenden Zeichenunterricht überall einzuführen': Zeichnen und Zeichenunterricht der Laien 1500–1870* (Frankfurt a. M., 1979); E. Schulze, *Nulla dies sine linea. Universitärer Zeichenunterricht – eine problemgeschichtliche Studie* (Stuttgart, 2004).

22. On the education of eighteenth-century draughtsmen, see for example H. Dickel, *Deutsche Zeichenbücher des Barock. Eine Studie zur Geschichte der Künstlerausbildung* (Hildesheim, 1987).

23. J. Sowerby, *A Botanical Drawing-Book, or an Easy Introduction to Drawing Flowers According to Nature* (London, 1788). On Sowerby, see also Nissen, *Die botanische Buchillustration*, vol. II, p. 173.

24. Sowerby, *Botanical Drawing-Book*, p. 4.

25. Similar procedures can be found in other contemporary drawing manuals, such as, to name only one further example, the German manual by J. C. Friedrich, *Anweisung zum Zeichnen und Blumenmalen* (Friedrichstadt, 1786). This drawing manual, published in Saxony, centres on the technical execution of drawing far more than Sowerby's index manual; the basic principle of training apprentices by having them copy standard examples of botanical objects, however, is the same.

26. C. J. Trew to J. J. Haid, 26 January 1750, Letter Collection of Erlangen, no. 266.

27. For further details, see K. Nickelsen, 'The challenge of colour: eighteenth-century botanists and the hand-colouring of illustrations', *Annals of Science*, 63 (2006), pp. 1–23.

28. Curtis, *Flora Londinensis*, preface.

29. W. Watson, 'Critical observations concerning the *Oenante aquatica* etc.', *Philosophical Transactions* 44 (1746), pp. 227–42, quoted pp. 239ff. For the illustration, see plate IV of the same volume.

30. See H.-W. Lack and D. J. Mabberley, *The Flora Graeca Story: Sibthorp, Bauer and Hawkins in the Levant* (Oxford, 1999), quoted on p. 50.

31. Two examples of photographed dried plants and the drawings made from them can be found in Lack and Mabberley, *The Flora Graeca Story*, figures 26 and 27 (pp. 78–9, *Campanula rupestris*) and figures 15 and 16 (pp. 56–7, *Ranunculus asiaticus*).

32. T. M. Fries (ed.), *Johann Beckmanns Schwedische Reise in den Jahren 1765–1766. Tagebuch* (Uppsala, 1911), pp. 146–7.

33. G. Saunders, *Picturing Plants: An Analytical History of Botanical Illustration* (Berkeley, 1995), p. 76.

34. G. Calman, *Ehret: Flower Painter Extraordinary* (Oxford, 1977), p. 68.

35. Sandberger's large collection of *c.*2,500 watercolours, presumably drawn by himself, is housed in the natural history collection of the Wiesbaden Museum, Germany, but has been little studied.

36. Many of these copying links are documented and analysed in K. Nickelsen, *Draughtsmen, Botanists and Nature: The Construction of Eighteenth-Century Botanical Illustrations,* (Dordrecht, 2006).

37. Kerner, *Beschreibung*, preface, fo. 2f.

38. S. Müller-Wille, 'Collection and collation: theory and practice of Linnaean botany', *Studies in History and Philosophy of Biological and Biomedical Sciences,* 38 (2007), pp. 541–62.

14   Botanical conquistadors

This chapter is drawn from D. Bleichmar, *Visible Empire: Botanical Expeditions and Visual Culture in the Hispanic Enlightenment* (Chicago, 2012). An earlier version appeared as 'Botanical conquistadors: plants and empire in the Hispanic Enlightenment', in Y. Batsaki, S. Burke Cahalane and A. Tchikine (eds.), *The Botany of Empire in the Long Eighteenth Century* (Dumbarton Oaks, 2017), pp. 35–60.

1. On imperial botany before the nineteenth century, see among others R. Drayton, *Nature's Government: Science, Imperial Britain and the 'Improvement' of the World* (New Haven, 2000) and L. Schiebinger and C. Swan (eds.), *Colonial Botany: Science, Commerce and Politics* (Philadelphia, 2004).

2. A. Lafuente and N. Valverde, 'Linnaean botany and Spanish imperial biopolitics', in Schiebinger and Swan (eds.), *Colonial Botany,* 134–47, and bibliography cited in Bleichmar, *Visible Empire.*

3. M. de los Angeles Calatayud Arinero, *Catálogo de las expediciones y viajes científicos españoles a América y Filipinas (siglos XVIII y XIX)* (Madrid, 1984), item 13.

4. J. E. McClellan and F. Regourd, *The Colonial Machine: French Science and Overseas Expansion in the Old Regime* (The Hague, 2012).

5. A. Lafuente and N. Valverde, *Los mundos de la ciencia en la ilustración española* (Madrid, 2003); J. Pimentel, *Testigos del mundo: ciencia, literatura y viajes en la Ilustración* (Madrid, 2003), 147–78; M. Villena, et al., *El gabinete perdido. Pedro Franco Dávila y la Historia Natural del Siglo de las Luces* (Madrid, 2009).

6. 'Carolus III Rex naturam et. artem sub uno tecto in publicam utilitatem consociavit.'

7. P. de Vos, 'The rare, the singular, and the extraordinary: natural history and the collection of curiosities in the Spanish Empire,' in D. Bleichmar, P. de Vos, K. Huffine and K. Sheehan (eds.), *Science in*

the Spanish and Portuguese Empires, 1500–1800 (Stanford, 2008), pp. 271–89.

8. Quoted in F. J. Puerto Sarmiento, *Ciencia de cámara. Casimiro Gómez Ortega (1741–1818), el científico cortesano* (Madrid, 1992), p. 54.

9. This cameralist approach to botany was also taken by Carl Linnaeus and Joseph Banks, among others. See Drayton, *Nature's Government;* J. Gascoigne, *Science in the Service of Empire: Joseph Banks, the British State and the Uses of Science in the Age of Revolution* (Cambridge, 1998); L. Koerner, *Linnaeus: Nature and Nation* (Cambridge, MA, 1999); D. P. Miller and P. H. Reill (eds.), *Visions of Empire: Voyages, Botany and Representations of Nature* (Cambridge, 1996).

10. C. Gómez Ortega, *Instrucción,* (Madrid, 1779), p. 22.

11. C. Gómez Ortega, *Historia natural de la Malagueta ó Pimienta de Tavasco* (Madrid, 1780), pp. 1–2.

12. C. G. Ortega to J. de Gálvez, Madrid, 23 February 1777, quoted in Puerto Sarmiento, *Ciencia de cámara,* pp. 154–6.

13. A. J. Cavanilles, 'Materiales para la historia de la Botánica', *Anales de Historia Natural,* 2:4 (1800; facs. edn 1993), pp. 3–57, at p. 24.

14. S. Varey (ed.), *The Mexican Treasury: The Writings of Dr Francisco Hernández* (Stanford, CA, 2000); S. Varey, R. Chabrán and D. V. Weiner (eds.), *Searching for the Secrets of Nature: The Life and Works of Dr Francisco Hernández* (Stanford, 2000).

15. F. Hernández, *Opera, cum edita, tum inedita, ad autographi fidem et integritatem expressa* (Madrid, 1790).

16. J. Pimentel, 'The Iberian vision: science and empire in the framework of a universal monarchy, 1500–1800', *Osiris,* 15 (2000), pp. 17–30.

17. A. Barrera-Osorio, *Experiencing Nature: The Spanish American Empire and the Early Scientific Revolution* (Austin, 2006); M. M. Portuondo, *Secret Science: Spanish Cosmography and the New World* (Chicago, 2009).

18. H. F. Cline (ed.), *Handbook of Middle American Indians, vol. XI: Guide to Ethnohistorical Sources* (Austin, 1972), pp. 183–542; F. de Solano (ed.), *Cuestionarios para la formación de las Relaciones Geográficas de Indias. Siglos XVI–XIX* (Madrid, 1988).

19. On Linnaean classification, see W. Blunt, *The Compleat Naturalist: A Life of Linnaeus* (1971; London, 2001); Koerner, *Linnaeus.* On Ehret, see W. Blunt and W. T. Stearn, *The Art of Botanical Illustration* (1951; Woodbridge, Suffolk, 1994), chapter 12.

20. On collective empiricism, see L. Daston and P. Galison, *Objectivity* (New York, 2007), pp. 19–23.

21. Bleichmar, *Visible Empire,* pp. 90–1. See also Daston and Galison, *Objectivity,* chapter 2, esp. pp. 84–98; K. Nickelsen, *Draughtsmen, Botanists and Nature: The Construction of Eighteenth-Century Botanical Illustrations* (Dordrecht, 2006), esp. pp. 1–68.

22. On scientific portraiture, see L. Jordanova, *Defining Features: Scientific and Medical Portraits* (London, 2000).

23. A. González Bueno, *Antonio José Cavanilles (1745–1804): la pasión por la ciencia* (Madrid, 2002).

24. A. Federico Gredilla, *Biografía de José Celestino Mutis* (1911; new edn Bogotá, 1982), pp. 42–3.

15   Bird sellers and animal merchants

1. Extract from the journal of Hester Piozzi Thrale (1741–1821) published in J. Black et al., *The Broadview Anthology of British Literature: The Restoration and the Eighteenth Century* (London, 2012), p. 949.

2. S. Chaplin, 'Nature dissected, or dissection naturalised? The case of John Hunter's museum', *Museums and Society*, 6 (2008), pp. 135–51; the story of 'Miss Poll' is given in detail in C. Plumb, *The Georgian Menagerie: Exotic Animals in Eighteenth-Century London* (London, 2015).

3. B. Silliman, *A Journal of Travels in England, Holland and Scotland and Two Passages across the Atlantic in 1805 and 1806*, 2 vols. (New York, 1810), vol. I, pp. 50, 105.

4. M. Senior (ed.), *A Cultural History of Animals in the Age of Enlightenment* (London, 2007); B. Boehrer (ed.), *A Cultural History of Animals in the Renaissance* (London, 2007); P. Findlen and P. Smith (eds.), *Merchants and Marvels: Commerce, Science and Art in Early Modern Europe* (London, 2002); E. Fudge (ed.), *Renaissance Beasts: Of Animals, Humans and Other Wonderful Creatures* (Chicago, 2004).

5. C. Gómez-Centurión, 'Treasures fit for a King: Charles III of Spain's Indian elephants', *Journal of the History of Collections*, 22 (2010), pp. 29–44; S. Bedini, *The Pope's Elephant* (Manchester, 1997); K. Walker-Meikle, *Medieval Pets* (Woodbridge, 2012); I. Tague, *Animal Companions: Pets and Social Change in Eighteenth-Century Britain* (University Park, PA, 2015); L. Robbins, *Elephant Slaves and Pampered Parrots: Exotic Animals in Eighteenth-Century Paris* (Baltimore, 2002); M. Morton and C. Bailey (eds.), *Oudry's Painted Menagerie: Portraits of Exotic Animals in Eighteenth-Century Europe* (Los Angeles, 2007); G. Ridley, *Clara's Grand Tour: Travels with a Rhinoceros in Eighteenth-Century Europe* (London, 2005).

6. M. R. Abbing, '"So Een Wunder heeft men hier nooijt gesien": de indische vrouwtjesolifant (1678/80–1706) van Bartel Verhagen', *Amstelodamum*, 106 (2014), pp. 12–39.

7. *Wonderen der natuur: In de menagerie van Blauw Jan te Amsterdam, zoals gezien door Jan Velten rond 1700/Wonders of Nature, in the Menagerie of Blauw Jan in Amsterdam, as seen by Jan Velten around 1700* (Berlin, 1998). The original album is held in the Artis Library, University of Amsterdam.

8. *Tuesday's Journal,* 24 July 1649.

9. Advertisements were regularly placed in newspapers like *The Post Boy, Flying Post* and *The Postman and Historical Account.* These are held in the Burney Collection, British Library.

10. R. Campbell, *The London Tradesman: Being a Compendious View of All the Trades* (London, 1747); T. Ward, *The Bird Fancier's Recreation, Including Choice Instructions for the Taking, Feeding, Breeding and Teaching of Them* (London, 1735).

11. *Gazetteer and New Daily Advertiser*, London, 15 May 1766; *Public Advertiser*, London, 24 December 1766; *Public Advertiser*, London, 28 September 1765.

12. J. Brookes, zoologist, handbill (British Library, L.23.c.3.(48.)); J. Brookes, bird merchant, trade card (British Museum, Heal, 14.1); P. Brookes, bird and animal seller, trade card (British Museum, Banks, 14.3).

13. J. Brookes Prob. 11/1386 and John Cross Prob. 11/1027, Public Record Office, the National Archives.

14. G. Pidcock's fire insurance, London Metropolitan Archives, 730081, 832885.

15. For more on the trade network and social connections of Brookes, see C. Grigson, *Menagerie: The History of Exotic Animals in England* (Cambridge, 2016).

16. W. Granger, *The New Wonderful Museum*, 6 vols. (London, 1804), vol. II, p. 711; G. Wilson, *The Eccentric Mirror: Reflecting a Faithful and Interesting Delineation of Male and Female Characters Ancient and Modern*, 4 vols. (London, 1807), vol. II, p. 32.

17. Robbins, *Elephant Slaves and Pampered Parrots.*

18. Advertisements for the camels appeared in *Mist's Journal* throughout 1758 (British Library, Lysons Collection, microfilm MC20452, frames 9 and 8e).

19. *A Journey to Paris in the Year 1698 by Dr. Martin Lister* (London, 1699), p. 177.

20. *Old Bailey Proceedings (OBP) Online* (www.oldbaileyonline.org, version 8.0, 7 March 2018), April 1768, trial of William Enoch (t17680413-54). OBP Online (www.oldbaileyonline.org, version 8.0, 7 March 2018), April 1789, trial of Thomas Andrews (t17890422-41).

16   Vegetable empire

I would like to thank the editors, and Charlie Jarvis and Julie Kim for comments on an earlier version of this chapter.

1. A. von Humboldt and A. Bonpland, *Essay on the Geography of Plants* (1807; Chicago, 2009), pp. 64–5, 69, 70.

2. L. Schiebinger and C. Swan (eds.), *Colonial Botany: Science, Commerce, and Politics in the Early Modern World* (Philadelphia, 2005).

3. L. Schiebinger, *Plants and Empire: Colonial Bioprospecting in the Atlantic World* (Cambridge, MA, 2004).

4. M. Ellis, R. Coulton and M. Mauger, *Empire of Tea: The Asian Leaf that Conquered the World* (London, 2015) and J. Walvin, *Fruits of Empire: Exotic Produce and British Taste, 1660–1800* (London, 1997).

5. R. H. Grove, *Green Imperialism: Colonial Expansion, Tropical Island Edens and the Origins of Environmentalism, 1600–1860* (Cambridge, 1995) and R. Drayton, *Nature's Government: Science, Imperial Britain, and the 'Improvement' of the World* (New Haven, 2000).

6. G. Williams, *Naturalists at Sea: Scientific Travellers from Dampier to Darwin* (New Haven, 2013).

7. J. Byron, *A Voyage Round the World, in His Majesty's Ship The Dolphin* (London, 1767), titlepage.

8. J. Gascoigne, *Joseph Banks and the English Enlightenment: Useful Knowledge and Polite Culture* (Cambridge, 1994).

9. D. Bleichmar, *Visible Empire: Botanical Expeditions and Visual Culture in the Hispanic Enlightenment* (Chicago, 2012), D. Mackay, *In the Wake of Cook: Exploration, Science and Empire, 1780–1801* (London, 1985) and E. C. Spary, *Utopia's Garden: French Natural History from Old Regime to Revolution* (Chicago, 2000).

10. D. P. Miller and P. H. Reill (eds.), *Visions of Empire: Voyages, Botany and Representations of Nature* (Cambridge, 1996).

11. N. Rigby, 'The politics and pragmatics of seaborne plant transportation, 1769–1805', in M. Lincoln (ed.), *Science and Exploration in the Pacific: European Voyages to the Southern Oceans in the Eighteenth Century* (Woodbridge, 1998), pp. 81–100.

12. J. Ellis, *A Description of the Mangostan and the Breadfruit* (London, 1775).

13. Mackay, *In the Wake of Cook*; National Library of Jamaica (NLJ) MS 208: West Indian Planters and Merchants Minutes, photostat, n. p.

14. S. Fuller to Lord Sydney (2 April 1787), NLJ MS 571.

15. G. Dening, *Mr Bligh's Bad Language: Passion, Power and Theatre on the Bounty* (Cambridge, 1994).

16. NLJ MS 553: letters on the transplanting of plants from Mauritius to Hispaniola, 1782 to 1783.

17. J.-N. Céré to G. de Bellecombe, Isle de France (12 August 1783 and 3 September 1783), NLJ MS 553 (my translation).

18. *Catalogue of Plants, Exotic and Indigenous, in the Botanical Garden, Jamaica* (St Jago de la Vega, 1792); Thomas Dancer to Edward Long, Jamaica (24 July 1789), British Library Add. MSS 22678.

19. Grove, *Green Imperialism*; Drayton, *Nature's Government*.

20. D. Hall, 'Planters, farmers and gardeners in eighteenth-century Jamaica', in B. L. Moore, B. W. Higman, C. Campbell and P. Bryan (eds.), *Slavery, Freedom and Gender: The Dynamics of Caribbean Society* (Kingston, 2001), pp. 97–114; J. Casid, *Sowing Empire: Landscape and Colonization* (Minneapolis, 2004).

21. BL Add. MSS 22678, fos. 35r–43r, at fo. 45r.

22. Thomas Dancer to Edward Long, Jamaica ([1790] and 20 July 1791), BL Add. MSS 22678, fos. 49v, 52r and 60v.

23. Quoted in D. Hall, 'Botanical and horticultural enterprise in eighteenth-century Jamaica', in R. A. McDonald (ed.), *West Indies Accounts* (Kingston, 1996), p. 119. See also Thomas Dancer, *Some Observations Respecting the Botanical Garden* (Jamaica, 1804).

24. Hall, 'Botanical and horticultural enterprise'; T. Thistlewood to E. Long, Jamaica (17 June 1776), BL Add. MSS 18275A, fo. 128v.

25. 'Plants growing in Thos. Thistlewood's garden at Bread Nutt Island Pen, June 1776', BL Add. MSS 18275A, fos. 120v–121r.

26. J. A. Carney and R. N. Rosomoff, *In the Shadow of Slavery: Africa's Botanical Legacy in the Atlantic World* (Berkeley, 2009).

27. Grove, *Green Imperialism*.

28. J. Browne, 'Biogeography and empire', in N. Jardine, J. A. Secord and E. C. Spary (eds.), *Cultures of Natural History* (Cambridge, 1996), pp. 305–24, at p. 314; Drayton, *Nature's Government*.

29. M. Dettelbach, 'Humboldtian science', in Jardine et al. (eds.), *Cultures of Natural History*, pp. 287–304.

30. Humboldt and Bonpland, *Essay*, pp. 64, 69, 71, 73.

31. Quoted in A. M. C. Godlewska, 'From Enlightenment vision to modern science? Humboldt's visual thinking', in D. N. Livingstone and C. W. J. Withers (eds.), *Geography and Enlightenment* (Chicago, 1999), p. 243.

32. Humboldt and Bonpland, *Essay*, p. 73.

33. Humboldt and Bonpland, *Essay*, p. 86.

34. Quoted in M. Dettelbach, 'Global physics and aesthetic empire: Humboldt's physical portrait of the tropics', in Miller and Reill (eds.), *Visions of Empire*, p. 261.

35. Humboldt and Bonpland, *Essay*, p. 87.

36. Humboldt and Bonpland, *Essay*, pp. 99, 149.

37. N. L. Stepan, *Picturing Tropical Nature* (London, 2001).

38. Humboldt and Bonpland, *Essay*, p. 72 and M. Dettelbach, 'Alexander von Humboldt between Enlightenment and Romanticism', *Northeastern Naturalist*, special issue 1 (2001), pp. 9–20.

39. M. Nicholson, 'Alexander von Humboldt and the geography of vegetation', in A. Cunningham and N. Jardine (eds.), *Romanticism and the Sciences* (Cambridge, 1990), pp. 169–85; M. L. Pratt, *Imperial Eyes: Travel Writing and Transculturation* (London, 1992).

40. J. Cañizares-Esguerra, 'How derivative was Humboldt? Microcosmic nature narratives in early modern Spanish America and the (other) origins of Humboldt's ecological sensibilities', in Schiebinger and Swan (eds), *Colonial Botany*, pp. 148–65.

41. See Dettelbach, 'Global physics and aesthetic empire' and Grove, *Green Imperialism*.

42. Browne, 'Biogeography and empire', pp. 305, 313.

43. Nicholson, 'Alexander von Humboldt and the geography of vegetation', pp. 171, 181 (Humboldt quotation).

17   Containers and collections

Some segments of this chapter have been previously published in 'Pressed into service: specimens, space, and seeing in botanical practice', in D. N. Livingstone and C. W. J. Withers (eds.), *Geographies of Nineteenth-Century Science* (Chicago, 2011), pp. 283–310. I am grateful to University of Chicago Press for permission to include this material here. Thanks are also due to Katarina Böhme Evengård of the SLU University Library, Sweden, Mikael Risedal (photographer), the British Library, Michael Wilson of Selwyn College Library, Cambridge, Domniki Papadimitriou of the Digital Content Unit, Cambridge University Library. Kind permission to quote from manuscripts has been granted by the Trustees of the Royal Botanic Gardens, Kew, the Trustees of the Natural History Museum, London, the Syndics of Cambridge University Library, Norfolk Museums Service (Norwich Castle Museum and Art Gallery), West Yorkshire Archive Service, Calderdale and Warrington Museum and Archives (Culture Warrington).

1. M. Foucault, *The Order of Things: An Archaeology of the Human Sciences*, translated from the French (London, 1970), p. 131. The book was originally published in French under the title *Les Mots et les Choses* in 1966.

2. M. L. Pratt, *Imperial Eyes: Travel Writing and Transculturation* (London, 1992), p. 33.

3. See J. E. and R. Cardinal (eds.), *The Cultures of Collecting* (London, 1994) and A. te Heesen, *The World in a Box: The Story of an Eighteenth-Century Picture Encyclopedia*, trans. A. M. Hentschel (Chicago, 2002).

4. S. Silver, *The Mind Is a Collection: Case Studies in Eighteenth-Century Thought* (Philadelphia, 2015).

5. British Library, Add. MS 31850, fos. 1–40.

6. D. Turner to E. Hutchins (24 July 1814), Royal Botanic Gardens, Kew (RBGK).

7. D. Turner, manuscript 'Journal of a Tour to France', 2 vols. (1814), Norfolk Museums Service (Norwich Castle Museum and Art Gallery), acc. no. NWHCM: 1970.483.11, vol. II, p. 190.

8. W. Withering, *A Botanical Arrangement of All the Vegetables Naturally Growing in Great Britain. With Descriptions of the Genera and Species, According to the System of the Celebrated Linnaeus. Being an Attempt to Render them Familiar to Those who are Unacquainted with the Learned Languages*, 2 vols. (Birmingham, 1776).

9. Withering, *Botanical Arrangement*, vol. I, pp. l–li. In recommending such a cabinet, Withering was following Linnaeus. See S. Müller-Wille, 'Linnaeus' herbarium cabinet: a piece of furniture and its function', *Endeavour*, 30 (2006), pp. 60–4.

10. Withering, *Botanical Arrangement*, vol. II, p. 838 and plate XII.

11. W. Mavor, *The Lady's and Gentleman's Botanical Pocket Book adapted to Withering's Arrangement of British Plants. Intended to Facilitate and Promote the Study of Indigenous Botany* (London, 1800), pp. viii–ix.

12. S. Stewart, *On Longing: Narratives of the Miniature, the Gigantic, the Souvenir, the Collection* (London, 1993), p. 155.

13. Stewart, *On Longing*, p. 162.

14. G. Gardner, *Musci Britannici, or Pocket Herbarium of British Mosses* (Glasgow, 1836); W. J. Hooker to W. H. F. Talbot (9 July 1836), *The Correspondence of William Henry Fox Talbot*, ed. L. J. Schaaf (available online at http://foxtalbot.dmu.ac.uk), document no. 353.

15. W. Gardiner, *Twenty Lessons on British Mosses*, 2nd edn (Edinburgh, 1846), p. 3.

16. Gardiner, *Twenty Lessons*, p. 49.

17. J. Dalton to W. H. F. Talbot (2 October 1816), *The Correspondence of William Henry Fox Talbot*, ed. L. J. Schaaf (available online at http://foxtalbot.dmu.ac.uk), document no. 720.

18. Gardiner's book went through two editions in 1846, and four editions had been produced by 1852; a second series of *Twenty Lessons* containing twenty-five specimens of mosses was produced in 1849.

19. For a discussion by librarians of the issues raised by the presence of loose specimens in books, see Katie Birkwood, 'Preserved flowers in books', available online at http://maedchenimmond .blogspot.co.uk/2014/11/preserved-flowers-in-books.html.

20. W. Wilson to W. J. Hooker (19 July 1831), RBGK, Directors' Correspondence, vol. VI, letter 346.

21. W. Wilson to W. J. Hooker (15 October 1831), RBGK, Directors' Correspondence, vol. VI, letter 347.

22. R. Buxton, *A Botanical Guide to the Flowering Plants, Ferns, Mosses, and Algae, found Indigenous within Sixteen Miles of Manchester ... Together with a Sketch of the Author's Life* (London, 1849), p. vi.

23. Buxton, *A Botanical Guide*, p. vi.

24. L. Daston, 'Speechless', in L. Daston (ed.), *Things that Talk: Object Lessons from Art and Science* (New York, 2004), pp. 9–24.

25. Mavor, *Botanical Pocket Book*, p. viii.

26. M. Allan, *The Hookers of Kew 1785–1911* (London, 1967), p. 38.

27. Allan, *The Hookers of Kew*, p. 141.

28. W. J. Hooker to M. J. Berkeley (8 December 1843), Natural History Museum, London, Botany Library, Berkeley Correspondence, vol. VII.

29. J. D. Hooker to W. Wilson (13 January 1844), Warrington Museum and Archives, Wilson Correspondence, MS 53.

30. When William Gilbertson, who possessed a fine fossil shell collection, 'declined housekeeping' in 1843, the notice of sale of his household goods included not only his furniture but also his books, his land and freshwater shells, and his fossil shells, together with the glass cases in which they were contained. West Yorkshire Archive Service, Calderdale, Roberts Leyland Natural History Correspondence, SH:7/JN/B/66/61.

31. G. A. Walker-Arnott to J. S. Henslow (9 April 1831), Cambridge University Library, Add. 8176, letter 196; R. K. Greville to M. J. Berkeley (1 April 1826), Natural History Museum, London, Botany Library, Berkeley Correspondence, vol. V.

32. J. Roberts to J. S. Henslow (15 October 1825), Cambridge University Library, Add. 8176, letter 33.

33. W. Wilson to J. D. Hooker (15 April 1847), RBGK, Directors' Correspondence, vol. CVI, fo. 177.

34. L. L. Merrill, *The Romance of Victorian Natural History* (Oxford, 1989), p. 9.

35. J. E. Winterbottom to W. Wilson (4 May 1852), Natural History Museum, London, Botany Library, Wilson Correspondence, vol. XII.

36. Stewart, *On Longing*, p. 162.

37. W. J. Hooker to W. Wilson (15 September 1828), RBGK, 'Letters from W. J. Hooker', fos. 31–3.

38. W. Wilson to W. J. Hooker (1 November 1828), RBGK, Directors' Correspondence, vol. I, letter 283.

39. W. Wilson to J. S. Henslow (10 February 1831), Cambridge University Library, Add. 8176, letter 180.

40. W. Wilson to W. J. Hooker (29 March 1832), RBGK, Directors' Correspondence, vol. VI, letter 352.

41. W. J. Hooker and Thomas Taylor had produced the first two editions of *Muscologia Britannica* (London, 1818 and 1827); Wilson published the third edition under the title *Bryologia Britannica* (London, 1855).

42. W. J. Hooker to W. Wilson ([March 1847]), Warrington Museum and Archives, Wilson Correspondence, MS 53.

43. W. Wilson to J. D. Hooker (27 October 1847), RBGK, Directors' Correspondence, vol. CVI, letter 187.

44. W. Wilson to M. J. Berkeley (3 April 1847), Natural History Museum, London, Botany Library, Berkeley Correspondence, vol. XI.

18   Natural history and the scientific voyage

1. G. Williams, *Naturalist at Sea: Scientific Travellers from Dampier to Darwin* (New Haven, 2013); D. MacKay, *In the Wake of Cook: Exploration, Science and Empire* (Victoria, 1985).

2. J. Browne, 'Biogeography and empire', in N. Jardine, J. A. Secord and E. C. Spary (eds.), *Cultures of Natural History* (Cambridge, 1996), pp. 305–21; E. C. Spary, *Utopia's Garden: French Natural History from Old Regime to Revolution* (Chicago, 2000), pp. 79–86; B. Smith, *European Vision and the South Pacific* (New Haven, 1985).

3. J. Dunmore, *French Explorers in the Pacific*, 2 vols. (Oxford, 1969), vol. II, pp. 316ff.

4. R. King, *Narrative of a Journey to the Shores of the Arctic Ocean in 1833, 1834, 1835*, 2 vols. (London, 1836), vol. I, pp. viii–ix.

5. J. Browne, *Charles Darwin, A Biography: Voyaging* (Princeton, 1996); J. Goodman, *The* Rattlesnake: *A Voyage of Discovery to the Coral Sea* (London, 2005); J. Dunmore, *From Venus to Antarctica: The Life of Jules Dumont D'Urville* (Auckland, 2007).

6. R. Cock, 'Sir Francis Beaufort and the coordination of British scientific activity 1829–1855', PhD dissertation, University of Cambridge (2003); J. Gascoigne, *Science in the Service of Empire: Joseph Banks, the British State and the Uses of Empire* (New York, 1998).

7. For example, P. Hatfield, *Lines in the Ice: Exploring the Roof of the World* (Montreal and Kingston, 2016); N. Rigby and P. van der Mer, *Pioneers of the Pacific: Voyages of Exploration, 1787–1810* (Berkeley, 2005); G. Williams, *The Search for the Northwest Passage in the Age of Reason* (New York, 2002).

8. R. Burkhardt, Jr, 'Naturalists' practices and nature's empire: Paris and the platypus 1815–1833', *Pacific Science*, 55 (2001), pp. 327–41.

9. J. Herschel (ed.), *A Manual of Scientific Enquiry* (London, 1849), p. iv.

10. A.-N. Vaillant, *Voyage autour du monde exécuté pendant les années 1836 et 1837 sur la corvette La Bonite* (Paris, 1840–52), vol. I, pp. xiv–xv; Dunmore, *French Explorers*, vol. II, pp. 269–70.

11. M. Sankey, P. Cowley and J. Fornasiero, 'The Baudin expedition in review: old quarrels and new approaches', *Australian Journal of French Studies*, 41 (2004), pp. 4–14.

12. King, *Narrative*, vol. II, p. 204.

13. J. MacDouall, *Narrative of a Voyage to Patagonia and Terra del Fuego* (London, 1833), pp. 140–1.

14. Capt. P. P. King to J. W. Croker, Secretary of the Admiralty (1 July 1830), ADM 1 2031, National Archives.

15. For example, *The South America Pilot* (London, 1832).

16. In a wide-ranging literature, see M. Cohen, *The Novel and the Sea* (Princeton, 2012) and R. Foulke, *The Sea Voyage Narrative* (New York, 2002). Work most relevant to historians of science includes: J. Fornasiero and J. West-Sooby, 'The narrative interruptions of science: the Baudin expedition to Australia (1800–1804)', *Forum for Modern Language Studies*, 49 (2013), pp. 457–71; D. Livingstone, 'Text, talk and testimony: geographical reflections on scientific habits: an afterword', *British Journal for the History of Science*, 38 (2005), pp. 93–100; M. Sankey, 'Writing the voyage of scientific exploration: the logbooks, journals and notes of the Baudin expedition (1800–1804)', *Intellectual History Review*, 20 (2010), pp. 401–13.

17. For its geographical reputation, see 'A Sketch of the Progress of Geography in 1836–7', *Journal of the Royal Geographical Society*, 7 (1837), p. 192.

18. Robert Fitzroy to his sister, Frances, on board HMS *Beagle* at St Helena (11 June 1836), Robert Fitzroy Letters to Family 1817–52, RP 2006, British Library.

19. C. Darwin to S. E. Darwin ([1 April 1838]), in F. Burkhardt et al. (eds.), *The Correspondence of Charles Darwin* (Cambridge, 1986), vol. II, p. 80.

20. R. Fitzroy, 'Memorandum', in *Narrative of the Surveying Voyages of His Majesty's Ships* Adventure *and* Beagle (London, 1839), vol. II, appendix, p. iv.

21. R. Yeo, *Encyclopaedic Visions: Scientific Dictionaries and Enlightenment Culture* (Cambridge, 2001).

22. G. S. Ritchie, *The Admiralty Chart, British Naval Hydrography in the Nineteenth Century* (London, 1967); M. Barford, 'The surveyor's St. Lawrence: route science and survey work', in K. Anderson and H. M. Rozwadowski (eds.), *Soundings and Crossings: Doing Science at Sea 1800–1970* (Sagamore Beach, MA, 2016), pp. 49–78; D. G. Burnett, 'Hydrographic discipline among the navigators: charting an empire of commerce and science in the nineteenth-century Pacific', in J. Akerman (ed.), *The Imperial Map: Cartography and the Mastery of Empire* (Chicago, 2009), pp. 185–259.

23. On the myth of Patagonian giants, see see P. Adams, *Travelers and Travel Liars* (Berkeley and Los Angeles, 1962).

24. P. P. King, *Narrative of the Surveying Voyages of His Majesty's Ships* Adventure *and* Beagle (London, 1839), vol. I, pp. 84–5; for navigation and visual expertise, see L. Martins, 'Navigating in tropical waters: maritime views of Rio de Janeiro', *Imago Mundi*, 50 (1998), pp. 141–55.

25. Fitzroy, *Narrative of HMS* Adventure *and* Beagle, vol. II, p. 134.

26. For a related discussion of the limits of control, see F. Driver, 'Distance and disturbance: travel, exploration and knowledge in the nineteenth century', *Transactions of the Royal Historical Society*, 14 (2004), pp. 73–92.

27. C. Darwin, 'Geology', in Herschel (ed.), *Manual* (1849), pp. 156–95, pp. 163–4, 183, 194–5.

28. A. Sponsel, 'An amphibious being: how maritime surveying re-shaped Darwin's approach to natural history', *Isis*, 107 (2016), pp. 254–81.

29. A. Du Petit-Thouars, *Voyage autour du monde dans la frégate Vénus*, 10 vols. (Paris, 1840–55), vol. V (1855), *Zoologie*, p. 2.

30. M.-N. Bourguet, C. Licoppe and H. O. Sibum (eds.), *Instruments, Travel and Science* (London, 2002).

31. Quoted in Dunmore, *French Explorers*, vol. II, p. 296.

32. S. Müller-Wille and I. Charmantier, 'Lists as research technologies', *Isis*, 103 (2012), pp. 743–52, p. 743.

33. See the discussion of the circulation of the specimen or its illustration as the 'dislocated global' in D. Bleichmar, 'Geography of observation: distance and invisibility in eighteenth-century

botanical travel', in L. Daston and E. Lunbeck (eds.), *Histories of Scientific Observation* (Chicago, 2011), pp. 373-95.

34. H. Rozwadowski, *Fathoming the Ocean* (Cambridge, MA, 2005), pp. 122-32, 150-73.

35. M. Deacon, *Scientists and the Sea 1650-1900* (London, 1971), pp. 306-65.

36. Rozwadowski, *Fathoming the Ocean*, pp. 183-4.

37. J. Murray with T. H. Tizard, H. N. Moseley, J. Y. Buchanan, *Narrative of the Cruise of HMS* Challenger, *with a General Account of the Scientific Results of the Expedition* (Edinburgh, 1885), art. 1, p. iv.

38. H. L. Burstyn, 'Science and government in the nineteenth century: the *Challenger* Expedition and its report', *Bulletin de l'Institut Océanographique*, special issue no. 2 (1968), pp. 603-11; L. K. Nyhart, 'Voyages and the scientific expedition report, 1800-1940', in R. D. Apple, G. J. Downey and S. L. Vaughn (eds.), *Science in Print: Essays on the History of Science and the Culture of Print* (Madison, 2012), pp. 65-86.

39. I am indebted in this conclusion to the argument of Nyhart on Victor Henson's plankton reports, cited in n. 38.

40. These were *Bulletin Statistique des Pêches Maritimes, Bulletin Hydrographique* (for dissemination of biological and physical data) and *Publications de Circonstance du Conseil Permanent International pour L'Exploration de la Mer*, renamed *Journal du Conseil* in 1926. See H. M. Rozwadowski, *The Sea Knows No Boundaries: A Century of Marine Science under ICES* (Seattle and London, 2002).

19   Humboldt's exploration at a distance

The research for this chapter was supported by a Marie Curie Grant awarded by the European Commission Research Executive Agency (AHumScienceNet, project number 327127, FP7-PEOPLE-2012-IOF), and carried out at the Huntington Library in San Marino, California.

1. See D. Outram, 'New spaces in natural history', in N. Jardine, J. A. Secord and E. C. Spary (eds.), *Cultures of Natural History* (Cambridge, 1996), pp. 259-60.

2. S. F. Cannon, *Science in Culture: The Early Victorian Period* (New York, 1978), chapter on Humboldtian science. See also M. Dettelbach, 'Humboldtian science', in Jardine et al. (eds.) *Cultures of Natural History*, pp. 287-304.

3. For Humboldt's preparations for his expedition through the Spanish colonies see M. A. Puig-Samper and S. Rebok, *Sentir y medir. Alexander von Humboldt en España* (Aranjuez, 2007);

S. Rebok, *Una doble mirada: Alexander von Humboldt y España en el siglo XX* (Madrid, 2009).

4. Humboldt to K. E. von Moll (5 June 1799), in U. Moheit (ed.), *Humboldt. Briefe aus Amerika. 1799–1804* (Berlin, 1993), p. 33. See also Humboldt to C. L. Willdenow (21 February 1801), in Moheit (ed.), *Humboldt*, p. 12.

5. Humboldt to J. Madison (19/20 June 1804), in I. Schwarz (ed.), *Alexander von Humboldt und die Vereinigten Staaten von Amerika. Briefwechsel* (Berlin, 2004), p. 94.

6. Humboldt to W. Thornton (20 June 1804), in Schwarz (ed.), *Humboldt und die Vereinigten Staaten*, p. 96. See also Humboldt to J. Vaughan (10 June 1805), in Schwarz (ed.), *Humboldt und die Vereinigten Staaten*, p. 105.

7. Humboldt was not the only source of information for the American president, but it allowed him to contrast the data provided with material that was provided through local agents, settlers, soldiers, merchants, such as James Wilkinson, Philip Nolan and John Sibley.

8. A. von Humboldt, 'Tablas geográfico-políticas del Reino de Nueva-España, en el año de 1803 . . .' *Boletín de geografía y estadística*, 1 (1869), pp. 635–57.

9. 'Tableau statistique du Royaume de la Nouvelle Espagne', in Schwarz (ed.), *Humboldt und die Vereinigten Staaten*, pp. 484–95.

10. M. Faak (ed.), *Alexander von Humboldt. Reise auf dem Rio Magdalena, durch die Anden und durch Mexiko*, (Berlin, 1986, new edition 2003) vol. VIII, pp. 329–30. A detailed list can also be found in his unpublished travel journals: M. Faak, *Alexander von Humboldts amerikanische Reisejournale: eine Übersicht* (Berlin, 2002), diaries VII b and c (fo. 278r–278v), VIII (fo. 64r, 89v).

11. S. T. Jackson and L. D. Walls (eds.), *Alexander von Humboldt. Views of Nature* (Chicago, 2014), pp. 49 and 151.

12. A. von Humboldt, *Cosmos: A Sketch of a Physical Description of the Universe*, 5 vols. (New York, 1858), vols. II and IV.

13. U. Leitner (ed.), *Alexander von Humboldt und Cotta: Briefwechsel* (Berlin, 2009); S. Panwitz and I. Schwarz (eds.), *Alexander von Humboldt. Familie Mendelsohn. Briefwechsel* (Berlin, 2011); I. Schwarz (ed.), *Alexander von Humboldt. Samuel Heinrich Spiker: Briefwechsel* (Berlin, 2007); *Briefwechsel Alexander von Humboldt's mit Heinrich Berghaus aus den Jahren 1825 bis 1858*, 3 vols. (Leipzig, 1863); U. Pässler, *Alexander von Humboldt. Carl Ritter. Briefwechsel* (Berlin, 2010).

14. H. R. Slotten, *Patronage, Practice, and the Culture of American Science: Alexander Dallas Bache and the US Coast Survey* (Cambridge, 1994), pp. 116–30.

15. I. Schwarz, 'Alexander von Humboldt's correspondence with Johann Gottfried Flügel', *Yearbook of German–American Studies*, 46 (2011), pp. 87–94.

16. See A. W. Whipple to Humboldt (8 August 1854), in Schwarz (ed.), *Humboldt und die Vereinigten Staaten*, pp. 334–5; Humboldt to A. W. Whipple (18 August 1855), in Schwarz (ed.), *Humboldt und die Vereinigten Staaten*, pp. 350–3.

17. J. G. Kohl to Humboldt (17 September 1856), in Schwarz (ed.), *Humboldt und die Vereinigten Staaten*, pp. 402–4.

18. J. B. Floyd to Humboldt (14 July 1858), in Schwarz (ed.), *Humboldt und die Vereinigten Staaten*, p. 457.

19. H. Stevens, *The Humboldt Library: A Catalogue of the Library of Alexander von Humboldt; with a Bibliographical and Biographical Memoir* (London, 1863).

20. A. von Humboldt, *On the Fluctuations in the Production of Gold* (New York, 1900). See also Humboldt's map 'Carte des divers routes par lequelles les richesses métalliques refluent d'un continent a l'autre', in his *Atlas de la Nouvelle-Espagne* (Paris, 1809).

21. See document 22 ('Über die Goldgewinnung in den Vereinigten Staaten', February 1833) and document 23 ('Goldwäsche am Altai', February 1833) in Schwarz (ed.), *Samuel Heinrich Spiker*, pp. 272–4.

22. Humboldt to J. L. Tellkampf (14 November 1843), in Schwarz (ed.), *Humboldt und die Vereinigten Staaten*, p. 229.

23. See Humboldt to Heinrich Berghaus (received 25 September 1849), in *Briefwechsel mit Heinrich Berghaus*, vol. III, p. 27.

24. Jackson and Walls (eds.), *Humboldt. Views of Nature*, p. 51.

25. Jackson and Walls (eds.), *Humboldt. Views of Nature*, p. 49 and pp. 151–2.

26. H. Berghaus and A. von Humboldt, 'Ethnographische Karte von Nordamerika', *Dr. Heinrich Berghaus' Physikalischer Atlas oder Sammlung von Karten*, vol. II, p. 8.17 (Gotha, 1846); H. Berghaus and A. von Humboldt, 'Bergketten in Nord-Amerika', in *Physikalischer Atlas*, vol. I, p. 3.6 (Gotha, 1842).

27. A. von Humboldt, *Political Essay on the Kingdom of New Spain*, 4 vols. (London, 1811), vol. I, p. civ.

28. Jackson and Walls (eds.), *Humboldt. Views of Nature*, p. 82.

29. S. Forry, *The Climate of the United States* (New York, 1842), pp. 37, 39, 102.

30. Jackson and Walls (eds.), *Humboldt. Views of Nature*, p. 85.

31. A. von Humboldt, *Zentral-Asien, Untersuchungen zu den Gebirgsketten und zur vergleichenden Klimatologie*, ed. O. Lubrich (Frankfurt am Main, 2009), pp. 30, 537–9.

32. See also J. B. de C. M. Saunders, *Humboldtian Physicians in California* (Davis, 1971).

33. 'Points de partage et communications projettées entre le Grand Océan et l'Océan Atlantique', in Humboldt, *Atlas de la Nouvelle-Espagne*.

34. Humboldt to J. Davis (24 Mar 1857), in Schwarz (ed.), *Humboldt und die Vereinigten Staaten*, pp. 417–18.

35. Humboldt to J. C. Frémont (7 October 1850), in Schwarz (ed.), *Humboldt und die Vereinigten Staaten*, pp. 275–6.

36. U. Leitner, *Alexander von Humboldt. Friedrich Wilhelm IV. Briefwechsel* (Berlin, 2013), pp. 74–8.

37. See Humboldt to Berghaus (3 March 1852), with material from Emory, Frémont, Nicollet, Albert, Agassiz and other books he had previously lent him, in *Briefwechsel Humboldt's mit Heinrich Berghaus*, vol. III, p. 307.

38. Humboldt, 'Metallurgische und geographische Nachrichten von Nord-Amerika. Aus einem Brief von Herrn Albert Gallatin', *Berlinische Nachrichten von Staats- und gelehrten Sachen*, n. 13 (16 January 1838), pp. 5–6.

39. A. Gallatin, 'Tabellarische Übersicht der Indianerstämme in den Vereinigten Staaten von Nordamerika, ostwärts von den Felsgebirgen (Stony Mountains), nach den Sprachen und Dialekten geordnet. 1826. Mitgetheilt von dem Freiherrn von Humboldt', *Hertha*, 8 (1827), pp. 328–34.

40. Humboldt and C. Möllhausen, 'Expedition zur wissenschaftlichen Erforschung des Rio Colorado, in den der Rio Gila einmündet', *Berlinische Nachrichten von Staats- und Gelehrten Sachen*, n. 299 (22 December 1857), pp. 3–4.

41. A list of all the places in the world named after Humboldt is in U.-D. Oppitz, '*Der Name der Brüder Humboldt in aller Welt*', in H. Pfeiffer (ed.), *Alexander von Humboldt, Werk und Weltgeltung* (München, 1969), pp. 277–429.

42. F. Baron and S. Seeger, 'Moritz Hartmann (1817–1900) in Kansas: a forgotten German pioneer of Lawrence and Humboldt', *Yearbook for German-American Studies*, 39 (2004), pp. 1–22, p. 9.

43. J. C. Fremont, *Geographical Memoir upon Upper California* (Philadelphia, 1849), p. 8.

44. J. B. Floyd to Humboldt (14 July 1858), in Schwarz (ed.), *Humboldt und die Vereinigten Staaten*, p. 457.

20   Publics and practices

1. See Richards, this volume, Chapter 32.

2. For a basic introduction to Linnaeus and Buffon, see P. R. Sloan, *Finding Order in Nature: The Naturalist Tradition from Linnaeus to E. O. Wilson* (Baltimore, 2000), chapter 1. On the popular

context for Buffon's *Histoire naturelle*, see L. E. Robbins, *Elephant Slaves and Pampered Parrots: Exotic Animals in Eighteenth-Century Paris* (Baltimore, 2002).

3. L. K. Nyhart, *Modern Nature: The Rise of the Biological Perspective in Germany* (Chicago, 2009).

4. J. Habermas, *The Structural Transformation of the Public Sphere*, trans. T. Burger (1962; Cambridge, MA, 1989). German original, 1962; T. H. Broman, 'The Habermasian public sphere and "science in the Enlightenment"', *History of Science*, 36 (1998), pp. 123–50.

5. R. Cooter and S. Pumfrey, 'Separate spheres and public places: reflections on the history of science popularisation and science in popular culture', *History of Science*, 32 (1994), pp. 237–67; B. Lightman, *Victorian Popularisers of Science: Designing Nature for New Audiences* (Chicago, 2007); A. W. Daum, 'Varieties of popular science and the transformations of public knowledge: some historical reflections', *Isis*, 100 (2009), pp. 319–32.

6. A. Daum, *Wissenschaftspopularisierung im 19. Jahrhundert* (Munich, 1998), table 2, pp. 91–5.

7. Daum, *Wissenschaftspopularisierung*, p. 101.

8. C. Kretschmann, *Räume öffnen sich: Naturhistorische Museen in Deutschland des 19. Jahrhunderts* (Berlin, 2006), pp. 179–84; Nyhart, *Modern Nature*, pp. 241–3; S. Köstering, *Natur zum Anschauen: Das Naturkundemuseum des deutschen Kaiserreichs 1871–1914* (Cologne, 2003), p. 198.

9. Köstering, *Natur zum Anschauen*, esp. pp. 24–30.

10. Köstering, *Natur zum Anschauen*, pp. 30–42; Nyhart, *Modern Nature*, pp. 203–14.

11. P. L. Martin, *Die Praxis der Naturgeschichte*, 3 vols. (Weimar, 1869–82), vol. II (1870), p. 2.

12. On European zoos, see Ash, this volume, Chapter 25, and sources cited therein; also E. Baratay and E. Hardouin-Fugier, *Zoo: A History of Zoological Gardens in the West* (London, 2002); on German zoos, A. Rieke-Müller and Lothar Dittrich, *Der Löwe brüllt nebenan: Die Gründung zoologischer Gärten im deutschsprachigen Raum 1833–1869* (Cologne, 1998); and Nyhart, *Modern Nature*, chapter 3.

13. M. Osborne, *Nature, the Exotic and the Science of French Colonialism* (Bloomington, 1994); W. Anderson, 'Climates of opinion: acclimatisation in nineteenth-century France and England', *Victorian Studies*, 35 (1992), pp. 135–57.

14. For more on zoos in the nineteenth and twentieth centuries, see Ash, this volume, Chapter 25.

15. See Plumb, this volume, Chapter 15.

16. D. F. Weinland, 'Ueber den Ursprung und die Bedeutung der neueren Zoologischen Gärten', *Der Zoologische Garten*, 3 (1862), pp. 1–3.

17. Both quoted in G. Kaselow, *Die Schaulust am exotischen Tier* (Hildesheim, 1999), pp. 143–4.

18. For example, A. E. Brehm, 'Von der Baustatte des Berliner Aquariums', *Gartenlaube* (1868), pp. 620–3; Brehm, 'Menschenaffen', *Gartenlaube* (1876), pp. 44–8, 160–3, 282–6; on Brehm's aquarium, see also Carl Nißle, 'Ein Wunderbau für die Thierwelt', *Gartenlaube* (1873), pp. 165–70.

19. Nyhart, *Modern Nature*, chapters 4 and 6 (on Möbius), chapter 3 (on Martin and Jaeger).

20. For a brief broad overview, see M. P. Winsor, 'Museums', in P. J. Bowler and J. V. Pickstone (eds.), *The Cambridge History of Science*, vol. VI, *The Modern Biological and Earth Sciences* (Cambridge, 2009), pp. 60–75; also Alberti, this volume, Chapter 21.

21. Nyhart, *Modern Nature*, pp. 223–40.

22. On dioramas, see Alberti, this volume, Chapter 21.

23. Köstering, *Natur zum Anschauen*, pp. 90–3, 116–22.

24. Nyhart, *Modern Nature*, pp. 251–92.

25. J. A. Williams, *Turning to Nature in Germany: Hiking, Nudism and Conservation, 1900–1940* (Stanford, 2007).

26. The 1899 figure: 'Die Junge-Feier im Kieler Lehrerverein', *Kieler Zeitung* (30 September 1899), p. 2; the 1915 figure: *Württembergischer Nekrolog für das Jahr 1919*, s. v. 'Lutz, Gottlob', reprinted in *Deutsches Biographisches Archiv*, Neue Folge, fiche 842, frames 56–9.

27. *Allgemeine Botanische Zeitschrift* (1906), pp. 206–7.

28. 'Der Ocean auf dem Tische', *Gartenlaube* (1855), p. 56.

29. E. A. Rossmässler, *Das Süßwasser-Aquarium. Eine Anleitung zur Herstellung und Pflege desselben* (Leipzig, 1857).

30. See, for example, K. G. Lutz, *Das Süsswasser-Aquarium und das Leben im Süsswasser* (Stuttgart, 1886); W. Hess, *Das Süßwasseraquarium und seine Bewohner* (Stuttgart, 1886). On thinking and teaching about natural communities, see Nyhart, *Modern Nature*.

31. See D. Hohl, 'Von den vivaristischen Anfängen bis zur VDA-Gründung', in Verband Deutscher Vereine für Aquarien- und Terrarienkunde e. V. (ed.), *Festschrift zum 90jährigen Jubiläum. Beiträge zur Geschichte der Aquaristik und Terraristik in Deutschland* (Bochum, 2001), pp. 13–72, for an overall picture, and on the guppy, p. 22; for descriptions of popular imported fish, see E. Bade, *Das Süßwasser-Aquarium*, 2nd edn (Berlin, 1898), pp.

374–5 (Veiltail [Schleierschwanz], Telescope), and p. 336 (climbing perch [Kletterbarsch]).

32. Hohl, 'Von den vivaristischen Anfängen', pp. 32–3, 51.

33. Activities tracked in *Blätter für Aquarien- und Terrarien-Freunde*, 1890–1914.

21   Museum nature

My thanks to Helen Anne Curry, Andrew Kitchener, Henry McGhie, Jim Secord, Emma Spary and Geoff Swinney.

1. J. V. Pickstone, *Ways of Knowing: A New History of Science, Technology and Medicine* (Manchester, 2000).

2. S. J. M. M. Alberti, 'Placing nature: natural history collections and their owners in nineteenth-century provincial England', *British Journal for the History of Science*, 35 (2002), pp. 291–311.

3. T. Barringer and T. Flynn (eds.), *Colonialism and the Object: Empire, Material Culture, and the Museum* (London, 1998); J. M. MacKenzie, *Museums and Empire: Natural History, Human Cultures and Colonial Identities* (Manchester, 2009).

4. W. H. Flower, *Essays on Museums and Other Subjects Connected with Natural History* (London, 1898), p. 14.

5. H. Ritvo, *The Platypus and the Mermaid, and Other Figments of the Classifying Imagination* (Cambridge, MA, 1997).

6. S. Sheets-Pyenson, *Cathedrals of Science: The Development of Colonial Natural History Museums during the Late Nineteenth Century* (Kingston, Ontario, 1988); W. T. Stearn, *The Natural History Museum at South Kensington: A History of the British Museum (Natural History) 1753–1980* (London, 1981).

7. S. Forgan 'Building the museum', *Isis*, 96 (2005), pp. 572–85; T. A. Markus, *Buildings and Power: Freedom and Control in the Origin of Modern Building Types* (London, 1993); C. Yanni, *Nature's Museums: Victorian Science and the Architecture of Display* (London, 1999).

8. S. Macdonald, *Behind the Scenes at the Science Museum* (Oxford, 2002).

9. I. Kopytoff, 'The cultural biography of things: commoditization as process', in A. Appadurai (ed.), *The Social Life of Things: Commodities in Cultural Perspective* (Cambridge, 1986), pp. 64–91; S. J. M. M. Alberti, 'Objects and the museum', *Isis*, 96 (2005), pp. 559–71.

10. S. Naylor and J. Hill, 'Museums', in J. Agnew and D. Livingstone (eds.), *The Sage Handbook of Geographical Knowledge* (London, 2011), pp. 64–75, p. 65; D. N. Livingstone, *Putting Science in its Place: Geographies of Scientific Knowledge* (Chicago, 2003).

11. G. N. Swinney, 'Appropriate and appropriated sites for elephants: a case-study of the making of museum objects', *Society and Animals* (forthcoming).

12. G. N. Swinney, 'Towards an historical geography of a "national" museum: the Industrial Museum of Scotland, the Edinburgh Museum of Science and Art and the Royal Scottish Museum, 1854–1939', PhD thesis, University of Edinburgh (2013).

13. 'Three inches taller than Jumbo', *Edinburgh Evening Dispatch* (19 October 1907).

14. R. D. Altick, *The Shows of London: A Panoramic History of Exhibitions, 1600–1862* (Cambridge, MA, 1978); Bullock Museum, *Sale Catalogue of the Bullock Museum 1819* (facsimile reprint; London, 1979), p. 70.

15. Scotch Education Department, *The Royal Scottish Museum, Edinburgh: A Guide to the Collections* (Glasgow, 1908), p. 28, original emphasis.

16. Following German naturalist Johann Friedrich Blumenbach; the species has since been categorised *Loxodonta africana*.

17. R. Poliquin, *The Breathless Zoo: Taxidermy and the Cultures of Longing* (University Park, PA, 2012).

18. B. Brenna, 'The frames of specimens: glass cases in Bergen Museum around 1900', in L. E. Thorsen, K. A. Rader and A. Dodd (eds.), *Animals on Display: The Creaturely in Museums, Zoos and Natural History* (University Park, PA, 2013), pp. 37–57.

19. S. C. Quinn, *Windows on Nature: The Great Habitat Dioramas of the American Museum of Natural History* (New York, 2006); K. Wonders, *Habitat Dioramas: Illusions of Wilderness in Museums of Natural History* (Uppsala, 1993).

20. Quoted in Rowland Ward to Percy Powell-Cotton (5 July 1907), Powell Cotton Museum, GB1711/3.1.1/259 AV 176 Q6/7.

21. J. Griesemer, 'Modelling in the museum: on the role of remnant models in the work of Joseph Grinnell', *Biology and Philosophy*, 5 (1990), pp. 3–36; P. A. Morris, *A History of Taxidermy: Art, Science and Bad Taste* (Ascot, Berkshire, 2010).

22. On blending of different continents' understandings and representations of elephants in a different context, see S. Sivasundaram, 'Trading knowledge: the East India Company's elephants in India and Britain', *Historical Journal*, 48 (2005), pp. 27–63.

23. Poliquin, *Breathless Zoo*.

24. R. M. Peck, 'Alcohol and arsenic, pepper and pitch: brief histories of preservation techniques', in S. A. Prince (ed.), *Stuffing Birds, Pressing Plants, Shaping Knowledge: Natural History in North America, 1730–1860* (Philadelphia, 2003), pp. 27–53.

25. H.-L. Chalk, 'Mobile stones: the uses and meanings of earth science teaching specimens', *Material Culture Review*, 74 (2012), pp. 149–60.

26. D. Haraway, *Primate Visions: Gender, Race and Nature in the World of Modern Science* (London, 1989).

27. J. A. Hendon, 'Having and holding: storage, memory, knowledge and social relations', *American Anthropologist*, 102 (2000), pp. 42–53; see also H. Geoghegan and A. Hess, 'Object-love at the Science Museum: cultural geographies of museum storerooms', *Cultural Geographies*, 22 (2015), pp. 445–65.

28. S. Everest, '"Under the skin": the biography of a Manchester mandrill', in S. J. M. M. Alberti (ed.), *The Afterlives of Animals: A Museum Menagerie* (Charlottesville, 2011), pp. 75–91.

29. R. V. Melville, *Towards Stability in the Names of Animals: A History of the International Commission on Zoological Nomenclature 1895–1995* (London, 1995).

30. J. C. Melvill and R. Standen, 'Description of *Conus (Cylinder) Clytospira* Sp. N. from the Arabian Sea', *Annals and Magazine of Natural History*, 4 (1899), pp. 461–3; F. W. Townsend, 'Notes on shell collecting in the northern parts of the Arabian Sea, including the Gulf of Oman and Persian Gulf in the years 1890–1914', *Proceedings of the Malacological Society*, 18 (1928), pp. 118–26.

31. See for example J. C. Melvill and R. Standen, *Catalogue of the Hadfield Collection of Shells from the Lifu and Uvea, Loyalty Islands*, 2 vols. (Manchester, 1895–7).

32. C. Gosden and C. Knowles, *Collecting Colonialism: Material Culture and Colonial Change* (Oxford, 2001).

33. R. E. Kohler, *All Creatures: Naturalists, Collectors and Biodiversity, 1850–1950* (Princeton, 2006).

34. J. A. Secord, 'Knowledge in transit', *Isis*, 95 (2004), 654–72.

35. S. J. M. M. Alberti, 'The museum affect: visiting collections of anatomy and natural history', in A. Fyfe and B. Lightman (eds.), *Science in the Marketplace: Nineteenth-Century Sites and Experiences* (Chicago, 2007), pp. 371–403.

22   Peopling natural history

1. C. Manias, *Race, Science and the Nation: Reconstructing the Ancient Past in Britain, France and Germany* (London, 2013).

2. The best summary is M. Rudwick, *Worlds before Adam: The Reconstruction of Geohistory in the Age of Reform* (Chicago, 2008), pp. 407–22.

3. Rudwick's works remain the authoritative guide.

4. A. Bowdoin von Riper, *Men among the Mammoths: Victorian Science and the Discovery of Human Prehistory* (Chicago, 1993).

5. M. Elshakry, *Reading Darwin in Arabic, 1860–1950* (Chicago, 2013) shows that some Muslim theologians embraced evolutionary ideas in an effort to prove that Islam was in accord with modern science, more accommodating than Christianity, and a rational religion against racialised accusations of superstitious ignorance.

6. C. Darwin, *The Descent of Man, and Selection in Relation to Sex* (London, 1871), p. 3.

7. D. N. Livingstone, *Adam's Ancestors: Race, Religion and the Politics of Human Origins* (Baltimore, 2008).

8. M.-J.-P. Flourens, 'On the natural history of man', *Edinburgh New Philosophical Journal*, 27 (1839), pp. 351–8, p. 353.

9. Darwin, *Descent*, vol. I, p. 226.

10. J. Endersby, 'Darwin on generation, pangenesis and sexual selection', in M. J. S. Hodge and G. Radick (eds.), *Cambridge Companion to Darwin* (Cambridge, 2003), pp. 69–91.

11. Manias, *Race, Science and the Nation*; G. Stocking, *Victorian Anthropology* (New York, 1987).

12. A. C. Haddon, 'President's address: anthropology, its position and needs', *Journal of the Anthropological Institute of Great Britain and Ireland*, 33 (1903), pp. 11–23, esp. pp. 19–22.

13. W. Rivers, 'Report on anthropological research outside America', in W. H. R. Rivers, A. E. Jenks and S. G. Morley (eds.), *Reports on the Present Condition and Future Needs of the Science of Anthropology*, Carnegie Publication 200 (Washington, DC, 1913), p. 7.

14. Older histories inaccurately describe the replacement of 'armchair anthropology' in favour of fieldwork. See F. Barth, et al., *One Discipline, Four Ways: British, German, French and American Anthropology* (Chicago, 2005) and G. W. Stocking, *After Tylor: British Social Anthropology, 1888–1951* (London, 1995). Compare with the continuities traced in E. Sera-Shriar, *The Making of British Anthropology, 1813 to 1871* (London, 2013).

15. S. Qureshi, *Peoples on Parade: Exhibitions, Empire and Anthropology in Nineteenth Century Britain* (Chicago, 2011).

16. See the timeline in C. Fusco, 'The other history of intercultural performance', in *English Is Broken Here: Notes on Cultural Fusion in the Americas* (New York, 1995), pp. 37–63.

17. A. T. Vaughan, 'Shakespeare's Indian: the Americanisation of Caliban', *Shakespeare Quarterly*, 39 (1988), pp. 137–53.

18. C. Crais and P. Scully, *Sara Baartman and the Hottentot Venus: A Biography and a Ghost Story* (Princeton, 2008), p. 70.

19. Crais and Scully, *Sara Baartman* and S. Qureshi, 'Displaying Sara Baartman, the "Hottentot Venus"', *History of Science*, 42 (2004), pp. 233–57.

20. Court Records, reprinted in Z. Strother, 'Display of the body Hottentot', in Bernth Lindfors (ed.), *Africans on Stage: Studies in Ethnological Showbusiness* (Bloomington, 1999), pp. 1–61, p. 43.

21. The body cast and skeleton were exhibited in Parisian museums until objections led to their removal in the 1980s. South Africans launched a repatriation campaign in 1995. Eight years later, Baartman's state funeral was held on National Women's Day in August 2002. Crais and Scully, *Sara Baartman*, 142–69.

22. Handbill for the 'Botocudos' (1822), 'Human Freaks', Box 4 (59) and entrance ticket for the 'Aztecs' (1853), all from University of Oxford, John Johnson Collection, Tickets Show Places (18).

23. 'Chit-Chat', *Theatrical Journal*, 8 (1847), p. 239.

24. 'Farini's Earthmen', *Era* (27 September 1884), p. 11.

25. On London's ethnic diversity see N. Merriman (ed.), *The Peopling of London: Fifteen Thousand Years of Settlement from Overseas* (London, 1993).

26. Anon., *Outlines of the Phrenological System of Drs. Gall and Spurzheim* (Edinburgh, 1819).

27. G. Catlin, *Adventures of the Ojibbeway and Ioway Indians in England, France and Belgium*, 3rd edn (London, 1852), vol. I, pp. 120–1.

28. Catlin, *Adventures of the Ojibbeway and Ioway Indians*, vol. I, pp. 247–50.

29. Anon., '*Now Exhibiting at the Egyptian Hall, Piccadilly: The Bosjesmans, or Bush People . . .*' (London, 1847).

30. R. Knox, *The Races of Men: A Fragment* (London, 1850), pp. 145–241.

31. R. Owen and R. Cull, 'A brief notice of the Aztec race, and a description of the so-called Aztec children', *Journal of the Ethnological Society of London*, 4 (1856), pp. 120–37.

32. W. H. Flower and J. Murie, 'Account of the dissection of the Bushwoman', *Journal of Anatomy and Physiology*, 1 (1867), pp. 189–208, p. 198.

33. Exhibition poster, British Library, Evanion Collection, Evan. 344.

34. Qureshi, *Peoples on Parade*, pp. 193–208.

35. 'A visit to the Crystal Palace', *Lady's Newspaper* (10 June 1854), p. 365.

36. R. G. Latham and E. Forbes, *A Hand Book to the Courts of Natural History Described* (London, 1854), p. 6.

37. 'Anthropological news', *Anthropological Review*, 5 (1867), pp. 240–56.

38. F. Galton, 'Opening remarks by the president', *Journal of the Anthropological Institute of Great Britain and Ireland*, 16 (1887), pp. 175–7.

39. T. Mitchell, 'The World as Exhibition', *Comparative Studies in Society and History*, 31 (1989), pp. 217–36; for the 1930s, see B. de l'Estoile, 'From the colonial exhibition to the museum of man: an alternative genealogy of French anthropology', *Social Anthropology*, 11 (2003), pp. 341–61.

40. P. Brinkman, 'Frederic Ward Putnam, Chicago's cultural philanthropists, and the founding of the Field Museum', *Museum History Journal*, 2 (2009), pp. 73–100, p. 76.

41. N. J. Parezo and D. D. Fowler, 'Taking ethnological training outside the classroom: the 1904 Louisiana Purchase Exposition as field school', *Histories of Anthropology Annual*, 2 (2006), pp. 69–102.

42. F. Galton, 'Opening remarks by the president', pp. 189–90, p. 189.

43. F. Galton, 'Address delivered at the anniversary meeting', *Journal of the Anthropological Institute of Great Britain and Ireland*, 16 (1887), pp. 387–402, p. 391.

44. W. H. Holmes, 'The World's Fair Congress of Anthropology', *American Anthropologist*, 6 (1893), pp. 423–34, p. 423.

45. Holmes, 'World's Fair Congress', p. 426.

46. Brinkman, 'Frederic Ward Putnam', p. 97.

47. N. J. Parezo and D. D. Fowler, *Anthropology Goes to the Fair: The 1904 Louisiana Purchase Exposition* (Lincoln, NE, 2008), pp. 307, 321, 348 and P. Blanchard, et al. (eds.), *Human Zoos: Science and Spectacle in the Age of Colonial Empires*, trans. Teresa Bridgeman (Liverpool, 2008).

48. J. Conolly, *The Ethnological Exhibitions of London* (London, 1855).

49. W. H. R. Rivers, '[Communication from Dr. W. H. R. Rivers regarding exhibition of human specimens]', *Journal of the Anthropological Institute of Great Britain and Ireland*, 30 (1900), pp. 6–7, p. 6.

50. A. Zimmerman, *Anthropology and Anti-Humanism in Imperial Germany* (Chicago, 2001) and E. Ames, *Carl Hagenbeck's Empire of Entertainments* (Seattle, 2009).

51. Qureshi, *Peoples on Parade*, pp. 155–82.

52. 'Chit-Chat', *Theatrical Journal*, 8 (1847), p. 239.

53. 'A South African native's picture of England', *Munger Africana Library Notes*, 9 (1979), pp. 8–19, p. 17.

23   The oils of empire

1. K. Pomeranz, *The Great Divergence: China, Europe and the Making of the Modern World Economy* (Princeton, 2000), pp. 66–7.

2. P. Parthasarathi, *Why Europe Grew Rich and Asia Did Not: Global Economic Divergence, 1600–1850* (Cambridge, 2011), p. 221.

3. J. McNeill, 'Energy, population, and environmental change since 1750: entering the Anthropocene', in J. R. McNeill and K. Pomeranz (eds.) *The Cambridge World History*, 7 vols. (Cambridge, 2015), vol. VII, part 2, pp. 51–82, p. 53.

4. For this approach, see S. Sivasundaram (ed.) 'Global histories of science', *Isis*, 101 (2010), pp. 95–158.

5. I build here on S. Sivasundaram, *Islanded: Britain, Sri Lanka and the Bounds of an Indian Ocean Colony* (Chicago, 2013), chapter 5, 'Gardens'.

6. H. Erni, *Coal, Oil and Petroleum: Their Origin, History, Geology and Chemistry* (Philadelphia, 1865); the table is on pp. 66–8.

7. Erni, *Coal, Oil and Petroleum*, p. 22.

8. 'Buchanan's Burmah Journal', British Library, IOR/H/87, p. 80.

9. R. Christison, 'Chemical examination of the petroleum of Rangoon', *Transactions of the Royal Society of Edinburgh*, 13 (1835), pp. 118–23, pp. 118, 120.

10. W. de La Rue and H. Müller, 'Chemical examination of Burmese naphtha, or Rangoon tar', *Proceedings of the Royal Society of London*, 8 (1856–7), pp. 221–8, p. 222.

11. M. V. Longmuir, *Oil in Burma* (Bangkok, 2001), pp. 58–60.

12. See for instance, Thant Myint-U, *The Making of Modern Burma* (Cambridge, 2001), p. 18.

13. 'Epistle from the courtier Son', in Maung Htin Aung (trans. and ed.), *Epistles Written on the Eve of the Anglo-Burmese War* (The Hague, 1968), pp. 31–3, p. 32.

14. 'Narrative of the Burmese War', *Museum of Foreign Literature and Science*, 10 (1821), pp. 353–63, p. 354.

15. J. J. Snodgrass, *Narrative of the Burmese War* (London, 1827), p. 106. T. A. Trant, *Two Years in Ava* (London, 1827), p. 40, notes that a 'great portion' of the British fleet was threatened by fire as a result of these fire rafts. On the destruction of fire rafts by the British forces, see also H. Lister Maw, *Memoir of the Early Operations of the Burmese War* (London, 1832), p. 19; also p. 95.

16. 'Marryat's private logbook and record of services', National Maritime Museum, Greenwich, MRY/6.

17. For the history of the *Diana*, see C. A. Gibson-Hill, 'The steamers in Asian waters, 1819–1839', *Journal of the Malayan Branch of the Royal Asiatic Society*, 27 (1954), pp. 120–62.

18. H. Havelock, *Memoir of Three Campaigns* (Serampore, 1828), pp. 210, 212.

19. H. Gouger, *A Personal Narrative of Two Years' Imprisonment in Burmah* (London, 1860), pp. 293–4.

20. J. Crawfurd, *Journal of an Embassy* (London, 1829), p. 445. For the visit of 'scientific men' to the coalfields of Burma, see Gouger, *A Personal Narrative*, p. 341.

21. Crawfurd, *Journal*, pp. 89, 321, 328.

22. M. Charney, 'Shallow-draft boats, guns, and the Aye-ra-wa-ti', *Oriens Extremus*, 40 (1997), 16–63; p. 60; M. Charney, *Powerful Learning: Buddhist Literati and the Throne in Burma's Last Dynasty, 1752–1885* (Ann Arbor, 2006), pp. 157–8.

23. *Documents Illustrative of the Burmese War: An Introductory Sketch* (Calcutta, 1827), p. xl; W. Buckland, 'Geological account of a series of animal and vegetable remains collected by J. Crawfurd', appendix no. XIII in Crawfurd, *Journal of an Embassy* (London, 1834), vol. II, pp. 143–62, pp. 145, 147 and 150.

24. Buckland, 'Geological account', pp. 154, 156.

25. *Documents Illustrative*, p. xli.

26. Crawfurd, *Journal* (1829) p. 442. On p. 441, he listed the mineral products of Burma: 'limestone and marble, gems, noble serpentine, iron, gold, silver, copper, tin, lead, antimony, amber, coal, petroleum, nitre, natron, and salt'.

27. For a debate about how to interpret Wallich's natural history, compare D. Arnold, 'Plant capitalism and company science: the Indian career of Nathaniel Wallich', *Modern Asian Studies*, 42 (2008), pp. 899–928 and R. Grove, *Green Imperialism: Colonial Expansion, Tropical Island Edens and the Origins of Environmentalism, 1600–1860* (Cambridge, 1995), pp. 407–15.

28. *Documents Illustrative*, p. xli.

29. Trant, *Two Years*, pp. 218–20.

30. H. Bell, *Narrative of the Late Military and Political Operations in the Birmese Empire* (Edinburgh, 1827), p. 64.

31. Nathaniel Wallich to the Secretary of the Navy Board (13 July 1831), British Library, IOR/P/13/1.

32. Crawfurd, *Journal*, p. 447.

33. N. Wallich, *Plantae Asiaticae Rariores* (London, 1830), vol. I, pp. 1–3.

34. F. B. Doveton, *Reminiscences of the Burmese War* (London, 1852), pp. 196–7.

35. E. H. Pascoe, 'The oil-fields of Burma', *Memoirs of the Geological Survey of India*, 40 (1912), plate 18.

36. See for instance M. Ridd and A. Racey, 'The historical background to Myanmar's petroleum industry', *Memoirs of the Geological Society of London*, 45 (2015), pp. 13–20, p. 18.

37. Pascoe, 'Oil-fields', p. 70. There were also oilfields elsewhere in Burma, for instance on the Arakan coast; for the oilfields of India, see A. Saikia, 'Imperialism, geology and petroleum: history of oil in colonial Assam', *Economic and Political Weekly*, 46 (2011), pp. 48–55.

38. Cited in Longmuir, *Oil in Burma*, p. 9; original from 1808.

39. See V. Lieberman, 'Secular trends in Burmese economic history, c.1350–1830, and their implications for state formation', *Modern Asian Studies*, 25 (1991), pp. 1–31 and E. Tagliacozzo, 'Ambiguous commodities, unstable frontiers: the case of Burma, Siam and Imperial Britain, 1800–1900', *Comparative Studies in Society and History*, 46 (2004), pp. 354–77.

40. Longmuir, *Oil in Burma*.

41. Longmuir, *Oil in Burma*, p. 24.

42. *An Account of an Embassy to the Kingdom of Ava* (London, 1800) p. 137; 'Buchanan's Burmah Journal', p. 211.

43. H. Cox, *Journal of a Residence in the Burmhan Empire* (London, 1821), p. 37.

44. Crawfurd, *Journal*, p. 55.

45. N. G. Cholmeley, 'The oil-fields of Burma', *Journal of the Royal Society of Arts*, 61 (1913), pp. 639–58, p. 648.

46. Pascoe, 'Oil-fields', pp. 75–6.

47. On the introduction of a 'diving-bell' at the jadeite mines of Burma, see 'Note by Fritz Noetling on the occurrence of jadeite in Upper Burma', F. Noetling, *Notes on the Mineral Resources of Upper Burma* (Rangoon, 1893), p. 3.

48. See British Library, IOR, photo 61/1. The tamarind tree is photo 61/1 (10), and is followed by 'Chatty Manufactory', photo 61/1 (11). This latter photograph is subtitled 'Petroleum is exported from Ye-nan-gyoung [whence its name river of fetid water] in pots such as represented above.'

49. Quotations from C. Grant, *Notes Explanatory of A Series of Views Taken . . . in 1855* (Calcutta, 1856[?]), pp. 9–10.

50. M. Symes, *An Account of an Embassy* (London, 1800), p. 199.

51. 'Buchanan's Burma Journal', IOR/H/Misc/687, p. 78.

52. Buckland, 'Geological account', p. 151.

53. For biography see, R. Struwe, 'An ambitious German in early twentieth century Tasmania', *Australian Archaeology*, 62 (2006), 31–7. For other work in Burma, see Noetling, *Notes on Mineral Resources*, parts 1–3.

54. F. Noetling, 'Report on the oil-fields of Twingoung and Beme, Burma', *Records of the Geological Survey of India*, 22 (1889), pp. 75–-136, pp. 75–6.

55. Noetling, 'Oil-fields of Twingoung', p. 86.

56. See tables in 'Record of the Wells', in Noetling, 'Oil-fields of Twingoung', pp. 112–36.

57. Noetling's papers are full of statistical tables connected to the oil industry. Another lengthy table which may be interpreted alongside that noted above is 'List of the Twinzas' oil-wells at Twingôn and Bemè in the Yenangyaung Subdivision, Magwe District, at the end of the Financial Year, 1890–91', *Report on the Petroleum Industry in Upper Burma* (Rangoon, 1892); this is followed by tables detailing, private, state and government oil wells, including attention to the names of the owners of wells.

58. Pascoe, *Oil-Fields,* p. 78; on desirable flow in a well, see p. 92.

59. A. M. Finlayson, 'Labour on the Burmese oilfields', *Mining Magazine,* 6 (1912), pp. 137–40, p. 137. I thank Michael Charney for this source.

60. N. Tate, *Petroleum and its Products* (London, 1863), pp. 2–4.

61. For an argument for the importance of the Burmese fields to the British Empire and how they could compete with oilfields in America and Russia, see C. Marvin, 'The oil wells of Burma', *National Review,* 14 (1889), pp. 341–8.

62. See, for instance, Charney, *Powerful Learning,* p. 168.

24   Global geology and the tectonics of empire

Earlier versions of this chapter were discussed at seminars at the European Central University in Budapest and the University of California at Los Angeles. I am particularly grateful to Jade Star Lackey, Eric Grosfils and their colleagues in the Department of Geology for the opportunity to deliver the 2016 Woodford–Eckis lecture at Pomona College; and to the Master and Fellows of Christ's College for their invitation to give a Lady Margaret Lecture in 2016.

1. Quoted in S. Herbert, *Charles Darwin, Geologist* (Ithaca, 2005), p. 175.

2. E. Suess, *The Future of Silver* (1892; trans. Washington, DC 1893), p. 95; A. Westermann, 'The end of gold? Monetary metals studied at the planetary and human scale during the classical gold standard era', in I. Borowy and M. Schmelzer (eds.), *History of the Future of Economic Growth* (London, 2017), pp. 69–90.

3. M. T. Greene, *Geology in the Nineteenth Century: Changing Views of a Changing World* (Ithaca, 1982), p. 157.

4. E. Suess, *The Face of the Earth,* 5 vols. (Oxford, 1904–24), vol. II, p. 556.

5. E. M. Forster, *A Passage to India* (London, 1924), chapter 12.

6. Quoted in P. Bowler, *Life's Splendid Drama: Evolutionary Biology and the Reconstruction of Life's Ancestry, 1860–1940* (Chicago, 1996), p. 375.

7. A. Noava and A. Levine, *From Man to Ape: Darwinism in Argentina, 1870–1920* (Chicago, 2010), pp. 109–11, 144–8;

I. Podgorny, 'Bones and devices in the constitution of paleontology in Argentina at the end of the nineteenth century', *Science in Context*, 18 (2005), pp. 249-83.

8. S. Ramaswamy, *The Lost Land of Lemuria: Fabulous Geographies, Catastrophic Histories* (Berkeley and Los Angeles, 2004).

9. Grace Yen Shen, *Unearthing the Nation: Modern Geology and Nationalism in Republican China* (Chicago, 2014), pp. 68-9, 105-6.

10. M. Greene, *Alfred Wegener: Science, Exploration, and the Theory of Continental Drift* (Ithaca, 2015), pp. 214-15.

11. A. Wegener to W. P. Köppen (6 December 1911), quoted and translated in Greene, *Alfred Wegener*, pp. 235-6.

12. C. Wessely, *Welteis: Eine Wahre Geschichte* (Berlin, 2013).

13. N. Oreskes, *The Rejection of Continental Drift: Theory and Method in American Earth Science* (New York and Oxford, 1999), p. 139.

14. Grace Yen Shen, 'Going with the flow: Chinese geology, international scientific meetings and knowledge circulation', in B. Lightman, G. McOuat and L. Stewart (eds.), *The Circulation of Knowledge between Britain, India and China: The Early-Modern World to the Twentieth Century* (Leiden, 2013), pp. 237-60, pp. 249-52.

15. É. Argand, *Tectonics of Asia*, trans. A. V. Carozzi and M. Carozzi (1924; New York, 1976), p. 165.

16. B. Willis, 'Continental drift, ein Märchen', *American Journal of Science*, 242 (1944), pp. 509-13.

17. Oreskes, *Rejection of Continental Drift*, p. 296. The later parts of this chapter are indebted to Oreskes's excellent study.

18. I purchased this copy at Princeton University in the 1970s. For Butterworth, see R. N. Nelson, 'Emerson McMillan Butterworth (1894-1961)', *Bulletin of the American Association of Petroleum Geologists*, 46 (1962), pp. 235-7.

19. J. H. F. Umbgrove, *The Pulse of the Earth* (The Hague, 1942), p. 2.

20. Margaret Robinson, quoted in N. Oreskes, '"*Laissez-tomber*": military patronage and women's work in mid-20th-century oceanography', *Historical Studies in the Physical and Biological Sciences*, 30 (2000), pp. 373-92.

21. R. E. Doel, 'Constituting the postwar earth sciences: the military's influence on the environmental sciences in the USA after 1945', *Social Studies of Science*, 33 (2003), pp. 635-66; Oreskes, *Rejection of Continental Drift*.

22. H. H. Hess, 'History of ocean basins', in *Petrologic Studies: A Volume to Honor A. F. Buddington* (Denver, 1962), pp. 599-620, p. 599.

23. M. D. Gordin, *The Pseudoscience Wars: Immanuel Velikovsky and the Birth of the Modern Fringe* (Chicago, 2012), p. 117.

24. J. Watson, 'Minds that live for science', *New Scientist* (21 May 1987), pp. 63-6, p. 63.

25. 'Marie Tharp bio', available online at www.whoi.edu/sbl/liteSite .do?litesiteid=9092&articleId=13407. See also H. Felt, *Soundings: The Story of the Remarkable Woman Who Mapped the Ocean Floor* (New York, 2012).

26. R. Siever, 'Doing earth science research during the Cold War', in N. Chomsky et al., *The Cold War and the University* (New York, 1997), pp. 147–70, p. 147.

27. N. Oreskes, *Science on a Mission* (Chicago, forthcoming).

28. H. E. Le Grand, *Drifting Continents and Shifting Theories: The Modern Revolution in Geology and Scientific Change* (Cambridge, 1988), pp. 4, 30, 238.

29. Oreskes, *Rejection of Continental Drift*, p. 127.

25   Zoological gardens

1. C. Wessely, *Künstliche Tiere. Zoologische Gärten und Urbane Moderne* (Berlin, 2008), p. 42.

2. Wessely, *Künstliche Tiere*, pp. 104–13.

3. Cited in H. Ritvo, *The Animal Estate: The English and Other Creatures in the Victorian Age* (Cambridge, MA, 1987), p. 218.

4. Cited in H. Ritvo, *The Platypus and the Mermaid and other Figments of the Classifying Imagination* (Cambridge, MA, 1998), p. 24.

5. I. Jahn, 'Zoologische Gärten-Zoologische Museen. Parallele ihrer Entstehung', *Bongo*, 24 (1994), pp. 7–30.

6. M. Osborne, *Nature, the Exotic and the Science of French Colonialism* (Bloomington, 1998), p. xv and chapter 3.

7. C. Darwin, Notebook C 79, cited in O. Hochadel, 'Watching exotic animals next door: "scientific" observations at the zoo (ca. 1870-1910)', *Science in Context*, 24 (2011), pp. 183–214, p. 200.

8. For this and a further example from Berlin, see Hochadel, 'Watching exotic animals', pp. 195–6.

9. Hochadel, 'Watching exotic animals', p. 205.

10. L. K. Nyhart, *Modern Nature: The Rise of the Biological Perspective in Germany* (Chicago, 2009), chapter 1.

11. D. F. Weinland, 'Was zu einem "ganzen" Thiere gehört und wie man vielleicht Tropenthiere gesunder erhalten könnte', *Der Zoologische Garten*, 2 (1860-1), pp. 185–7, pp. 185, 186.

12. Nyhart, *Modern Nature*, p. 80.

13. C. O. Whitman, 'A biological farm. For the experimental investigation of heredity, variation and evolution and for the study of life-histories, habits, instincts and intelligence', *Biological Bulletin*, 3 (1902), pp. 214–24, p. 221, cited in Hochadel, 'Watching exotic animals', p. 206. Emphasis in the original.

14. E. Ames, *Carl Hagenbeck's Empire of Entertainments* (Seattle, 2008), pp. 145ff, 150ff.

15. See, for example, N. Rothfels, *Savages and Beasts: The Birth of the Modern Zoo* (Baltimore, 2002).

16. Ames, *Hagenbeck's Empire*, p. 170.

17. Ames, *Hagenbeck's Empire*, pp. 160, 172.

18. A. Sokolowsky, 'Ein neuer Tierpark nach biologischem Prinzip', *Wild und Hund*, 13 (1907), p. 22, cited in L. Dittrich and A. Rieke-Müller, *Carl Hagenbeck (1844–1913). Tierhandel und Schaustellungen im Deutschen Kaiserreich* (Frankfurt am Main, 1998), p. 208.

19. Cited in Ames, *Hagenbeck's Empire*, p. 176.

20. A. Sokolowsky, *Beobachtungen über die Psyche der Menschenaffen. Mit einem Vorwort von Ernst Haeckel* (Frankfurt am Main, 1908).

21. O. Pfungst, *Zur Psychologie der Affen. Bericht über den 5. Kongress für experimentelle Psychologie in Berlin* (Jena, 1912), pp. 200–5.

22. Sokolowsky, *Beobachtungen*, p. 76.

23. For examples, see Dittrich and Rieke-Müller, *Hagenbeck*, pp. 233–4.

24. Dittrich and Rieke-Müller, *Hagenbeck*, pp. 243–51; see also Ames, *Hagenbeck's Empire*, pp. 188ff.

25. M. Gretzschel and O. Pelc, *Hagenbeck: Tiere Menschen, Illusionen* (Hamburg, 1998), p. 89, esp. the image on that page, showing the main display at Hagenbeck's Animal Park from the back with visiting American zoo directors.

26. G. Mitman, *Reel Nature: America's Romance with Wildlife on Film* (Cambridge, MA, 1999), p. 133.

27. R. W. Burckhardt, Jr, *Patterns of Behaviour: Konrad Lorenz, Niko Tinbergen, and the Founding of Ethology* (Chicago, 2005), pp. 138–9.

28. 'Ziele und Programm der neuen Zeitschrift "Der Zoologische Garten"', *Der Zoologische Garten*, new series, 1 (1929), cited in G. Heindl, 'Otto Antonius – ein Wissenschaftler als Tiergärtner', in H. Pechlaner, D. Schratter and G. Heindl (eds.), *Otto Antonius: Wegbereiter der Tiergartenbiologie* (Vienna, 2010), pp. 1–90, p. 49.

29. O. Antonius, *Gefangene Tiere* (Salzburg, 1933), pp. 31ff. Antonius, a folkish German nationalist, joined the Austrian branch of the National Socialist party in 1932, and later participated in Nazi-era discussions about efforts by German zoo directors like Lutz Heck to recreate ancient, supposedly 'Germanic' equine and bovine species by back-breeding from existing species. For details see Heindl, *Otto Antonius*, p. 41.

30. The following is derived from H. Hediger, *Wildtiere in Gefangenschaft. Ein Grundriss der Tiergartenbiologie* (Basel, 1942). See also V. Hofer, 'Wissenschaft und Authentizität. Der Schönbrunner Tiergarten in der ersten Hälfte des 20. Jahrhunderts und die Anfänge der Tiergartenbiologie', in M. G. Ash (ed.), *Mensch, Tier und Zoo. Der Tiergarten*

Schönbrunn im internationalen Vergleich vom achtzehnten Jahrhundert bis heute (Vienna, 2008), pp. 251–80.

31. C. R. Schmidt, 'Entstehung und Bedeutung der Tiergartenbiologie', in L. Dittrich, D. von Engelhardt and A. Rieke-Müller (eds.), *Die Kulturgeschichte des Zoos* (Berlin, 2001), pp. 117–28.

32. For an excellent analysis of such exhibits see I. Braverman, 'Looking at zoos', *Cultural Studies*, 25:6 (2009), pp. 809–42.

33. For a detailed analysis of 'theme worlds', including zoos, see J. Steinkrüger, *Thematisierte Welten. Über Darstellungspraxen in Zooligschen Gärten und Vergnügungsparks* (Bielefeld, 2013).

34. Concealed from view are the holding rooms where animals are kept overnight; see Braverman, 'Looking at zoos', pp. 820ff.

35. The classic statement of these positions is C. Tudge, *Last Animals at the Zoo: How Mass Extinction Can Be Stopped* (London, 1991).

36. K. Kawata, 'The profession of zookeeper', in M. D. Irwin, J. B. Stoner, A. M. Cobaugh (eds.), *Zookeeping: An Introduction to Science and Technology* (Chicago, 2013), pp. 3–12.

37. For references to works by Morris, see www.desmond-morris.com/research.php.

38. P. C. Watts, K. R. Buley, S. Sanderson et al., 'Parthenogenesis in Komodo dragons', *Nature*, 444 (2006), pp. 1021–2.

39. For examples, see G. R. Hosey, V. Melfi and S. Pankhurst, *Zoo Animals: Behaviour, Management, and Welfare* (Oxford, 2009), chapter 14.

40. R. Pizzi, 'Slap me with a dead penguin: what we can learn from 1001 post-mortems at Edinburgh Zoo', in C. MacDonald (ed.), *Proceedings of the Sixth Annual Symposium on Zoo Research* (Edinburgh, 2004), pp. 253–5, cited in Hosey et al., *Zoo Animals*, pp. 507–8.

41. AZA and WAZA documentation are available online at: www.aza.org/species-survival-plan-programs; www.waza.org/en/site/conservation/waza-conservation-projects.

42. For a sceptical account, see J. E. Fa, S. M. Funk and D. O'Connell, *Zoo Conservation Biology* (Cambridge, 2011), section 5.8.

43. For the following, see Hosey et al., *Zoo Animals*, esp. section 14.4, and Fa et al., *Zoo Conservation Biology*, chapters. 5, 7 and 9.

44. T. Milstein, '"Somethin' tells me it's all happening at the zoo": discourse, power and conservationism', *Environmental Communication*, 3 (2009), pp. 25–48, p. 40.

26 Provincialising global botany

I would like to thank the Max Planck Institute for the History of Science, as the major part of the work was done during my post-doctoral fellowship there.

1. The image is available online at www.jsdi.or.jp/~taka-lib/page045 .html; the author of the website does not want it to be reprinted.
2. J. Matsumura, 'Cerasi Japonicæ duæ species novæ', *Botanical Magazine, Tokyo* (*BMT* hereafter), 15 (1901), pp. 99–101.
3. R. H. Myers and M. R. Peattie (eds.), *The Japanese Colonial Empire, 1895–1945* (Princeton, 1984).
4. Bernhard Adalbert Emil Koehne, an expert on the Rosaceae family, made this claim after examining the specimens sent by Emile Taquet, who collected plants to finance his missionary activities.
5. E. H. Wilson, *The Cherries of Japan* (Cambridge, MA, 1916). It is the current consensus that it is a cultivar.
6. Y. Komori, *Postcolonial: Colonial Sub-consciousness and Imperial Consciousness* [in Korean], trans. Song Tae-uk (Seoul, 2002).
7. For more details, see J. Lee, 'Between universalism and regionalism', *British Journal for the History of Science*, 48 (2015), pp. 661–84.
8. R. Yatabe, 'A few words of explanation to European botanists', *BMT*, 4 (1890), pp. 355–6, p. 355.
9. Yatabe, 'A few words', p. 355.
10. T. Nakai, 'An outline of Dr Matsumura Jinzo's achievement' [in Japanese], *BMT*, 29 (1915), pp. 342–8.
11. Myers and Peattie (eds.), *Japanese Colonial Empire*.
12. T. Nakai, *A Synoptical Sketch of Korean Flora* (Tokyo, 1952), p. 1; H. Ohashi, 'Bunzo Hayata and his contributions to the flora of Taiwan', *Taiwania*, 54 (2009), pp. 1–27.
13. T. Nakai, *Flora Sylvatica Koreana*, 22 vols. (Seoul, 1930), vol. XVIII p. 52.
14. Ohashi, 'Bunzo Hayata'.
15. B. Hayata, 'On *Taiwania* and its affinity to other genera', *BMT*, 21 (1907) pp. 21–28, p. 21.
16. Hayata, 'On *Taiwania* and its affinity', pp. 23–27.
17. B. Hayata, *Icones of the Plants of Formosa*, 10 vols. (Taiwan, 1911–20), vol. I, pp. 2–3, 11–12.
18. Hayata, *Icones*, vol. IV, pp. v–vi.
19. Hayata, *Icones*, vol. IV, pp. ii–v; vol. V, p. iii.
20. Hayata, *Icones*, vol. IV, p. iv.
21. Hayata, *Icones*, vol. IV, p. 6.
22. B. W. Ogilvie, *The Science of Describing* (Chicago, 2006); L. Daston and P. Galison, *Objectivity* (New York, 2007). See also Nickelsen, this volume, Chapter 13.

23. J. Endersby, *Imperial Nature: Joseph Hooker and the Practices of Victorian Science* (Chicago, 2008).

24. T. Nakai, 'An observation on Japanese Aconitum I', *BMT*, 22 (1908), pp. 127–33, pp. 127–8; T. Nakai, 'An observation on Japanese Aconitum II', *BMT*, 22 (1908), pp. 133–40. J. Endersby, '"From having no herbarium." Local knowledge versus metropolitan expertise: Joseph Hooker's Australasian correspondence with William Colenso and Ronald Gunn', *Pacific Science*, 55 (2001), pp. 343–58.

25. Nakai, *Flora Sylvatica Koreana* (Seoul, 1927), vol. XVI.

26. T. Hayata, 'The natural classification of plants, according to the dynamic system', in Hayata, *Icones*, vol. X, pp. 97–234; G. E. Du Rietz, 'The fundamental units of biological taxonomy', *Svensk Botanisk Tidskrift*, 24 (1930), pp. 333–428, pp. 406, 413; D. Hull, *Science as a Process* (Chicago, 1988), pp. 81–109.

27. Hayata, *Icones*, vol. X, pp. 98, 114.

28. Hayata, *Icones*, vol. X, pp. 2–3, 75–6.

29. Hayata, *Icones*, vol. X, pp. 114, 115, 115–28, 99.

30. Hayata, *Icones*, vol. X, pp. 98, 108–9.

31. Hayata, *Icones*, vol. X, p. 52.

32. Hayata, *Icones*, vol. X, p. 52.

33. 'The history of [the] Engler and Prantl era [still] remains to be written': F. A. Stafleu, 'The volumes on cryptogams of "Engler und Prantl"', *Taxon*, 21 (1972), pp. 501–11, p. 501; F. A. Stafleu, 'Engler und seine Zeit', *Botanische Jahrbücher für Systematik*, 102 (1981), pp. 21–38.

34. Nakai, *Flora Sylvatica Koreana*, vol. XXII.

35. T. Hayata, 'The succession and participation theories and their bearings upon the objects of the Third Pan-Pacific Science Congress', in *Proceedings of the Third Pan-Pacific Science Congress*, 2 vols. (Tokyo, 1928), vol. II, pp. 1869–75, pp. 1874, 1872, 1870.

36. Hayata, 'The succession and participation theories', pp. 1875, 1876.

37. Hayata, 'The succession and participation theories', p. 1870.

38. Hayata, *Icones*, vol. X, p. 80; Anesaki Masaharu, *History of Japanese Religion* (London, 1930), p. v.

39. Hayata, 'The succession and participation theories', p. 1875.

40. T. Hayata, 'Discussing the concept of systematics' [in Japanese], *Botany and Zoology*, 2 (1934), pp. 79–88, pp. 87, 88; H. D. Harootunian, *Overcome by Modernity* (Princeton, 2000).

27   Descriptive and prescriptive taxonomies

1. Hansard HL Deb (series 5), vol. DXXXVIII, cc1295–342 (9 July 1992) (Dainton, Frederick Sydney, 1992). All the quotes from Dainton here are from the Hansard record of this debate.

2. Hansard HL Deb.

3.   J. Endersby, *Imperial Nature: Joseph Hooker and the Practices of Victorian Science* (Chicago, 2008), pp. 137–69.

4.   K. Vernon, 'Desperately seeking status: evolutionary systematics and the taxonomists' search for respectability 1940-60', *British Journal for the History of Science*, 26:2 (1993), pp. 207-27, pp. 221-2.

5.   E. J. H. Corner et al., *Taxonomy: Report of a Committee Appointed by the Council of the Royal Society* (London, 1963), p. 1.

6.   Corner et al., *Taxonomy*, p. 10.

7.   K. Johnson, 'Natural history as stamp collecting: a brief history', *Archives of Natural History*, 34:2 (2007), pp. 244–58.

8.   My thinking has been much influenced by the work of Keith Vernon, particularly 'Desperately seeking status' (fn. 4).

9.   Corner et al., *Taxonomy*, p. 8.

10.   Both the terms 'numerical phenetics' and 'cladistics' were coined by Ernst Mayr, and the proponents of these rival schemes did not initially adopt them: E. Mayr, 'Numerical phenetics and taxonomic theory', *Systematic Zoology*, 14:2 (1965), pp. 73-97, p. 78; D. L. Hull, *Science as a Process: An Evolutionary Account of the Social and Conceptual Development of Science* (Chicago, 1988), pp. 132-3. However, they are now so widespread that alternative terms would be unnecessarily confusing.

11.   C. Darwin, *On the Origin of Species by Means of Natural Selection: Or the Preservation of Favoured Races in the Struggle for Life* (London, 1859), p. 421.

12.   Darwin, *Origin of Species*, p. 484.

13.   P. F. Stevens, 'Metaphors and typology in the development of botanical systematics 1690-1960, or the art of putting new wine in old bottles', *Taxon*, 33:2 (1984), pp. 169-211, p. 169; J. Endersby, 'Classifying sciences: systematics and status in mid-Victorian natural history', in M. Daunton (ed.), *The Organisation of Knowledge in Victorian Britain* (Oxford, 2005), pp. 61-86; J. Endersby, 'Lumpers and splitters: Darwin, Hooker and the search for order', *Science* 326:5959 (2009), pp. 1496-9.

14.   F. A. Bather, 'Biological classification: past and future', *Quarterly Journal of the Geological Society of London*, 83 (1927), pp. lxii-civ. All quotes from Bather are from this source.

15.   See J. Endersby, *A Guinea Pig's History of Biology: The Plants and Animals Who Taught Us the Facts of Life* (London, 2007); J. Endersby, 'Mutant utopias: evening primroses and imagined futures in early twentieth-century America', *Isis*, 104:3 (2013), pp. 471-503.

16.   Vernon, 'Desperately seeking status', pp. 207-8.

17.   F. N. Egerton, 'History of ecological sciences, part 48: formalizing plant ecology, about 1870 to mid-1920s', *The Bulletin of the Ecological Society of America*, 94:4 (2013), pp. 341-78.

18. Much of the historiography of twentieth-century biology has been built around a narrative shaped by the supposed contest between naturalists and experimentalists, but the history of taxonomy illustrates why this is an oversimplification. See J. B. Hagen, 'Experimentalists and naturalists in twentieth-century botany: experimental taxonomy, 1920–1950', *Journal of the History of Biology*, 17:2 (1984), pp. 249–70, p. 250.

19. Hagen, 'Experimentalists and naturalists', p. 251.

20. F. E. Clements, *Research Methods in Ecology* (Lincoln, NE, 1905), p. 13.

21. J. Clausen, D. D. Keck and W. M. Hiesey, *Experimental Studies on the Nature of Species*, 6 vols. (Washington, DC, 1940), vol. I, p. 4; A. Müntzing, 'Göte Wilhelm Turesson', *Taxon*: 20:5/6 (1971), pp. 773–5; Hagen, 'Experimentalists and naturalists', pp. 251–4.

22. V. B. Smocovitis, 'Botany and the evolutionary synthesis, 1920–1950', in M. Ruse (ed.), *The Cambridge Encyclopedia of Darwin and Evolutionary Thought* (Cambridge, 2013), pp. 313–21, p. 317.

23. Clausen et al., *Experimental Studies*, pp. 15–18.

24. Clausen et al., *Experimental Studies*, pp. 6–7, p. 25.

25. V. B. Smocovitis, 'G. Ledyard Stebbins, Jr. and the evolutionary synthesis (1924–1950)', *American Journal of Botany* 84:12 (1997), pp. 1625–37; 'Darwin's botany in the *Origin of Species*', in M. Ruse and R. J. Richards (eds.), *The Cambridge Companion to the 'Origin of Species'* (Cambridge, 2009), pp. 216–36; V. B. Smocovitis, 'The "Plant Drosophila": E. B. Babcock, the genus *Crepis*, and the evolution of a genetics research program at Berkeley, 1915–1947', *Historical Studies in the Natural Sciences*, 39:3 (2009), pp. 300–55.

26. J. W. Gregor, V. McM. Davey and J. M. S. Lang, 'Experimental taxonomy I: experimental garden technique in relation to the recognition of the small taxonomic units', *The New Phytologist*, 35:4 (1936), pp. 323–6; P. A. Rydberg, 'Scylla or Charybdis?', in B. M. Duggar (ed.), *Proceedings of the International Congress of Plant Sciences, Ithaca, New York, August 16–23, 1926* (Menasha, WI, 1929), pp. 1539–51.

27. Vernon, 'Desperately seeking status', pp. 210–11.

28. K. Fægri, 'Some fundamental problems of taxonomy and phylogenetics', *Botanical Review*, 3:8 (1937), pp. 400–23, p. 401.

29. E. Mayr and W. Provine (eds.), *The Evolutionary Synthesis: Perspectives on the Unification of Biology*, 2nd edn (Cambridge, MA, 1998); V. B. Smocovitis, *Unifying Biology: The Evolutionary Synthesis and Evolutionary Biology* (Princeton, 1996).

30. J. S. Huxley, 'Foreword', in J. S. Huxley (ed.), *The New Systematics* (Oxford, 1940), n.p.

31. 'Introductory: towards the new systematics', in Huxley (ed.), *The New Systematics*, pp. 1–46, pp. 1–2.

32. Huxley 'Introductory', p. 3.

33. M. P. Winsor, 'The English debate on taxonomy and phylogeny, 1937–1940', in *History and Philosophy of the Life Sciences*, 17 (1995), pp. 227–52; Vernon, 'Desperately seeking status', p. 212.

34. For example, W. T. Calman, 'The meaning of biological classification', *Nature*, 136: 3427 (1935), pp. 9–10, p. 10.

35. J. S. Lennox Gilmour and S. M. Walters, 'Two early papers on classification', *Plant Systematics and Evolution*, 167 (1936, 1937 [reprinted 1989]), pp. 97–107, p. 100.

36. R. E. Blackwelder and A. A. Boyden, 'The nature of systematics', *Systematic Zoology* (1952), pp. 26–33, p. 30, quoted in Vernon, 'Desperately seeking status', p. 223.

37. Stevens, 'Metaphors and typology', pp. 169–70.

38. Hull, *Science as a Process*, pp. 120–5; K. Vernon, 'The founding of numerical taxonomy', *British Journal for the History of Science*, 21:2 (1988), pp. 143–59, p. 148; M. P. Winsor, 'Setting up milestones: Sneath on Adanson and Mayr on Darwin', in D. M. Williams and P. L. Forey (eds.), *Milestones in Systematics* (Boca Raton, 2004), pp. 1–17.

39. P. H. A. Sneath, 'Some thoughts on bacterial classification', *Journal of General Microbiology*, 17 (1957), pp. 184–200; P. H. A. Sneath, 'The application of computers to taxonomy', *Journal of General Microbiology*, 17 (1957), pp. 201–26; P. H. A. Sneath, 'Early experience with computers', *Binary: The Newsletter of the Society for General Microbiology Computer Club*, 1 (1985), pp. 5–7.

40. Hull, *Science as a Process*, p. 119; Vernon, 'The founding of numerical taxonomy', p. 151.

41. K. Vernon, "A truly taxonomic revolution? Numerical taxonomy, 1957–1970', *Studies in the History and Philosophy of the Biological and Biomedical Sciences*, 32:2 (2001), pp. 315–41, pp. 315–16.

42. Hull, *Science as a Process*, p. 121.

43. See E. Mayr, E. G. Linsley and R. L. Usinger, *Methods and Principles of Systematic Zoology* (New York, 1953); also Hull, *Science as a Process*, p. 107; Vernon, 'Truly taxonomic revolution', pp. 208–9.

44. Mayr, 'Numerical phenetics and taxonomic theory', pp. 73–4.

45. Mayr, 'Numerical phenetics and taxonomic theory', p. 76, emphasis added.

46. See note 8.

47. Hull, *Science as a Process*, p. 131; R. Craw, 'Margins of cladistics: identity, difference and place in the emergence of phylogenetic systematics, 1864–1975', in P. E. Griffiths (ed.), *Trees of Life: Essays*

in *Philosophy of Biology* (Dordrecht, 1992), pp. 65–106, p. 67; P. L. Forey, *Cladistics: A Practical Course in Systematics* (Oxford, 1992); H. Gee, *Deep Time: Cladistics, the Revolution in Evolution* (London, 2001); B. Hennig and A. Kluge, 'Willi Hennig', available online at https://cladistics.org/willi-hennig/.

48. W. Hennig, *Phylogenetic Systematics* (Champaign, IL, 1966); Hull, *Science as a Process*, pp. 144–6.

49. Craw, 'Margins of cladistics'.

50. I discuss some of the rhetorical aspects of recent taxonomy in J. Endersby, '"The realm of hard evidence": novelty, persuasion and collaboration in botanical cladistics', *Studies in History and Philosophy of Biological and Biomedical Sciences*, 32:2 (2001), pp. 343–60.

51. Dean argued that controversy in classification was an ideal case to test the premises of the strong programme in the sociology of scientific knowledge, because there are no 'facts of the matter' that can resolve taxonomic disputes: J. P. Dean, 'Controversy over classification: a case study from the history of botany', in B. Barnes and S. Shapin (eds.), *Natural Order: Historical Studies of Scientific Culture* (Beverly Hills, 1979), pp. 1–30.

52. Select Committee appointed to consider Science and Technology, 'What on Earth? The threat to the science underpinning conservation', *House of Lords Session 2002–3, 3rd Report* (London, 2003).

53. Vernon, 'Desperately seeking status', p. 209.

54. A further key question not tackled here, for want of space, is the gendered nature of systematics. Systematics seems even more male dominated than comparable sciences at the same time, but the reasons for this require further research.

55. Bather, 'Biological classification', p. lxiv.

28   Imperilled crops and endangered flowers

1. 'Beautiful new waterlily species discovered by Kew plant hunter Carlos Magdelena', available online at www.kew.org/science/news/beautiful-new-waterlily-species-discovered-by-kew-plant-hunter-carlos-magdalena.

2. Overviews of 'plant hunting' include A. Coats, *The Plant Hunters: Being a History of the Horticultural Pioneers, their Quests, and their Discoveries from the Renaissance to the Twentieth Century* (New York, 1969); C. Lyte, *The Plant Hunters* (London, 1983); T. Musgrave, C. Gardner and W. Musgrave, *The Plant Hunters: Two Hundred Years of Adventure and Discovery Around the World* (London, 1998); C. Fry, *The Plant Hunters: The Adventures of the World's Greatest Botanical Explorers* (Chicago, 2013).

3. Histories of botanical gardens that emphasise their role in the expansion of empire include: L. Brockway, *Science and Colonial Expansion: The Role of the British Royal Botanic Gardens* (New York, 1979); R. Drayton, *Nature's Government: Science, Imperial Britain and the 'Improvement' of the World* (New Haven, 2000); D. P. McCracken, *Gardens of Empire: Botanical Institutions of the Victorian British Empire* (London, 1997).

4. M. R. Finlay, 'Science, practice and politics: German agricultural experiment stations in the nineteenth century', PhD dissertation, Iowa State University (1992).

5. H. L. Hyland, 'History of US plant introduction', *Environmental Review*, 2:4 (1977), pp. 26–33; K. A. Williams, 'An overview of the US National Plant Germplasm System's exploration program', *HortScience*, 40:2 (2005), pp. 297–301.

6. Quoted in I. G. Loskutov, *Vavilov and his Institute: A History of the World Collection of Plant Genetic Resources in Russia* (Rome, 1999), p. 18.

7. O. Elina, S. Heim and N. Roll-Hansen, 'Plant breeding on the front: imperialism, war and exploitation', *Osiris*, 20 (2005), pp. 161–79, p. 166; M. Flitner, 'Genetic geographies: a historical comparison of agrarian modernization and eugenic thought in Germany, the Soviet Union and the United States', *Geoforum*, 34:2 (2003), pp. 175–85, pp. 178–9.

8. Flitner, 'Genetic geographies', p. 126.

9. See N. I. Vavilov, *Origin and Geography of Cultivated Plants*, ed. V. F. Dorofeyev, trans. D. Löve (Cambridge, 1992). On the emergence of an idea of 'genetic resources', see C. Bonneuil, 'Life diversity as a "universal store of genes": the emergence of *genetic resources* as a modernization/conservation category', presented at the workshop 'Seeds for Survival', July 2017, Cambridge, UK.

10. L. U. Wingen et al., 'Establishing the A. E. Watkins landrace cultivar collection as a resource for systematic gene discovery in bread wheat', *Theoretical and Applied Genetics*, 127:8 (2014), pp. 1831–42.

11. J. G. Hawkes, 'The Commonwealth Potato Collection', *American Potato Journal*, 28:1 (1951), pp. 465–71.

12. Elina et al., 'Plant breeding on the front'; Flitner, 'Genetic geographies'; T. Saraiva, 'Breeding Europe: crop diversity, gene banks, and commoners', in N. Disco and E. Kranakis (eds.), *Cosmopolitan Commons: Sharing Resources and Risks across Borders* (Cambridge, 2013), pp. 185–212.

13. I. Bonnin et al., 'Explaining the decrease in the genetic diversity of wheat in France over the 20th century', *Agriculture, Ecosystems and Environment*, 195 (2014), pp. 183–92; C. Bonneuil and F. Thomas, 'Purifying landscapes: the Vichy regime and the

genetic modernization of France', *Historical Studies in the Natural Sciences*, 40:4 (2010), pp. 532–68.

14. H. V. Harlan and M. L. Martini, 'Problems and results in barley breeding', in *Yearbook of Agriculture 1936* (Washington, DC, 1936), pp. 303–46, p. 317.

15. One such example is a maize-collecting programme of the 1950s; see H. A. Curry, 'Breeding uniformity and banking diversity: the genescapes of industrial agriculture, 1935–1970', *Global Environment*, 10:1 (2017), pp. 83–113.

16. Saraiva, 'Breeding Europe', pp. 196–8; H. A. Curry, 'From working collections to the World Germplasm Project: agricultural modernization and genetic conservation at the Rockefeller Foundation', *History and Philosophy of the Life Sciences*, 39:5 (2017), doi:10.1007/s40656-017-0131-8.

17. M. Fenzi and C. Bonneuil, 'From "genetic resources" to "ecosystems services": a century of science and global policies for crop diversity conservation', *Culture, Agriculture, Food and Environment*, 38:2 (2016), pp. 72–83.

18. E. Bennett (ed.), *Record of the FAO/IBP Technical Conference on the Exploration, Utilisation and Conservation of Plant Genetic Resources*, PL/FO: 1967/M/12 (Rome, 1968), p. 4.

19. On the history of this partnership, see R. Pistorius, *Scientists, Plants and Politics: A History of the Plant Genetic Resources Movement* (Rome, 1997), chapters 1–2.

20. O. H. Frankel and E. Bennett, 'Genetic resources', in O. H. Frankel and E. Bennett (eds.), *Genetic Resources in Plants: Their Exploration and Conservation*, IBP Handbook no. 11 (Oxford, 1970), pp. 7–17, pp. 13–15.

21. H. Ito and K. Kumagai, 'The National Seed Storage Laboratory for genetic resources in Japan', *Japan Agricultural Research Quarterly*, 4: 2 (1969), pp. 32–8, p. 32.

22. National Plant Genetic Resources Board, *Plant Genetic Resources: Conservation and Use* (Washington, DC, 1979).

23. FAO, *Plant Introduction Newsletter*, 10 (July 1962), pp. 5–6.

24. Saraiva, 'Breeding Europe', pp. 198–9.

25. IBPGR Secretariat, *International Board for Plant Genetic Resources, Annual Report 1974* (Rome, 1975).

26. Committee on Managing Global Genetic Resources, National Research Council, *Managing Global Genetic Resources: Agricultural Crop Issues and Policies* (Washington, DC, 1993), chapter 2. On the conflicts involved in the creation of IBPGR, see Pistorius, *Scientists, Plants and Politics*.

27. M. Holdgate, *The Green Web: A Union for World Conservation* (London, 1999).

28. R. Melville, 'Plant conservation and the Red Book', *Biological Conservation*, 2: 3 (1970), pp. 185–8, p. 188; J. Tinker, 'One flower in 10 faces extinction', *New Scientist and Science Journal*, 13 May 1971, pp. 408–13.

29. 'The function of living plant collections in conservation and in conservation-orientated research and public education', Report of a Conservation Conference, Royal Botanic Gardens, Kew, 2–6 September 1975, available online at https://issuu.com/kewguild journal/docs/v9s80p378-all/32.

30. See, for example, H. Synge and H. Townsend (eds.), *Survival or Extinction: The Practical Role of Botanic Gardens in the Conservation of Rare and Threatened Plants* (Kew, 1979); D. Bramwell et al. (eds.), *Botanic Gardens and the World Conservation Strategy: Proceedings of an International Conference, 26–30 November 1985* (London, 1987); V. H. Heywood, *The Botanic Gardens Conservation Strategy* (Gland, Switzerland, 1989).

31. J. Merson, 'Bio-prospecting or bio-piracy: intellectual property rights and biodiversity in a colonial and postcolonial context', *Osiris*, 15 (2000), pp. 282–96.

32. FAO, International Undertaking on Plant Genetic Resources, Resolution 8/83, available online at www.fao.org/docrep/x5563E/X5563e0a.htm.

33. FAO, International Treaty on Plant Genetic Resources for Food and Agriculture, available online at www.planttreaty.org/content/texts-treaty-official-versions.

34. United Nations, Convention on Biological Diversity (1992), available online at www.cbd.int/doc/legal/cbd-en.pdf.

35. Brockway, *Science and Colonial Expansion*; J. R. Kloppenburg, Jr, *First the Seed: The Political Economy of Plant Biotechnology 1492–2000*, 2nd edn (Madison, 2004).

36. D. Godrej, '8 things you should know about patents', *New Internationalist*, 349 (September 2002), newint.org/features/2002/09/05/keynote/.

29   Networks of natural history in Latin America

Diane Grosklaus Whitty translated this text from Portuguese into English. The author also thanks Wilson Picado, Tom, CNPq, FAPEMIG, and the editors of this book.

1. J. Lavery, C. Crady, E. Walh and E. J. Emanuel, *Issues in International Biomedical Research, a Casebook* (New York, 2007), pp. 21–42; United Nations, *Convention on Biological Diversity*, 1992, www.cbd.int.

2. As proposed by J. Adelman, 'Latin American and world histories: old and new approaches to the *pluribus* and the *unum*', *Hispanic American Historical Review*, 84:3 (2004), pp. 399-409. On regional/global approaches in Latin America history, see L. Benton, 'No longer odd region out: repositioning Latin America in world history', *Hispanic American Historical Review*, 84:3 (2004), pp. 423-30. On networks in Latin American science, see L. López-Ocón, 'La Comisión Científica del Pacífico: de la ciencia imperial a la ciencia federative', *Bulletin des Institutes Françaises de Études Andines*, 32:3 (2003), pp. 479-514, p. 486.

3. N. Safier, 'Global knowledge on the move: itineraries, Amerindian narratives, and deep histories of science', *Isis*, 101:1 (2010), pp. 133-45.

4. J. M. Gallegos, 'En la Baja California', *Magazine de Geografía Nacional*, 26 (1926), pp. 3-46. On the field notes from Huey's 1924 expedition, see *L. M. Hueys's Coronado Island Field Notes*, www.sdnhm.org/about-us/history/coronado-islands.

5. A. L. Herrera, *Zoología* (Ciudad de Mexico, 1924), pp. 9-17, 538-40. See also A. Herrera, *Nociones de Biología* (México, 1904), pp. 243-4. On Herrera, see C. Cuevas and I. Ledesma, 'Alfonso L. Herrera: controversia y debates durante el inicio de la biología en México', *Historia Mexicana*, 55 (2006), pp. 973-1013.

6. C. Onelli, 'El Jardín Zoológico en 1912', *Revista del Jardín Zoológico de Buenos Aires* (hereafter, *RJZBA*), 8:32 (1912), pp. 351-8; C. Onelli, 'Idiosincracias individuales de los pensionistas del Jardín', *RJZBA*, 1:4 (1905), pp. 327-42. See also Onelli, *Aguafuertes del Zoológico* (Buenos Aires, 1916). On the Buenos Aires zoo, see D. A. del Pino, *Historia del Jardín Zoológico Municipal* (Buenos Aires, 1979).

7. For the correspondence between Onelli and Frank Baker, and Onelli and Hagenbeck, see 'Las últimas adquisiciones por compras y por canje', *RJZBA*, 1:4 (1905), pp. 424-45.

8. N. Safier, 'Itineraries of Atlantic science: new questions, new approaches, new directions', *Atlantic Studies*, 7:4 (2010), pp. 357-64, p. 358. On the idea of people-in-between, see Sujit Sivasundaram, 'Sciences and the global: on methods, questions, and theory', *Isis*, 101:1 (2010), pp. 146-58, p. 158.

9. C. de Mello Leitão, *Zoogeografia do Brasil*, 2nd edn (São Paulo, 1947), pp. 7-8; C. de Mello Leitão, 'A gênese dos continentes e oceanos segundo Wegener', *Revista Nacional de Educação*, 2:15 (1933), pp. 49-54.

10. See R. Horta Duarte, *Activist Biology: The National Museum, Politics, and Nation Building in Brazil*, trans. D. G. Whitty (Tuscon, 2016), pp. 149-50.

11. IUCN, 'About', www.iucn.org/about.

12.   *International Union for the Protection of Nature: Established at Fontainebleau 5 October 1948* (Brussels, 1948), pp. 23–9.

13.   On 'Creole science', see S. McCook, *States of Nature: Science, Agriculture and Environment in the Spanish Caribbean, 1760–1940* (Austin, 2002), p. 6. On 'criollo culture' as an expression of nationalism, see C. Rama, *Historia de las Relaciones Culturales entre España y America Latina* (Mexico, 1982), p. 122.

14.   F. Vuilleumier, 'In Memoriam: William H. Phelps Jr', *The Auk*, 107 (1990), pp. 181–3.

15.   Á. Cabrera and J. Yepes, *Mamiferos Sud-Americanos (Vida, Costumbres y Descripción)* (Buenos Aires, 1940). On nationalism and science in Latin America, see E. Beltran, 'La historia de la ciencia en América Latina', *Quipu*, 1:1 (1984), pp. 7–23; M. Cueto, 'Andean biology in Peru: scientific styles on the periphery', *Isis*, 80:4 (1989), pp. 640–58; J. Saldaña, 'Acerca de la historia de la ciencia nacional', *Cuadernos de Quipu*, 4 (1992), pp. 9–54.

16.   J. A. Vellard, *Le Venin des Araignées* (Paris, 1936); J. Vellard, 'Preparation du curare par les Ñambikwara', *Journal de la Societé des Américanistes*, 31:1 (1939), pp. 211–22, p. 221. On Vellard's anthropological studies, see O. Dollfus, 'Jean-Albert Vellard', *Bulletin de L'Institut Francais d'Etudes Andines*, 25:2 (1996), pp. 165–7.

17.   A. Crosby, *The Columbian Exchange: Biological and Cultural Consequences of 1492* (Westport, CT, 1972); S. McCook, 'The neo-Columbian exchange: the second conquest of the greater Caribbean, 1720–1930', *Latin American Research Review*, 46:4 (2011), pp. 11–31.

18.   On INBio's history, see R. Gámez Lobo, *De Biodiversidad, Gentes y Utopias: Reflexiones en los 10 Años del INBio* (Heredia, 1999); J. M. Feinsilver, 'Prospección de la biodiversidad: potencialidades para los países en desarrollo', *Revista CEPAL*, 60 (1996), pp. 111–28. For unfavourable views, see I. Rojas, 'Mercantilización de la biodiversidad: la actividad de bioprospección del INBio en Costa Rica', *Economía y Sociedad*, 33/34 (2008), pp. 21–38; S. Rodríguez, 'El talon de Aquiles del INBio', *Biodiversidad*, 83 (2015), pp. 15–16, www.grain.org/e/5158.

19.   P. Koleff and J. Laso, *Conabio: Informe 2002–2004* (México, DF, 2005), pp. 8–13, 14, 46.

20.   'Instituto Humboldt y Colciencias se refieren a controversia por Expediciones BIO Putuamayo', *Boletin de Prensa Instituto Humboldt*, Bogotá, 31 May 2016. Colombia has a Sistema de Informaciones sobre Biodiversidad (SIB); on SIB, see Decreto 1375, 27 June 2013, Bogotá, Colombia, Ministerio de Ambiente

y Desarrollo Sostenible. On the collections, see www.humboldt
.org.co/es/investigacion/programas/colecciones-biologicas.

21. J. S. Pearse, 'The promise of integrative biology: resurrection of the
naturalist', *Integrative and Comparative Biology* 43 (2003),
pp. 276-7. On 'labscapes' and 'border crossings', see R. E. Kohler,
*Landscapes and Labscapes: Exploring the Lab-Field Border in
Biology* (Chicago, 2002), pp. 1-5; R. Kohler, 'Place and practice in
field biology', *History of Science*, 40 (2002), pp. 189-210. On the
centrality of collections to practice and theory in integrative biology, see M. Sunderland, K. Klitz and K. Yoshihara, 'Doing natural
history', *BioScience* 62:9 (2012), pp. 824-9.

22. M. C. da Cunha and M. B. de Almeida (eds.), *Enciclopédia da
Floresta. O Alto Juruá: Práticas e Conhecimentos das Populacões*
(São Paulo, 2002).

23. K. Brown, Jr, and A. V. L. Freitas, 'Diversidade biológica no Alto
Juruá: avaliação, causas e manutenção', in M. C. da Cunha and
A. V. L. Almeida (eds.), *Enciclopédia da Floresta*, pp. 33-42.

24. F. Olmos et al., 'Correção política e biodiversidade: a crescente
ameaça das "populações tradicionais" à Mata Atlântica', in
J. L. Berger Albuquerque et al., *Ornitologia e Conservação: da
Ciência às Estratégias* (Tubarão, 2001), pp. 279-312.

25. M. N. L. Pastana and W. M. Ohara, 'A new species of
Hyphessobrycon Durbin (Characiformes: Characidae) from rio
Aripuanã, rio Madeira basin, Brazil', *Zootaxa*, 4161:3 (2016),
pp. 386-98; T. H. Condez et al., 'A new species of flea-toad
(Anura: Brachycephalidae) from southern Atlantic Forest,
Brazil', *Zootaxa*, 4083:1 (2016), pp. 40-56.

26. On the impact of the activities of so-called traditional peoples on
protection areas in Amazonia and Atlantic Forest reserves, see
J. Drummond and J. L. de A. Franco, *Terras de Quilombolas
e Unidades de Conservação: Uma Discussão Conceitual e Política,
com Ênfase nos Prejuízos para a Conservação da Natureza* (São
Paulo, 2009).

30 The unnatural history of human biology

1. G. Hardin, *Nature and Man's Fate* (New York, 1959), pp. 289-90.
2. R. Dubos, *Man Adapting* (New Haven, 1966), p. xx.
3. J. D. Hamblin, *Arming Mother Nature: The Birth of Catastrophic
Environmentalism* (New York, 2013).
4. M. Midgley, *Science as Salvation: A Modern Myth and its Meaning*
(New York, 1992).

5. R. Pearl, from the Preface to *Studies in Human Biology* (1924), cited in M. A. Little and R. M. Garruto, 'Raymond Pearl and the shaping of human biology', *Human Biology*, 82:1 (2010), pp. 77–102.

6. R. Pearl, *Natural History of Population* (New York, 1939).

7. W. B. Provine, *The Origins of Theoretical Population Genetics* (Chicago, 1971).

8. D. F. Roberts, and G. A. Harrison (eds.), *Natural Selection in Human Populations* (New York, 1959), p. 60.

9. J. Bangham and S. de Chadarevian, 'Human heredity after 1945: moving populations centre stage', *Studies in History and Philosophy of Biological and Biomedical Sciences*, 47 (2014), pp. 45–9.

10. S. Washburn, 'The new physical anthropology', *Transactions of the New York Academy of Sciences*, 13:7 (1951), pp. 298–304.

11. G. A. Harrison, J. S. Weiner, J. M. Tanner and N. A. Barnicot (eds.), *Human Biology: An Introduction to Human Evolution, Variation, and Growth* (New York, 1964).

12. E. B. Worthington, *The Evolution of IBP*, vol. I (Cambridge, 1975).

13. These projects are described in K. J. Collins and J. S. Weiner, *Human Adaptability: A History and Compendium of Research in the International Biological Programme* (London, 1977).

14. Collins and Weiner, *Human Adaptability*, p. 4.

15. J. S. Weiner and J. A. Lourie, *Human Biology: A Guide to Field Methods* (Oxford, 1969).

16. H. Kuklick, 'Islands in the Pacific: Darwinian biogeography and British anthropology', *American Ethnologist*, 23:3 (1996), pp. 611–38.

17. Damon to Leahy, 25 June 1971, Harvard Peabody Museum (hereafter, HPM), Box 4: Solomon Islands Expedition, Damon and Howells, Grant Applications, Reference Materials, Folder: Grant Applications.

18. P. Worsley, *The Trumpet Shall Sound: A Study of 'Cargo' Cults in Melanesia* (London, 1957).

19. Department of Health, Education and Welfare Public Health Service, Grant Application for 'Medical-anthropological studies in Solomon Islands' for January 1972 to December 1976, HPM, Box 4: Solomon Islands Expedition, Damon and Howells, Grant Applications, Reference Materials, Folder: Grant Applications.

20. 'Proposal for use of R/V Alpha Helix', HPM, Box 2 Peabody Museum Archives – Solomon Islands Expedition (W. W. Howells files), Folder: Travel Corr and Grant Appl, 1971–1972.

21. Damon to Howells, 24 July 1966, HPM, Box 2, Peabody Museum Archives – Solomon Islands Expedition (W. W. Howells files), Folder: Corr and Study, 1966–1976.

22. Damon to Levy, 18 July 1972, HPM, Box 2: 1964–1972 Solomon Islands, Exp. Albert Damn Corresp., Notes: Budget info; grant applications, Folder: Levy, 2004.1.197, 2.17.

23. Ross to Lubin and Ellis, 12 April 1974, HPM, unprocessed portion of Harvard Solomon Islands Project papers.

24. J. S. Friedlaender, W. W. Howells and J. G. Rhoads, *The Solomon Islands Project: A Long-Term Study of Health, Human Biology, and Culture Change* (New York, 1987), p. 12.

25. D. A. Merriwether, 'Freezer anthropology: new uses for old blood', *Philosophical Transactions: Biological Sciences*, 354:1379 (1999), pp. 121–9.

26. R. Ventura Santos and C. E. A. Coimbra, 'On the (un)natural history of the Tupi-Monde Indians: bioanthropology and change in the Brazilian Amazon', in A. H. Goodman and T. L. Leatherman (eds.), *Building a New Biocultural Synthesis: Political-Economic Perspectives on Human Biology* (Ann Arbor, 1998), pp. 269–94.

27. W. Anderson, 'The possession of Kuru: medical science and bio-colonial exchange', *Comparative Studies in Society and History*, 42 (2000), pp. 713–44.

28. D. Oliver, *Aspects of Modernization in Bougainville, Papua New Guinea*, Working Papers Series, Pacific Island Studies (Honolulu, 1981).

29. N. Thomas, *Entangled Objects: Exchange, Material Culture and Colonialism in the Pacific* (Cambridge, 1991), p. 9.

30. A. M. Kakaliouras, 'An anthropology of repatriation', *Current Anthropology*, 53, suppl. 5 (2012), pp. S210–21.

31. Quoted in R. Borofsky, *Yanomami: The Fierce Controversy and What We Might Learn from It* (Berkeley, 2005).

31   Fieldwork out of place

1. E. O. Wilson, *Naturalist* (Washington, DC, 1994), pp. 11–12; see discussion in P. L. Farber, *Finding Order in Nature: The Naturalist Tradition from Linnaeus to E. O. Wilson* (Baltimore, 2000), p. 109. On the importance of place in the history of natural history and the field sciences, see R. de Bont, *Stations in the Field: A History of Place-Based Animal Research, 1870–1930* (Chicago, 2015); T. F. Gieryn, 'City as truth-spot: laboratories and field-sites in urban studies', *Social Studies of Science*, 36:1 (2006), pp. 5–38; R. E. Kohler, *Landscapes and Labscapes: Exploring the Lab–Field Border in Biology* (Chicago, 2002); R. E. Kohler and J. Vetter, 'The field', in B. Lightman (ed.), *A Companion to the History of Science* (Hoboken, NJ, 2016), chapter 20; J. Lachmund, *Greening*

Berlin: The Co-production of Science, Politics and Urban Nature (Cambridge, MA, 2013); S. Kingsland, *The Evolution of American Ecology, 1890–2000* (Baltimore, 2005); J. Vetter (ed.), *Knowing Global Environments: New Historical Perspectives on the Field Sciences* (New Brunswick, NJ, 2011); J. Vetter, *Field Life: Science in the American West during the Railroad Era* (Pittsburgh, 2016).

2. For a case study that addresses related issues with a focus on governance, see H. B. Stokland, 'Field studies in absentia: counting and monitoring from a distance as technologies of government in Norwegian wolf management (1960s–2010s)', *Journal of the History of Biology*, 48:1 (2015), pp. 1–36.

3. M. V. Barrow, *A Passion for Birds: American Ornithology after Audubon* (Princeton, 1998), pp. 58–9; P. L. Farber, *Discovering Birds: The Emergence of Ornithology as a Scientific Discipline* (Baltimore, 1997), pp. 27–48; D. Worster, *Nature's Economy: A History of Ecological Ideas*, 2nd edn (New York, 1994), pp. 6–10.

4. J. A. Jackson, W. E. Davis, Jr and J. Tautin (eds.), *Bird Banding in North America: The First Hundred Years*. Memoirs of the Nuttall Ornithological Club, no. 15 (Cambridge, MA, 2008); Barrow, *Passion for Birds*, pp. 169–71; R. M. Wilson, *Seeking Refuge: Birds and Landscapes of the Pacific Flyway* (Seattle, 2010), pp. 72–5; K. Whitney, 'Domesticating nature? Surveillance and conservation of migratory shorebirds in the "Atlantic Flyway"', *Studies in History and Philosophy of Biology and Biomedical Sciences*, 45 (2014), pp. 78–87.

5. L. J. Cole, 'The early history of bird banding in America', *Wilson Bulletin*, 34:2 (1922), pp. 108–44.

6. E. W. Nelson, 'Bird banding work being taken over by the Biological Survey', *Wilson Bulletin*, 32 (1920), pp. 63–4; F. C. Lincoln, 'The history and purposes of bird banding', *Auk*, 38 (1921), pp. 217–28; F. C. Lincoln, *The Migration of American Birds* (New York, 1939); Wilson, *Seeking Refuge*, pp. 73–5.

7. E. S. Benson, 'A centrifuge of calculation: managing data and enthusiasm in early twentieth-century bird banding', *Osiris*, 32 (2017), pp. 286–306.

8. R. MacLeod, '"Strictly for the birds": science, the military and the Smithsonian's Pacific Ocean Biological Survey Program, 1963–1970 ', *Journal of the History of Biology*, 34:2 (2001), pp. 315–52; E. Benson, *Wired Wilderness: Technologies of Tracking and the Making of Modern Wildlife* (Baltimore, 2008), pp. 5–51.

9. L. D. Mech, *Handbook of Animal Radio-Tracking* (Minneapolis, 1983); R. E. Kenward, *A Manual for Wildlife Radio Tagging* (London, 2001).

10. W. W. Cochran, 'Long-distance tracking of birds', in S. R. Galler, K. Schmidt-Koenig, G. J. Jacobs and R. E. Belleville (eds.), *Animal Orientation and Navigation* (Washington, DC, 1972), pp. 39–59.

11. W. W. Cochran, H. Mouritsen and M. Wikelski, 'Migrating songbirds recalibrate their magnetic compass daily from twilight cues', *Science*, 304:5669 (2004), pp. 405–8.

12. D. W. Warner, 'Space tracks: bioelectronics extends its frontiers', *Natural History*, 62 (1963), pp. 8–15; Benson, *Wired Wilderness*, pp. 79–83; E. S. Benson, 'One infrastructure, many global visions: the commercialization and diversification of Argos, a satellite-based environmental surveillance system', *Social Studies of Science*, 42 (2012), pp. 846–71.

13. J. R. Varney, J. J. Craighead and J. S. Sumner, 'An evaluation of the use of Erts-1 satellite imagery for Grizzly bear habitat analysis', *Bears: Their Biology and Management*, 3 (1976), pp. 261–73; W. B. Cohen and S. N. Goward, 'Landsat's role in ecological applications of remote sensing', *BioScience*, 54:6 (2004), pp. 535–45; P. E. Mack, *Viewing the Earth: The Social Construction of the Landsat Satellite System* (Cambridge, MA, 1990).

14. D. J. Haraway, 'Situated knowledges: the science question in feminism and the privilege of partial perspective', *Feminist Studies*, 14:3 (1988), pp. 575–99; D. N. Livingstone, *Putting Science in its Place: Geographies of Scientific Knowledge* (Chicago, 2003); S. Shapin, 'Placing the view from nowhere: historical and sociological problems in the location of science', *Transactions of the Institute of British Geographers*, 23:1 (1998), pp. 5–12.

32   Wild visions

1. Landmarks are multipart documentary series focusing on academic subjects and authored by a single knowledgeable on-screen presenter. Attenborough's landmarks began with *Life on Earth* (1979) and continued to *David Attenborough's Great Barrier Reef* (2016). Other BBC series such as *The Blue Planet* (2001), *Planet Earth* (2006) and *Planet Earth II* (2016) are variations on the landmark format, with Attenborough acting as a narrator rather than on-screen presenter and writer; in *Frozen Planet* (2011), Attenborough appeared on-screen only in the final episode.

2. C. James, *The Crystal Bucket: Television Criticism from the Observer* (London, 1981), p. 164.

3. James, *Crystal Bucket*, p. 163.

4. D. Attenborough, *Life on Air: Memoirs of a Broadcaster* (London, 2010), p. 34.

5. K. Clark, *A Self-Portrait: The Other Half* (London, 1986), p. 212.

6. Attenborough, *Life on Air*, p. 10.

7. D. Attenborough, 'Oral history transcription', interview by C. Parsons, 31 August 2000, Richmond, UK, www.wildfilmhistory .org/oh/3/David+Attenorough.html.

8. Attenborough, *Life on Air*, p. 201.

9. Attenborough, *Life on Air*, p. 246.

10. C. Parsons, *True to Nature* (London, 1982), pp. 12, 20.

11. Parsons, *True to Nature*, p. 314.

12. Attenborough, *Life on Air*, p. 37.

13. D. Hawkins, BBC Natural History Unit: Report by Head of West Regional Programmes, 18 September 1962, courtesy of the BBC Written Archives Centre.

14. M. Richards, 'Greening wildlife documentary', in L. Lester and B. Hutchins (eds.), *Environmental Conflict and the Media* (New York, 2013), p. 178.

15. M. Jeffries, 'BBC Natural History versus science paradigms', *Science as Culture*, 12:4 (2003), pp. 527–45, pp. 527, 543.

16. Attenborough, *Life on Air*, p. 279.

17. Attenborough, *Life on Air*, p. 280.

18. D. Attenborough, narration in *Life on Earth*, TV series (London, 1979).

19. D. Attenborough, interview with the author, 2 February 2004, Richmond, UK.

20. Parsons, *True to Nature*, pp. 11–20.

21. Parsons, *True to Nature*, pp. 318–19.

22. Attenborough, *Life on Earth*.

23. Attenborough, *Life on Earth*, pp. 290–1.

24. J. Burt, *Animals in Film* (London, 2002), p. 47.

25. Burt, *Animals in Film*, pp. 47–8. Yet the same scene has inspired vastly different reactions. Donna Haraway, for example, writes about her sense of alienation, dismissing Attenborough's interaction with the gorilla as part of 'the theatre of male exhibition'. Referring to Attenborough's role in whispering the audience into the documentary, she writes, 'It is not the drama of touch that fills the screen; it is Attenborough, the master of ceremonies.' D. Haraway *Primate Visions: Gender, Race, and Nature in the World of Modern Science* (New York, 1989), p. 401.

26. Parsons, *True to Nature*, p. 349.

27. Attenborough, 'Oral history transcription'.

28. Attenborough, interview with the author, 3 February 2012, Richmond, UK

29. For a more thorough exploration of the exclusion of environmental issues in wildlife documentaries, see Richards, 'Greening wildlife documentary', pp. 171–85.

30. A. Hough, '"Frozen planet": controversial BBC climate change episode to air in America', *The Telegraph*, 7 December 2011, available online at www.telegraph.co.uk/news/earth/earthnews/8939 592/Frozen-Planet-controversial-BBC-climate-change-episode-t o-air-in-America.html.

Natural history and its histories in the twenty-first century

1. For example, G. Ceballos et al., 'Accelerated modern human-induced species losses: entering the sixth mass extinction', *Science Advances*, 1:5 (19 June 2015), e1400253. There are many varied estimates of current extinction rates, which are difficult to determine for a number of reasons; see F. Pearce, 'Global extinction rates: why do estimates vary so wildly?', *Yale Environment 360* (17 August 2015), available online at e360.yale.edu/ features/global_extinction_rates_why_do_estimates_vary_so_ wildly.

2. D. S. Wilcove and T. Eisner, 'The impending extinction of natural history', *Chronicle of Higher Education* (15 September 2000), p. B24.

3. For a few prominent examples, see H. W. Greene and J. B. Losos, 'Systematics, natural history and conservation: field biologists must fight a public-image problem', *BioScience*, 38:7 (1988), pp. 458–62; R. F. Noss, 'The naturalists are dying off', *Conservation Biology*, 10:1 (1996), pp. 1–3; A. V. Suarez and N. D. Tsutsui, 'The value of museum collections for research and society', *BioScience*, 54:1 (2004), pp. 66–74.

4. Noss, 'The naturalists are dying off', p. 2.

5. Wilcove and Eisner, 'The impending extinction of natural history', p. B24.

6. B. Strasser, 'The experimenter's museum: GenBank, natural history, and the moral economies of biomedicine', *Isis*, 102:1 (2011), pp. 60–96, pp. 62–63, 96.

7. C. Hine, *Systematics as Cyberscience: Computers, Change, and Continuity in Science* (Cambridge, MA, 2008), p. 245.

8. B. M. Thiers, M. C. Tulig and K. A. Watson, 'Digitisation of the New York Botanical Garden herbarium', *Brittonia*, 68:3 (2016), pp. 324–33.

9. iDigBio (Integrated Digitized Biocollections), www.idigbio.org.

10. E. O. Wilson, 'My wish: build the Encyclopedia of Life', 2007 TED Prize speech transcript, available online at www.ted.com/talks/ e_o_wilson_on_saving_life_on_earth.

11. L. Odling-Smee, 'Encyclopedia of Life launched', *Nature News* (9 May 2007), available online at www.nature.com/news/2007/07 0508/full/news070508-7.html.

12. 'Encyclopedia of Life reaches historic "one million species pages" milestone', press release from Encylopedia of Life (9 May 2012), available online at eol.org/info/May_9.

13. Encyclopedia of Life, Global Partners, eol.org/info/global_partners.

14. E. O. Wilson, 'The Encyclopedia of Life', *Trends in Ecology and Evolution*, 18:2 (2003), pp. 77–80.

15. 'UF, FSU receive $10 million for project to digitize US biology collections', iDigBio, available online at www.idigbio.org/content/uf-fsu-receive-10-million-project-digitize-us-biology-collections.

16. 'Notes from Nature (Bird Collection)', Natural History Museum (London), available online at www.nhm.ac.uk/take-part/citizen-science/notes-from-nature.html.

17. J. Beckmann, 'The Swedish Taxonomy Initiative: managing the boundaries of "Sweden" and "taxonomy"', in K. H. Nielsen, M. Harbsmeier and C. J. Ries (eds.), *Scientists and Scholars in the Field: Studies in the History of Fieldwork and Expeditions* (Aarhus, 2012), pp. 395–414; G. Miller, 'Linnaeus's legacy carries on', *Science*, 307:5712 (2005), pp. 1038–9.

18. Swedish Species Information Centre, 'Species observations', available online at www.artdatabanken.se/en/species-observations.

19. Chinese Field Herbarium, www.cfh.ac.cn/default-en.html; other recent citizen science initiatives in China are documented in A. Hsu, A. Weinfurther and C. Yan, 'The potential for citizen-generated data in China', *Yale Data-Driven* (January 2017), available online at datadriven.yale.edu/wp-content/uploads/2017/01/Third_Wave_Citizen-Science_FINAL.pdf.

20. R. Monastersky, 'Anthropocene: the Human Age', *Nature*, 519 (2015), pp. 144–7.

21. J. Hannon, 'Earth in the Age of Humans', *Carnegie* (Winter 2016), pp. 22–7.

22. 'Mining the museum', https://miningthemuseumblog.wordpress.com.

23. Information about the exhibition is available via the Deutsches Museum web archive, www.deutsches-museum.de/en/exhibitions/special-exhibitions/archive/2015/anthropocene.

24. Center for Postnatural History, 'About', www.postnatural.org/Specimen-Vault/About. Emphasis in original.

25. B. Valentine, 'A museum for our postnatural age', interview of R. Pell, 24 February 2016, hyperallergic.com/271709/a-museum-for-our-postnatural-age.

26. The displays can be viewed at www.naturkundemuseum.berlin/en/museum/exhibitions/world-dinosaurs. For the excavations, see G. Maier, *African Dinosaurs Unearthed: The Tendaguru Expeditions* (Bloomington, IN, 2003).

27. Deutsches Historisches Museum, *German Colonialism: Fragments Past and Present* (Berlin, 2016).

Index

Printed in the United States
by Baker & Taylor Publisher Services